CAMBRIDGE MONOGRAPHS ON PHYSICS

INTRODUCTION TO THE THEORY OF LIQUID METALS

INTRODUCTION TO THE THEORY OF LIQUID METALS

T. E. FABER

University Lecturer in Physics and Fellow of Corpus Christi College, Cambridge

CAMBRIDGE
AT THE UNIVERSITY PRESS
1972

CAMBRIDGE UNIVERSITY PRESS
Cambridge, New York, Melbourne, Madrid, Cape Town, Singapore,
São Paulo, Delhi, Dubai, Tokyo, Mexico City

Cambridge University Press
The Edinburgh Building, Cambridge CB2 8RU, UK

Published in the United States of America by Cambridge University Press, New York

www.cambridge.org
Information on this title: www.cambridge.org/9780521154499

First published 1972
First paperback edition 2010

A catalogue record for this publication is available from the British Library

Library of Congress Catalogue Card Number: 76-184903

ISBN 978-0-521-08477-2 Hardback
ISBN 978-0-521-15449-9 Paperback

CONTENTS

Preface *page* xi

1 Potentials and pseudo-potentials 1

1.1 Prologue 1
1.2 The free-electron model 6
1.3 Exchange and correlation 8
1.4 Screening: the dielectric constant 14
1.5 Partial waves, phase shifts and virtual bound states 22
1.6 Friedel oscillations 28
1.7 Scattering of electrons by potential wells 30
1.8 Pseudo-potentials in the ion core 33
1.9 Coherent scattering by assemblies of ions 42
1.10 Effective masses for conduction electrons 48
1.11 Dispersion theory of effective mass 54
1.12 Energy gaps and pseudo-gaps 58
1.13 Ion–ion interactions 63
1.14 The rigid-sphere model 70
1.15 The energy for vacancy formation 72

2 The structure of simple liquids 74

2.1 Prologue 74
2.2 Models 76
2.3 Entropy 79
2.4 Specific heat 84
2.5 Surface tension and surface entropy 86
2.6 Molecular dynamics 91
2.7 An equation of state for the rigid-sphere model 97
2.8 The melting temperature 103
2.9 Distribution functions 106
2.10 The interference function 110
2.11 The interference function for a solid 114
2.12 Mixtures 121

2.13 Experimental methods for the determination of $a(q)$ *page* 124
2.14 Discussion of results 128
2.15 Temperature dependence of $a(q)$ 134
2.16 The Born–Green theory 137
2.17 The Percus–Yevick theory 140
2.18 Results for $w(R)$ 144
2.19 Compressibility 147

3 Molecular motion in liquids 154
3.1 Prologue 154
3.2 Diffusion as a random walk process 158
3.3 The friction coefficient for rigid spheres 162
3.4 A rigid-sphere theory of viscosity 165
3.5 Isotope effects in lithium 168
3.6 The velocity auto-correlation function 169
3.7 The back-scatter correction for rigid spheres 174
3.8 The frequency spectrum of molecular motion in liquids 177
3.9 Inelastic scattering 181
3.10 Quantum-mechanical corrections 186
3.11 Time-dependent pair distribution functions 187
3.12 Moment theorems 191
3.13 Approximate methods for the determination of G_s 194
3.14 Longitudinal 'phonons' in liquid metals 198
3.15 Visco-elasticity and shear waves 204
3.16 Relaxation times for conduction electrons 207
3.17 Nuclear quadrupole interaction in liquid metals 213

4 Electron states in disordered structures 221
4.1 Prologue 221
4.2 Survey of the literature 224
4.3 Hot solids in one dimension 232
4.4 The dislocated solid 236
4.5 Energy gaps in one-dimensional liquids 238

4.6 Localisation of states in one dimension *page* 241
4.7 Perturbation theory for wave groups 245
4.8 Comparison of results in one dimension 254
4.9 Extension to real metals in three dimensions 263
4.10 Localisation and band gaps in three dimensions 271
4.11 Beyond the Born approximation 276
4.12 Second thoughts on screening 285
4.13 Positron annihilation 288
4.14 Compton scattering 293
4.15 Magnetic susceptibilities 294
4.16 The Knight shift 302

5 Electronic transport processes 309
5.1 Prologue 309
5.2 Ziman's theory 312
5.3 The role of the effective mass 317
5.4 Comparison with experiment: the magnitude of ρ_L 322
5.5 Temperature dependence of ρ_L 330
5.6 Thermo-electric properties 334
5.7 Thermal conductivity 337
5.8 The effects of compression 339
5.9 Cs at high densities 343
5.10 Results at high temperatures and low densities 345
5.11 A different approach to the calculation of conductivity 353
5.12 Evaluation of the formula 356
5.13 Corrections when the mean free path is short 358
5.14 Conduction at high frequencies: the Drude equations 360
5.15 The anomalous skin effect 365
5.16 Dynamic screening 367
5.17 Many-body effects 370
5.18 Residual inter-band absorption 371
5.19 The absorption sum rule 375
5.20 Core polarisability 381

5.21 Optical properties of liquid metals: experimental methods *page* 382
5.22 Optical properties of liquid metals: results 385
5.23 Optical properties of liquid metals: mercury 390
5.24 Absorption and emission processes involving core electrons 394
5.25 Photo-emission and the work function 401
5.26 Plasma oscillations 404
5.27 The Hall effect 407
5.28 Magneto-resistance 414
5.29 Appendix 415

6 Liquid alloys 421
6.1 Prologue 421
6.2 Pseudo-potentials in alloys 424
6.3 The NFE model for alloys 428
6.4 The rigid-sphere model 433
6.5 Partial interference functions for rigid spheres 435
6.6 The volume and energy of mixing 441
6.7 The elastic continuum model 444
6.8 The entropy of mixing 446
6.9 Compound formation 448
6.10 Electrical transport properties: the Faber–Ziman theory 455
6.11 Results of diffraction experiments 461
6.12 Thermodynamic properties: normal systems 469
6.13 Thermodynamic properties: compound-forming systems 474
6.14 Surface tension 477
6.15 Other surface properties 481
6.16 Diffusion 482
6.17 Viscosity 484
6.18 Electromigration, thermal diffusion and related effects 485
6.19 Magnetic susceptibility 492

6.20 Nuclear magnetic resonance *page* 499
6.21 The Hall effect 508
6.22 Optical properties 513
6.23 Resistivity: normal systems 515
6.24 Resistivity: amalgams 521
6.25 Resistivity: systems with miscibility gaps 524
6.26 Resistivity: compound-forming systems 526
6.27 Thermo-electric power 530
6.28 Liquid semiconductors 535

Index of principal symbols 543

Bibliography and author index 549

Subject index 581

PREFACE

I first started to think about liquid metals in 1958, when I was awarded a generous Fellowship by the Worshipful Company of Armourers and Brasiers to work on the subject. I knew virtually nothing about the theory of liquids at the time and my understanding of the properties of metals was distinctly limited. Nevertheless, within a week of the public announcement of the award I was approached by a publisher suggesting that I should write this book.

To begin with it was to be a very slender monograph. I had the privilege to be associated with John Ziman in 1961–2, when he was producing the papers that made the theory of liquid metals respectable for the first time. Ziman's theory, using the concept of the *pseudo-potential* that was just becoming fashionable, seemed so beautifully simple that I felt myself competent to provide the brief account of it in book form which was clearly needed, and a period of leave in Berkeley in 1962–3 gave me the leisure to start writing. But the more I wrote the more I realised how many essential points there were that I did not really understand. I became fascinated by all sorts of problems peripheral to the theory: the justification for the nearly-free-electron (NFE) model, the two-way interaction between the conduction electrons and the ionic structure of a liquid metal, the influence of thermal motion of the ions, the complex behaviour of liquid alloys, and so on. I became fascinated to the point of obsession by the task of expounding the work of Edwards, which was recognised as being of the first importance as soon as it appeared but which others beside myself found obscure. And then I had to absorb a mass of fresh facts and ingenious new theories, which the papers of Ziman and Edwards soon stimulated. My manuscript grew fatter and fatter.

Inside every fat book is a thin one struggling to get out, and the essential bones and muscle of this one could still be compressed, by ruthless cutting, into a hundred pages or two. But the general reader, who wants no more than a quick birds-eye view of the subject is already provided for, e.g. by March's book on *Liquid Metals* (Pergamon, 1968) or by the chapter which I contributed to *The Physics of Metals: I Electrons* (Cambridge University Press,

1969). What is lacking is a detailed guide to help the experimental physicist or metallurgist, with rather little in the way of mathematical equipment, who is obliged to struggle at ground level through the rapidly expanding jungle of literature. That is what I have attempted to provide.

Much of the background knowledge that will be needed by the readers I have in mind is not readily to be found in books; it has to be culled from review articles written by active researchers, who are often too close to their own work to appreciate the problems that it presents to the uninitiated. I have therefore felt it necessary in many places to go right back to first principles: to give a detailed account in chapter 1, for example, of the basic assumptions behind the NFE model, or of distribution function theory in chapter 2, or of inelastic scattering in chapter 3. Furthermore, I have digressed in many places to explain the nature of phenomena (viscoelasticity, for example, or the Knight shift or plasma oscillations) which are not confined to liquid metals but which may nevertheless be unfamiliar. The mathematics is spelt out in sufficient detail to enable those who wish to do so to reproduce the results without excessive effort; familiarity with elementary perturbation theory in quantum mechanics and the ability to perform simple contour integrations are about all that is required. This mathematical detail, and the fact that I have preferred to avoid advanced methods involving Green functions and the other fashionable tools of many-body theory, will make much of the book tedious for the professional theorist, though I hope that he may find the physics of the subject more clearly illuminated here than it is by more sophisticated treatments.

The book presents a personal viewpoint throughout and contains a good deal of original work which has not been published elsewhere. It is intended, however, as a comprehensive and critical introduction to the work of many others besides myself. I have not referred in detail to all the older papers, partly because they can easily be traced through reviews and bibliographies that are already available, and partly because, in some cases, they are better forgotten. But I have struggled to keep up to date with more recent publications, say to the end of 1971, and I hope that the references provide an adequate guide to sources of experimental data and not just to theories. Being reluctant to write about what I do not understand, I have said little about liquid transition metals or about liquids such as Te which are half way towards being semiconductors. In these areas, where my book overlaps with *Electronic Processes in Non-Crystalline Materials* by Mott & Davis (Oxford, 1971), it makes no claim to be comprehensive.

Each of the six long chapters has been written as a more or less continuous narrative, with a 'prologue' to prepare the reader for the more important conclusions. I have endeavoured to ensure, nevertheless, that those who wish to use the book for reference only should be able to consult individual sections with enlightenment. There are frequent cross-references between one section and another and, indeed, between chapters. I have experienced the familiar shortage of symbols, so that the same symbol is sometimes used to represent several different quantities in different sections. The list of symbols on p. 543 may help to prevent confusion. Except where otherwise stated (e.g. where Bohr units are used) the formulae are expressed throughout in the unrationalised Gaussian system which is still more popular in the literature than any other: electric charges, field strengths, current densities and conductivities are in e.s.u., while magnetic field strengths and moments are in e.m.u.

I owe a deep debt of gratitude to Brian Pippard, who first directed my attention to liquid metals; to Sir Nevill Mott for his constant encouragement; to several friends besides John Ziman who have explained difficult points to me, especially Sam Edwards, Volker Heine and Norman March; to all my fellow-members of the British Liquid Metals Forum, especially Norman Cusack and John Enderby; to Walter Knight for his hospitality in Berkeley during 1962–3 and to Bill Spicer for hospitality in Stanford in 1970, where my first draft was heavily revised; and lastly to all past and present members of the Liquid Metals group at Cambridge – Clive Bradley, Guy Wilson, Richard Aldridge, Neville Smith, George Turner, Jane Miller, Peter Shackle, Neville Comins, Séan McAlister, Juliet Valiant and Walter Undrill.

The book is dedicated, however, to my wife.

December 1971 T. E. FABER

CHAPTER 1

POTENTIALS AND PSEUDO-POTENTIALS

1.1. Prologue

To reach an understanding of the properties of metals, whether liquid or solid, the traveller must choose at the start between two roads. They are signposted 'via Tight Binding' and 'via Jellium'. The territory between them is uncharted and perilous.

He who adopts the tight binding approach will see the metal as an assembly of *atoms*, brought together from infinity. He will see the discrete energy levels of these atoms broadened into *bands* by the interaction between neighbours. He will see the possibility of electrical conductivity arising when the bands have broadened to such an extent that they are overlapping.

The traveller via jellium has a quite different perspective. To him the metal consists of a sea of conduction electrons which are always free to move, with a background of positive charge which is not. In the early stages of his journey the positive background is treated as continuous, and the electrons may then be described by the free electron model; they can be labelled by well-defined wave vectors **k**, lying within a Fermi surface which is itself sharply defined and which is spherical. Later, he explores the consequences of supposing the positive charge to be concentrated into *ions*. If these are arranged to form a regular lattice, the electrons are liable to be scattered by them in a coherent fashion and must be described by *Bloch waves*. To label the electrons by their **k**-vectors is still legitimate, however, and a 'nearly-free-electron', or NFE, model can be used. The electron energy spectrum has gaps along certain planes in **k**-space, the Fermi surface may be distorted in consequence, and the electrons may acquire an effective mass m^* which is different from m. These 'band structure' characteristics of the conduction electrons influence many of the observable properties of the metal as a whole.

Much of the work that has been done on solid metals over the last twenty years has been devoted to filling in the pot-holes and eliminating the corners on the jellium road and this is the one that we shall take. At the risk of boring those readers who have been along it before, the basic theory of the *free-electron model*, of the *exchange and correlation* effects that keep electrons apart from one another and of their power to *screen* a perturbing potential, is developed in §§1.2–1.4 of the present chapter. At the risk of intimidating other readers, a good many points of difficulty that are sometimes glossed over are here exposed. One such is the distinction between the two dielectric constants ϵ and ϵ'; it has consequences which are not trivial and we cannot afford to ignore it.

Sections 1.5–1.7 contain a number of important results of a general nature concerning the scattering of electrons in a jellium by perturbing potential wells, and they lead up to §1.8 in which the essential concept of a *pseudo-potential*, which may be used to describe the scattering by a single ion, is first introduced. A detailed explanation of how pseudo-potentials are calculated in practice would require another chapter as long again as this, but the provenance of some which have been applied to liquid metal theory is briefly outlined.

In §1.9 we leave jellium behind us and face the problem of how to describe the scattering by an assembly of ions in a real metal. The philosophy adopted is what Ziman (1964), who was the first to formulate it clearly, has christened the 'method of neutral pseudo-atoms'. It appears that the information which is needed about the ionic arrangement, if the coherence between wavelets scattered off adjacent ions is to be properly allowed for, is contained in the *structure factor*

$$F(\mathbf{q}) = \sum_i \exp(-i\mathbf{q}\cdot\mathbf{r}_i),$$

where $\mathbf{q}$ is the scattering vector, and $\mathbf{r}_i$ denotes the position of the ith ion.

The structure factor for a perfect crystal has sharp spikes whenever $\mathbf{q}$ coincides with one of the reciprocal lattice vectors and is zero in between. The coherent scattering associated with each of these spikes is responsible (see §1.12) for the energy gaps and for the Fermi surface distortion. But what about the liquid phase, in which there is no long-range order? One can imagine circumstances in which the short-range order would be so pronounced as to make the structure look like that of a polycrystalline solid, in which case we could hardly accept the free-electron model in its simplest form. We would have to use Bloch waves and distorted Fermi surfaces to describe the electrons within each microscopic region of order and find some way to describe the effects of interaction between one region and the next. A theory of amorphous conductors has been constructed along these lines by Gubanov and is set out in his book (1965). There is growing circumstantial evidence, however, that in real liquid metals the situation is not so complicated. The evidence may be summarised as follows.

(*a*) Fermi surfaces have now been mapped out for a great number of solid metals by measurements at low temperatures on the de Haas–van Alphen effect etc., and the distortions have proved to be smaller than was at one time supposed. It would not seem to require very much disorder to wash them out entirely. A few 'semi-metals' such as Ga, Sb and Bi may be exceptions, but these are anomalous in that they tend to shrink when they melt. There is little doubt that melting brings them into line with the metals proper. The same is true for Si and Ge, which are semiconductors when solid.

(*b*) Low-temperature measurements of the electronic specific heat for solid metals indicate that the density of states at the Fermi level is substantially greater than the free-electron model predicts. It is now known, however, that the discrepancies are mainly due to an electron–phonon enhancement which is relevant only at low temperatures. Except in the semi-metals just referred to, the influence of the electron band structure on the density of states seems to be slight (see §1.10).

(*c*) Band structure effects in solid metals show up in optical absorption experiments; there are absorption edges in the visible and near ultra-violet associated with the excitation of conduction electrons to vacant bands above the Fermi level. On melting, the *intra*-band absorption increases but the *inter*-band part disappears. The absorption spectrum becomes essentially free-electron-like (see §5.22).

(*d*) Likewise the Hall coefficient, which shows peculiarities in the solid phase that are presumably due to band structure, becomes free-electron-like on melting (see §5.27).

(*e*) Experimental and theoretical work on the structure of simple liquids has shown (see chapter 2) that the range of solid-like order, if such order exists at all, must be extremely small – a few atomic spacings at the most. But consider the estimates for the electronic mean free path L which are shown in table 1.1. (This table, incidentally, which is repeatedly referred to through the book, will serve to indicate the range of elements with which we shall mainly be concerned.) In the monovalent metals L is so long compared with the atomic radius R_A that it must also be long compared with the range of order. Each electron therefore traverses many 'crystallites' of different orientations between one collision and the next. To describe the propagation characteristics of the electrons we must surely average the $(E,\mathbf{k})$ characteristics for a perfect crystal over all directions for $\mathbf{k}$ and an isotropic Fermi surface is bound to result. In some of the polyvalent liquid metals L is much shorter. Then, however, the uncertainty principle prevents us from defining the $\mathbf{k}$-vector of an electron with precision, and to talk of local distortions of the Fermi surface becomes meaningless on this account.

Nowhere else in this book are these arguments spelt out again in detail, but they are very important nevertheless. They justify a premise that lies beneath much of the detailed work that is to come: that the electrons in liquid metals may be handled by the methods of *perturbation theory*, using the plane waves of the free-electron model, isotropically distributed in $\mathbf{k}$-space, as a set of basis states. In principle, of course, there is nothing to stop one from adopting this approach whatever the range of order in the specimen, but it seems that to resurrect the band structure of a polycrystalline solid with its aid one must go beyond the second order in the perturbing pseudo-potential and the method then becomes too complicated to be useful (§4.11). We hope to get by with second-order calculations in practice.

Even in the first order, the pseudo-potential can affect the density of states of the conduction electrons. Despite their spherical Fermi surface we must recognise that they are only *nearly* free; m^* is not necessarily equal to m. The theory of the effective mass is discussed from two different points of view in §§1.10 and 1.11. We discover in passing why the curve of density of states versus

TABLE 1.1. *Electronic relaxation times and mean free paths in liquid metals*

Metal	Valency	$\hbar/2\tau K_F$ $(= 1/Lk_F)$	L/R_A
Li	1	0.022	24
Na	1	0.0071	73
K	1	0.0078	67
Rb	1	0.013	41
Cs	1	0.019	28
Cu	1	0.023	23
Ag	1	0.017	31
Au	1	0.030	17
Mg	2	0.027	15
Zn	2	0.047	9
Cd	2	0.038	11
Hg (25 °C)	2	0.101	4.1
Al	3	0.032	11
Ga	3	0.036	10
In	3	0.040	9
Tl	3	0.087	4.1
Si	4	0.12	2.7
Ge	4	0.10	3.0
Sn	4	0.064	5.1
Pb	4	0.12	2.7
Sb	5	0.16	1.9
Bi	5	0.18	1.7
Te	6	(0.7)	(0.4)

Note: The figures in this table have been deduced from the measured conductivity σ at the melting point (except in the case of Hg) using the free-electron formulae

$$\sigma = ne^2\tau/m = ne^2L/\hbar k_F,$$

which are largely substantiated by the arguments in chapter 5. The difference between the relaxation time τ and the lifetime τ_l, or between the mean free path L and the range of coherence l, should in most cases be rather trivial (see p. 322).

energy for a liquid transition metal may have a narrow peak in it, sufficient to accommodate the extra d electrons. Some speculations are advanced in § 1.12 concerning the troughs or *pseudo-gaps* which may open in the density of states curve for any metal, whether ordered or disordered, when it is expanded into the vapour phase.

Finally, in § 1.13, we turn to consider the forces of interaction between one ion and another. It is shown that, provided a second-order calculation is sufficient, the term in the energy of a metal which depends upon its structure may be calculated from an ion–ion pair potential $w(R)$. The most important feature of this is its strong repulsive core, which should hold the ions apart so that their cores do not overlap. Its relatively long-range tail, which is thought to be oscillatory, is too weak compared with $k_B T_M$, where T_M is the melting temperature, to be of much significance in the liquid phase. Indeed, it may prove an adequate approximation at high temperatures to represent the ions in a liquid metal by rigid spheres, though their effective diameter should decrease slowly on heating (see §1.14). They must be held together by a term in the energy which depends upon volume but not upon structure, but the magnitude of this term is not seriously discussed below.

The energy required to form a vacancy in a metal can be estimated once $w(R)$ is known. The necessary theory is briefly outlined in §1.15.

1.2. The free-electron model

To begin at the very beginning, let us consider the simplest of all models for a metal, in which the positive charge is regarded not as concentrated in ions but as spread in a uniform jelly throughout the specimen, 'without form and void'. The equilibrium state of the conduction electrons in this model must surely be likewise one of uniform charge density so as to preserve charge neutrality throughout, and one does not have to hunt very far to find wave functions for these electrons which are *self-consistent* in the Hartree sense: they are plane waves of uniform amplitude of the form

$$\psi_s = \Omega^{-\frac{1}{2}} \exp(i\mathbf{k}_s \cdot \mathbf{r}), \tag{1.1}$$

where Ω is the total volume. The Hartree philosophy is that an electron in a state ψ_s contributes a term

$$e^2 \int \frac{\psi_s^*(\mathbf{r}')\,\psi_s(\mathbf{r}')}{|\mathbf{r}-\mathbf{r}'|}\,d\mathbf{r}' \tag{1.2}$$

to the potential† experienced by the other electrons at the point $\mathbf{r}$. The total potential at $\mathbf{r}$, $V(\mathbf{r})$, includes a term of this sort for every

† 'Potential' is really an abbreviation for 'potential energy', which accounts for the factor e^2 in (1.2); it is measured in electron volts rather than volts.

occupied state, but since no electron can act upon itself the potential experienced by the sth electron is not V but

$$U_s(\mathbf{r}) = V(\mathbf{r}) - e^2 \int \frac{\psi_s^*(\mathbf{r}')\psi_s(\mathbf{r}')}{|\mathbf{r}-\mathbf{r}'|}\, \mathrm{d}\mathbf{r}'. \tag{1.3}$$

In a free atom U_s may differ substantially from V and an elaborate iterative procedure is required to find a self-consistent form for ψ_s. But if equation (1.1) is assumed to hold for a metal it immediately follows that the difference between U_s and V is proportional to $\Omega^{-\frac{1}{3}}$ and can be neglected for a macroscopic specimen. Not only that, but for the *jellium model* which we are considering U_s and V are uniformly zero. The self-consistency of (1.1) is then apparent.

Whatever boundary conditions we apply to the specimen there is always some restriction on the values of $\mathbf{k}_s$ that are permitted, the general rule being that there are $\Omega/(2\pi)^3$ such values per unit volume of $\mathbf{k}$-space. Since the energy of an electron in the state ψ_s is given for this model by

$$E_s = \hbar^2 k_s^2/2m, \tag{1.4}$$

the *density of states per unit energy with a given spin orientation* is given by

$$\mathscr{N}(E) = \frac{\Omega}{(2\pi)^3}\, 4\pi k^2 \frac{\mathrm{d}k}{\mathrm{d}E} = \frac{m\Omega}{2\pi^2\hbar^3}(2mE)^{\frac{1}{2}}. \tag{1.5}$$

At a temperature T, the probability that a given state is occupied should be described by the Fermi–Dirac function

$$f(E) = [\exp((E-E_{\mathrm{F}})/k_{\mathrm{B}}T)+1]^{-1}, \tag{1.6}$$

where the *Fermi energy* E_{F} is fixed by the total number of electrons that have to be accommodated. If there are n of them per unit volume, then

$$E_{\mathrm{F}} = K_{\mathrm{F}} = \frac{\hbar^2}{2m}(3\pi^2 n)^{\frac{2}{3}}\left(1 - \frac{\pi^2}{12}\left(\frac{k_{\mathrm{B}}T}{K_{\mathrm{F}}}\right)^2 + \ldots\right); \tag{1.7}$$

K_{F} is used in this book to represent the part of the total Fermi energy which is kinetic rather than potential. If there are N atoms in volume Ω, each with valency z, we may assume n to be given by (Nz/Ω), and for a typical metal this is about 10^{23} cm^{-3}. Hence a typical value for K_{F} would be 7 eV. Since $k_{\mathrm{B}}T$ is still only about

0.2 eV at a temperature of 2000 °K, it follows that even in a liquid metal at a high temperature the conduction electrons should remain highly *degenerate*; the *Fermi surface*, which divides the occupied from the unoccupied states in **k**-space and which for the present model is a sphere with a radius

$$k_F = (3\pi^2 n)^{\frac{1}{3}}, \tag{1.8}$$

remains quite sharp. Moreover, the temperature-dependent correction term in equation (1.7) for the Fermi energy remains insignificant.

Equations (1.4) to (1.8) are the basic equations of any *free-electron* model for a metal and have been known since the time of Sommerfeld. They are collected here for ease of reference and to introduce some notation that is used throughout this book.

1.3. Exchange and correlation

One of the limitations of the Hartree method is that it makes no allowance for the *exchange effects* which tend to keep electrons of like spin apart. Fock showed that if we include these in the conventional way by using many-electron wave functions which are Slater determinants of the various ψ_s, then the potential which should be adopted in calculating ψ_s (the *Hartree–Fock* potential) is given by

$$U_s(\mathbf{r}) = V(\mathbf{r}) - e^2 \sum_{t \neq s} \int \frac{\psi_t^*(\mathbf{r}')\psi_t(\mathbf{r})}{|\mathbf{r}-\mathbf{r}'|} \frac{\psi_s(\mathbf{r}')}{\psi_s(\mathbf{r})} \, d\mathbf{r}', \tag{1.9}$$

where the sum is to be taken over occupied states of like spin to the one under consideration. This is very different from the Hartree formula and the distinction between U and V is no longer trivial. It remains true that U is independent of $\mathbf{r}$, of course, so that (1.1) remains a self-consistent solution, and this allows us to rewrite (1.9) in the form

$$U_s(\mathbf{r}) = V(\mathbf{r}) - \frac{e^2}{\Omega} \int \frac{\exp(-i\mathbf{k}_s \cdot \mathbf{R})\, d\mathbf{R}}{R} \int \frac{\Omega}{(2\pi)^3} \exp(i\mathbf{k} \cdot \mathbf{R})\, d\mathbf{k},$$

where $\mathbf{R} = \mathbf{r} - \mathbf{r}'$ and the sum over t in (1.9) has been converted into an integral in **k**-space over the interior of the Fermi sphere (which it is convenient to evaluate before integrating over **R** in real space). After performing the integration over **k** and integrating also over all possible directions for **R** one obtains

$$U_s(\mathbf{r}) = V(\mathbf{r}) - \frac{2e^2}{\pi k_s}\int_0^\infty \left(\frac{\sin k_s R \sin k_F R}{R^3} - \frac{k_F \sin k_s R \cos k_F R}{R^2}\right) dR. \tag{1.10}$$

Thus the difference between U and V is *the potential that would be set up by a distribution of charge around the point* $\mathbf{r}$ *given by*

$$-\frac{2e}{\pi k_s}\int_0^\infty \left(\frac{\sin k_s R \sin k_F R}{R^2} - \frac{k_F \sin k_s R \cos k_F R}{R}\right) dR$$
$$= -\frac{e}{\pi}\int_0^\infty \left(\frac{\sin (k_F + k_s)R}{R} + \frac{\sin (k_F - k_s)R}{R}\right) dR$$
$$= -e(\tfrac{1}{2} + \tfrac{1}{2}). \tag{1.11}$$

We shall be mainly concerned with the potential experienced by electrons on the Fermi surface for which $k_s = k_F$. The physical meaning of equation (1.11) in such a case is clear enough. Wherever the electron is situated in the metal it is to be thought of as surrounded by an *exchange hole* in the sea of other electrons, whose radius is of order $(2k_F)^{-1}$ and is therefore comparable with the wavelength of the electrons at the Fermi surface; from this hole a net charge equal to half the electronic charge is excluded. The excluded $\frac{1}{2}e$ must be distributed in a uniform fashion through the rest of the specimen. Since no electron can act upon itself, however, we must remember in evaluating U to make the correction that Hartree suggested, which means subtracting from the sea a uniformly distributed charge e. The combined effects of exchange and of the Hartree correction, therefore, ensure that $\frac{1}{2}e$ is missing from the exchange hole and a further $\frac{1}{2}e$ is missing from the rest of the specimen, which is just what (1.11) says.

It is a straightforward matter to evaluate U_s for the jellium model from equation (1.10); the result is

$$U_s^X = -\frac{e^2 k_F}{\pi}\left(1 + \frac{k_F^2 - k_s^2}{2k_F k_s}\log\left|\frac{k_F + k_s}{k_F - k_s}\right|\right). \tag{1.12}$$

The superscript X has been added as a reminder that this equation describes the potential energy of a single electron due to *exchange*. To find the total energy of a single electron we must add U^X to the kinetic energy, and for an electron at the Fermi level we get

$$E_F = K_F + U_F^X = \frac{3.68}{r_s^2} - \frac{1.22}{r_s} \text{ rydberg,} \tag{1.13}$$

where r_s is the conventional measure of the mean distance between the conduction electrons, defined by

$$n(4\pi r_s^3/3) = 1 \tag{1.14}$$

and expressed for the purposes of (1.13) in Bohr units.† So long as r_s exceeds about 3 Bohr units (a condition which is not in fact satisfied by all real metals) E_F as given by (1.13) is safely negative; the attractive potential due to the positive jelly which is exposed around each electron due to the operation of exchange is sufficient to counterbalance the kinetic energy and to prevent an electron at the Fermi level from escaping from the specimen.

Actually, we ought not to take it for granted that the energy required to extract an electron from the Fermi level of the jellium and remove it to infinity, where its energy is zero, is necessarily given by $-E_F$. This energy – the so-called *work function* of the metal – could be perturbed by two corrections. First, there is the possibility that a *dipole layer* exists on the surface of the jellium, which would shift the potential of all the electrons inside by a uniform term which is not included in (1.13). Secondly, there is the possibility that the extraction of one electron changes the energy of all the other electrons which are left behind. It is worth a short digression to expound the basis of *Koopman's theorem*, which assures us that in a large specimen of jellium the correction required on this second count is infinitesimal, though it might be important for a single atom.

The first point to be clear about is that the total energy of the jellium is *not* just the sum over all the electrons of the single-particle energies $(K_s + U_s)$; such a sum would include twice over the coulomb interaction e^2/R_{st} between each pair of electrons and

† The Bohr unit of length is a_H, the radius of the first Bohr orbit of the H atom, = 0.53 Å. The rydberg is the ionisation energy of the H atom, = 13.6 eV.

would also omit the self energy of the positive jelly. In fact the total energy may be expressed as

$$\sum_{s=1}^{N_e} (K_s + U_s) - \sum_{s>t} \frac{e^2}{R_{st}} + \text{self energy of positive jelly,}$$

where in evaluating the second term each pair of electrons is to be counted only once; in the Hartree approximation, of course, the second and third terms exactly cancel. Now if there are N_e electrons initially and the N_eth one is removed from the Fermi level, the change in the total energy is

$$-(K_F + U_F) + \sum_{s=1}^{N_e-1} \Delta U_s + \sum_{s=1}^{N_e-1} \frac{e^2}{R_{sN_e}},$$

where ΔU_s is the change in U_s which the removal brings about. But the missing electron occupied a plane wave state which extended through the whole specimen and whose amplitude at any point was infinitesimal. Its removal therefore constitutes an infinitesimal perturbation which leaves the wave functions for all the other electrons effectively unchanged. It is not difficult to see that in these circumstances ΔU_s must be equal and opposite to e^2/R_{sN_e}, in which case the change in the total energy does reduce to $-(K_F + U_F)$ as required.

An alternative approach which is instructive is to write the total energy of the jellium as

$$\sum_s (K_s + \tfrac{1}{2}U_s) = \sum_s K_s + \tfrac{1}{2}N_e\bar{U},$$

where the factor $\frac{1}{2}$ takes care that no interaction energy is counted twice. The change in the total energy when an electron is extracted is then given by

$$-K_F - \frac{d}{dN_e}(\tfrac{1}{2}N_e\bar{U}) = -K_F - \tfrac{1}{2}\bar{U} - \tfrac{1}{2}N_e \frac{d\bar{U}}{dN_e}.$$

But it is readily shown by averaging (1.12) over all values of k_s up to k_F that $\frac{1}{2}\bar{U}^X = \frac{3}{4}U_F^X$. Moreover, since U^X is inversely proportional to r_s it is proportional to $N_e^{\frac{1}{3}}$, whence

$$\tfrac{1}{2}N_e \frac{d\bar{U}}{dN_e} = \tfrac{1}{6}\bar{U} = \tfrac{1}{4}U_F^X.$$

Once again the change in energy reduces to $-(K_F+U_F)$, at any rate within the limits of the Hartree–Fock approximation.

The trouble with the Hartree–Fock method is that it fails to allow in detail for the effects of coulomb repulsion between individual electrons; in this respect it is no better than the Hartree method upon which it is based. The repulsion tends to keep electrons apart from one another irrespective of their spin, but there is nothing in (1.3) or (1.9) to suggest this. The truth is that we ought to represent the whole electron system by a *many-body* wave function, much more complex than a Slater determinant of single-particle wave functions, if we wish to describe the subtle correlations between the coordinates of the particles that must in reality exist. No attempt is made in this book to discuss the results of many-body theory in detail, but one semi-quantitative idea of value will emerge at the end of the following section. If we are to fix our attention on a single electron then we must regard it as accompanied wherever it goes not just by an *exchange hole* in the surrounding electron sea but by a combined *exchange + correlation hole* for which the coulomb repulsion is at least partly responsible. From this hole, as it turns out, a charge of e rather than $\frac{1}{2}e$ is missing.

To take account of correlation, some of the results obtained above must be modified by adding to U^X an appropriate correction U^C. In particular, (1.13) becomes

$$E_F = \frac{3.68}{r_s^2} - \frac{1.22}{r_s} + U_F^C \text{ rydberg,} \tag{1.15}$$

while the total energy of the jellium becomes

$$N_e(\tfrac{3}{5}K_F + \tfrac{1}{2}\bar{U}^X + \tfrac{1}{2}\bar{U}^C) = N_e\left(\frac{2.21}{r_s^2} - \frac{0.916}{r_s} + \tfrac{1}{2}\bar{U}^C\right) \text{ rydberg.} \tag{1.16}$$

Perhaps the best values of U_F^C and $\bar{U}^C$ are those calculated by Hubbard and conveniently tabulated by Animalu & Heine (1965). Like U^X, U^C increases in magnitude as the density of electrons is increased, though somewhat less rapidly; there are grounds for expecting U^C to be proportional to the screening parameter q_s

which is introduced in the next section and hence to be proportional to $N_e^{\frac{1}{3}}$ and this is roughly true of Hubbard's results. Since

$$\tfrac{1}{2}\bar{U}^{X,C}+\tfrac{1}{2}N_e\frac{d\bar{U}^{X,C}}{dN_e} = U_F^{X,C}$$

whether we are concerned with exchange or correlation effects, it follows that $\bar{U}^C$ is roughly $(12/7)U_F^C$.

Given a reliable set of values for $\bar{U}^C$ as a function of r_s we can find the equilibrium value of r_s for a jellium by minimising (1.16). The result, using Hubbard's values, is 4.13 Bohr units. The two real metals which come closest to this conduction electron density are Na and K, for which r_s is 3.96 and 4.87 Bohr units respectively, and these are the metals for which the predictions of the free-electron model are most accurately obeyed. For $r_s = 4.13$, (1.15) becomes

$$E_F = 0.216-0.296-0.082 = -0.162 \text{ rydberg},$$

which implies that the so-called *inner* work function (i.e. the work function in the absence of any surface dipole layer) should be 2.20 eV. The measured work functions of Na and K are 2.15 and 2.27 eV respectively. The close agreement indicates that for these two metals at any rate the effect of the surface dipole layer is not a large one. According to Lang & Kohn (1971), however, it increases as r_s diminishes and may contribute as much as 4 eV to the measured work function at the values of r_s which are typical for polyvalent metals.

Finally we should consider what effect exchange and correlation corrections are liable to have upon the density of states. This quantity is inversely proportional to the rate of change of the single-particle energy with k, and the result given in (1.5) must therefore be amended by a factor

$$\left[1+\frac{m}{\hbar^2 k}\frac{\partial}{\partial k_s}(U_s^X+U_s^C)\right]^{-1} = \frac{m_{XC}^*}{m}. \qquad (1.17)$$

It is apparent from (1.12) that $\partial U_s^X/\partial k_s$ is always positive and in fact it becomes infinite at the Fermi level. Hence, if the Hartree–Fock equations were exact the density of states at the Fermi level would be zero! When correlation effects are properly included this singularity disappears, and the best recent calculations (see fig. 4.21

on p. 296) suggest that (1.5) as it stands its not too far from the truth.

In the next section we go on to consider a jellium which is perturbed in some way so that the true electrostatic potential is no longer everywhere zero. We shall write

$$U = V + U^{XC}, \tag{1.18}$$

where U^{XC} is the attractive potential due to the exchange+correlation hole, hitherto expressed as the sum of two separate terms. We shall need to remember that U^{XC} may depend not only on the k value of the individual electron under consideration but also, in cases where this is liable to vary from place to place through the specimen, on the local density of the whole electron gas.

1.4. Screening: the dielectric constant

We have seen that the free-electron model, with perhaps a modified density of states, represents a self-consistent solution for a jellium in which the potential V is uniformly zero. Now we must consider what happens when the potential is perturbed in some way, perhaps by the introduction of charges from outside. Suppose, for example, that we introduce a point charge Q of positive sign and hence a *bare* potential†

$$V_b = -Qe/R. \tag{1.19}$$

The attractive field is surely bound to start electrons flowing towards Q, and space charges will build up to *screen* the perturbation. What is the *screened* potential V when equilibrium is established?

It is convenient to answer questions such as this by breaking the initial perturbation into its Fourier components and discussing the screening of each of them separately. It is also convenient, as with a dielectric medium, to describe the screening by means of a *dielectric constant.* Suppose that we introduce a continuous distribution of charge with density ρ so as to set up a bare potential‡

† The symbol **R** rather than **r** is used whenever we have to deal with a spherically symmetric perturbation centred about the point **R** = 0. In later sections **R** represents a coordinate measured from the centre of an ion.

‡ The symbol Δ attached to V_b, V, U^{XC} etc. implies that we are to consider only a small perturbation in the quantity concerned, from the value that it would have in the unperturbed jellium.

ΔV_b which is periodic in space, proportional to cos $(\mathbf{q}\cdot\mathbf{r})$. After the jellium has been allowed to polarise we can still define the electric induction by means of the equation

$$\operatorname{div} \mathbf{D} = 4\pi\rho,$$

which means that $\mathbf{D}$ is given by $-\operatorname{grad}(\Delta V_b)$. The electric field $\mathbf{E}$ on the other hand, is $-\operatorname{grad}(\Delta V)$. Hence

$$\frac{\Delta V_b}{\Delta V} = \frac{\mathbf{D}}{\mathbf{E}} = \epsilon(q). \tag{1.20}$$

The respect in which a jellium differs from a non-conducting medium is that this dielectric constant $\epsilon(q)$ turns out to vary with q. For long wavelengths it tends to infinity like q^{-2}, i.e. the screening becomes almost perfect, but for short wavelengths it tends to unity.

The reason for this variation is straightforward enough. When the perturbing charge is introduced the mean density of electrons, which is given by n in the absence of any perturbation, acquires a periodic component Δn. Once equilibrium has been re-established the electrons see the screened potential ΔV rather than ΔV_b and if this is weak enough for their response to be linear – a cardinal assumption throughout this discussion – we may write

$$\Delta n = \text{constant} \times \Delta V,$$

where the constant of proportionality should be a negative quantity if, as seems likely, the electrons tend to cluster where ΔV is attractive. But Poisson's equation tells us that the screening field which they set up must be such that

$$q^2(\Delta V - \Delta V_b) = 4\pi e^2 \Delta n. \tag{1.21}$$

It follows at once that
$$\epsilon = 1 + \frac{q_s^2}{q^2}, \tag{1.22}$$

where the *screening parameter* q_s is defined by the equation

$$q_s^2 = 4\pi e^2(-\Delta n/\Delta V). \tag{1.23}$$

The screening parameter may be estimated by the *Thomas–Fermi* method, i.e. by making the assumptions (*a*) that in circumstances where the potential is varying it is possible to define a local value of K_F, determined by the local value of n, and (*b*) that the total

Fermi energy E_F remains uniform in space. If we forget about exchange and correlation for the moment, it follows from these assumptions that

$$\Delta K_F = \frac{dK_F}{dn}\Delta n = \frac{\Omega \Delta n}{2\mathcal{N}(E)_F} = -\Delta V,$$

and hence that $$q_s^2 = 8\pi \mathcal{N}(E)_F e^2/\Omega. \tag{1.24}$$

When the free-electron equation (1.5) is adequate to describe the density of states this answer can be expressed as

$$q_s^2 = 6\pi n e^2/K_F, \tag{1.25}$$

but (1.24) is the more fundamental formula.

The approximations of the Thomas–Fermi method become impossible to justify when q is comparable with k_F; physical intuition suggests that a very short wavelength perturbation is unlikely to modulate the electron density at all, in which case an accurate expression for q_s^2 should tend to zero for large q. This is confirmed by a calculation due originally to Bardeen, in which q_s^2 is derived by perturbation theory, starting from the plane wave states of the free electron-model. Consider an electron in the unperturbed state

$$\psi_0 = \Omega^{-\frac{1}{2}} \exp(i\mathbf{k}\cdot\mathbf{r})$$

with energy $E(k)$, subjected to a perturbation ΔV which is proportional to $\cos(\mathbf{q}\cdot\mathbf{r})$. This mixes components of the form $\exp(i(\mathbf{k}\pm\mathbf{q})\cdot\mathbf{r})$ in with ψ_0 and to first-order in ΔV the result is a periodic modulation in the amplitude of the wave function such that

$$\begin{aligned}\psi\psi^* &= \psi_0\psi_0^*\left(1+\frac{\Delta V}{E(\mathbf{k})-E(\mathbf{k}+\mathbf{q})}+\frac{\Delta V}{E(\mathbf{k})-E(\mathbf{k}-\mathbf{q})}\right)\\ &= \psi_0\psi_0^*\left(1+\frac{4m\Delta V}{\hbar^2(4k_\parallel^2-q^2)}\right),\end{aligned} \tag{1.26}$$

where $k_\parallel$ is the component of $\mathbf{k}$ in the direction of $\mathbf{q}$. To obtain q_s^2 we must average (1.26) over all the occupied states within the Fermi sphere. (It is to be noted that it is only for states such that $|2k_\parallel|$ is less than q that the electrons tend to avoid the regions where ΔV is positive. Fast electrons, with $|2k_\parallel|$ greater than q, behave in just the opposite fashion; far from screening the perturbation they tend to accentuate it. This may seem surprising at first sight, but it is in fact an elementary result of the WKB method for solving Schrödinger's equation that the amplitude of a wave function for a fast electron is inversely proportional to its local momentum – because the electron spends more time in a region where it travels slowly – and is therefore bound to be bigger where the potential is high. Of course when one averages over the Fermi sphere the screening effect of the slow electrons outweighs the anti-screening effect of the fast ones.) The average may be conducted in two stages, first over all possible orientations of **k** with respect to **q** and then over all values of k up to k_F. At the end of the first stage one has

$$\overline{\psi\psi^*} = \psi_0\psi_0^* \left(1 - \frac{m\Delta V}{\hbar^2 qk} \log\left|\frac{2k+q}{2k-q}\right|\right). \tag{1.27}$$

The final average yields

$$\begin{aligned}\Delta n/n &= (\overline{\psi\psi^*} + \psi_0\psi_0^*)/\psi_0\psi_0^* \\ &= -\Delta V \frac{3m}{\hbar^2 k_F^2}\left\{\frac{1}{2} + \frac{1-y^2}{4y}\log\left|\frac{1+y}{1-y}\right|\right\},\end{aligned}$$

so that

$$q_s^2 = \frac{6\pi Nze^2}{\Omega K_F}\left\{\frac{1}{2} + \frac{1-y^2}{4y}\log\left|\frac{1+y}{1-y}\right|\right\}, \tag{1.28}$$

where $y = q/2k_F$. This expression for q_s^2 has the expected properties; it reduces to (1.25) when y is small, but for large y tends to zero. There is a weak singularity at $y = 1$ where the slope of q_s^2 becomes infinite.

The problem is further complicated, alas, by the effects of exchange and correlation. Once the electron density is modulated a periodic term, which may be a function of the wave vector **k** of the electron upon which it acts, appears in the hole potential U^{XC}. An electron at the Fermi level, for example, experiences a perturbation which is not ΔV but

$$\Delta U_F = \Delta V + \Delta U_F^{XC},$$

and for many purposes we shall find that an effective dielectric constant ϵ', defined by the equation

$$\epsilon'(q) = \Delta V_b/\Delta U_F \tag{1.29}$$

is of more use than the $\epsilon(q)$ of equation (1.20).† Equation (1.21) becomes

$$q^2\Delta V_b\left(\frac{1}{\epsilon}-1\right) = -q_s^2\overline{\Delta U},$$

where q_s^2 is still given by (1.24) in the limit of small q and may be supposed to vary for large q in the way that (1.28) suggests. Here $\overline{\Delta U}$ is some appropriate average of ΔU over values of k from zero up to k_F. We can probably ignore the distinction between $\overline{\Delta U}$ and ΔU_F in which case

$$\epsilon' = \frac{\epsilon'}{\epsilon}+\frac{q_s^2}{q^2} \tag{1.30}$$

or

$$\epsilon = 1+\frac{\epsilon}{\epsilon'}\frac{q_s^2}{q^2}.$$

In the limit of large q, when screening is relatively unimportant, we can safely replace ϵ'/ϵ by unity; in this limit there is no difference between ϵ' and ϵ and both of them are adequately described by our previous expression (1.22). But for small q, as the following argument shows, ϵ'/ϵ is rarely more than 0.5.

Let us go back to the Thomas–Fermi calculation, which can readily be modified to allow for exchange and correlation (Heine, Nozières & Wilkins, 1966). In the limit $q \to 0$ we can treat the specimen as an assembly of macroscopic slices (cut perpendicular to $\mathbf{q}$) within each of which V is uniform, and there is no problem in defining a Fermi radius separately for each. Since the total Fermi energy is the chemical potential of the electrons, it must certainly be the same in every slice, which means that

$$\Delta K_F+\Delta V+\left(\frac{\partial U^{XC}}{\partial n}\right)_k \Delta n+\left(\frac{\partial U^{XC}}{\partial k}\right)_n \Delta k_F = 0. \tag{1.31}$$

The quantity in which we are interested, however, is the periodic potential that would be experienced by an electron travelling through the specimen with a *uniform* wave-vector $\bar{k}_F$, and this is given by

$$\begin{aligned}\Delta U_F &= \Delta V+\left(\frac{\partial U^{XC}}{\partial n}\right)_k \Delta n\\ &= -\Delta K_F-\left(\frac{\partial U^{XC}}{\partial k}\right)_n \Delta k_F\\ &= -(\Omega/2\mathcal{N}(E)_F)\,\Delta n,\end{aligned} \tag{1.32}$$

† The two dielectric constants ϵ and ϵ' have been labelled ϵ_1 and ϵ_2 by Heine & Weaire (1966) and ϵ_p and ϵ_e by Heine & Weaire (1970).

where $\mathcal{N}(E)$ is the density of states for one spin only in the unperturbed specimen, including now the correction for exchange and correlation. Alternatively, we could write (1.31) as

$$\Delta V = -\left(\frac{\mathrm{d}K_{\mathrm{F}}}{\mathrm{d}n} + \Omega\frac{\mathrm{d}U_{\mathrm{F}}^{\mathrm{XC}}}{\mathrm{d}N_e}\right)\Delta n$$

$$= -(\Omega/2\mathcal{N}^0(E)_{\mathrm{F}})\left(1 - \frac{1}{2}\frac{r_s}{K_{\mathrm{F}}}\frac{\mathrm{d}U_{\mathrm{F}}^{\mathrm{XC}}}{\mathrm{d}r_s}\right)\Delta n, \qquad (1.33)$$

$\mathcal{N}^0(E)$ being the density of states without this correction. Thus so long as the Thomas–Fermi approach is justified we have (see (1.13))

$$\frac{\epsilon'}{\epsilon} = \frac{\Delta V}{\Delta U_{\mathrm{F}}} = \frac{m_{\mathrm{XC}}^*}{m}\left(1 - \frac{r_s}{6.04} - \frac{r_s^3}{7.36}\frac{\mathrm{d}U_{\mathrm{F}}^{\mathrm{C}}}{\mathrm{d}r_s}\right). \qquad (1.34)$$

If one assumes the effective mass ratio, first introduced in (1.17), to be close to unity and takes $U_{\mathrm{F}}^{\mathrm{C}}$ from Hubbard's calculations, one finds

$$\frac{\epsilon'}{\epsilon} \simeq (1.03 - 0.19\, r_s) \quad (q \to 0) \qquad (1.35)$$

over the range of r_s that concerns us. For the electron density that is typical of polyvalent metals this amounts to about 0.5, while in alkali metals it is substantially less – in Cs, indeed, it is negative! Fortunately, its magnitude does not affect the low q limit of ϵ', which is always given correctly by q_s^2/q^2; it is ϵ rather than ϵ' which is negative in Cs for low q.

We have pursued the theory of the dielectric constant far enough for the moment – the reader who wishes to go further may consult Heine & Weaire (1970) or, for a more sophisticated analysis, Pines & Nozières (1966) or Hedin & Lundqvist (1969). Let us go back to the problem with which we started, concerning the screening of a point charge. To solve it we express the bare potential (1.19) in Fourier components and assume, the response being linear, that each of them is separately screened. Before screening we have

$$\Delta V_{\mathrm{b}} = \frac{4\pi eQ}{\Omega}\sum_{\mathbf{q}}\frac{\cos(\mathbf{q}\cdot\mathbf{R})}{q^2} = \frac{eQ}{2\pi^2}\int\frac{\cos(\mathbf{q}\cdot\mathbf{R})}{q^2}\,\mathrm{d}\mathbf{q} \qquad (1.36)$$

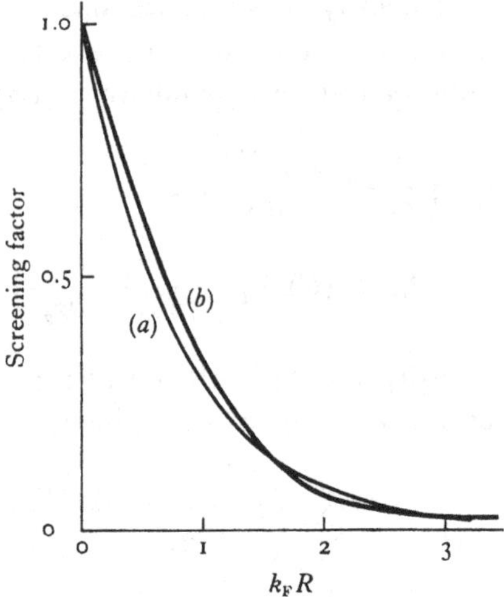

Fig. 1.1. The factor by which screening reduces the potential round a point charge in a jellium for which $r_s = 2.4$ Bohr units: (*a*) Thomas–Fermi approximation, (*b*) Bardeen approximation.

(the equivalence of (1.36) to (1.19) is an elementary example of Fourier transform theory). After screening, and after integrating over all directions for $\mathbf{q}$ we have

$$\Delta U_F = \frac{2eQ}{\pi}\int_0^\infty \frac{\sin(qR)}{qR}\frac{dq}{\epsilon'(q)}. \tag{1.37}$$

The integration is readily performed by contour methods if one is prepared to use the simple Thomas–Fermi expression (1.22) for ϵ', with q_s independent of q. The integrand has only two poles, at $q = \pm iq_s$, and so

$$\Delta U_F = -\frac{eQ}{R}\exp(-q_s R). \tag{1.38}$$

This is often referred to as a screened coulomb potential, the *screening length* being q_s^{-1}. If the Bardeen expression (1.28) is preferred one has to resort to numerical methods, but fig. 1.1 shows that the end result is very much the same as (1.38) over a substantial range of R; the singularity in (1.28) does make itself felt, as we shall see in the following section, but only when R is large. The fact is that the main part of ΔU_F still arises from the poles on the imaginary axis for q and it can readily be shown that

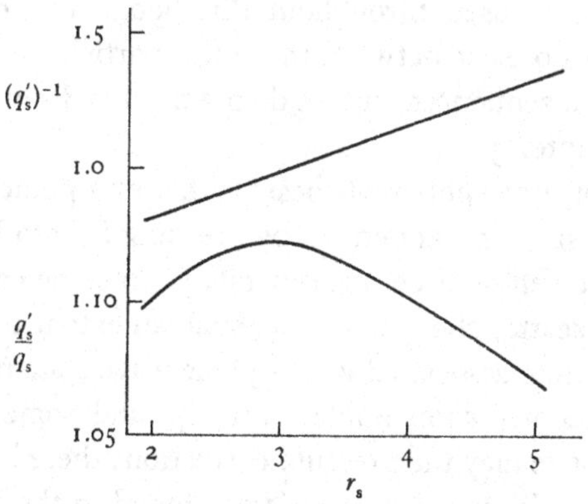

Fig. 1.2. The screening length $(q_s')^{-1}$ in Bohr units as a function of electron density. The lower curve shows, on a different scale, the ratio between q_s' and q_s. (1 Bohr unit of length = 0.53 Å.)

although the use of Bardeen's expression shifts these poles it does not shift them very far.

We can use a similar argument to guess the form that ΔU_F will take, or ΔV for that matter, if allowance is made for exchange and correlation. These effects turn out to shift the principal poles in both $(\epsilon')^{-1}$ and $(\epsilon)^{-1}$ to $\pm iq_s'$, where the ratio between q_s' and the Thomas–Fermi value for q_s is shown graphically as a function of r_s in fig. 1.2; the curve there derives from expressions quoted by Heine & Weaire (1970). Hence both ΔU_F and ΔV may be expected to approximate to the screened coulomb form, with $(q_s')^{-1}$ as the screening length. A second curve in fig. 1.2 shows the absolute magnitude of $(q_s')^{-1}$ derived from (1.25), i.e. on the assumption that the density of states is given by the free-electron value. It is smaller by a factor of 3 or so than r_s itself.

Just as electrons screen any charge introduced into a jellium from outside they presumably screen each other, and this is the justification for the statement in the previous section that the total charge excluded from the exchange + correlation hole must be a full e and not just $\frac{1}{2}e$ as suggested by the Hartree–Fock argument.

It is the efficiency of the screening that allows us in so many contexts to treat the electrons, together with their attendant exchange + correlation holes, as *independent quasi-particles*. This philosophy has been adopted without question in the argument so far and will be freely used throughout this book. The residual interactions which do exist between the quasi-particles have one or two important consequences, but we do not need to discuss them until we get to chapter 5.

The screening of other potentials besides that of a point charge can be treated in a similar manner, i.e. by breaking V_b into Fourier components and screening each one separately. In some cases the perturbation may scatter the electrons to such an extent that $\hbar/\tau_l$, where τ_l is the lifetime associated with a plane wave state near the Fermi surface, becomes comparable with K_F, and some of the results obtained above may then require correction; there is reason to believe, for example, that equation (1.24) for q_s^2 in the limit of low q ought to be modified by a factor

$$1-(\hbar/2\tau K_F) \tag{1.39}$$

(see §4.12). The figures in table 1.1 suggest that for a typical liquid metal a correction such as (1.39) describes can usually be ignored.

1.5. Partial waves, phase shifts and virtual bound states

The dielectric constant method is a very convenient way of treating the screening problem *but it is valid only when the perturbation is weak.* To take the specific case of a perturbation due to a point charge Q, the method is certainly sound when $|Q/e|$ is very much less than unity. When $|Q/e|$ is equal to or greater than unity its success, in a typical metal, is far from assured.

An alternative method of solution exists when, as in this specific case, the perturbation is spherically symmetric and localised. The procedure is very similar to that which is adopted in calculations of wave functions and potentials for free atoms. One starts with a plausible guess at the potential, solves the Schrödinger equation exactly for each electron, computes the resulting charge distribution, and, if this does not come out to be consistent with the potential initially assumed, iterates the calculation until it does. The procedure is of course laborious and it is often tempting to take a short cut to the answer required, via the *Friedel sum rule.*

The philosophy behind this sum rule (Friedel, 1954) is that if, when one adds a perturbing charge Q, one simultaneously adds $-(Q/e)$ electrons to provide the necessary charge to do the screening and to leave the specimen neutral, then the momentum distribution of the electrons in a region far from Q must remain unchanged; the screening is so effective that there can be no way of telling from a distance that the perturbation has been introduced. This means that the kinetic energy at the Fermi level remains unchanged far from the perturbation and that the total Fermi energy, E_F, is also unchanged. So what we do is count up the number of *extra* states with energy less than E_F which are generated by the perturbation and equate it to $-(Q/e)$.

It is well known that the exact solutions of Schrödinger's equation in a spherical potential are partial waves of the form

$$\psi = f(R)\, P_l(\cos\theta)$$

where $P_l(\cos\theta)$ is a spherical harmonic. If the specimen is a sphere of radius R_0 with the perturbing potential localised near its centre their asymptotic form for large R is

$$\psi \propto R^{-1}\sin(k_R R - \tfrac{1}{2}l\pi + \eta_l)\, P_l(\cos\theta), \tag{1.40}$$

where the effect of the perturbation lies in the *phase shift*, η_l; in the absence of any perturbation all the phase shifts necessarily vanish. Over any small volume of the specimen a long way from the centre a solution of this form is indistinguishable from a plane wave, with momentum $\hbar k_R$ in the radial direction and angular momentum $[l(l+1)]^{\frac{1}{2}}\hbar$ about the centre. Associated with the angular momentum there is of course some kinetic energy $l(l+1)\hbar^2/2mR^2$. If one wishes to concentrate upon the radial motion this rotational kinetic energy may be interpreted as a fictitious centrifugal potential, which prevents electrons with large angular momentum from approaching the region near $R = 0$. Since it thereby prevents them from experiencing the perturbation, one may infer that η_l is non-zero only for small values of l; the more localised the potential the fewer the phase shifts that have to be considered.

For a particular value of l, the value of k_R which corresponds to a total energy E_F is given by

$$k_{R,F}^2 = k_F^2 - l(l+1)/R^2, \tag{1.41}$$

but for the small values of l which are all we need to bother about it is sufficient to set

$$k_{R,F} = k_F, \tag{1.42}$$

at any rate near $R = R_0$. It is convenient to assume boundary conditions such that ψ has to vanish at R_0. This restricts k_R to values that satisfy the equation

$$k_R R_0 + \eta_l = (n + \tfrac{1}{2}l)\,\pi, \tag{1.43}$$

where n (not to be confused with the electron density) is an integer as well as l. It follows that for a given (small) value of l the value of n which describes a state with total energy E_F rises by

$$\eta_l(k_R = k_F)/\pi$$

when the perturbing potential is introduced. Since there is a $(2l+1)$-fold multiplicity associated with the orientation of the angular momentum and a further two-fold multiplicity associated with the orientation of the electron spin, the number of extra states with energy less than E_F for which the perturbation is responsible is

$$2(2l+1)\,\eta_l(k_F)/\pi \tag{1.44}$$

for this particular l. A summation over l yields Friedel's equation,

$$-(Q/e) = (2/\pi)\sum_{l=0}^{\infty}(2l+1)\,\eta_l(k_F). \tag{1.45}$$

Now for many purposes a detailed knowledge of how the electronic wave functions behave in the neighbourhood of the perturbing potential is not required, it is only their asymptotic form that matters. This, of course, is completely determined by the phase shifts. Since, when the potential is due to something as small as a metallic ion, the phase shifts are normally insignificant beyond say η_3, there are only four numbers to be calculated, η_0, η_1, η_2 and η_3. While the sum rule expressed by (1.45) is clearly not sufficient by itself to fix all of these, it may be of great help in checking that the values chosen for them are internally consistent.

An example of its use in this way is provided by the work of Meyer, Nestor & Young (1967), who set out to calculate phase shifts for the alkali and noble metals. Their calculations apply, in effect, to a single atom of the metal in question, inserted into a jellium with the appropriate charge density and Fermi energy. In this situation the perturbing charge Q is strictly speaking that of the nucleus and $-(Q/e)$ is the atomic number. For all but the lighest elements, however, the nuclear charge is surely strong enough to force a number of electrons to occupy bound states, thus forming an *ion core* around it. Such bound states have an energy which is negative, relative to a zero corresponding to $k = 0$, and it is well known that every time a new one is created (e.g. by increasing the atomic number) one of the phase shifts at $k = 0$ jumps suddenly upwards by another unit of π. It follows that if we insert on the right of (1.45) the *reduced* phase shifts, obtained from the true phase shifts by subtracting integral multiples of π until each one goes to zero at $k = 0$, we may use for Q the net positive charge on the ion core, so that $-(Q/e)$ on the left of (1.45) becomes the valency z. Meyer *et al.* (1967) take this to be unity, of course.†

The details of their subsequent procedure hardly concerns us here – the interested reader who wishes to assess the reliability of the results obtained must refer to the original paper. But briefly, they assume the potential inside the perturbing atom to be the same before screening as the Hartree–Fock potential which acts upon the valence electron in the free atom, obtaining this from tables computed by previous workers. Screening is supposed to be effected by a thin spherical shell of charge centred about each ion with a radius R_{sc} which lifts the potential uniformly inside the shell by an amount e^2/R_{sc} and eliminates entirely the coulomb tail outside. Reduced phase shifts (up to η_4) are then computed by a method that relies to some extent on arguments developed in the next section, and R_{sc} is adjusted until the sum rule is satisfied. The values of R_{sc} which are needed turn out to be within 5% of the mean atomic radius R_A.

Meyer *et al.* have computed their phase shifts for a range of values of k_F, which enables them to discuss the effects of alloying and compression. They have also computed them, in some cases, as a function of k for fixed k_F, and fig. 1.3 shows their results for metallic Cs at zero pressure. The behaviour shown is characteristic

† It is arguable that phase shifts of more relevance to some of the properties of the metals concerned would be obtained by supposing that at the same time as the atom is introduced into the jellium a sphere of positive jelly is cut out to make way for it. In this case $-(Q/e)$ would be zero. Some justification for the procedure used by Meyer *et al.*, however, is implicit in equation (1.73) below.

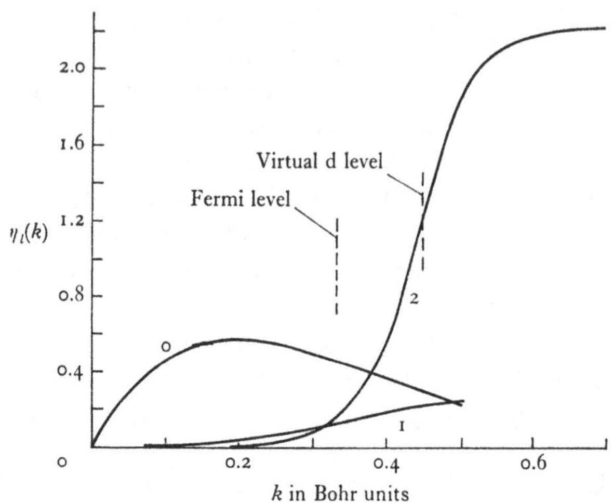

Fig. 1.3. Reduced phase shifts for Cs. (Redrawn by permission from Meyer, Nestor & Young, 1967.)

for attractive potential wells: the reduced phase shifts rise from zero at $\mathbf{k} = 0$, pass through a maximum and then steadily decrease. Now it can be shown (e.g. Kittel, 1963, p. 342) that the normalisation factor omitted from the asymptotic formula (1.40), which would be $(1/2\pi R_0)^{\frac{1}{2}}$ in the absence of any perturbing potential, requires to be corrected when η_l is not zero by a further factor

$$(1 + \tfrac{1}{2}R_0^{-1}\partial\eta_l/\partial k)^{-1}. \tag{1.46}$$

Where $(\partial\eta/\partial k)$ is positive, therefore, for low-energy electrons, the amplitude of the electronic eigenfunction is reduced at large distances from the potential well, which implies that in the neighbourhood of the well it is enhanced. For high-energy electrons, such that $(\partial\eta/\partial k)$ is negative, the reverse is true. Evidently, as we have noted already on p. 17, it is the low-energy ones which are most effective for screening purposes.

In some cases it happens that $(\partial\eta/\partial k)$ is large and there is a particularly marked tendency for the eigenfunction to be localised in the well, which is then said to have a *virtual bound state* at the energy concerned. Virtual bound states are normally a consequence of the centrifugal potential and therefore arise mainly for $l \geqslant 1$. One is used to thinking that in atoms the centrifugal potential is

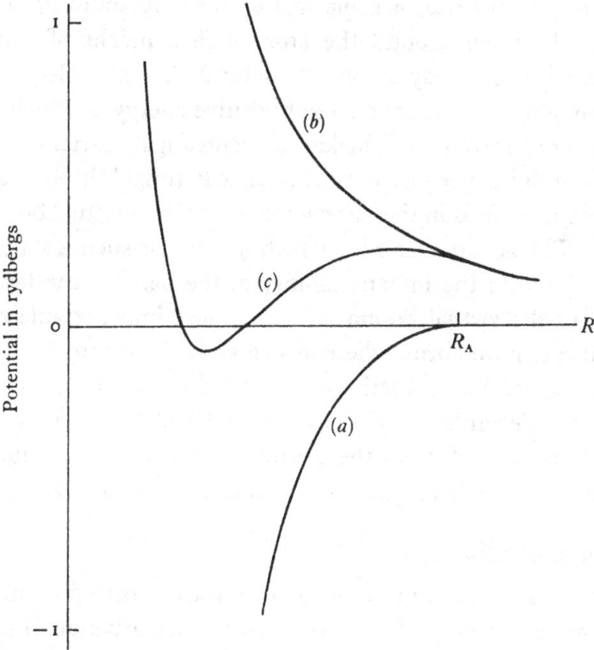

Fig. 1.4. Potentials in an atom surrounded by jellium: (*a*) true potential, (*b*) centrifugal, (*c*) total. (These curves are for the case $l = 1$, $z = 2$, $R_A = 3$ Bohr units. Since (*a*) has been computed from (1.69), however, and therefore includes no allowance for screening or core effects, they should be regarded as essentially schematic.)

important chiefly near $R = 0$, where, because it varies like R^{-2}, it ultimately dominates the true potential and hence prevents electrons of large angular momentum from approaching the nucleus. But in a metal as opposed to a free atom it can also be important for large R. In a free atom the Hartree or Hartree–Fock potential seen by a valence electron has an attractive coulomb tail for large R which is stronger than the repulsive centrifugal term. If the atom is surrounded by a matrix of jellium, however, into which the wave function of the valence electron is free to spread, the potential which the electron sees is altered; within the framework of the Hartree approximation it becomes identical with the true potential (see §1.2) and this, because the atom as a whole is neutral, reaches zero at the atomic radius and does not have a coulomb tail stretching beyond. Hence the centrifugal potential when $l \geqslant 1$ may give rise, as suggested by the diagram in fig. 1.4, to a potential barrier around the atom with a height of order $\hbar^2 l(l+1)/2mR_A^2$ which may amount to several electron volts. It is therefore possible to envisage states of positive energy in which an electron is very nearly bound inside the centrifugal barrier – very nearly but not quite, because it must be able to tunnel through the barrier given time. It is in these circumstances that a virtual bound state exists. The abruptness with which η rises for such a state is of course related to the impenetrability of the barrier; the lower the energy of the virtual bound state the more impenetrable the barrier and the more abrupt the rise. To go back now to fig. 1.3, the curves suggest that in Cs there is a virtual bound d state some way above the Fermi level. Meyer *et al.* find that Li has a virtual bound p state quite close to the Fermi level, and they attribute many of the anomalous properties of metallic Li to this feature.

1.6. Friedel oscillations

We can use (1.40) to compute the change in the electron density, Δn, at large values of R due to the central perturbation. From (1.44) it follows that the number of states per unit range of k_R with a given value of l is

$$2(2l+1)(R_0+\mathrm{d}\eta_l/\mathrm{d}k_R)/\pi, \tag{1.47}$$

but $\mathrm{d}\eta_l/\mathrm{d}k_R$ is negligible here by comparison with R_0. Hence for large R

$$\begin{aligned}\Delta n &= \frac{1}{2\pi R^2 R_0}\sum_{l=0}^{\infty}\frac{2(2l+1)R_0}{\pi}\int_0^{k_{R,\mathrm{F}}}\{\sin^2(k_R R-\tfrac{1}{2}l\pi+\eta_l) \\ &\qquad\qquad -\sin^2(k_R R-\tfrac{1}{2}l\pi)\}\,\mathrm{d}k_R \\ &= \frac{1}{2\pi^2 R^3}\Sigma 2(2l+1)\int_0^{k_\mathrm{F}}\sin(2k_R R-l\pi+\eta_l)\sin\eta_l\,\mathrm{d}(2k_R R) \\ &= \frac{1}{2\pi^2}\Sigma(2l+1)(-1)^{l+1}\sin\eta_l\,\frac{\cos(2k_\mathrm{F}R+\eta_l)}{R^3}.\end{aligned} \tag{1.48}$$

Once again it is only the lowest values of l which contribute, which justifies the replacement of $k_{R,\mathrm{F}}$ by k_F in the upper limit of the integral. Hence at large distances from the perturbation there are liable to be oscillations in the electron density and associated oscillations, of course, in the potential. These are known as the

Friedel oscillations. Since their amplitude falls off like R^{-3} the potential round a point charge does not decay quite as rapidly as the screened coulomb law, with its exponential factor $\exp(-q_s' R)$ would suggest. The Friedel oscillations may still be apparent at a range of 2 or 3 atomic spacings, where $q_s' R$ may be 5 or 10.

There is nothing in our derivation of these oscillations to suggest that they arise only when the perturbation is strong, so that they should be implied also by calculations based upon the dielectric constant method. It is often claimed (e.g. Harrison, 1966, p. 52) that they are a consequence of the singularity in the dielectric constant at $q = 2k_F$. Except in the limit of *very* large R this is not strictly true because the slightest degree of blurring of the sharp Fermi surface, due to thermal excitation or to lifetime broadening, is enough to remove the singularity but it does not necessarily remove the oscillations: one can see immediately from (1.48) that to damp out the oscillations at a radius R requires a blurring over a range of k of order $\pi/2R$. The truth seems to be that the oscillations which undoubtedly occur in practice at a range of 2 or 3 atomic spacings arise from the general shape of the curve for ϵ rather than from its infinite slope at one point. It is interesting to observe that some sign of them is visible in fig. 1.1 where the curve computed using (1.28) is already oscillating with the expected periodicity, even at these relatively small values of R, about the exponentially screened potential; the computation was far too crude for the results to have been sensitive to the fine details of the dielectric constant's behaviour.

Quantitative treatments of the effects of thermal excitation and lifetime broadening on the amplitude of the oscillations have been attempted by Flynn & Odle (1963) and Gaskell & March (1963). The latter authors conclude that if the probability of occupation of the state k be written (compare the Fermi–Dirac function (1.6)) as

$$[\exp((k^2 - k_F^2)/k_F^2 \Delta) + 1]^{-1}, \qquad (1.49)$$

then values of Δ of at least 0.1 are needed before the oscillations at a range of 1 or 2 atomic spacings are thoroughly dampened. Since the value of Δ due to thermal excitation alone should be only $(k_B T/K_F)$ it is clear that the damping due to thermal excitation should be quite negligible at ordinarily accessible temperatures.

As for lifetime broadening, the arguments in §4.12 below suggest that

$$\tfrac{1}{2}-\frac{1}{\pi}\tan^{-1}\left(2\tau_l K_F(k^2-k_F^2)/\hbar k_F^2\right) \qquad (1.50)$$

may be a more appropriate formula to use for the occupation probability than (1.49), but for small $(k^2-k_F^2)$ at any rate the two are equivalent with $\Delta = \pi(\hbar/2\tau_l K_F)/4$. Hence, unless the rather longer tail of (1.50) for large $(k^2-k_F^2)$ is unexpectedly significant, the oscillations should not be damped out unless $(\hbar/2\tau_l K_F)$ is more than about 0.15. The figures in table 1.1 suggest that it does not exceed this limit in practice, except perhaps for the pentavalent liquid metals.

1.7. Scattering of electrons by potential wells

The phase shifts discussed in §1.5 play a crucial role in determining the scattering cross-section which a potential well presents to an incident electron. Suppose that the electron is described by a plane wave of the form $\exp(i\mathbf{k}_0\cdot\mathbf{r})$. It is well known that this can be decomposed into a set of partial waves which collapse inwards onto the well and a similar set which radiate outwards from it. In the presence of the perturbing potential the outgoing waves are shifted in phase by $2\eta_l$ and the wave function at large R becomes

$$\exp(i\mathbf{k}_0\cdot\mathbf{r})+\frac{\exp(ikR)}{2ikR}\sum_{l=0}^{\infty}(2l+1)P_l(\cos\theta)(\exp(2i\eta_l)-1),$$

where k is the magnitude of $\mathbf{k}_0$. Hence the amplitude of the wave scattered elastically through an angle θ, relative to the incident amplitude, is

$$\frac{1}{kR}\sum_{l=0}^{\infty}(2l+1)P_l(\cos\theta)\sin\eta_l\exp(i\eta_l). \qquad (1.51)$$

The larger the pase shifts the stronger the scattering. Evidently it is liable to be especially strong if the energy of the incident electron happens to coincide with that of a virtual bound state, in which case one of the η_l may pass through $\tfrac{1}{2}\pi$.

It is so particularly important to understand the way in which electrons are scattered in a metal, that it is worth a detour to compare the partial wave treatment with the alternative method based upon the *Born approximation.* In the latter, one sets out to calculate by perturbation theory the rate at which the amplitude of the scattered wave $\psi_1 \propto \exp(i\mathbf{k}_1 \cdot \mathbf{r})$ increases when the potential $\Delta U(\mathbf{r})$ is switched on. This rate is proportional to the matrix element

$$\int \psi_1^* \Delta U \psi \, d\mathbf{r}, \tag{1.52}$$

where $\psi(\mathbf{r})$ is the whole wave function at $\mathbf{r}$, i.e. it includes all the scattered waves as well as the incident wave ψ_0. These scattered waves have the effect of modulating the amplitude and phase of ψ_0 to some extent, and we may express this formally by writing

$$\psi(\mathbf{r}) = (1 + \gamma(\mathbf{r}))\,\psi_0(\mathbf{r}). \tag{1.53}$$

So long as the perturbation is weak, however, γ should be small and in the Born approximation it is ignored completely. The matrix element which determines the scattering is then proportional to

$$\int \exp(-i\mathbf{k}_1 \cdot \mathbf{r}) \Delta U \exp(i\mathbf{k}_0 \cdot \mathbf{r})\, d\mathbf{r} = \int \exp(-i\mathbf{q} \cdot \mathbf{r}) \Delta U \, d\mathbf{r} = \Delta U(\mathbf{q}), \tag{1.54}$$

i.e. to the $\mathbf{q}$th Fourier component of ΔU, where $\mathbf{q}$ is the *scattering vector*. For elastic scattering, as may be seen from fig. 1.5,

$$q = 2k \sin \tfrac{1}{2}\theta. \tag{1.55}$$

The condition for the Born approximation to give the same answer as (1.51) then turns out to be that

$$\Delta U(\mathbf{q}) = -(\Omega/\pi \mathcal{N}(E)) \sum_{l=0}^{\infty} (2l+1) P_l(\cos\theta) \sin \eta_l \exp(i\eta_l). \tag{1.56}$$

To check this result, consider the particular case of scattering by a point charge Q embedded in a jellium, for an electron at the Fermi level with $k = k_F$. For the Born approximation to be valid we must let Q be small, in which case the phase shifts are also small and we may replace $\sin \eta_l$ by η_l and $\exp(i\eta_l)$ by unity. If we further restrict ourselves to the limit when θ and q tend to zero, we can replace $P_l(\cos\theta)$ by unity and the right-hand side of

(1.56) can then be evaluated with the aid of the sum rule: it is $(\Omega Q/2e\mathcal{N}(E))$. But if Q is small the dielectric constant method is valid and the left-hand side (cp. (1.36)), is $(4\pi eQ/q^2\epsilon')$. These two answers do turn out to be the same if (1.30) is used for ϵ' and (1.24) for q_s^2.

If the perturbation is not weak and γ is not negligible, it may still be possible to use the machinery of the Born approximation – which is often convenient in practice – provided that ΔU is replaced by a suitable *effective* potential $\Delta U'$. One recipe which will give exactly the right answer, as may be seen by inserting (1.53) into (1.52), is to use

$$\Delta U'(\mathbf{r}) = \Delta U(\mathbf{r})(1+\gamma(\mathbf{r})). \tag{1.57}$$

However, it is evident from (1.51) that any change in ΔU whose only effect upon the phase shifts is to add or subtract some multiple of 2π to or from each of them leaves the scattering amplitudes the same, and it follows that there is an infinite set of ΔU's which cannot be distinguished by their scattering behaviour. Each of these may be plugged into (1.57) to give a different $\Delta U'$. Hence there is also an infinite set of effective potentials, all of which, when used in conjunction with the Born approximation, should work equally well. With luck, one of them may have Fourier components which are relatively weak and which do not extend out to large values of q so that a calculation which involves an integration over all q will rapidly converge.

Ziman (1964) has pointed out that if the phase shifts are known equation (1.56), with $\Delta U'(q)$ instead of $\Delta U(q)$ on the left-hand side, may be used to *define* an effective potential if desired. However, this equation only tells us how to calculate $\Delta U'$ when k_0 and k_1 are equal in magnitude. We shall have occasion later to consider matrix elements of $\Delta U'$ between states of unequal k (see fig. 1.5) and to calculate these we should have to generalise (1.56). That $\Delta U'(\mathbf{q})$ is not necessarily independent of the direction of $\mathbf{q}$ relative to k_0 follows from the fact that $\Delta U'(\mathbf{r})$ is not necessarily spherically symmetric in real space, even when $\Delta U(\mathbf{r})$ is; the lack of symmetry of $\Delta U'(\mathbf{r})$ is a consequence of lack of symmetry in the modulation factor $\gamma(\mathbf{r})$ in (1.57). Another complication is that since γ (or else the phase shifts, if one prefers to base the

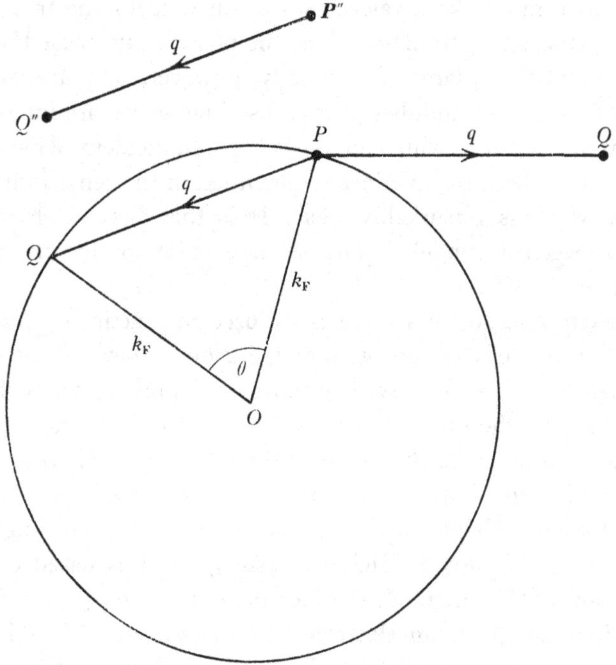

Fig. 1.5. Representation in **k**-space of three scattering processes for which $\Delta U'$ may be different although q is the same. P and Q lie on the Fermi sphere and the process linking these two states is elastic.

calculation of $\Delta U'$ upon (1.56)) is liable to vary with k_0, the effective potential may depend upon the wave number and energy of the incident electron. We cannot treat $\Delta U'$ as an ordinary potential which is single-valued, and it is sometimes described as a *non-local operator* in consequence.

1.8. Pseudo-potentials in the ion core

In a real metal, as opposed to a jellium, the positive charge is, of course, concentrated in ions, and the conduction electrons penetrate the ion cores to experience an attractive potential that becomes, near the nucleus, very strong indeed compared with K_F. If we want to describe the conduction electrons by plane wave functions we must therefore admit that the modulation factor γ introduced in the previous section is liable to be large inside a core and to oscillate rapidly, making the true wave function in this region look more like a valence wave function for the free atom than a plane wave. In other words the phase shifts caused by the ionic potential are large. Fortunately, however, they are usually close to an integral number of π and so long as we are interested only in the extent to which an ion scatters an incident plane wave we can describe it by an effective potential, in the sense indicated above, which is remarkably weak. It is this fact which makes the free-electron model a good starting point for the theory of metals.

The effective potentials which are used in practice to describe the interaction of electrons with metallic ions – they are known as *pseudo-potentials* – are derived by a variety of approximate methods, of which only the method of Meyer *et al.* (1967) for monovalent metals corresponds to the recipe outlined in the previous section. It would be out of place to discuss these exhaustively (see Heine, 1970; Cohen & Heine, 1970) but it is instructive to go through the argument of Phillips & Kleinman (1959). This is based on the well-known OPW method, the idea being to decompose the true wave function ψ for an electron into a *pseudo* wave function ϕ, which describes its asymptotic behaviour at large distances but varies *smoothly* inside the ion cores, and a residue which contains the rapid oscillations; this residue is then expressed in terms of the core eigenfunctions ψ_c. It is because these core eigenfunctions form an almost complete set of orthonormal functions in the region occupied by the core that one may reasonably hope to expand the oscillatory component of ψ in terms of them. Strictly the core functions we should use, and the associated eigenvalues of the energy, are those which would be appropriate if the core electrons experienced the same Hartree potential, including any correction for exchange and correlation that may be necessary, as do the conduction electrons; it should be apparent from the discussion in §1.2 above that the Hartree potential experienced by one of the actual core electrons will be noticeably different. This reservation is necessary to ensure that ψ_c is an eigenfunction of the same Hamiltonian, H, that acts on ψ.

Let us therefore write

$$\psi = \phi - \sum_c a_c \psi_c, \qquad (1.58)$$

and suppose for the moment that ψ is one of the eigenfunctions of H. Then ψ must be orthogonal to all the core functions, so that†

$$\langle\psi_c|\phi\rangle - \sum_{c'} a_{c'}\langle\psi_c|\psi_{c'}\rangle = 0,$$

or

$$a_c = \langle\psi_c|\phi\rangle \tag{1.59}$$

since the core functions are orthogonal to one another. Substitution of (1.58) into the Schrödinger equation

$$H\psi = E\psi \tag{1.60}$$

therefore leads to

$$\left[H + \sum_c (E-E_c)\frac{\langle\psi_c|\phi\rangle}{\phi}\psi_c\right]\phi = E\phi. \tag{1.61}$$

The pseudo wave function ϕ is therefore an eigenfunction of a *pseudo-Hamiltonian*, with the same eigenvalue for the energy as (1.60). The pseudo-Hamiltonian contains an extra term which is essentially positive (since E is always greater than E_c) and which plays the role of a repulsive potential confined to the ion core. It is to be hoped that this will cancel most of the strongly attractive potential in the true Hamiltonian H, leaving only a relatively weak pseudo-potential to be considered. Reasons why the cancellation should indeed be fairly effective have been given by Heine (1970) and Pendry (1971).

Equation (1.61) is exact so far as it goes, but we have made the assumption in deriving it that ψ is an eigenfunction, in which case ϕ is a partial wave such as equation (1.40) describes. If we want to determine the pseudo-potential experienced by an electron whose asymptotic wave function is a plane wave, we ought really to decompose this plane wave into partial waves and evaluate the repulsive term in (1.61) differently for each. A reasonable approximation when the resultant pseudo-potential is weak may be to write it as

$$\sum_c (E-E_c)\frac{\langle\psi_c|\exp(i\mathbf{k}\cdot\mathbf{r})\rangle}{\exp(i\mathbf{k}\cdot\mathbf{r})}\psi_c, \tag{1.62}$$

† The convention of the notation here is that

$$\langle x|O|\rangle y> = \int x^* O y \,d\mathbf{r},$$

the integral being taken over the whole volume Ω in which the wave functions are normalised. O may be any operator, in this case the unit one.

but this cannot be exact. One could approximate still further by saying that, for values of k near k_F, $\exp(i\mathbf{k}\cdot\mathbf{r})$ should not vary much over the region where the ψ_c are significant; (1.62) can then be simplified to

$$\sum_c \langle\psi_c|1\rangle\psi_c(E-E_c),$$

which has the great advantage that it is independent of k. We are forced to recognise, however, that this last step may be going too far. The more perfect the cancellation between the attractive and repulsive terms that make up the pseudo-potential the more likely is the result to be a *non-local* operator in the sense defined in the previous section.

The normalisation of the pseudo wave functions is a matter of some importance. If we suppose that

$$\phi = (C/\Omega)^{\frac{1}{2}} \exp(i\mathbf{k}\cdot\mathbf{r}), \tag{1.63}$$

it is readily shown that the normalisation condition

$$\langle\psi|\psi\rangle = 1$$

implies

$$C^{-1} = 1-\Omega^{-1}\sum_c |\langle\psi_c|\exp(i\mathbf{k}\cdot\mathbf{r})\rangle|^2. \tag{1.64}$$

This is for a single ion surrounded, as we have tacitly assumed throughout the above discussion, by jellium. If there are N ions present the correction term on the right should be N times bigger. A knowledge of how the core functions behave enables C to be estimated from (1.64), and a typical value for it in a real metal seems to be about 1.1 for $k = k_F$; the difference from unity is usually small and in most calculations is completely ignored.

The fact that C, as given by (1.64), is always greater than unity is a mathematical consequence of the OPW procedure of little physical significance. It does *not* imply, as may appear at first sight, that the conduction electrons in a metal are necessarily repelled from the ion cores. To find the distribution of the electronic charge we must of course go a step further and allow the pseudo-potential to perturb the pseudo wave functions. This perturbation is bound, for small k at any rate, to overcome the repulsive effects of orthogonalisation and to draw the electrons back into the core, for otherwise there would be no screening.

Micah, Stocks & Young (1969) have discussed the issue of normalisation using the language of partial waves and phase shifts and with the aid of (1.46) they obtain a result which, for a single ion immersed in jellium, is equivalent to

$$C^{-1} = 1+(2\pi/\Omega k^2)\sum_l (2l+1)\frac{\partial \eta_l}{\partial k}.$$

This appears to be in flagrant disagreement with (1.64), since the correction term on the right is certainly positive for small k (see fig. 1.3) and, if the phase shifts of Meyer *et al.* are to be trusted, is still positive for $k = k_F$ in all the alkali metals. In the partial-wave analysis, however, the effect of the potential in drawing in the electrons is included from the start.

It is possible to generalise the Phillips–Kleinman approach and to produce any number of different formulae from which a pseudo-potential of sorts may be calculated (Cohen & Heine, 1970). They are none of them to be trusted to better than a few tenths of an eV, however, and for practical purposes it is probably more accurate, and certainly more convenient, to assume some simple and plausible model for the pseudo-potential and to determine the one or more adjustable parameters that characterise it by reference to experimental data. Heine and his colleagues, for example, have assumed the model illustrated by fig. 1.6(*a*). Outside the core one has, if the ion is supposed to be immersed in jellium, an ordinary coulomb potential $-ze^2/R$; inside some model radius R_M one has only a weak potential A that is uniform in space. It can be shown that if a potential of this form is used in a 'pseudo' Schrödinger equation like (1.61), it is capable of giving all the correct eigenvalues and scattering matrix elements for a conduction electron, provided that one always takes the trouble to split up an incident plane wave into partial waves and allows A to depend not only on the energy of the incident electron but also on the quantum number l. Heine *et al.* determine appropriate values of A up to $l = 2$ by appeal to experimental data at a rather fundamental level: they use spectroscopic term values for the free ion, as in the quantum defect method. Their calculations are for an energy corresponding to $k = k_F$ and their final results are expressed as a table of Fourier components of the pseudo-potential, it being understood that

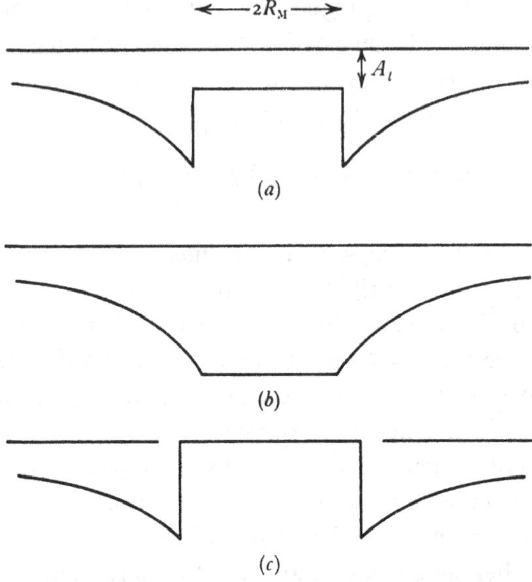

Fig. 1.6. Three model pseudo-potentials (unscreened): (*a*) Heine, (*b*) Shaw, (*c*) Ashcroft.

$(\mathbf{k}+\mathbf{q})$ lies on the Fermi surface as well as $\mathbf{k}$ when $q \leqslant 2k_F$ (see fig. 1.5) or is anti-parallel to $\mathbf{k}$ when $q > 2k_F$. It is an inevitable feature of the model, with its sharp change at the radius R_M, that the pseudo-potential oscillates in q space; R_M is chosen so that the oscillations decay as rapidly as possible. Shaw (1968) has achieved a more rapid convergence using the model illustrated by fig. 1.6(*b*): he allows R_M as well as A to vary with energy and l.

For some purposes it may be sufficient to ignore the variation of both R_M and A (e.g. Cohen, 1962). Indeed, Ashcroft (1966*a*, 1968) and Ashcroft & Langreth (1967*a*,*b*) have set A uniformly zero as in fig. 1.6(*c*), leaving only one adjustable parameter R_M, which they suggest may be fixed by reference to the resistivity of the liquid metal. Other authors have adjusted their semi-empirical pseudo-potentials so as to give the right values for the band gaps in the electronic energy spectrum of the solid phase (see § 1.12) and hence the right shape for the Fermi surface (e.g. Stark & Falicov, 1967), or else so as to fit the observed phonon dispersion curves for the solid. Schneider & Stoll (1966), for example, are able to determine up to six adjustable parameters for their relatively complicated model by reference to dispersion curves; but, like Ashcroft's, theirs is a model in which the non-locality and energy dependence of the pseudo-potential are lost from view.

It is a feature common to all the models so far suggested that well outside the core, for a single ion immersed in jellium, the pseudo-potential is that of a point charge and varies like R^{-1}. This means that for small q its Fourier components diverge like q^{-2}, and the non-divergent contribution due to the core becomes negligible when q is small enough. Thus the 'empty core' model suggested by Ashcroft leads to the result†

$$\begin{aligned}\Delta V_{\mathrm{b}}(q) &= -\frac{4\pi z e^2}{q^2}\cos qR_{\mathrm{M}}\\ &= -4\pi z e^2(q^{-2}-\tfrac{1}{2}R_{\mathrm{M}}^2+\ldots).\end{aligned} \tag{1.65}$$

Other models lead to comparable formulae in the small-q region.

It is arguable, however, that the usual valency of the ion, z, should be replaced in (1.65) by an effective valency z^* which is somewhat greater. We have seen that, before screening is taken into account, the effect of orthogonalisation to the core functions is to exclude conduction electrons from the core, and on this account the apparent positive charge carried by the core may be enhanced. From (1.64) one has

$$z^*-z \sim n\sum_{\mathrm{c}}|\langle\psi_{\mathrm{c}}^*|\exp(\mathrm{i}\mathbf{k}\cdot\mathbf{r})\rangle|^2, \tag{1.66}$$

where the right-hand side should be averaged over values of k from 0 to k_{F}. The correction is not a large one and by many authors it has been disregarded.

Equation (1.65) describes the bare, or unscreened, pseudo-potential. It is calculated on the assumption that the true conduction electron wave functions have developed oscillations inside the ion core, but that their *pseudo* wave functions are still unperturbed

† At the risk of confusion we shall use the same symbols V_{b} and U to denote respectively a bare and a screened *pseudo*-potential as we have used previously for true potentials. The addition of Δ implies that we are considering a small perturbation of a primitive jellium. The Δ will be dropped in the following section, where we abandon the jellium model.

plane waves of uniform amplitude. When the pseudo wave functions are pertubed by the pseudo-potential there must be a shift of charge density such as to screen ΔV_b. Granted that the perturbation is weak, the resultant (screened) ΔU, including a correction for the exchange + correlation hole, may in principle be determined by the dielectric constant method, though the fact that ΔV_b is non-local somewhat complicates the computation of the appropriate dielectric constant (Heine & Weaire, 1970). An analytic expression may be obtained in the limit of small q from (1.65), (1.29), (1.22) and (1.24):

$$\Delta U(q \to 0) = -4\pi z^* e^2/q_s^2 = -z\Omega/2\mathcal{N}(E)_F. \qquad (1.67)$$

The true valency appears in the final answer rather than z^* because the screening efficiency of the conduction electrons (i.e. q_s^2) should be enhanced by the normalisation factor C, and this is just z^*/z.†

The expected behaviour of $\Delta U(q)$ for large q is shown in fig. 1.7. The curve for Sn is typical of the results computed by Animalu (1965) using the model of Heine *et al.* and those for Al and Na, based upon the models of Shaw and Ashcroft respectively, are not very different. The curves for Cu (due to Harrison (1969) and Moriarty (1970)), Cs (due to Bortolani & Calandra (1970)) and Hg (due to Evans, Greenwood, Lloyd & Ziman (1969)) represent three rather special cases. Hg is of interest because the topmost core states (5d) are unusually close to the conduction band (6s), which seems to affect may of its electrical properties, as we shall see below. According to Evans *et al.* the phase shift η_2 is exceptionally large on this account (0.2 instead of about 0.04 as in most similar metals), and it is this feature which makes their curve for Hg bend downwards as it does. Incidentally, the Evans pseudo-potential for Hg is a highly non-local one; the values plotted in the figures are those appropriate when the two states linked by $\mathbf{q}$ both lie on the Fermi sphere. In the noble metal Cu the 3d band actually overlaps the conduction band and it used to be feared that this would invalidate the pseudo-potential approach. The pseudo-potentials calculated

† For a more careful discussion of this point see Shaw & Harrison (1967). Readers who are concerned with the distinction between z and z^* should also consult Ballentine (1968).

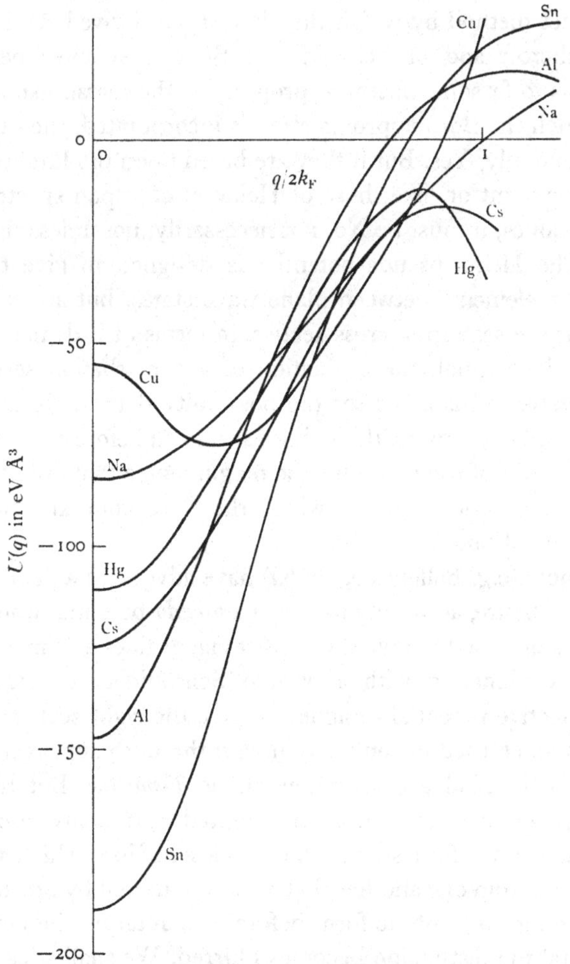

Fig. 1.7. Screened pseudo-potentials or form factors for scattering round the Fermi sphere.

by Moriarty for Cu, Ag and Au, however, seem to be quite capable of explaining the band gaps for the solid phase and we shall see later that they account for the resistivity of the liquid as well.

How accurately do these model pseudo-potentials correspond to the effective potential advocated in §1.7? If they are used in conjunction with the Born approximation do they necessarily give the correct scattering cross-section for an ion, or at least a good approximation to it? The answer to this question rather depends upon the exact method by which they have been derived. If, like those of Ashcroft and of Schneider & Stoll, they have been adjusted so as to fit some observed property of the metal, using a theory in which the Born approximation is incorporated, then the answer is, hopefully, Yes. But if they are based upon the Phillips–Kleinman argument or, like those of Heine *et al.*, upon spectroscopic term values, it must be No, not necessarily, not unless they are weak. The Heine pseudo-potential is designed to give the correct matrix elements between plane wave states, but an exact calculation of the scattering cross-section in terms of it demands, as with a real potential, the evaluation of a perturbation series whose successive terms allow for the possibility that an electron may be scattered once, twice, three times and so on before escaping from the influence of the ion. There is no guarantee that only the first term in this series, the one which the Born approximation describes, is significant.

Some authors (e.g. Ballentine, 1966*a*) have advocated a distinction of nomenclature, according to which a *pseudo*-potential would always be one designed to give the correct eigenvalues and matrix elements in conjunction with a pseudo Schrödinger equation, while any effective potential designed to give the right scattering cross-section when used in conjunction with the Born approximation would be labelled a *quasi*-potential, or *T-matrix*. But this distinction has not been systematically adopted, and in any case it is only a clear-cut one for a single ion or for a set of ions which are so far separated from one another that a wave scattered by one has always reached its asymptotic form before encountering the next. In a real metal the distinction becomes blurred. We shall stick to the name 'pseudo-potential' in what follows.

1.9. Coherent scattering by assemblies of ions

The time has now come to abandon the simple jellium model with which we have been largely concerned so far and to think of a metal in more realistic terms, i.e. as an assembly of positive ions with a good deal of space between them, filled only by the conduction electron sea. How do we calculate the pseudo-potential which is seen by an electron wave as it travels through an assembly of this sort?

It is best to start by considering the bare pseudo-potential, to be denoted by V_b.† By this is meant, as above, the pseudo-potential that would be seen if the pseudo wave functions of all the conduction electrons were frozen in their unperturbed condition of uniform amplitude, i.e. if the total charge density of the conduction electron sea were uniform, apart from the slight deficiency inside each ion core which converts the effective charge on each ion from $-ze$ to $-z^*e$. V_b should consist of two parts, one due to the ions alone, and the other to the electrons. In order to make progress we need to assume that the contributions made by separate ions are additive, in which case we may write

$$V_b(\mathbf{r}) = \sum_i v_b(\mathbf{r}-\mathbf{r}_i) + V_{b,\,elec},$$

$v_b(\mathbf{r}-\mathbf{r}_i)$ being the pseudo-potential due to the ith ion alone, whose centre is at $\mathbf{r}_i$. The assumption of additivity is certainly legitimate at a point well outside any of the ion cores, where $v_b(\mathbf{r}-\mathbf{r}_i)$ is just the coulomb potential $-z^*e^2/|\mathbf{r}-\mathbf{r}_i|$. Inside an ion core, however, we shall need to use some model pseudo-potential for v_b, of the sort discussed in the previous section, and this may bear no very direct relation to the true electrostatic potential. It is far from certain that the v_b remain additive inside an ion core. We must just hope that the ions stay far enough apart in practice to prevent the pseudo-potential of one atom from overlapping too seriously the core regions of its neighbours. In the end, of course, it is the overlap of screened rather than unscreened, pseudo-potentials which really matters, and the assumption of additivity then looks more plausible.

The coulomb potential is a long-range one and the ionic contribution to $V_b(\mathbf{r})$ necessarily diverges as the size of the specimen is increased. Since the metal as a whole is electrically neutral, however, the divergent term in V_b due to distant ions is bound to be cancelled by an equal and opposite term due to distant conduction electrons, so that V_b as a whole is independent of the specimen size. Indeed, V_b, in the neighbourhood of $\mathbf{r}_i$, should be dominated by the effects of the charge on the ith ion and on the part of the electron sea which immediately surrounds it. A convenient first

† See footnote regarding notation on p. 39.

approximation in this neighbourhood is to ignore *all* the other ions and simultaneously to ignore all the conduction electrons for which

$$R = |\mathbf{r}-\mathbf{r}_i| > R_A$$

where R_A is the mean atomic radius defined by

$$\Omega/N = \tfrac{4}{3}\pi R_A^3. \tag{1.68}$$

Then for a value of R which is less than R_A but greater than the core radius one has

$$V_b(R) \sim z^*e^2\left(-\frac{1}{R}+\frac{3}{2R_A}-\frac{R^2}{2R_A^3}\right). \tag{1.69}$$

This potential goes to zero at $R = R_A$ because the 'atom' as a whole is neutral; evidently the conduction electrons do a good deal, even before screening has been taken into account, to mitigate the potential due to the ion alone. This consideration encourages one to suppose that V_b is everywhere weak enough for the effects of screening to be calculated in due course by the dielectric constant method.

To obtain a more exact representation of V_b it is helpful to concentrate upon its Fourier transform, $V_b(\mathbf{q})$. It turns out that *so long as we restrict attention to those Fourier components for which q is greater than the inverse of the specimen size it is legitimate to ignore the contribution due to the electrons.* This is a vital point in the argument, which requires careful justification. Consider, for simplicity, the case of a spherical specimen of radius R_0, in which $V_{b,\,elec}$ varies like R^2 about the centre. To calculate its qth Fourier component one may use Poisson's equation to relate this to the corresponding Fourier component in the charge density of the conduction electron sea. Thus

$$\begin{aligned} V_{b,\,elec}(\mathbf{q}) &= \frac{4\pi ne^2}{q^2}\int \exp{(i\mathbf{q}\cdot\mathbf{r})}\,d\mathbf{r} \\ &= 16\pi^2ne^2(\sin qR_0 - qR_0\cos qR_0)/q^5. \end{aligned} \tag{1.70}$$

For values of q so small that qR_0 is less than unity, $V_{b,\,elec}(q)$ comes out proportional to R_0^3 and is certainly not negligible. When qR_0 is large, however, it increases only as R_0, whereas $V_b(q)$ (as will emerge more fully in chapter 2) is proportional to the square root of the number of ions present, i.e. to $R_0^{\frac{3}{2}}$.

With this justification we may neglect $V_{\mathrm{b,\,elec}}$ for finite q in large specimens and write

$$\begin{aligned} V_{\mathrm{b}}(\mathbf{q}) &= \int \exp(-i\mathbf{q}\cdot\mathbf{r}) \Sigma v_{\mathrm{b}}(\mathbf{r}-\mathbf{r}_i)\,\mathrm{d}\mathbf{r} \\ &= \sum_i \int \exp(-i\mathbf{q}\cdot\mathbf{R}) v_{\mathrm{b}}(\mathbf{R})\,\mathrm{d}\mathbf{R}\cdot\exp(-i\mathbf{q}\cdot\mathbf{r}_i) \\ &= v_{\mathrm{b}}(\mathbf{q})F(\mathbf{q}), \end{aligned} \tag{1.71}$$

where F is the *structure factor* defined by

$$F(\mathbf{q}) = \sum_i \exp(-i\mathbf{q}\cdot\mathbf{r}_i), \tag{1.72}$$

the sum being taken over all the ions in the specimen.† The factorisation of (1.71) into a term which depends only on the properties of a single ion and a term which depends only on the structure hinges, of course, on the assumption that all the ions are identical; the argument is generalised to cover alloys in §2.12.

Provided that the pseudo-potential is indeed weak enough to justify the dielectric constant method, we may now obtain the Fourier transform of the *screened* pseudo-potential U by dividing (1.71) by $\epsilon'(q)$. It is best to absorb this factor on the right-hand side into the term which refers to a single ion, i.e. to write

$$U(\mathbf{q}) = (v_{\mathrm{b}}(\mathbf{q})/\epsilon'(q))F(\mathbf{q}) = u(\mathbf{q})F(\mathbf{q}). \tag{1.73}$$

The quantity $u(\mathbf{q})$ here is the same thing as the $\Delta U(\mathbf{q})$ of §1.7. Thus, as Ziman (1964) was the first to stress, we can calculate the screened pseudo-potential of an assembly of ions by calculating $u(\mathbf{q})$ for a single ion first, as though it were immersed in a jellium, and then multiplying by the structure factor F. An elegant and powerful result.

If we choose to discuss the Fourier transform of the pseudo-potential in this way, the argument to justify the use of the dielectric constant method which was based on equation (1.69)

† To the X-ray crystallographer the term 'structure factor' really implies a sum over all the atoms or ions of a unit cell of a solid lattice. Its use as above is excusable for liquids on the grounds that no periodic lattice exists. In some work on liquid metals it has been denoted by $S(\mathbf{q})$, but it seems preferable to avoid this symbol since it has been applied by other authors to the interference function.

above needs to be rephrased. The equivalent argument in $\mathbf{q}$-space is that $N^{-\frac{1}{2}}F(\mathbf{q})$ is much less than unity, as we shall discover in chapter 2, over the range where q is small, which is where $u(q)$ tends to be particularly large. This does a lot to weaken the resultant pseudo-potential. And for large q, where $N^{-\frac{1}{2}}F$ becomes comparable with unity, the screening is so ineffective anyway (i.e. ϵ' is so close to unity) that it should matter little just how one sets out to estimate it. But whether these qualitative arguments are really good enough has not been seriously checked by quantitative calculations.

We shall find later that $U(\mathbf{q})$ is what we most often need to know, rather than $U(\mathbf{r})$. If, for example, we want to calculate how an electron wave is scattered as it travels through a liquid metal we shall use perturbation theory, and so long as the perturbation is weak enough for a first-order theory to be sufficient, i.e. for the Born approximation to be valid, we shall find the scattered amplitude to be directly proportional to $U(\mathbf{q})$, where $\mathbf{q}$ is the scattering vector as in (1.54).† It is worth recalling the usual method whereby the scattering of say X-rays by an assembly of ions or atoms is discussed. One starts with a *form factor* f, which describes the amplitude of the X-ray scattered by an individual particle in the assembly, and then adds the contribution of all the particles *coherently*, taking due account of their phase. A wave which emanates from a distant source at $\mathbf{r}_S$ and is scattered from the ith particle arrives at some distant detector at $\mathbf{r}_D$ with a phase

$$\mathbf{k}_0 \cdot (\mathbf{r}_i - \mathbf{r}_S) + \mathbf{k}_1 \cdot (\mathbf{r}_D - \mathbf{r}_i) = -\mathbf{q} \cdot \mathbf{r}_i + \text{constant}.$$

Hence from this point of view the total scattered amplitude at the detector should be proportional to

$$\sum_i f(\mathbf{q}) \exp(-\mathrm{i}\mathbf{q} \cdot \mathbf{r}_i) = f(\mathbf{q}) F(\mathbf{q}). \qquad (1.74)$$

The analogy with (1.73) is obvious. *For an electron wave $u(\mathbf{q})$ is the analogue of the X-ray form factor.*

It must be emphasised that even if U is weak enough for the

† The theory of scattering is more fully discussed in chapters 3 and 4 where, in particular, the possibility that the scattering may be *inelastic* is examined. Throughout the present chapter we make the simplifying assumption that the ions can be regarded as at rest, in which case the scattering occurs without any Doppler shift of frequency.

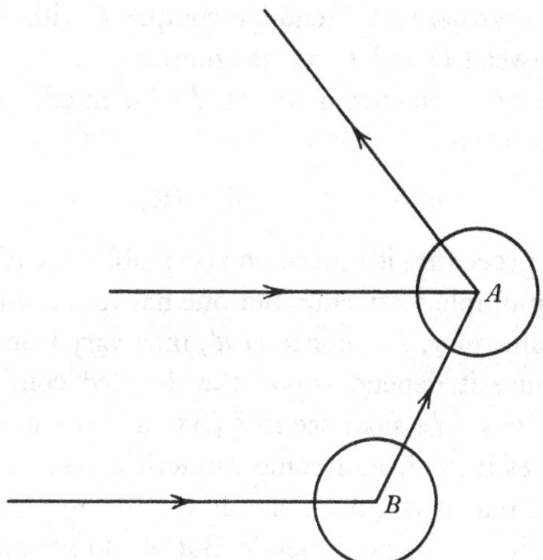

Fig. 1.8. Multiple scattering.

dielectric constant method to provide an adequate account of screening, it is not necessarily weak enough for the Born approximation to be valid in a scattering calculation. The wavelet scattered from the ion at A in fig. 1.8 includes a component which has previously been scattered from another ion at B, and in principle there are an infinite number of multiple scattering processes involving more than one ion to be considered, even when the ionic pseudo-potential has been adjusted so that it fully describes the multiple scattering within a single ion. An argument in § 1.7 suggests that we may preserve the formalism of the Born approximation if we replace U by an effective pseudo-potential

$$U'(\mathbf{r}) = (1+\gamma(\mathbf{r}))\,U(\mathbf{r}), \tag{1.75}$$

the amplitude of the wavelet that arrives at A from B, for example, being allowed for in the modulation factor $\gamma(\mathbf{A})$. Now it is just the modulation of the electrons' pseudo wave functions which is responsible for screening. Faith in the dielectric constant method implies that γ can be calculated with sufficient accuracy by first-order perturbation theory, i.e. that it is linearly proportional to the strength of U. It does not follow that for an electron at the Fermi surface (the screening is done by all the electrons but in most scattering calculations it is only the ones at the Fermi surface that matter) γ is everywhere small enough compared with unity for the distinction between U and U' to be ignored.

It would be a convenience if we could allot an effective pseudo-potential to each ion,

$$u'(\mathbf{R}) = (1+\gamma(\mathbf{R}))u(\mathbf{R}).$$

Formally this procedure is indeed an acceptable way of expressing the effects of multiple scattering, but one has to recognise that the appropriate value of γ, and hence of u', may vary from one ion to the next because it depends upon the detailed configuration of neighbouring ions. We shall see in §4.11 how the mean value of $\gamma(\mathbf{R})$ may be estimated, and some numerical results are quoted which suggest that it could be significant for an electron at the Fermi surface in many liquid metals. But we do not know enough about their structure to take into account the fluctuations of γ, and this makes it impossible to go beyond the Born approximation without a good deal of guesswork.

1.10. Effective masses for conduction electrons

It was emphasised in §1.3 that the free-electron formula for the density of states may be inaccurate even for a primitive jellium, and it looks still more unreliable for a real metal. To first order in the pseudo-potential the potential energy of an electron in the state $\mathbf{k}$ is

$$\Omega^{-1}\langle \exp(i\mathbf{k}\cdot\mathbf{r})|\,U(\mathbf{r})\,|\exp(i\mathbf{k}\cdot\mathbf{r})\rangle, \tag{1.76}$$

and it will be clear from what has been said about pseudo-potentials that this is liable to depend upon k, quite apart from whether or not the exchange + correlation correction is k-dependent.

If the pseudo-potential is a straightforward local one, then (1.76), apart from the Ω^{-1}, is just its zero q Fourier component, i.e. the integral of $U(\mathbf{r})\,d\mathbf{r}$ over the whole specimen. It is important to note that this is *not* the same quantity as the apparent limit of (1.73) when q goes to zero, because (1.73) is not valid unless q is still large compared with the inverse of the specimen size. In fact the evaluation of (1.76) is a task of some complexity. It is sometimes assumed (e.g. Heine, 1969; Ashcroft & Langreth, 1967*a*) that one may proceed by integrating the *bare* pseudo-potential $V_b(\mathbf{r})$, together with some uniform correction for the exchange +

correlation hole, on the grounds, presumably, that the shift of electronic charge which accompanies screening merely adds some oscillatory Fourier components to $V_{\mathrm{b, elec}}$ which do not alter its mean value. This procedure is not entirely trustworthy, however; a shift of electronic charge which gives rise to a Fourier component of the charge density with wave number q generates a Fourier component in $V_{\mathrm{b, elec}}$ which is inversely proportional to q^2, and because q^{-2} diverges it is necessary to be careful in proceeding to the limit $q = 0$. In fact it is readily verified, by consideration of some simple examples, that the mean value of $V_{\mathrm{b, elec}}$ is *not* necessarily unaffected by displacement of the electronic charge.

Leaving this criticism on one side, and neglecting also the exchange + correlation correction, the part of (1.76) which is liable to depend upon k comes only from the region inside each ion core, where we need to invoke some model for the pseudo-potential. With the model of Heine *et al.*, for example, the rate of change of (1.76) with k involves the energy dependence of the well-depths A; the necessary formulae are given by Weaire (1967). Unless the ion cores overlap, the answer should not depend upon the way in which they are arranged, so that to first-order in the pseudo-potential the density of states should be the same in a liquid metal as in the solid at the same volume.

We shall find it convenient to represent the effect of the pseudo-potential on the density of states at the Fermi level by means of an *effective mass* for the conduction electrons. The effective mass is usually introduced to describe the rate at which electrons are accelerated by a field, as in the equation

$$\Omega \frac{\mathrm{d}j_x}{\mathrm{d}t} = \frac{e^2 \mathsf{E}_x}{m_{xx}}$$

for the current density j per electron induced by an electric field E; and in this case, as is well known,

$$\frac{1}{m_{xx}} = \hbar^{-2} \frac{\partial^2 E}{\partial k_x^2}.$$

If there is spherical symmetry in $\mathbf{k}$-space, however, then

$$\frac{\partial^2 E}{\partial k_x^2} = \frac{\partial^2 E}{\partial k^2}\frac{k_x^2}{k^2} + \frac{1}{k}\frac{\partial E}{\partial k}\left(1 - \frac{k_x^2}{k^2}\right),$$

and an average over all the occupied states within the Fermi sphere yields

$$\overline{\left(\frac{1}{m_{xx}}\right)} = \frac{1}{\hbar^2 k_F^3}\int_0^{k_F}\frac{1}{3}\left(\frac{\partial^2 E}{\partial k^2}+\frac{2}{k}\frac{\partial E}{\partial k}\right)k^2\,dk = \frac{1}{\hbar^2 k_F}\left(\frac{\partial E}{\partial k}\right)_F.$$

If, therefore, we choose to define an effective mass m^* by the equation

$$m^*/m = \mathcal{N}(E)_F/\mathcal{N}(E)_{F,\,FE}, \tag{1.77}$$

where $\mathcal{N}(E)_{F,\,FE}$ is the free-electron density of states, its inverse is just the average of the inverse of the more conventional m_{xx}. We may use (m_1^*/m) to represent the effect of the pseudo-potential to first-order only.

It is when the calculation of m^* is carried beyond the first order the structure-dependent terms are liable to appear. Higher order terms will be discussed in detail in chapter 4, but there is no difficulty in writing down straight away the contribution to the energy associated with the state $\mathbf{k}$ which is of order U^2. Orthodox perturbation theory shows this to be

$$\frac{2m}{\hbar^2\Omega^2}\sum_{q\neq 0}\frac{U(\mathbf{q})U(-\mathbf{q})}{k^2-|\mathbf{k}+\mathbf{q}|^2};$$

its average (over an ensemble of possible configurations for the specimen) can be expressed with the aid of (1.73) in the form

$$\frac{2mN}{\hbar^2\Omega^2}\sum_{\mathbf{q}\neq 0}\frac{u(\mathbf{q})u(-\mathbf{q})a(\mathbf{q})}{k^2-|\mathbf{k}+\mathbf{q}|^2}, \tag{1.78}$$

where $a(\mathbf{q})$ is a quantity known as the *interference function* (see §2.10) defined by

$$a(\mathbf{q}) = N^{-1}\,\overline{F(\mathbf{q})F(-\mathbf{q})} = N^{-1}\overline{\sum_{i,j}\exp i\mathbf{q}\cdot(\mathbf{r}_j-\mathbf{r}_i)}. \tag{1.79}$$

Clearly, this second-order contribution is just as likely as the first-order one to depend upon k and hence to make m^* differ from m. Equally clearly, the structure enters through the quantity $a(\mathbf{q})$. In the second order, therefore, m^* for the liquid and solid need not be identical.

Since we have introduced in the previous section the idea of an effective pseudo-potential u', it may be worth remarking that a calculation of $\mathcal{N}(E)_F$ or of m^*/m to second order in u is the same as a calculation to first order in u', if the modulation factor γ that distinguishes the two is itself expressed to the first order only in u. The interested reader may verify this statement by reference to §4.11. We shall consider in some detail in chapter 4 how the calculation can in principle be made more accurate, e.g. by using energies $E(k)$ that have an imaginary component to allow for scattering and by computing γ self-consistently in terms of u' rather than u. In all this work, however, we shall be circumscribed by a lack of detailed knowledge about the local order of the ions in liquid metals and about how this order varies from point to point. We shall be restricted, in fact, to what has sometimes been called (e.g. by Ham, 1962) the *spherical approximation*,† in which the environment of all the ions is assumed to be identical and to be spherically symmetric. Evidently we are bound, with this approximation, to land up with spherical symmetry in **k**-space also, and a spherical Fermi surface. In principle, as soon as third-order terms in u become significant one ought to contemplate the possibility of a distorted Fermi surface, the axes of the distortion having different orientations in different parts of the specimen, as in a polycrystalline solid. Fortunately perhaps there is no indication that in real liquid metals the situation is ever so complex.

Theoretical estimates for m^*/m in solid metals vary quite widely according to the method employed in the calculation. The figures quoted in the first three columns of table 1.2, which are taken from (*a*) Ham (1962), (*b*) Ashcroft & Wilkins (1965), Ashcroft & Lawrence (1968) and Allen, Cohen, Falicov & Kasowski (1968), and (*c*) Shaw & Smith (1969), illustrate the sort of predictions that have been made for some of the relatively 'free-electron-like' metals, i.e. for metals whose Fermi surfaces are known from work on the de Haas–van Alphen effect to be almost spherical. The next two columns show some recent estimates, based on different model pseudo-potentials, for the effect of the first-order contribution alone; these are due to (*d*) Weaire (1967) and (*e*) Shaw & Smith (1969) and Shaw (1969). A quick comparison of the entries in

† Ham's quantum defect method for calculating effective masses in the spherical approximation and older methods such as that of Bardeen (1938) appear to be equivalent to the modern pseudo-potential approach.

TABLE 1.2. *Theoretical effective mass ratios for solid metals*

Metal	(m^*/m) (a)	(b)	(c)	(m_1^*/m) (d)	(e)	Phonon enhancement (f)	γ/γ(FE)
Li	1.66	—	1.27	1.19	1.14	(1.5)	2.3
Na	1.00	1.06	—	1.00	0.99	1.18	1.26
K	1.09	—	—	0.99	0.97	1.14	1 20
Rb	1.21	—	—	0.97	0.89	1.14	2.14, 1.26
Cs	1.76	—	—	0.98	—	1.12	2.82, 1.43
Be	—	—	—	1.28	1.13	1.26	0.5
Mg	—	—	—	1.01	1.04	1.31	1.3
Zn	—	0.59	—	0.93	0.95	1.42	0.9
Cd	—	0.54	0.94	0.87	0.95	1.40	0.8
Hg	—	—	—	0.80	—	(1.9)	2.1
Al	—	1.05	—	1.04	1.04	1.5	1.5
Ga	—	—	—	0.96	—	1.25	0.6
In	—	0.91	0.90	0.89	0.93	(1.9)	1.4
Tl	—	—	—	0.82	—	(2.1)	1.1
Sn	—	—	—	0.93	—	(1.9)	1.3
Pb	—	1.12	—	0.86	—	(2.3)	2.1
Bi	—	—	—	0.87	—	—	0.01

Note. References for the figures in columns (*a*) to (*f*) are given in the text. Two experimental values of γ/γ(FE) are quoted for Rb and Cs, for which the measurements of Martin, Zych & Heer (1964) and of Lien & Phillips (1964) are in gross disagreement. Martin (1961) has found γ to be the same for the two isotopes Li^6 and Li^7 to within 5%, though the phonon enhancement factor presumably depends upon ionic mass.

columns (*a*), (*b*), and (*c*) with those in (*d*) and (*e*) suggests that in a typical 'free-electron-like' metal the structure-dependent effects of the pseudo-potential are unlikely to modify m^*/m by more than 20% or so (Ham's figure of 1.76 for Cs is almost certainly too large). Hence the change of m^*/m on melting may well be rather trivial. But metals such as Be, Ga and Bi are not free-electron-like in the solid phase and there is little doubt that in these m^*/m is substantially reduced by distortion of the Fermi surface; parts of the Fermi surface which are in contact with Brillouin zone faces contribute nothing, of course, to the density of states available for excitation. Hence in Be, Ga and Bi the increase of m^*/m on melting could be considerable.

Unfortunately, m^*/m cannot be determined directly from experiment, whether the metal is solid or liquid. The obvious way to try to measure it for a solid is to find the electronic specific heat coefficient, which should be directly proportional to $\mathcal{N}(E)_F$; this, of course, requires experiments at very low temperatures where the linear term in the specific heat, γT, is no longer swamped by the T^3 term from the lattice. It is now realised, however, that at very low temperatures it is not only the electrons which contribute to the linear term; the zero-point energy of the ions turns out to include a small correction proportional to T, which also contributes. Indirectly, it is true, the electrons are responsible, since the extra zero-point energy arises from the thermal broadening of the Fermi–Dirac distribution function and its magnitude depends upon the strength of the electron–phonon interaction, but it is certainly not allowed for in the conventional theory of the electronic specific heat. This so-called *phonon enhancement* of γ is important only at temperatures below the Debye temperature, Θ_D, because above Θ_D the thermal energy stored in the lattice vibrations goes over to the classical value of $3RT$, irrespective of the magnitude of the zero-point energy. Reliable calculations of the strength of the enhancement require a good pseudo-potential and a knowledge of the phonon dispersion characteristics of the metal concerned; the most comprehensive estimates available are those of Allen & Cohen (1969) and the figures listed in column (*f*) of the table are taken from their paper, which includes the results of much previous work.

The experimental results for $\gamma/\gamma(\text{FE})$ which are shown in the final column are indeed greater than the predicted values of m^*/m in nearly every case where a comparison is possible, and the difference can safely be attributed to phonon enhancement. If the figures in column (*b*) are multiplied by those in column (*f*) the agreement with experiment is excellent, so it looks as though the value of 0.54 for m^*/m in solid Cd is closer to the truth than 0.94, at any rate at very low temperatures; there is evidence (see p. 308) that m^*/m in Cd increases on heating to the melting point, and the same may be true of Zn. The results for $\gamma/\gamma(\text{FE})$ in Be, Ga and Bi, which presumably set an upper limit for m^*/m, confirm the view that in these metals m^*/m is significantly less than unity.

We shall consider other evidence relating to the effective mass in §§4.15 and 4.16 below.

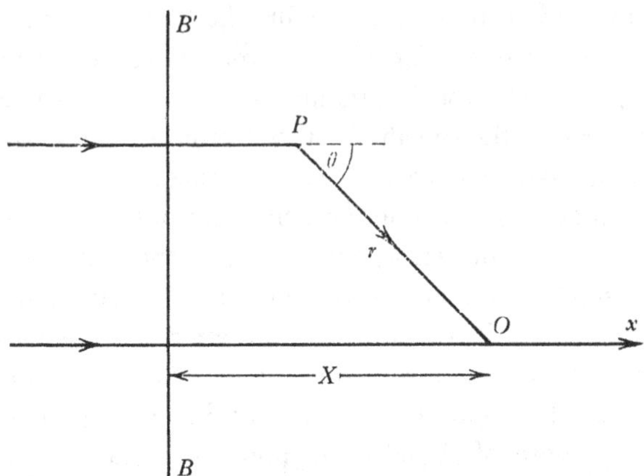

Fig. 1.9. BB' is the front face of the specimen.

1.11. Dispersion theory of effective mass

It is instructive to see how m^* may be estimated from a different point of view, using the language of phase shifts rather than pseudo-potentials. The argument adds little that is essentially new, for the chief results can be derived from (1.56) and (1.76) if preferred. We shall find in §6.3, however, that it helps to clarify the situation in alloys, and it leads on to a discussion of how the d electrons in a liquid transition metal are perhaps to be described.

Imagine a uniform jellium, with the charge density appropriate for the metal of interest, containing a small number ΔN of ions, randomly disposed. Whereas in a pseudo-potential calculation one fixes k first and then analyses the shift of E, we shall here consider an electron propagating in the x direction with a fixed energy – one that would correspond in the absence of any ions to a wave number k_0 – and show how the effect of scattered wavelets is to make the mean amplitude of its wave function vary like exp (ikx), where $k = k_0 + \Delta k$. We shall prove a standard result relating Δk to $f(0)$, where $f(\theta)$ is the 'scattering amplitude' of an individual ion for angle θ, given in terms of the phase shifts by (see (1.51))

$$f(\theta) = \frac{1}{k_0} \sum_l (2l+1) \sin \eta_l \exp (i\eta_l) P_l(\cos \theta).$$

To make the final answer realistic, incidentally, it is probably desirable to choose phase shifts that satisfy the Friedel sum rule with $Q = 0$ in (1.45); that is to say, we should imagine that before inserting each ion, we scoop out an equivalent amount of positive jelly to make room for it. The reduced phase shifts calculated by Meyer *et al.* (see p. 24 and the footnote on p. 25) are not necessarily the best ones to use when it is the scattering angle for $\theta = 0$ that matters.

It is convenient to suppose the medium to be semi-infinite as in fig. 1.9, with the electron wave entering normally from the left. Evidently, to ensure self-consistency at a point such as O, where the mean amplitude is A, we must satisfy the equation

$$A = A\exp(-\mathrm{i}kX)\exp(\mathrm{i}k_0X) + A\frac{2\pi(\Delta N)}{\Omega}\int_0^{\pi} f(\theta)\sin\theta\,\mathrm{d}\theta \times \int_0^{X\sec\theta,\,\infty} r\exp(\mathrm{i}r(k_0 - k\cos\theta))\,\mathrm{d}r; \quad (1.80)$$

the first term on the right describes the amplitude that the incident wave would have had on reaching O were it not for the intervening scatterers, and the second term describes the mean contribution made by scattered wavelets. Note that each wavelet must be treated as propagating from its source, at P say, to O with the wave number k_0; to use k in this context would be to allow twice over for the effects of scattering by other ions that lie between P and O.

Provided that $k_0 \gg \Delta k$, $k_0X \gg 1$ it is legitimate to ignore the whole range of θ between some small angle θ' and π; θ' may be small enough, in fact, to justify the replacement of $f(\theta)$ by $f(0)$ in (1.80). Then integration over r and θ leads to

$$1 - \exp(-\mathrm{i}(\Delta k)X) = \frac{2\pi(\Delta N)xf(0)}{\Omega k_0}\left|\frac{1-\exp(\mathrm{i}y)}{y}\right|_{-(\Delta k)X}^{-(\Delta k)X+\frac{1}{2}k_0X\theta'^2},$$

where $y = (k_0\sec\theta - k)X$, and by letting $k_0X\theta'^2$ tend to infinity one arrives at the required result,

$$\Delta k = \Delta N\frac{2\pi f(0)}{\Omega k_0} = \Delta N\frac{2\pi}{\Omega k_0^2}\sum_l (2l+1)\sin\eta_l\exp(\mathrm{i}\eta_l). \quad (1.81)$$

It may be seen that Δk includes an imaginary term. This describes the attenuation of the incident wave for which the scattering is responsible; there is a general theorem (e.g. Schiff, 1955, p. 105) which relates the imaginary part of $f(0)$ for any scatterer to its total scattering cross-section.

Presumably the density of states is determined by $\partial k'/\partial E$, where k' is the *real* part of k, and this is confirmed by the analysis in §4.7 below, at any rate so long as the imaginary part, k'', is small. Hence it follows from (1.81) that in the presence of the scatterers

$$\begin{aligned}\frac{m^*}{m} &= \frac{\hbar^2 k_F}{m}\left(\frac{\partial k_0}{\partial E}\right)_F\left(\frac{\partial k'}{\partial k_0}\right)_F \\ &= \left(\frac{m^*_{XC}}{m}\right)\left(1+\Delta N\frac{2\pi}{\Omega k_F^3}\sum_l(2l+1)\left(k_0\frac{\partial\eta_l}{\partial k_0}\cos 2\eta_l-\sin 2\eta_l\right)_F\right),\end{aligned} \tag{1.82}$$

where m^*_{XC}/m expresses the effect of exchange and correlation on the density of states of the jellium. This formula tells us the influence that each scattering ion has upon the effective mass, for a particularly crude form of the 'spherical approximation' where everything outside a sphere of radius R_A with the ion at its centre is replaced by jellium. Within this approximation the effective mass for the actual metal is independent of its structure and may be obtained from (1.82) by replacing ΔN by N. If the phase shifts are small, we have

$$\left(\frac{m^*_1}{m}\right) = \left(\frac{m^*_{XC}}{m}\right)\left(1+\frac{2}{3\pi z}\sum_l(2l+1)\left(k_0\frac{\partial\eta_l}{\partial k_0}-2\eta_l\right)_F\right).$$

If they obey the Friedel sum rule with $Q = 0$, as suggested above, then

$$\left(\frac{m^*_1}{m}\right) = \left(\frac{m^*_{XC}}{m}\right)\left(1+\frac{2}{3\pi z}\sum_l(2l+1)\left(k_0\frac{\partial\eta_l}{\partial k_0}\right)_F\right). \tag{1.83}$$

It is hardly surprising, in the light of this result, that the theoretical values for (m^*_1/m) which are quoted in table 1.2 are mostly close to unity.

To extend this dispersion theory argument to cases where the phase shifts are *not* small is far from easy, as the reader who consults the papers of Ziman (1965) and of Anderson & McMillan (1967) will soon appreciate. A careful discussion of multiple scattering is required, and the structure of the specimen is bound to enter the problem, as it did in the pseudo-potential calculations of the previous section when we tried to go beyond the first-order. But just as one can hope to make the 'spherical approximation' more accurate within the pseudo-potential framework by replacing u by some appropriate u', so one may hope to improve upon (1.81) by replacing the phase shifts η_l by effective phase shifts η_l'; in computing them one should imagine the jellium surrounding the central ion to be distorted in a spherically symmetric fashion, so as to represent the reaction upon the ion of scattering by its immediate neighbours. This is the philosophy behind the methods which Ziman, Anderson & McMillan, and later workers such as Morgan (1969), Gyorffy (1970) and Schwartz & Ehrenreich (1971) have explored.

The particular problem of concern to Ziman and to Anderson & McMillan was the effect upon the electronic density of states of resonance with a virtual bound state in the scattering ions. If such a resonance occurs, we must suppose that one of the reduced phase shifts η_l increases rather rapidly with k_0, passing through $\frac{1}{2}\pi$. The quantity $(k'-k_0)$, according to (1.81), should vary like $\sin 2\eta_l$ and the curve of k' as a function of k_0^2 (or E) is therefore liable to be distorted in the manner shown schematically in fig. 1.10(*a*). There is general agreement between this naive prediction and the curves which Anderson & McMillan have generated by their more subtle analysis, using phase shifts appropriate for liquid Fe, which are reproduced in fig. 1.10(*b*). A wave whose amplitude varies like $\exp(\mathrm{i}k'x)\exp(-k''x)$ has a Fourier transform which is peaked at the wave number k' but spreads over a range of order k'' on either side, and it is this spread which is indicated by the shading in fig. 1.10(*b*).

In the circumstances represented by fig. 1.10 it is clearly inappropriate to calculate the density of states from $\partial k'/\partial E$, for in the resonance region this slope is negative. The electronic wave functions in this region are heaped up inside each ion, and on this account their Fourier transforms extend out to large values of k in a manner of which fig. 1.10(*b*), which describes only the Fourier transform of the enveloping plane wave, gives no indication. The greater the volume of **k**-space (or phase space) they cover, the more discrete states there must be. The resonance therefore produces a

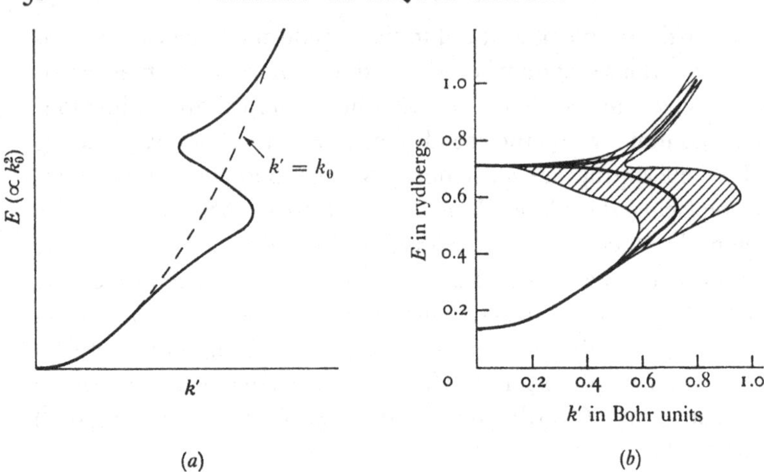

(*a*) (*b*)

Fig. 1.10. Effect of a virtual bound state on the density of states. (Curve (*b*) after Anderson & McMillan, 1967.)

narrow band with a high density of states, overlying the normal conduction band.† In the case of Fe, of course, it is η_2 which passes through $\frac{1}{2}\pi$. The states in the narrow band must then have a d-like character around each ion, and there must presumably be enough of them to accommodate 10 electrons per atom.

1.12. Energy gaps and pseudo-gaps

The distortion of the Fermi surface in solid metals, referred to briefly in § 1.10, is associated with the gaps that exist in the energy spectrum of the conduction electrons for values of $\mathbf{k}$ that lie on the faces of the Brillouin zones. We shall need an expression for the width of a gap, and one may readily be obtained in terms of the pseudo-potential. From (1.73) and (1.72) we know that in a perfectly regular crystal, the Fourier components $U(\mathbf{q})$ are zero except where $\mathbf{q}$ is equal to one of the *reciprocal lattice vectors*, $\mathbf{G}$, when‡

$$U(\mathbf{G}) = Nu(\mathbf{G}).$$

These strong Fourier components tend to scatter an electron in

† According to Schwartz & Ehrenreich (1971) the density of states passes through *two* well-defined maxima, and there is an abrupt minimum at the resonance energy itself.

‡ It involves no essential loss of generality to choose the origin of $\mathbf{r}$, as we do here, to lie at the centre of one of the ions.

the state $\mathbf{k}$ to the state $\mathbf{k}-\mathbf{G}$ and vice versa and if these happen to be degenerate, which is the case when $\mathbf{k}$ lies on a Brillouin zone face, the resultant pseudo-eigenfunctions have to be equal mixtures of the form

$$\begin{aligned}\phi &= (2\Omega)^{-\frac{1}{2}}(\exp(\mathrm{i}\mathbf{k}\cdot\mathbf{r}) \pm \exp(\mathrm{i}(\mathbf{k}-\mathbf{G})\cdot\mathbf{r}))\\ &= (2/\Omega)^{\frac{1}{2}}\exp(\mathrm{i}(\mathbf{k}-\tfrac{1}{2}\mathbf{G})\cdot\mathbf{r})\begin{matrix}\cos\\ \sin\end{matrix}(\tfrac{1}{2}\mathbf{G}\cdot\mathbf{r}).\end{aligned}$$

The difference of energy between the two is just

$$\begin{aligned}&2\Omega^{-1}\int(\cos^2(\tfrac{1}{2}\mathbf{G}\cdot\mathbf{r})-\sin^2(\tfrac{1}{2}\mathbf{G}\cdot\mathbf{r}))\,U(\mathbf{r})\,\mathrm{d}\mathbf{r}\\ &\quad= \Omega^{-1}\int(\exp(\mathrm{i}\mathbf{G}\cdot\mathbf{r})+\exp(-\mathrm{i}\mathbf{G}\cdot\mathbf{r}))U(\mathbf{r})\,\mathrm{d}\mathbf{r}\\ &\quad= 2N\Omega^{-1}\,\mathrm{Re}\,(u(\mathbf{G})). \qquad (1.84)\end{aligned}$$

In a typical metal this gap width might be 1 or 2 eV, reasonably small compared with the Fermi energy K_F, which is why the distortion of the Fermi surface is often relatively slight. The gaps would need to be a good deal wider to prevent any overlap in energy between one zone and the next and to turn the substance – assuming the number of valence electrons to be just sufficient to fill one or more zones completely – from a metal to an insulator or semiconductor.

But if a metallic specimen could be uniformly expanded until the atoms were virtually out of range of one another it would clearly become an insulator; the electronic energy spectrum would collapse into bands centred about the energy levels of a free atom which would be much too narrow to overlap at all. Presumably there is a one-to-one correspondence between the eigenstates of the dense and the expanded crystals, and we usually picture the transition between them as a continuous one, brought about by a continuous enlargement of the energy gap on each Brillouin zone face.

One should not be misled by this picture into supposing that the separation of the energy spectrum into discrete bands depends in any very essential way upon the Brillouin zone structure. A liquid metal is too disordered for the concept of a zone to retain any real

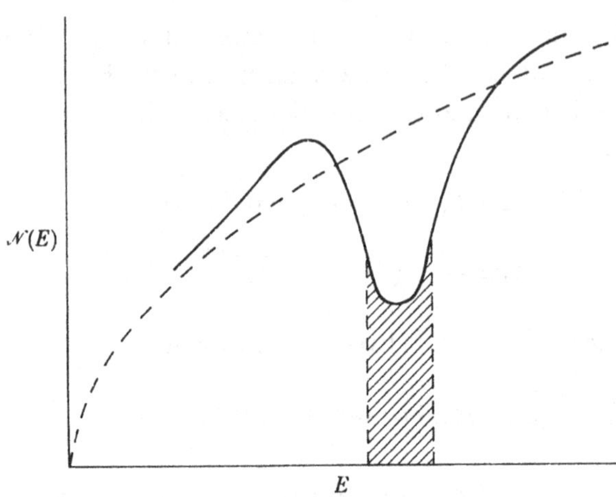

Fig. 1.11. A Mott pseudo-gap in the density of states. The broken curve is the free-electron parabola. Shading indicates the energy range where, according to Mott, the electron states are liable to be localised.

meaning; its interference function $a(q)$ does have peaks, as we shall see below, which reflect some degree of local order among the ions and which enhance the scattering through certain values of q, but it turns out that they are unlikely to be sufficient by themselves to open up any energy gaps. Gaps must nevertheless appear if the liquid is expanded, for there is no doubt that the energy levels in the vapour phase are perfectly discrete; presumably they must appear at some stage however great the disorder, even if there are no peaks in $a(q)$ at all. In §5.10 we shall examine some results obtained with Hg at temperatures in the neighbourhood of, and above, its critical point, where its structure is highly disordered and its volume is two or three times greater than at normal temperatures and pressures. Under these conditions the DC conductivity is still finite and the energy spectrum is presumably still continuous, but it is probable that a deep trough is developing in the density of states near the Fermi level – what Mott (1969) has christened a *pseudo-gap* – as shown schematically in fig. 1.11. Further expansion would split the spectrum completely, into an occupied valence band of s-like states and an empty conduction band.

One of the changes that occur on expansion is a reduction of the kinetic energy K_F, and it is possible that the point at which the pseudo-gap begins to form, in a disordered metal such as Hg, is the point at which the Fermi level has fallen to such an extent that electrons must resort to *tunnelling* to get from one atom to the next. The sort of model that we have used so far, based on freely-propagating plane waves, becomes inappropriate in such circumstances; the *tight-binding* model constitutes a more plausible starting point. The tight-binding model inevitably leads, whatever the structure, to a set of atomic-like energy bands. Initially they may overlap, but the bands will be found to get narrower as the ions are moved further apart and sooner or later pseudo-gaps and gaps are bound to appear.

Fig. 1.12(*a*) shows, in a schematic fashion, the screened potential experienced by the valence electrons in a typical dense metal. At points such as P and P', midway between atoms, the true electrostatic potential is zero (because the atoms are neutral) though an electron sees a potential of U^{XC} which is negative because of exchange and correlation. Relative to the PP' level, which is sometimes referred to as the zero of the 'muffin-tin' potential, the bottom of the conduction band BB' is slightly negative; one may estimate by averaging (1.69) over the volume of an atom that the mean potential energy of a conduction electron is of the order of $U^{\mathrm{XC}} - 0.3\, ze^2 R_{\mathrm{A}}^{-1}$. But since

$$K_{\mathrm{F}} = (\hbar^2/2m)(9\pi z/4)^{\frac{2}{3}} R_{\mathrm{A}}^{-2} > 0.3 z e^2 R_{\mathrm{A}}^{-1},$$

i.e. since

$$z^{\frac{1}{3}} R_{\mathrm{A}} < 3.3\ \text{Å} \tag{1.85}$$

at normal densities, the Fermi level FF' should be well above PP' and electrons at the Fermi level are free to move without tunnelling. The majority of metals, including Hg, will tolerate a volume expansion by a factor of something like 3 before the inequality in (1.85) breaks down.

That cannot be quite the whole story, however, for the eigenfunctions predicted by the usual tight-binding model still extend throughout the specimen. In the isolated atom represented by fig. 1.12(*b*), or in any assembly of atoms which are completely separated from one another, the electrons are essentially *localised*. In the latter case, as we shall see in §4.10, the DC conductivity is bound to vanish, whereas according to the tight-binding model a

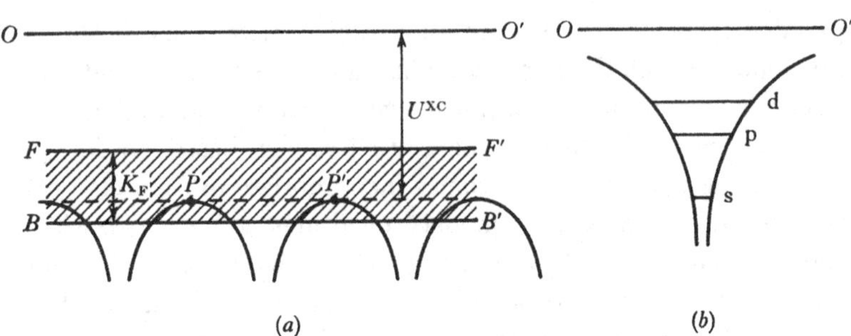

Fig. 1.12. (*a*) Position of the conduction band in a typical metal, relative to the 'muffin-tin zero', *PP′*. On expansion the Fermi level *FF′* may fall below *PP′*. (*b*) An isolated atom in which the valence electrons occupy discrete states.

substance with an odd number of valence electrons per unit cell should in principle remain a conductor however narrow the bands.

It is now widely accepted that localisation can be brought about by the effects of correlation when the ratio between the screening length $(q_s')^{-1}$ and the Bohr radius a_H becomes larger than some critical value. The simplest arguments suggest an abrupt transition to an insulating state – the so-called *Mott transition* – at an electron density such that

$$n^{\frac{1}{3}}a_H \sim 0.2\text{–}0.25, \tag{1.86}$$

and there is experimental evidence to support this, e.g. in the behaviour of heavily doped semiconductors (see Doniach, 1970, for a recent review). In a semiconductor the Bohr radius needs to be calculated from the formula $a_H = \hbar^2\epsilon/m^*e^2$, using a dielectric constant which is often substantially greater than unity and an effective mass m^* which is less than m; a_H may therefore be as much as 10^{-6} cm, say, so that the Mott transition can be observed when n is only about 10^{16} cm^{-3}. If we set $\epsilon = 1$ and $m^* = m$, equation (1.86) suggests

$$n > 10^{23}\ \text{cm}^{-3}$$

as a criterion for metallic behaviour, but this is clearly too stringent a condition for there are many perfectly good metals (Na for example) which fail to satisfy it. The Mott transition incidentally, need not be an abrupt one at finite temperatures.

In *disordered* systems, and especially for states that are situated in the middle of a pseudo-gap, it is thought that localisation may set in, perhaps at an earlier stage, for reasons that have nothing whatever to do with correlation. References to the literature concerning this still somewhat controversial point are given in §4.11.

1.13. Ion–ion interactions

For the purpose of calculating cohesive energies, equilibrium densities, compressibilities and such-like, it seems necessary at first sight to have some recipe for calculating the total energy of a metal. The discussion of the ideal jellium in §1.3 above shows that this is a tricky business; it is certainly not correct to sum the energies of the individual electrons, using the methods outlined above to evaluate $E(k)$, because an important part of $E(k)$ arises from the coulomb interaction between the electron of interest and other electrons and this part would be counted twice over in effecting the sum. With care it is possible to generalise the arguments in §1.3, and to obtain expressions for the total energy of a real metal in terms of its pseudo-potential, with corrections for correlation and exchange included. There is a lot to be said, however, for by-passing such a calculation where possible.

A by-pass exists if one's ultimate goal is, say, the compressibility. It is not essential to go through the total energy to reach this quantity. All that is really needed is the *change* in the energy when the metal is disturbed by a wave which compresses it in some places and expands it in others but leaves its total volume unchanged. Our object in this section is to see how, so long as the total volume is unaffected, changes of the energy due to any sort of rearrangement of the ionic structure may be calculated with relative ease. It turns out that, provided a calculation up to second order in the pseudo-potential is sufficient, these changes may be expressed rather neatly in terms of an effective potential acting between the ions. The necessary theory is discussed e.g. by Harrison (1966) and March (1968); the elementary argument below is due to Heine & Weaire (1966).

It is helpful to start by considering how much work would be required to distort the positively-charged background in a jellium so as to set up a potential ΔV, oscillating sinusoidally with wave number q. If ρ is the charge density of this background we know from Poisson's equation that

$$4\pi e\Delta\rho = q^2\Delta V_b = q^2\epsilon(q)\Delta V, \tag{1.87}$$

ϵ being defined as in (1.20). Since the background is acted on, the whole time that it is being distorted, by the *screened* potential ΔV rather than ΔV_b, the work required is clearly the integral over the volume of the specimen of

$$\frac{1}{2e}\Delta\rho\Delta V = \frac{q^2\epsilon}{8\pi e^2}(\Delta V)^2, \tag{1.88}$$

and this is the extra energy of the jellium due to the distortion. In the absence of any electrons the work required would of course be

$$\frac{1}{2e}\Delta\rho\Delta V_b = \frac{q^2\epsilon^2}{8\pi e^2}(\Delta V)^2, \tag{1.89}$$

and we may regard this as the energy stored in the self-energy of the positive charge. The remainder of (1.88), i.e.

$$\frac{q^2\epsilon(1-\epsilon)}{8\pi e^2}(\Delta V)^2 = \frac{q^2\epsilon'^2}{8\pi e^2}\left(\frac{1}{\epsilon}-1\right)(\Delta U_F)^2, \tag{1.90}$$

which may also be written, with the aid of (1.30), as

$$-\frac{q_s^2\epsilon'}{8\pi e^2}(\Delta U_F)^2, \tag{1.91}$$

represents energy of interaction of the electrons with the positive charge or with each other. Equation (1.91) should be entirely accurate up to second order in the perturbation – not beyond that, because of limitations in the validity of the dielectric constant approach – and in fact it is possible (Heine & Weaire, 1970) to obtain it by second-order perturbation theory if care is taken not to count certain terms in the total energy twice over.

The argument is easily extended to cover more general distortions of a jellium at constant volume. Whatever the form of ΔV it may be split up into Fourier components and the energy contribution evaluated separately for each. The total energy contains no cross-terms from different Fourier components because the product of $\Delta\rho(\mathbf{q}_1)$ and $\Delta V(\mathbf{q}_2)$ is zero when integrated over the volume of the specimen if $\mathbf{q}_1$ and $\mathbf{q}_2$ are not equal.

The fact that (1.91) can be obtained by perturbation theory is proof, if proof were needed, that it can be applied equally well to a real metal, provided that one uses for ΔU_F the appropriate pseudo-potential,† i.e. the proper matrix element for scattering of an electron from a state $\mathbf{k}$ on the Fermi surface to the state $\mathbf{k}+\mathbf{q}$. Using (1.71) and (1.79), and summing over all Fourier components other than $q = 0$, one may arrive immediately at a formula for the structure-dependent term in the total electronic energy at constant volume up to second order in u: it is just

$$-\frac{N}{8\pi e^2 \Omega} \sum_{\mathbf{q} \neq 0} q_s^2 \epsilon'(q) a(\mathbf{q}) u(\mathbf{q}) u(-\mathbf{q}). \tag{1.92}$$

Now had we set out to describe the energy in terms a pair interaction w' with Fourier components $w'(\mathbf{q})$ we would have written it as

$$\frac{1}{2} \sum_{i \neq j} w'(\mathbf{r}_j - \mathbf{r}_i) = \frac{1}{2\Omega} \sum_{i \neq j} \sum_{\mathbf{q}} \exp i\mathbf{q}\cdot(\mathbf{r}_j - \mathbf{r}_i) w'(\mathbf{q}),$$

and if w' is the same for every pair of ions this reduces with the aid of (1.79) to

$$\frac{N}{2\Omega} \sum_{\mathbf{q}} w'(\mathbf{q})(a(\mathbf{q}) - 1). \tag{1.93}$$

So long as we are concerned only with the part of the energy which is structure-dependent we may forget the 1 inside the bracket in (1.93) and its equivalence to (1.92) is then apparent; we must clearly use

$$w'(\mathbf{q}) = -\frac{q_s^2 \epsilon'}{4\pi e^2} u(\mathbf{q}) u(-\mathbf{q}) = -\frac{q_s^2}{4\pi e^2 \epsilon'} v_b(\mathbf{q}) v_b(-\mathbf{q}), \tag{1.94}$$

from which the variation of this effective pair interaction in real space is readily constructed if required.

The quantity w' is an *indirect* interaction energy between ions, that allows for the energy taken up by the conduction electrons. There is also of course a *direct* interaction w'' to be considered, which is the ordinary coulomb energy shared between two bare ions carrying a charge z^*e plus a repulsive term that rises rapidly when the separation between the ions becomes so small that their

† It is readily shown that if this is complex $(\Delta U_F)(\Delta U_F)^*$ has to replace $(\Delta U_F)^2$ in (1.91).

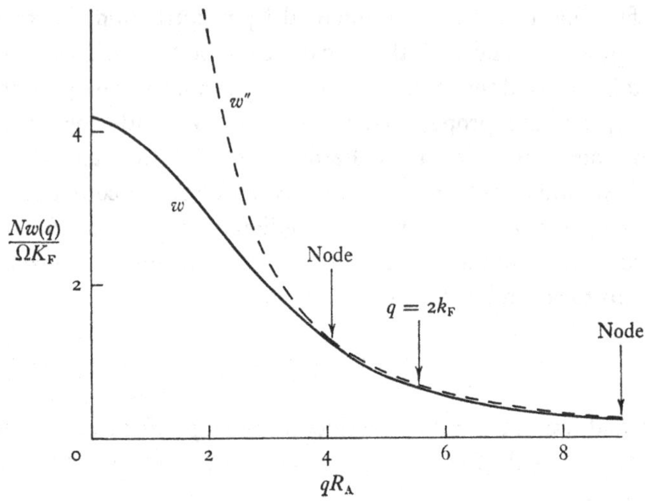

Fig. 1.13. Fourier transform of the ion–ion potential for Al. The curve is based upon the Heine–Animalu pseudo-potential and an approximate expression for ϵ', so it may not be completely accurate. But it illustrates how small $w'(q)$ is compared with $w''(q)$ for large values of q. Arrows indicate the position of nodes in $u(q)$ where $w'(q)$ vanishes.

cores begin to overlap. If the ions are always far enough apart for repulsive forces between the cores to be ignored, then

$$w''(q) \simeq 4\pi(z^*e)^2/q^2. \tag{1.95}$$

This must be added to (1.94) to get the total effective interaction, w.

Insight into the nature of w may be obtained by consideration of Ashcroft's 'empty core' model for the ionic pseudo-potential, which involves the single adjustable parameter R_M (see p. 38). From (1.65), (1.54) and (1.95), if the distinction between z and z^* may be ignored in the interests of simplicity, it follows that

$$w(q) = \frac{4\pi(ze)^2}{q^2}\left(1 - \frac{q_s^2}{q^2\epsilon'}\cos^2 qR_M\right). \tag{1.96}$$

The limiting value of this expression for small q is

$$w(0) = 4\pi(ze)^2\left(R_M^2 + \frac{\epsilon'}{q_s^2\epsilon}\right), \tag{1.97}$$

and its behaviour for large q is sketched in fig. 1.13; for small q the indirect interaction plays an essential role in eliminating the divergence which would be displayed by the direct interaction alone, but for large q, where $\epsilon' \simeq 1$ and q_s is tending to zero, it is only the direct part that matters. Use will be made of these results to estimate compressibilities of liquid metals in §2.19.

To find the variation of $w(R)$ in real space we must take the Fourier transform of (1.96). A simple analytical solution may be obtained by making the crude assumption that ϵ' is given by the Thomas–Fermi expression (1.22), with q_s a constant independent of q; for $R > 2R_M$ (but not otherwise) it is

$$w(R) \simeq \frac{(ze)^2}{R} \cosh^2(q_s R_M) \exp(-q_s R). \qquad (1.98)$$

If, following the argument on p. 21, we replace q_s by q_s' this is perhaps a reasonable first approximation to the true answer. It suggests that the ions repel one another by means of a screened coulomb interaction, and the surprising effect of the finite core radius R_M is simply to scale up the strength of the repulsion without altering its form. Let us put numbers into the expression for a typical metal such as Al. According to Ashcroft the best value to take for R_M in Al is 0.59 Å, since the model pseudo-potential which corresponds to this will explain the shape of the Fermi surface in solid Al as determined experimentally. The screening parameter q_s' may be read off from fig. 1.2, if $m^*/m = 1$, and should be 2.3 Å^{-1}. Then for $R = 3$ Å which is about the mean distance between adjacent ions at temperatures close to the melting point T_M, one finds $w(R) \simeq 0.20$ eV, while $w(2R_M)$ turns out to be about 30 eV. The increase of energy required to make two ion cores overlap is therefore vast compared with $k_B T_M$ ($\simeq 0.08$ eV) which justifies the initial assumption that core overlap does not occur.

To obtain an accurate picture of $w(R)$, of course, we need a better model for the pseudo-potential than Ashcroft's and a detailed knowledge of the q-dependence of ϵ'. Many calculations exist (Harrison, 1966; Ashcroft & Langreth, 1967*b*; Schneider & Stoll, 1967*b*; Shyu *et al.*, 1967, 1968, 1969, 1971*a*, *b*; Bortolani & Magnaterra, 1968; Shaw, 1969; Braunbek *et al.*, 1970; Hasegawa & Watabe, 1972), but the three curves for Al which are plotted in fig. 1.14 show that the results are far from consistent. The latest

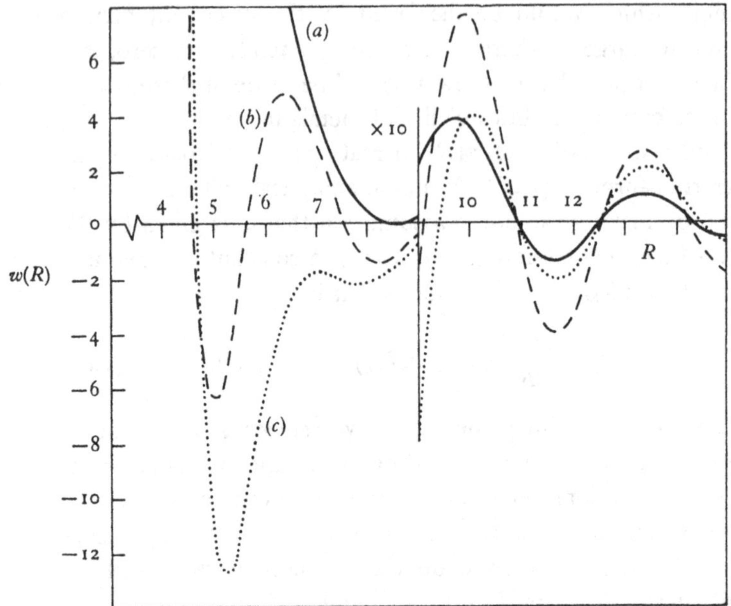

Fig. 1.14. Theoretical ion–ion potentials for Al in real space: (*a*) Shaw, (*b*) Harrison, (*c*) Animalu. On the scale used for R (Bohr units) the mean separation is about 5.6. The units for w are rydberg × 10^{-3}, so that $k_B T_M$ is about 6. The oscillations on the right of the diagram are expanded by a factor 10. (Redrawn by permission from Shaw, 1969.)

of these three is matched with fair success by (1.98). Still later work suggests, however, that Shaw's allowance for the effects of exchange and correlation is not sufficient, and curves which are not everywhere repulsive but which pass, like those of Harrison and Animalu in fig. 1.14, through an attractive minimum of some sort are currently more favoured. It is worth stressing, perhaps, that a minimum is *not* essential, since the cohesion of metals can be attributed to the part of their energy which depends upon volume but not upon structure and which is not described by $w(R)$ at all. The situation is very different in a simple non-metallic liquid such as Ar, for which an inter-atomic potential is sketched in fig. 1.15.

Another distinction between metals and rare gases is that $w(R)$ for metals appears to have an oscillatory tail extending over several atomic spacings. This is a common feature of all three curves in fig. 1.14 and it appears in curves calculated for other metals

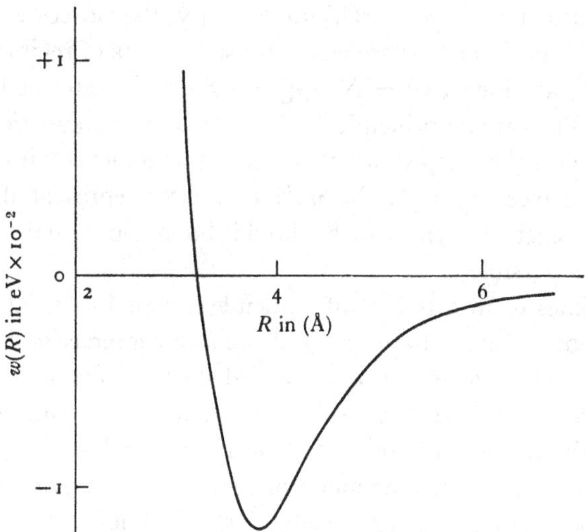

Fig. 1.15. The interatomic pair potential for Ar.

besides Al. It arises from the singularity in the slope of q_s^2 at $q = 2k_F$. These are in fact the *Friedel oscillations* discussed in §1.6, appearing now in the interaction between ions and other ions rather than between ions and electrons. In the limit of large R their wavelength is expected to be (π/k_F).

Because the interatomic potential in Ar is a short-range one, which is scarcely felt beyond the first coordination shell, the crystal structure of solid Ar is free to adapt itself until each atom has the largest possible number of neighbours. A close-packed structure is the inevitable result. Many metals, however, adopt structures in the solid phase which are severely distorted versions of hexagonal or cubic close-packing, and for several of them (e.g. the b.c.c. alkali metals) the coordination number is less than 12. Presumably the relatively long-range tail of $w(R)$ for metals plays an essential part in determining just what structure is the most stable one at low temperatures.

Attempts have been made to explain the lattic structures of solid metals in terms of theoretical curves for $w(R)$, and a brief account of the arguments of Heine & Weaire (1966, 1970) and Weaire (1968) is appropriate here because they have been applied to liquid metals too. The analysis is conducted entirely in **q**-space, to make a transformation from $w(q)$ to $w(R)$ unnecessary; the lattices whose energies are to be compared are characterised by sets of reciprocal lattice vectors at which $a(\mathbf{q}) = N$, $a(\mathbf{q})$ being zero for intermediate values of $\mathbf{q}$. The general principle invoked is that it is energetically unfavourable for the reciprocal lattice vectors to coincide with the nodes of $u(q)$ (see fig. 1.7); the *indirect* energy represented by (1.92) is a negative term which should be made as large in magnitude as possible.

Two weaknesses in this elegant principle impair its usefulness for our purposes. One is the neglect of the *direct* interaction term in the energy. How far it pays a metal to distort so as to shift one of its reciprocal lattice vectors $\mathbf{G}$ depends upon the magnitude of $\partial w(q)/\partial q$ in the neighbourhood of $\mathbf{G}$, and it is clear from fig. 1.13 that this is determined just as much or more by w'' as by w' (the first reciprocal lattice vector usually lies at about $qR_{\mathrm{A}} = 5$). Secondly, the equilibrium of a lattice is determined by its free energy, and at high temperatures it is necessary to compare not only the *energies* of two rival structures but also their *entropies*.

The magnitude of $k_{\mathrm{B}} T_{\mathrm{M}}$ for Al, where T_{M} is the melting temperature, is large compared with the oscillations in $w(R)$ which are suggested by theory, and this seems to be the case for other metals besides Al. In the liquid state above the melting point, therefore, the long-range tail of $w(R)$ may be of rather slight importance.

1.14. The rigid-sphere model

There are several properties of liquid metals which can be explained quite successfully by treating the ions as *rigid spheres*; examples include the change of volume on melting (§2.7), the pair distribution function (§2.14) and the self-diffusion coefficient (§3.7). Rigid-sphere models are scarcely appropriate for metals at low temperatures, but they become more plausible above the melting point, largely because the long-range oscillations in $w(R)$ are so small compared with $k_{\mathrm{B}} T_{\mathrm{M}}$. What value should we use for the effective rigid-sphere diameter σ?

Presumably we ought to calculate the closest distance of approach for two ions that collide head-on, with kinetic energies of order $k_{\mathrm{B}} T$, and the prescription that is usually recommended is to equate $k_B T$, or some multiple of it, to $w(\sigma)$. It is readily shown from (1.98), with q_s' in place of q_s, that when $q_s' R_M > 1$ an answer of the form

$$\sigma \simeq 2R_M + \text{constant } (q_s')^{-1}$$

is suggested by this argument, the numerical constant lying in the range between 3 and 5. It looks a sensible result and it illustrates once again that core overlap, which requires $R \leqslant 2R_M$, is virtually out of the question.

It should be remembered, however, that we are dealing with a dense system, very different from the dilute gas of conventional kinetic theory. When two ions collide they do not approach each other from some large distance, where $w(R)$ is virtually zero, and then separate completely; all that happens is that R changes from say $\bar{R}$, the mean distance between two adjacent ions, to σ and then back again. A more honest prescription for estimating σ might therefore be to set

$$w(\sigma) - w(\bar{R}) = \text{constant} \times k_B T. \qquad (1.99)$$

Now evidence is presented in chapter 2 that the effective *packing fraction* y is usually about 0.46 for liquid metals on the point of solidification. This quantity is defined as the fraction of the total volume occupied by the rigid spheres, so

$$y = \frac{N\pi\sigma^3}{6\Omega} = \frac{\pi\sqrt{2}}{6}\left(\frac{\sigma}{\bar{R}}\right)^3 \qquad (1.100)$$

for a structure with close-packed coordination. For Al, therefore, in which $\bar{R} \simeq 2.95$ Å, it seems that $\sigma \simeq 2.5$ Å. To reconcile these figures with (1.98) and (1.99), and with the values for R_M and q_s' which were quoted on p. 67, the constant in (1.99) would need to be about 6.6.

We may now enquire how y is likely to vary with temperature above the melting point. Evidently σ decreases on heating, according to (1.99), while thermal expansion makes $\bar{R}$ increase. It may be shown by differentiation of (1.99) after some approximations that are justified by the orders of magnitude involved (the effect of thermal expansion on q_s', for example, should be negligible in this context), that

$$y \propto T^{-\nu},$$

where
$$\nu \simeq \frac{w(\sigma)-w(\bar{R})}{w(\sigma)}\left(\frac{3}{1+q_s'\sigma}+\alpha T\right). \tag{1.101}$$

For Al, which is probably typical, ν should be about 0.39 at constant pressure. At constant volume the term in α, which is the thermal expansion coefficient, should be omitted and 0.31 is the result.

However, equation (1.100) is no more than a rough rule of thumb. In practice we had better treat ν as an adjustable parameter, to be chosen so as to reconcile the predictions of the rigid-sphere model to some experimental observation. In view of the obvious limitations of the model, we should not be surprised if different values of ν turn out to be needed in different situations.

1.15. The energy for vacancy formation

It forms a simple application of the methods developed in the previous section to calculate the energy required to form a vacancy in a metal, at the absolute zero of temperature and under zero applied pressure. A more significant application, to the calculation of compressibility, is deferred to §2.19.

We can imagine the vacancy to be formed in two stages. First the metal is expanded *uniformly* through one atomic volume, Ω/N. It is shown on p. 113 below that the interference function changes during this process from $a_0(q)$ to

$$a_0(q)+\frac{q}{3N}\frac{\mathrm{d}}{\mathrm{d}q}a_0(q). \tag{1.102}$$

Secondly, the ions are rearranged at constant volume so as to produce a single vacancy, the surrounding metal being restored to its original density. It is shown on p. 115 that the interference function at the end of this second process is

$$a_0(q)+N^{-1}(1-a_0(q)). \tag{1.103}$$

Now if the metal during stage 1 does not depart from the equilibrium configuration appropriate to its volume at any particular moment, which seems a plausible assumption, then its energy during this stage increases by an infinitesimal amount: the rate of change of internal energy with volume is zero in equilibrium at zero T and p. The whole energy required to form the vacancy must therefore be provided during stage 2 and must be due to the change in $a(q)$ which then occurs. This change is just the difference between (1.103) and (1.102), and the required answer for the energy is therefore given by

$$\frac{N}{2\Omega}\frac{4\pi\Omega}{(2\pi)^3}\int_0^\infty N^{-1}\left(1-a_0-\frac{q}{3}\frac{\mathrm{d}}{\mathrm{d}q}a_0\right)w(q)q^2\,\mathrm{d}q;$$

the sum over q in e.g. (1.93) is here converted into an integral on the assumption that $a(\mathbf{q})$ and $w(\mathbf{q})$ are isotropic. Integration by parts, with the knowledge (see §2.10) that a_0 tends rapidly to unity for large q, simplifies this to

$$\frac{1}{12\pi^3}\int_0^\infty (a_0-1)\frac{\mathrm{d}w}{\mathrm{d}q}q^3\,\mathrm{d}q. \tag{1.104}$$

This result does not seem to have been exploited in the literature, though there would be little difficulty in computing an answer from it, for a liquid metal at any rate, given a knowledge of $a_0(q)$ and $w(q)$. Previous attempts to calculate the energy for vacancy formation (e.g. March, 1966) have been based on rather crude models and have suggested an answer simply proportional to zK_F. It is possible to cast (1.104) into a similar form, but only after approximations that seem altogether too drastic to be realistic.

The argument could be extended so as to give the energy of formation of a hole from which more than one atom is missing, and it is tempting to apply it to a *macroscopic* hole and hence to calculate the *surface tension* for a liquid metal. The whole argument rests, however, on the assumption that the screening can properly be taken into account by the dielectric constant method, and it is quite certain that the perturbation represented by a macroscopic hole is far too strong for this to be legitimate; it may indeed break down even for a single vacancy. Surface tension and the surface dipole effect referred to briefly in §1.3 are phenomena which it does not seem possible to analyse quantitatively by the methods examined in this chapter

CHAPTER 2

THE STRUCTURE OF SIMPLE LIQUIDS

2.1. Prologue

In this chapter we are concerned with the arrangement of the ions in liquid metals, and how this compares with the arrangement of the atoms in a simple non-metallic liquid such as Ar. We shall need a word to describe either an ion or an atom according as the context dictates and 'molecule' seems to be the best available.

There are two ways to approach the problem of molecular structure. The first is in terms of *models* and we begin with these in §2.2. Quasi-crystalline models have sometimes been favoured in the past, though it is clear that if crystallites do ever exist in the liquid phase they must be extremely small; otherwise it is next to impossible to explain the entropy of melting. Entropy, specific heat and surface entropy are discussed in §§2.3–2.5 and the important point emerges that if the liquid phase at its melting point is compared not with the solid but with a perfect gas of the same density, then the so-called *excess* entropy of the liquid is much the same for all the simple substances that concern us. Here is a preliminary indication that their structures are also much the same, which surely undermines the quasi-crystalline approach.

The non-crystalline models constructed by Bernal and others are of little help to us, however, because they are *static*. Fortunately it is now possible to construct *dynamic* models, in which the molecules are endowed with kinetic energy, with the aid of a computer. In principle one should be able to obtain an exact picture of a liquid in this way, provided that the pair potential $w(R)$ is accurately known. The very interesting results obtained by Alder & Wainwright (1957) for a system composed of non-attracting rigid spheres are presented in §2.6. The system gets continuously denser as the external pressure is increased, without passing through any vapour–liquid transition, until the volume reaches about 1.63 times the close-packed volume; at this stage the *packing fraction* y, i.e. the fraction of the total volume which is occupied by the spheres themselves, is about 0.46. Any further increase of pressure forces the spheres to crystallise, and the volume shrinks abruptly by a further 5%. This 5% shrinkage, incidentally, is markedly less than the 14% suggested by Bernal's model.

We have seen in chapter 1 that it may be legitimate at high temperatures to treat the ions in a metal as rigid spheres, confined in a box whose volume is determined not so much by the external pressure as by the term in the energy of the metal which depends upon volume but not upon structure. The effect of the volume-dependent term on the equation of state computed by Alder & Wainwright is discussed in §2.7. The argument explains why the change of volume on solidification is if anything less than 5% for metals but substantially greater than 5% for Ar. It also explains the famous empirical rule established by Lindemann, which relates the melting temperature to the characteristic Debye temperature of the solid (see §2.8).

In §2.9 the attempt to think in terms of models is abandoned and the language of *distribution functions* introduced instead; one defines a pair distribution function g_2, a three-body distribution function g_3 and so on. These, as it turns out, describe just those average properties of the ensemble of possible configurations for a liquid that are needed in order to calculate its macroscopic properties. An argument about whether or not the local order is quasi-crystalline is really an argument about the nature of g_3. It is shown in §2.10 that $(g_2(R)-1)$, which is sometimes referred to as the *pair correlation function*, is a Fourier transform of $(a(q)-1)$, where $a(q)$ is the interference function introduced in chapter 1. Since $a(q)$ may be determined by the methods of X-ray or neutron diffraction, it follows that g_2 comprises all the information we can learn about the molecular structure by the application of these techniques. There is no direct way to find g_3, let alone any of the higher-order distribution functions.

Lack of information about g_3 is not going to hamper us, when we come to consider in detail the effect of the ionic pseudo-potential upon the conduction electrons in a liquid metal, provided that the pseudo-potential is weak. In these circumstances we can use perturbation theory and a calculation to the second order only should be sufficient. In this order, as we have seen already in chapter 1, it is $a(q)$ that matters. But a calculation to third order is bound to involve g_3, and so on.

A number of significant theoretical results concerning $a(q)$ are proved in §2.10 and the properties of this important function are further illustrated in §2.11 by consideration of a crystalline solid. In §2.12 the argument is extended to mixtures; in a binary mixture, for example, three *partial* interference functions a_{00}, a_{11} and a_{01} need to be defined. All this material is a necessary preliminary to a discussion in §§2.13, 2.14 and 2.15 of how $a(q)$ and therefore g_2 are measured in practice and of the nature of the results obtained.

The results are indeed much the same for all liquid metals at their melting points, as the excess entropy had led us to expect, and they correspond closely to the results predicted for a rigid-sphere fluid with a packing fraction y of 0.46; on further heating, of course, the value of y decreases, because the effective rigid-sphere diameter σ goes down while the volume of the specimen goes up. Differences of detail do occur, however, and presumably these are related to differences in the form of $w(R)$ from one metal to another. Approximate theories of the liquid state exist which enable $w(R)$ to be inferred from $g_2(R)$, and a simple account of two of the more important is given in §§2.16 and 2.17. When used to analyse the data available for liquid metals, alas, they generate curves for $w(R)$ of quite unexpected form. The reason for this failure is still not apparent. To test whether the theory of $w(R)$ is at fault it is employed in the final section in an attempt to explain the compressibility of liquid metals. This attempt is on the whole a success.

In so far as the *thermodynamic* properties of pure metals are discussed at all in this book they are discussed in the present chapter. The reader should appreciate that many more data are available in the literature than there is space to summarise (Wilson, 1965), while fresh results are steadily being accumulated from experiments to measure density, sound velocity, specific heat etc. at high temperatures and pressures (Ross & Greenwood, 1969).

2.2. Models

What does a liquid really look like on the microscopic scale? One thing that is certain is that all the molecules must be in rapid motion, with a mean kinetic energy of $\frac{3}{2}k_B T$ so that the environment of each one must be continuously changing.† But suppose that one could sit on a single molecule and record the position of its neighbours at successive instants of time, well separated from one another, what sort of picture of the average environment would emerge?

The crystallographer prefers to approach this problem in terms of a model which he can visualise or even build out of balls and spokes, and a whole variety of models have at one time or another been proposed (Barker, 1963). At one extreme are what may broadly be termed the *quasi-crystalline* models, in which the local coordination just above the melting point is treated as very similar to that which prevails in the solid phase just below. We shall be referring below to evidence which suggests that the mean separation between neighbouring molecules changes less on melting than might be expected from the increase in volume which occurs; it is the mean number of nearest neighbours – the so-called *coordination number* – which alters. This is the sort of evidence that has been used to justify the quasi-crystalline approach, the extra volume in the liquid phase being regarded as used up in 'holes' in the liquid structure. In the crudest sort of model the holes may be treated as actual vacancies in a crystal lattice, but more generally they could be dislocations, grain boundaries between crystallites or any sort of misfit that arises when the long-range order of the solid is destroyed.

One of the difficulties facing any quasi-crystalline model is to explain the observed increase of entropy on melting, which, as shown in table 2.1 below, is typically of order k_B per atom in simple monatomic substances. This difficulty is examined in greater detail in the following section, where it emerges that if crystallites do indeed exist in liquids they must be very small – say less than 5 molecules across. It is sometimes argued that they

† The result quoted for the mean kinetic energy is a classical one, derived from the principle of equipartition. Quantum corrections are never likely to be significant in liquids except for elements of low atomic weight which remain liquid at low temperatures: in He, of course, they are of overwhelming importance. Throughout this chapter and most of the next we shall be using classical language, describing the molecules as distinguishable objects whose coordinates of position and velocity can be precisely defined.

could not be even as big as this, because the presence of nuclei 5 molecules across would make it impossible for a liquid to supercool – whereas in ideal circumstances (Turnbull & Cech, 1950) almost all liquid metals can be persuaded to supercool through as much as 20% of their melting temperature. It is possible to get round this argument by various special hypotheses: if the crystallites in a liquid are to be regarded as in a state of free rotation, for example, there may be a potential barrier inhibiting the growth of one at the expense of another. But it is a good deal more plausible to suppose that, if crystallites exist at all, they are so small and so irregular and so continuously swapping molecules between themselves and changing their orientations that an observer sitting on a molecule in a liquid would scarcely recognise their existence. All he would detect over long periods of time would be certain average correlations between the positions of neighbouring molecules, reminiscent of the correlations that hold exactly in the solid lattice.

At the opposite extreme from the quasi-crystalline picture lies the model of *random close-packed* (RCP) spheres which Bernal (1965; see also Bernal & Finney, 1967; Finney, 1970) in particular has investigated. The RCP state can be studied on the macroscopic scale by pouring a lot of rigid ball bearings, for example, into a container with irregular surfaces (smooth sides encourage 'crystallisation') and shaking them together until they can be compressed no further. Bernal has devised a number of ingenious methods for revealing the coordination patterns that are characteristic of it. Briefly, his conclusions are the following. (*a*) The mean number of neighbours which virtually make contact with some central sphere, i.e. whose separation from it is less than 5% of the sphere diameter, is about 8. (*b*) If, however, the entire volume of the system is divided up into *Voronoi polyhedra*,† one polyhedron being associated with each sphere, then the mean number of faces per polyhedron, i.e. the mean number of 'geometrical' rather than 'structural' neighbours, is close to 13.6; nearly all these geometrical neighbours lie within $\sqrt{2}$ diameters of the central sphere. (*c*) It follows from Euler's theorem that the mean number of edges per face on the polyhedra should be very close to 5. A high

† These polyhedra are constructed by drawing planes to bisect at right angles the lines that join adjacent molecules.

incidence of 5-sided faces is indeed observed, and associated with these are 5-fold rings of neighbouring atoms. This is a feature which is foreign to *any* crystal lattice with translational symmetry, so that an essential difference between the RCP state and any arrangement that could legitimately be described as quasi-crystalline does exist.

The increase of volume needed to pass from a truly close-packed arrangement of rigid spheres to an RCP one is 14%, which is encouragingly close to the expansion that is observed to occur on melting for the rare gases (see table 2.1). We shall find in §2.6, however, that the agreement is fortuitous. The weakness of the Bernal model is its *static* character. When we come to construct *dynamic* models with the aid of a computer, we discover that the increase of volume on melting is greatly affected by the kinetic energy of the molecules, which ensures that in the solid just before melting they are already quite well separated from one another. For a system of non-attracting rigid spheres, in fact, $\Delta\Omega/\Omega$ turns out to be only 5%. That it is as much as 15% in practice for rare gases but rather less than 5% for most metals can be understood, as we shall see, once the effects of the attractive forces that hold these substances together are properly appreciated.

2.3. Entropy

Some secrets about the structure of a liquid must be locked up in its entropy, and it is worth considering whether the data for this quantity favour one model rather than another. If we are inclined to believe in quasi-crystalline models it is obviously of interest to look at the increase of entropy on melting, ΔS, which can of course be determined from the latent heat of melting. It is equally illuminating, however, to compare the total entropy of the liquid phase, as determined from specific heat measurements down to very low temperatures and tabulated e.g. by Hultgren, Orr, Anderson & Kelley (1963), with the quantity

$$S_{\text{PG}} = Nk_{\text{B}} \log \left(\text{e} \frac{\Omega}{N} \left(\frac{2\pi \text{e} m_{\text{A}} k_{\text{B}} T}{h^2} \right)^{\frac{3}{2}} \right),$$

which describes the entropy that a perfect gas would have at the

TABLE 2.1. *The volume and entropy of simple liquids*

Substance	Increase of volume on melting (%) $100(\Delta\Omega/\Omega)$	Increase of entropy on melting $\Delta S/Nk_B$	'Excess' entropy at melting point $-S_E/Nk_B$
A	14.3	1.7	3.6
Kr	15.1	1.7	—
Xe	15.3	1.7	—
Li	1.6	0.80	3.7
Na	2.5	0.85	3.55
K	2.5	0.84	3.61
Rb	2.5	0.87	—
Cs	2.6	0.87	—
Cu	4.4	1.16	3.6
Ag	3.8	1.13	3.85
Au	5.1	1.11	4.0
Be	7.0	1.13	3.7
Mg	3.1	1.17	3.45
Zn	4.2	1.28	3.9
Cd	4.0	1.24	4.15
Hg	3.6	1.18	5.1
Al	6.0	1.39	3.6
Ga	−3.2 (+1.5)†	2.2 (1.1)†	4.75
In	2.0	0.92	4.4
Tl	2.2	0.85	4.05
Si	−9.6	3.6	(2.5)
Ge	−5.0	3.2	(1.0)
Sn	2.6	1.67	4.2
Pb	3.5	0.96	4.1
Sb	(−0.9)	2.61	3.2
Bi	−3.4	2.40	3.9
Te	4.9	3.0	(2.6)

† The figures in parentheses for Ga apply to the metastable solid phase Gaβ, which tends to crystallise out of the melt at −16.3°C if liquid Ga is supercooled past the melting point (30°C) of the stable phase, GaI. (Curien, Rimsky & Defrain, 1962; Bosio & Defrain, 1964.)

same temperature and volume. The difference, $(S - S_{PG})$, is known as the *excess entropy* and denoted by S_E. It is, of course, a negative quantity, since the spatial coordinates of the molecules are bound to be more ordered in the liquid phase than they would be for a perfect gas. Values for ΔS and for S_E just above the melting point are shown in table 2.1.

We have already noted that ΔS is about k_B per atom for many monatomic substances, but evidently there are systematic variations in its magnitude from one group of substances to the next. It is $1.7k_B$ for the rare gases, $0.85k_B$ for the b.c.c. metals (Li, Na, K, Rb, Cs, Tl), and say $1.15k_B$ for most of the close-packed ones (Cu, Ag, Au, Be, Mg etc.). For Ga, Si, Ge, Sb, Bi and Te, which are semi-metals or semiconductors in the solid phase, it is exceptionally large. By comparison, the variations in S_E seem to be unsystematic and relatively small.

Surely the moral is that when these simple substances melt they tend to lose the structural features that distinguish them from one another and to conform to a common pattern? Even Sb and Bi seem to conform to this pattern; they are semi-metals in the solid Phase, with relatively open crystal structures of low symmetry, but the contraction that occurs on melting seems to bring them into line with the metals proper. Ga is left with a rather large value for $-S_E$, and it is possibly that Ga, together with Hg, has a slightly more ordered structure just above the melting point than the average liquid metal, though the distinction is clearly not a large one. Si and Ge, which are semiconductors in the solid phase but contract a great deal on melting and become quite ordinary metals, would seem if anything to be *less* ordered in the liquid phase than the average; the specific heat data for these elements are unreliable, however, and the entries for $-S_E$ are probably erroneous.

The same must be said of Te, which is a case of special interest. It is known that in solid Te the atoms lie on continuous spiral chains, the distance between two neighbours on the same chain being less than the distance between adjacent chains. Solid Te is in fact half way to being a molecular solid like S, bound together by directional covalent bonds, and its electrical properties are those of a semi-conductor. Obviously much of the structural order is lost on melting, for ΔS is certainly large, but Te is not so unambiguously metallic in the liquid phase as Si and Ge: its Knight shift looks metallic, but its resistivity is about five times higher than that of any of the liquid metals proper and its Hall coefficient, relative to the free-electron value, is about three times higher (see tables 4.3, 5.2 and 5.8 respectively). It is clear that chain-like molecules exist in liquid S, though they break up on heating above the melting point, and it is tempting to associate the somewhat peculiar electrical properties of liquid Te with a similar tendency for the atoms to associate in chains. But the evidence from viscosity measurements which has been quoted in support of this view (e.g. Glazov, Chizhevskaya & Glagoleva, 1969) is not by itself convincing. If the unreliable entry for liquid Te in table 2.1 is any guide, the chain model for the structure of liquid Te is scarcely tenable. Other evidence against it has been summarised by Cabane & Friedel (1971).

The above considerations scarcely favour the quasi-crystalline approach, but a brief account of some efforts which have been made to reconcile it with the data for ΔS may nevertheless be of interest. One early theory was that of Mott (1934). He observed that the thermal entropy of an Einstein solid is given in terms of its characteristic vibration frequency ν_S by the formula

$$S = 3Nk_B \log (k_B T/h\nu_S) + O(h\nu/k_B T)^2. \qquad (2.1)$$

On the assumption that a similar formula can be applied to the liquid one has immediately

$$\Delta S \simeq 3Nk_B \log (\nu_S/\nu_L), \qquad (2.2)$$

so that an entropy increase of order Nk_B can be explained by the hypothesis that ν falls by some 40% in going from the solid to the liquid phase. In support of this hypothesis it was noted that the electrical resistivity of solid metals at high temperatures (Mott & Jones, 1936, p. 244) should be proportional to the mean square amplitude of vibration of each ion and hence to ν_S^{-2}. If this principle can be extended to the liquid phase one should be able to write

$$\Delta S \simeq \tfrac{3}{2}Nk_B \log (\rho_L/\rho_S), \qquad (2.3)$$

a relation which does indeed hold with reasonable accuracy in a number of cases.

Unfortunately, this elegant theory will not stand up to scrutiny in the light of later work. The suggested increase of ν on melting is too large to be reconciled with other evidence about molecular motion in liquids (see chapter 3), and ρ_L and ρ_S depend upon more factors than just ν_L and ν_S (see chapter 5). The real trouble is that in attributing all the entropy of a liquid to the thermal agitation of its molecules Mott ignored the *configurational entropy* arising from disorder in the rest positions about which the molecules are instantaneously vibrating. Even the most primitive of quasi-crystalline models, in which all the extra volume introduced on melting is supposed to be used up in the creation of vacancies, suggests a configurational entropy equal to

$$Nk_B(\Delta\Omega/\Omega)(1 - \log(\Delta\Omega/\Omega)), \tag{2.4}$$

and this is not trivial – it amounts to $0.2Nk_B$ if $\Delta\Omega/\Omega$ is 0.05. For any realistic model the configurational entropy may well be greater.

A different approach to the calculation of ΔS was suggested by Mott & Gurney (1939), again resting on a quasi-crystalline model. These authors pictured the liquid as composed of a number of crystallites, each containing say s^3 molecules, and they estimated the configurational entropy on the assumption that every crystallite was free to take up at random one of about s^2 orientations. If changes in the entropy due to thermal agitation are now ignored, this suggests

$$\Delta S \sim Nk_B(\log s^2)/s^3, \tag{2.5}$$

which is clearly incapable of fitting the observed results, however small the choice of s. A slightly different answer may be obtained by supposing the crystallites to be freely rotating; after making allowance for the 3 modes that must be lost from the vibrational spectrum of the solid whenever free rotation of a crystallite sets in, this picture leads to

$$\Delta S \simeq \tfrac{3}{2}Nk_B \log(4k_B m_A R_A^2 \Theta_D^2 s^3/5\hbar^2 T_M)/s^3, \tag{2.6}$$

where m_A is the atomic mass, R_A the atomic radius, Θ_D the characteristic (Debye) temperature of the solid and T_M its melting temperature. The combination of parameters inside the brackets is much the same for all metals (see p. 103) and the result can be expressed as

$$\Delta S \simeq \tfrac{3}{2}Nk_B \log(10^3 s^2)/s^3, \tag{2.7}$$

which would fit the observations with a value of s between 2 and 3.

It can be argued that both the above calculations underestimate the configurational entropy because they ignore the freedom of each crystallite to exchange molecules with its neighbours. Various attempts have been made to estimate an extra term due to this (e.g. Ookawa, 1960). A simple argument is to say that each crystallite has about $6s^2$ molecules in its surface layer, and that each of these is free to jump to a neighbouring crystallite without much increasing the energy. This suggests an extra term of order $6Nk_B(\log 2)/s$, which amounts to $0.8Nk_B$ when s is as big as 5.

The crudity of all these attempts to estimate the entropy of melting from quasi-crystalline models will be very apparent to the reader, but the general moral does seem to emerge, as mentioned in §2.1, that if crystallites did exist in liquids they would have to be very small. Once one decides to abandon the quasi-crystalline approach, of course, the problem changes; it is the magnitude of S_E rather than ΔS which requires explanation. An approximate method for estimating S_E that might be applicable to Bernal's RCP model has been suggested by Collins (1965), but we shall see how the problem can be tackled in a quite general fashion, using the language of distribution functions, in §2.9. Further discussion of the data presented in table 2.1 must be postponed till then.

2.4. Specific heat

Accurate data for the specific heat of monatomic liquids are not plentiful, but the results for liquid Na which are reproduced in table 2.2 are fairly typical.† At the melting point the specific heat at constant volume, C_Ω, is quite close to the Dulong & Petit value of $3Nk_B$, as it is in monatomic solids. This may occasion some surprise, since a good part of the specific heat of solids is attributed to the excitation of transverse vibrational modes which an ideal fluid is unable to support. Real liquids are viscous, however, and their viscosity at the melting point is sufficient for transverse modes whose wavelength is comparable with atomic dimensions to propagate with little attenuation (see §3.15). On heating, the

† A small correction for the electronic specific heat, which should be linear in T (see §1.10), has been subtracted from these results. The conduction electrons, incidentally, are liable to affect the entropy of a metal as well as its specific heat, and in principle there could be an electronic term in the entropy of melting ΔS if m^*/m were to change significantly on passing from the solid to the liquid phase. In practice it could never amount to more than $0.1Nk_B$; its influence on the figures for ΔS in table 2.1 is negligible.

TABLE 2.2. *The specific heat of liquid Na*†

T(°C)	100	200	300	400	500
$C_\Omega/3Nk_B$	1.14	1.06	1.00	0.91	0.81

† See footnote on p. 84.

TABLE 2.3. *Specific heat ratios for liquids*

Ar	2.2	Zn	1.21
		Cd	1.22
		Hg	1.12
Na	1.10	Al	1.25
K	1.12	Ga	1.10
Rb	1.15	In	1.12
Cs	1.18	Tl	1.18
Cu	(1.45)	Sn	1.15
Ag	(1.45)	Pb	1.18
		Sb	1.14
		Bi	1.15

Note. These data are for temperatures close to T_M.

viscosity rapidly falls and the transverse modes do become hindered. It is no doubt for this reason, as Wannier & Piroue (1956) have pointed out, that C_Ω diminishes on heating in the way that it does. It generally tends to a value of about $2Nk_B$ at the critical point T_C, consistent with the idea that by this stage the transverse vibrations are no longer contributing. In a monatomic gas, which can sustain no vibrational modes of any sort, C_Ω is of course only $\frac{3}{2}Nk_B$.

Since C_Ω exceeds $(C_\Omega)_{PG}$, the excess entropy defined in the previous section must rise on heating; it changes from say $-4Nk_B$ at the melting point to something less than

$$-4Nk_B + \tfrac{3}{2}Nk_B \log (T_C/T_M) \simeq -2Nk_B$$

at the critical point. It is bound to remain negative, of course, as long as the molecular structure retains some degree of order.

The ratio between the specific heats at constant pressure and constant volume may be deduced from the well-known thermodynamic formula

$$\gamma = C_p/C_\Omega = 1 + \alpha^2 \Omega T/\beta C_\Omega, \tag{2.8}$$

where α is the volume expansion coefficient and β the isothermal compressibility. Having discovered that the increase of volume on melting is some four or five times greater for the rare gases than for a typical metal, it may not surprise the reader to learn that (αT_M) is also four or five times greater; $(\alpha\Omega/\beta C_\Omega)$ on the other hand is much the same for all monatomic liquids (see table 2.5 on p. 96). Hence it turns out that there is a big distinction between rare gases and metals in the magnitude of γ for the liquid phase. Figures are given in table 2.3. The distinction has been further discussed by Bratby, Gaskell & March, (1970).

2.5. Surface tension and surface entropy

The more sophisticated discussion on which we are shortly to embark concerning the structure and thermodynamic properties of liquids in the bulk will not help us to understand their surface properties. It is for this reason that the present section is inserted where it is, though it constitutes a digression from the main argument.

The data available for the surface tension, ℓ,† of liquid metals have been discussed by Semenchenko (1961) and they are presented graphically in fig. 2.1, in such a way as to bring out the fair correlation that exists between ℓ and the heat of vaporisation. Compared with many non-metallic liquids the surface tension is large, but this is not unexpected; quite primitive theories based upon the jellium model are capable of giving answers of the right order of magnitude for alkali metals and pseudo-potential theories are yielding promising results for polyvalent ones (Lang & Kohn, 1970). But exact calculations are not easy (see §1.15) and until more theoretical work has been done there is little to be learnt about the structure of the liquid surface from the magnitude of ℓ alone. Its temperature variation, however, may be more informative.

In the normal course of events ℓ is a decreasing function of temperature; it goes to zero at the critical temperature like $(T_C - T)^n$, where n seems generally to be a little greater than unity. This means that the *surface entropy* per unit area, which is given

† The more usual symbols for surface tension have been pre-empted for other purposes; ℓ is not inappropriate, since the surface tension is also the surface free energy per unit area.

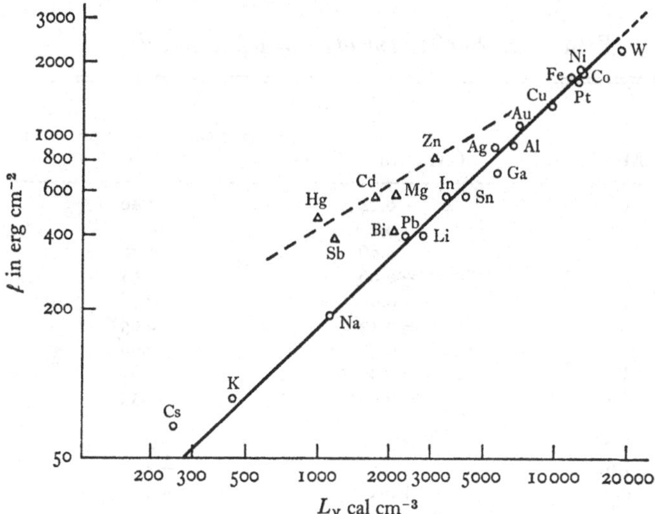

Fig. 2.1. The surface tension of liquid metals at the melting point compared with the latent heat of vaporisation. (Redrawn by permission from Grosse (1964), *J. Inorg. Nucl. Chem.* **26**, 1349.)

by $-(d\ell/dT)$, is a positive quantity. The surface entropy corresponding to the area occupied by a single atom in the surface layer is typically rather less than k_B – figures are given in table 2.4 for a number of metals which have been investigated recently and for which the results for $d\ell/dT$ should be fairly reliable (see Wilson, 1965; Germer & Mayer, 1968; White, 1966, 1968 and private communication). There are some liquid metals, however, for which $d\ell/dT$ is reported to be positive over a temperature range of 100 °C or more above the melting point, notably Cu, Zn and Cd. White, who has carried out particularly careful measurements with extremely pure materials, claims that the tendency for $d\ell/dT$ to be positive is enhanced if steps are taken to prevent the specimen from continuously evaporating while measurements are being made, by enclosing it in a box filled with saturated vapour. It seems that an evaporation rate which corresponds to a loss of only about 100 atomic layers per second is sufficient to effect a marked change in ℓ and $d\ell/dT$, a surprising result since one would expect the surface to reach equilibrium in a time of order 10^{-11} sec at most (see chapter 3). There seems no reason to doubt White's results, however, which carry with them the implication that for some way

TABLE 2.4. *Surface entropies of liquid metals*

Metal	$\frac{d\mathscr{T}}{dT}$ (dyne cm^{-1} °C^{-1})	Surface entropy per atom (units of k_B)
Li	−0.14	0.80
Na	−0.1	0.85
K	−0.06	0.78
Rb	−0.06	0.94
Cs	−0.05	0.78
Ag	−0.13	0.65
Al	−0.135	0.68
In	−0.096	0.61
Sn	−0.083	0.55
Cu	+0.75	−2.9
Zn	+0.5	−2.2
Cd	+0.5	−2.9
Hg (25 °C)	−0.20	+1.2

Note. These data are for temperatures close to T_M, except in the case of Hg.

above the melting point the surface entropy in Cu, Zn and Cd is *negative*. The surface entropy per atom seems to be less than it is for other liquid metals by at least $3.5k_B$.

To convince ourselves that it is Cu, Zn and Cd which are anomalous, rather than the metals above them in table 2.4, let us consider how much surface entropy we might expect to arise from the thermal excitation of *ripples*. An approximate answer may readily be obtained from a two-dimensional analogue of the familiar Debye theory, which tells us the entropy associated with the excitation of vibrational modes in the bulk of the specimen. Ripples, at any rate when their wavelength is much less than 1 cm and yet large compared with atomic dimensions, obey the dispersion relation

$$\omega^2 = q^3\mathscr{T}/\rho, \tag{2.9}$$

where ω is the angular frequency for wave vector q and ρ is the density. It follows that the 'density of states', i.e. the number of distinct ripple modes per unit range of ω for an area of surface A is given by

$$\mathscr{N}(\omega) = \frac{A}{3\pi}\left(\frac{\rho}{\mathscr{T}}\right)^{\frac{2}{3}}\omega^{\frac{1}{3}}. \tag{2.10}$$

In the spirit of the Debye model let us suppose that (2.9) and (2.10) are accurate right up to some cut-off frequency ω_{max}, determined by the condition

$$\int_0^{\omega_{max}} \mathcal{N}(\omega)\, d\omega = A/a, \tag{2.11}$$

where a is the area of surface occupied by a single molecule, which would be $1.09\,(m_A/\rho)^{\frac{2}{3}}$ for a close-packed structure with close-packed molecules in the surface layer. We may define a characteristic surface temperature Θ_S by the equation

$$\Theta_S = \frac{\hbar\omega_{max}}{k_B} = \frac{\hbar}{k_B}\left(\frac{4\pi}{1.09}\right)^{\frac{3}{4}}\left(\frac{\ell}{m_A}\right)^{\frac{1}{2}}, \tag{2.12}$$

to play the role of the Debye temperature of the bulk, Θ_D.

It is now an elementary exercise in statistical mechanics to show that at high temperatures ($> \Theta_S$)

$$\ell(T) = \ell(0) - \frac{k_B T}{A}\int_0^{\omega_{max}} \mathcal{N}(\omega)\left\{\log\left(\frac{k_B T}{\hbar\omega}\right) - \frac{1}{24}\left(\frac{\hbar\omega}{k_B T}\right)^2 + \ldots\right\} d\omega,$$

where $\ell(0)$ is the surface free energy in the absence of ripples, presumably independent of temperature. This suggests a surface entropy per atom in the surface layer of

$$-a\frac{d\ell}{dT} = k_B\left\{\log\left(\frac{T}{\Theta_S}\right) + \frac{7}{4} + \frac{1}{60}\left(\frac{\Theta_S}{T}\right)^2 + \ldots\right\}. \tag{2.13}$$

We must remember, however, not to over-specify the motion of the specimen as a whole; if we wish to count some surface ripple modes we must deduct an equal number of bulk modes from the usual Debye spectrum. The entropy associated with these bulk modes can be estimated by more or less the same argument, though we must now use a density of states proportional to ω^2 rather than ω. It turns out that

$$k_B\left\{\log\left(\frac{T}{\Theta_D}\right) + \frac{4}{3} + \frac{1}{40}\left(\frac{\Theta_D}{T}\right)^2 + \ldots\right\}$$

should be subtracted from the right-hand side of (2.13). The final answer is therefore

$$-a\frac{d\ell}{dT} = k_B\left\{\log\left(\frac{\Theta_D}{\Theta_S}\right) + \frac{5}{12}\right\}; \tag{2.14}$$

the terms of order $(\Theta/T)^2$ are negligible above the melting point.

Experimental values of Θ_D are listed in table 2.6, while Θ_S is readily evaluated from (2.12). For liquid metals the ratio Θ_D/Θ_S lies in the range between 1.4 and 2.1, with 1.8 as about the mean value, so the surface entropy per atom is expected to be close to $+k_B$. The theory is therefore capable of explaining the entries in the top half of table 2.4 with reasonable precision. Detailed agreement was scarcely to be expected, in view of the assumptions and approximations involved.

So Cu, Zn and Cd are clearly the anomalous ones, and to explain their negative surface entropies just above the melting point we must apparently postulate an unusual degree of ordering in the arrangement of the ions near the surface. The crudest explanation is that the liquid there is virtually crystalline. Since the entropy of melting of Cu, Zn and Cd is about $1.2k_B$ per atom (see table 2.1) we would need to regard the top three layers of ions as crystalline in order to account for the results. Curiously enough, a suggestion that the surface of a liquid might be crystalline has been made previously by Bernal (private communication) on the basis of his observations regarding the packing of rigid spheres. It would be of great interest to confirm the hypothesis by electron diffraction techniques. Such experiments have been attempted and provide some evidence for solid-like structure near the surface of liquid Pb (Croxton, 1969), a metal which according to White shows an anomalously small value of $d\ell/dT$ just above the melting point, though one that is not actually positive. The difficulties of the experiment and of interpreting the results are, however, severe.

The surface structure of Hg is of particular interest in connection with its optical properties (see §5.23) but the situation as regards its surface tension is still obscure. Most of the measurements of ℓ reported for this metal are naturally for temperatures above 0 °C and they indicate a positive surface entropy. It is in fact a rather large one, as shown in table 2.4. There are reports that below 0 °C, however, at temperatures close to its melting point, Hg is another metal for which $d\ell/dT$ becomes positive (Semenchenko, 1961).

From a theoretical point of view it may prove easier to calculate the surface tension for a liquid–solid interface than for a liquid–vapour interface; the former constitutes a much less severe perturbation of the conduction electrons than the latter and the methods of chapter 1 may be applicable. Values of the liquid–solid surface tension have been estimated by Turnbull (1950) from data concerning the nucleation rate of supercooled droplets. It seems to be about 10% of ℓ, though it is naturally greater than this for substances like Ge and Sb where severe misfit between the liquid and solid phases is to be expected. In fact the liquid–solid surface tension correlates better with the heat of melting than with the heat of vaporisation.

The ripple model may be used, by the way, to estimate the *roughness* of a free liquid surface in thermal equilibrium. According to the law of equipartition of energy we may equate the mean potential energy in each ripple mode to $\frac{1}{2}k_B T$ at high temperatures, and the mean square displacement for which it is responsible is therefore given by

$$\overline{\xi^2} = k_B T/\ell A q^2 = (k_B T/\ell A)(\ell/\rho\omega^2)^{\frac{2}{3}}.$$

The total mean square displacement of any point on the surface may be obtained by summing the contributions due to all modes between ω_{max}, as given by (2.11), and some ω_{min}, determined by the configuration of the specimen, at which the dispersion law (2.9) breaks down. In a large specimen, whose depth is several centimetres at least, (2.9) breaks down when

$$q^2 < \rho g/\ell,$$

since ripples of long wavelength are controlled by gravity rather than surface tension. Hence in these circumstances we may expect

$$\Sigma\overline{\xi^2} = \frac{k_B T}{3\pi\ell}\log\left(\frac{\omega_{max}}{\omega_{min}}\right) \simeq \frac{k_B T}{4\pi\ell}\log\left(\frac{4\pi\ell}{\rho g a}\right). \qquad (2.15)$$

The total r.m.s. displacement suggested by this expression is not more than about 3 Å for a typical liquid metal.

2.6. Molecular dynamics

Bernal's RCP model for liquids has been criticised above for being static. The method of *molecular dynamics*, or the virtually equivalent technique of *Monte Carlo* calculation (consult Barker, 1963; Wood, 1968), enables one to construct a model in which the molecules are continually on the move. Results obtained by this method have revolutionised our attitude to liquids in recent years.

The idea is to simulate the behaviour of a limited number of molecules, confined to a given volume and interacting with one another through a given pair potential, with the aid of a computer. The molecules are endowed with more or less ordered coordinates of position at the start, so that none overlap, but with random coordinates of velocity, and the computer is then left to follow their trajectories by a detailed solution of Newton's laws of motion. It is usual to assume periodic boundary conditions, so that a molecule which leaves the volume across one face re-enters across the face opposite. In principle one should be able to construct an exact picture of a liquid in this way, provided that the appropriate pair potential is accurately known. In practice there are limitations set by the capacity of the computer; with the machines at present available it takes at least 1 hour of computer time to follow the motion of a system composed of 10^3 molecules for a time of 10^{-11} sec.

The first computations were for rigid spheres (Alder & Wainwright, 1957), but the method has since been extended to systems in which the spheres interact by means of the 'Lennard–Jones' potential† which is often used for calculations on Ar (Rahman, 1964*a*, 1966; Verlet, 1967, 1968; Levesque & Verlet, 1970; McDonald & Singer, 1967 – though these last authors use Monte Carlo techniques). Results have also been obtained (Paskin & Rahman, 1966; Schiff, 1969) using potentials with long-range oscillatory tails, such as may be appropriate for liquid metals (see §1.13), but our expectation that oscillations whose amplitude is a small fraction of $k_B T_M$ should be of rather little importance appears to be confirmed; Schiff has emphasised that where one set of results differs from another it is usually because different assumptions have been made about the steepness of the repulsive core. For the moment we shall concentrate upon the behaviour shown

† The Lennard–Jones potential has an attractive term proportional to R^{-6} and a repulsive one proportional to R^{-12}. Whether it is an accurate representation of the true potential between two Ar atoms has been strongly questioned by Guggenheim (1960) and others. There is evidence from the results of X-ray diffraction work on Ar near its critical point (Mikolaj & Pings, 1967) that the strength of the two terms may depend to some extent on density.

by rigid spheres, and particularly upon their thermodynamic properties. The information which molecular dynamics has so far provided about the detailed structure of the liquid phase can be better discussed at a later stage, using the language of distribution functions. It may be said at once, however, that none of the calculations provides support for any of the more extreme quasi-crystalline models. Nor do they verify the details of Bernal's RCP model; they suggest that the mean number of geometrical neighbours in liquid Ar, for example, is 14.45 rather than 13.6.

Perhaps the most striking single result is that a liquid–solid phase transition always occurs when the volume is diminished, even when there are no attractive forces between the spheres at all. It may be detected not only by extracting coordinates for the molecules and verifying that long-range crystalline order has set in, but also by programming the computer to calculate the mean pressure exerted by the spheres against the walls of their container. Since the internal energy $\mathcal{U}$ of a rigid-sphere fluid is entirely kinetic it does not depend upon volume at constant temperature, so that

$$(\partial\mathcal{U}/\partial\Omega)_T = -p+T(\partial S/\partial\Omega)_T = 0.$$

It follows that

$$p/T = (\partial p/\partial T)_\Omega, \tag{2.16}$$

and hence that (p/T) is a unique function of volume; all the isotherms for the fluid may be made to coincide by plotting the dimensionless quantities $(p\Omega_{CP}/Nk_BT)$ against (Ω/Ω_{CP}), where Ω_{CP} is the volume that the spheres would occupy if they were truly close-packed. A convenient parameter to use instead of Ω/Ω_{CP} is the *packing fraction*, y, defined as the fraction of the total volume which is occupied by the spheres themselves. Evidently

$$y = \pi N\sigma^3/6\Omega = (\sqrt{2}\pi/6)(\Omega/\Omega_{CP})^{-1}. \tag{2.17}$$

Alder & Wainwright's results are shown in this way in fig. 2.2, the two non-intersecting curves representing the pressure exerted by the liquid and solid phases. Although the number of spheres in the system investigated was sufficient for these curves to provide a reasonably accurate picture of how an infinite *homogeneous* system would behave, it was insufficient to allow *inhomogeneous* states, with both phases coexisting side by side, to persist for any length of time. Consequently, it was not possible to compute directly the

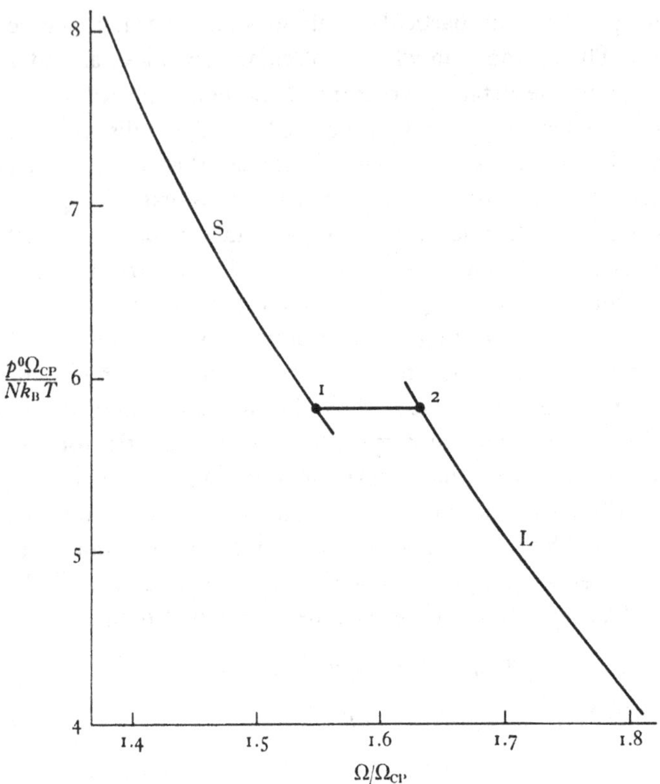

Fig. 2.2. The solid (S) and liquid (L) isotherms for non-attracting rigid spheres

position of the horizontal tie-line which joins the solid and liquid isotherms, and there is still some uncertainty as to where this should be drawn. The level of the tie-line in fig. 2.2 is copied from Longuet-Higgins & Widom (1964) and is rather lower than the one recommended by Ross & Alder (1966) and Hoover & Ree (1968). For the argument that follows to yield the best agreement with experiment it would need to be somewhat lower still.

From the figure the following conclusions can at once be drawn.

(*a*) A rigid-sphere fluid is compelled by lack of space to solidify when Ω/Ω_{CP} reaches about 1.63 or y reaches about 0.46.

(*b*) In the solid phase just below the melting point Ω/Ω_{CP} is still about 1.55, i.e. the thermal motion of the spheres keeps them well apart from one another still.

(*c*) The change of volume on melting is only about 5%, as mentioned in §2.2.

(*d*) The melting pressure is given by

$$p_M \simeq 5.8Nk_B T/\Omega_{CP}. \tag{2.18}$$

(*e*) Since the Gibbs free energy is the same at the points 1 and 2 in the figure we know that

$$p_M(\Omega_2 - \Omega_1) = T(S_2 - S_1),$$

and it follows that the change of entropy on melting is

$$\Delta S \simeq 5.8Nk_B \Delta\Omega/\Omega \simeq 0.5Nk_B. \tag{2.19}$$

It is a somewhat surprising feature of the rigid-sphere fluid that $(\Delta S/\Delta\Omega)$ at the melting point is the same as $(\partial S/\partial\Omega)_T$ in the liquid and solid phases just above and below it.

If we are to be justified in using rigid-sphere models to explain the properties of real liquids, we must first ascertain whether their entropy depends upon volume in the way that fig. 2.2 suggests. The dimensionless quantity

$$\Omega\left(\frac{\partial S}{\partial\Omega}\right)_T = \Omega\left(\frac{\partial p}{\partial T}\right)_\Omega = \frac{\alpha\Omega}{\beta}$$

is easily determined from experimental data for real liquids. Does it equal

$$\frac{\Omega p_M^0}{T} \simeq 5.8Nk_B \times 1.63 = 9.4Nk_B$$

just above the melting point as it should? The figures in table 2.5 show that for polyvalent metals at any rate the agreement is not unsatisfactory. By lowering the tie-line it could be improved.

To make the rigid-sphere model realistic, of course, we are obliged to include some attractive forces between the spheres. Without these the fluid has no cohesion; external pressure has to be applied to keep it in the form of a dense liquid or gas,† while to solidify it at a temperature of say 500 °K, when Ω_{CP}/N is say 10^{-23} cm^3, would require a pressure of at least 40 kbar according to (2.18). Longuet-Higgins & Widom (1964) have suggested an elegant way in which one may take the effects of attraction into account without embarking on a completely fresh computation.

† In the absence of attractive forces there is no real distinction between liquid and gas, since the critical temperature goes to zero.

TABLE 2.5. *Thermal expansion of liquids*

Substance	Volume expansion coefficient $10^4\alpha$ (°K^{-1})	αT_M	Volume per gram atom Ω (cm^3)	$\frac{\alpha\Omega}{Nk_B\beta}$
Ar	45	0.38	28.5	5.9
Na	2.75	0.10	24.7	4.5
K	2.9	0.095	47.2	4.4
Cu	1.0	0.14	7.96	6.3
Ag	0.97	0.12	11.6	6.5
Mg	1.66	0.15	15.3	7.6
Zn	1.50	0.10	9.7	7.0
Cd	1.51	0.09	14.0	8.0
Hg (25°C)	1.82	0.04	14.6	8.5
Al	1.22	0.11	11.4	6.9
Ga	1.26	0.04	11.5	7.9
In	1.2	0.05	16.4	7.9
Tl	1.4	0.08	18.1	8.0
Sn	0.88	0.045	17.0	6.8
Pb	1.27	0.075	19.6	8.4
Sb	0.96	0.09	18.8	4.5
Bi	1.1	0.06	20.9	6.5

Note. These data are for temperatures close to T_M, except in the case of Hg.

Their argument was originally framed with rare gases such as Ar in mind, but a modification which is appropriate for metals is presented in the following section.

For some of the calculations that follow it will prove convenient to have in mind an analytical formula to represent Alder & Wainwright's isotherm for the liquid phase. According to Carnahan & Starling (1969), the empirical equation

$$\frac{p^0\Omega}{Nk_BT} = f(y) = \frac{1+y+y^2-y^3}{(1-y)^3} \tag{2.20}$$

is the most successful of a number that have been suggested. Throughout the range of y which concerns us, the mismatch between Carnahan & Starling's formula and the computed curve is small enough for most purposes to be ignored.

2.7. An equation of state for the rigid-sphere model

The basic assumptions made by Longuet-Higgins & Widom are (*a*) that attractive forces do not affect the way in which the entropy of a rigid-sphere system depends upon volume, and (*b*) that they add to the initial energy a term which is a smooth function of volume at constant temperature, the same function for the liquid and solid phases. The validity of assumption (*a*) has been tested in table 2.5. Assumption (*b*) is plausible enough, especially for metals, whose cohesion can be attributed to the uniformly distributed sea of conduction electrons rather than to short range forces between neighbouring atoms.

From these assumptions it follows immediately that

$$p = p^0 - (\partial \mathscr{U}/\partial \Omega)_T, \tag{2.21}$$

where p^0 is the rigid-sphere pressure, to be taken from the curves in fig. 2.2; the superscript zero will be used throughout this section to denote properties of the rigid-sphere system in the absence of attractive forces. Longuet-Higgins & Widom suggest that in Ar the potential energy of each atom is proportional to the density of its neighbours, i.e. to Ω^{-1}, and they therefore arrive at an equation of state of the form

$$p = p^0 - a/\Omega^2;$$

this is reminiscent of Van der Waals's equation except that p^0 is no longer the pressure to be expected of a perfect gas. *For metals, however,* $(\partial \mathscr{U}/\partial \Omega)_T$ *is likely to increase with volume instead.*

This key statement can be justified both theoretically and by examination of experimental data for the compressibility. To appreciate the theoretical argument consider first an ideal jellium, for which the internal energy may be calculated from (1.16). It is a moderately complicated function of volume, but all that matters for our purposes is that it passes through a minimum at some volume Ω_0, and may be expanded about this minimum in a Taylor series of the form

$$\mathscr{U}(\Omega) = \mathscr{U}(\Omega_0) + C(\Omega - \Omega_0)^2 + \ldots, \tag{2.22}$$

the constant C being necessarily positive. For a real metal the values of Ω_0 and C are somewhat different, since they are affected by the details of the ionic pseudo-potential, but it must still be true that the part of the internal energy which depends upon volume rather than structure has a minimum somewhere, and equation (2.22) may still be used if $(\Omega-\Omega_0)$ is not too big. The data suggest, as we shall see, that it is not more than 15% of Ω for a typical metal at its melting point.

As for the compressibility, it follows by differentiation of (2.21) that

$$(\partial^2\mathscr{U}/\partial\Omega^2)_T = (\beta\Omega)^{-1}-(\beta^0\Omega)^{-1}. \tag{2.23}$$

Let us consider the liquid phase at its melting point, just above T_M. We shall discover shortly that the packing fraction y is more or less unaffected by the attractive forces, i.e. it is close to 0.46 at the melting point, so β^0 is given according to Carnahan & Starling's equation (2.20) by

$$\Omega_M/Nk_BT_M\beta^0 = f(0.46)+0.46f'(0.46) \simeq 39. \tag{2.24}$$

Hence
$$\frac{\beta}{\beta^0} \simeq \frac{39Nk_BT_M\beta}{\Omega_M} = 39a(0). \tag{2.25}$$

The quantity $a(0)$ defined by (2.25) occurs frequently in later sections and experimental values for it are listed in table 2.7. For liquid Ar $a(0)$ is *greater* than 1/39 (= 0.026) so β is greater than β^0 and $\partial^2\mathscr{U}/\partial\Omega^2$ is negative, as postulated by Longuet-Higgins & Widom. For every liquid metal except Li and Cs, however, $a(0)$ is *less* than 0.026, so $\partial^2\mathscr{U}/\partial\Omega^2$ must be positive.

If we take $\partial^2\mathscr{U}/\partial\Omega^2$ from (2.23), (2.24) and (2.25), we may readily deduce how the volume should change on melting. The problem may be discussed in terms of fig. 2.3, which shows the effect of the substantial correction $-(\partial\mathscr{U}/\partial\Omega)_T$ on the rigid-sphere isotherms for the two phases. Since this correction is not, like p^0, proportional to T, the isotherms can no longer be scaled into a single curve. The ones in the figure are drawn to represent the situation at the temperature T_M, where melting occurs for zero applied pressure. The points labelled 5 and 6 are therefore such that the free energies $\mathscr{F}_5$ and $\mathscr{F}_6$ are equal, just as $\mathscr{G}_1^0$ and $\mathscr{G}_2^0$ are equal. We may use these equalities to show, after a trivial analysis in which (2.21) is used to express $(\mathscr{U}_6-\mathscr{U}_5)$ as

$$\tfrac{1}{2}(p_4^0+p_3^0)(\Omega_4-\Omega_3),$$

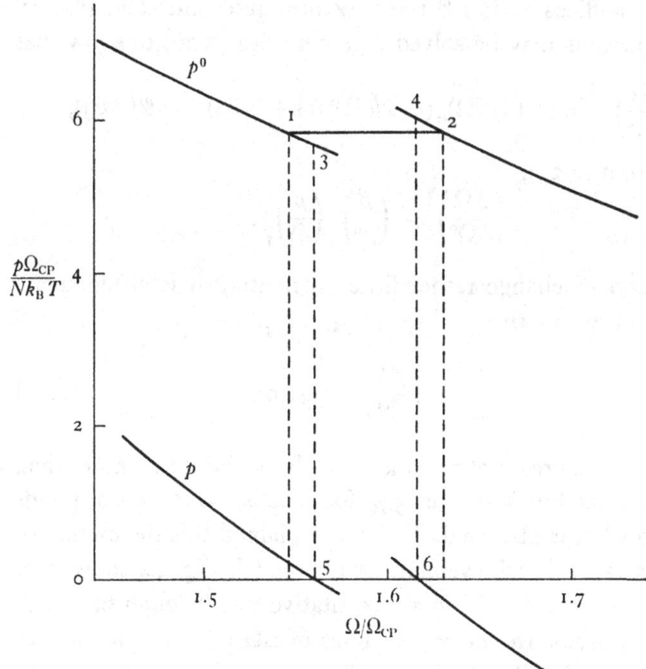

Fig. 2.3. Attractive forces lower the rigid-sphere isotherms and reduce the melting pressure to zero.

that $$(p_4^0 - p_M^0)(\Omega_2 - \Omega_3) = (p_M^0 - p_3^0)(\Omega_4 - \Omega_1). \tag{2.26}$$

This result proves that whatever the dependence of $\mathscr{U}$ on Ω the points 3 and 4, on the rigid-sphere isotherms for the solid and liquid phases respectively, must lie on opposite sides of the horizontal tie-line at the pressure p_M^0. Hence the attractive forces cannot shift the solid–liquid transition to a quite different range of volume; the value of y for the liquid phase at T_M should remain close to 0.46, as stated above.

If the change of volume on melting, $\Delta\Omega$ ($= \Omega_4 - \Omega_3$), remains comparable with $\Delta\Omega^0$ ($= \Omega_2 - \Omega_1$) we may safely write

$$\begin{aligned}(p_4^0 - p_M^0) &= (\beta^0\Omega)_L(\Omega_2 - \Omega_4),\\ (p_M^0 - p_3^0) &= (\beta^0\Omega)_S(\Omega_3 - \Omega_1),\\ (p_4^0 - p_3^0) &= (\partial^2\mathscr{U}/\partial\Omega^2)(\Omega_4 - \Omega_3),\end{aligned}$$

where the suffices L and S refer to the liquid and solid phases. These equations may be solved together with (2.26) to show that

$$\left(\frac{\Delta\Omega}{\Delta\Omega^0}\right)^{-2} = (1+(\beta^0\Omega)_S(\partial^2\mathscr{U}/\partial\Omega^2))(1+(\beta^0\Omega)_L(\partial^2\mathscr{U}/\partial\Omega^2)),$$

so that, from (2.23),

$$\left(\frac{\Delta\Omega}{\Delta\Omega^0}\right)^{2} = \left(\frac{\beta}{\beta^0}\right)_S\left(\frac{\beta}{\beta^0}\right)_L.$$

Since β and β^0 change rather little on melting it is an adequate approximation to write

$$\frac{\Delta\Omega}{\Delta\Omega^0} \simeq \left(\frac{\beta}{\beta^0}\right)_L \simeq 39a(0). \tag{2.27}$$

Evidently the increase of volume on melting should be more than 5% for the Ar but less than 5% for metals, which corresponds exactly to what is observed. To have explained this distinction so neatly is a valuable achievement for the model. Fig. 2.4 shows that (2.27) is not unsuccessful in a quantitative way, though of course it cannot describe the negative values of $\Delta\Omega$ observed for metals such as Sb or Bi. Some of the other metals for which it works badly, such as Na and K, are ones for which the model is already suspect because $(\alpha\Omega/Nk_B\beta)$ just above the melting point differs rather substantially from 9.4.

The argument may be extended to give the change of entropy on melting. Since (S_6-S_5) should be the same as $(S_4^0-S_3^0)$, and since

$$\left(\frac{\partial S^0}{\partial\Omega}\right)_S = \left(\frac{\partial S^0}{\partial\Omega}\right)_L = \frac{\Delta S^0}{\Delta\Omega^0},$$

it suggests that

$$\Delta S/\Delta S^0 = \Delta\Omega/\Delta\Omega^0,$$

i.e. that

$$\Delta S \simeq 20a(0)Nk_B. \tag{2.28}$$

This is capable of explaining why ΔS for Ar is greater than for a typical metal by something approaching Nk_B (see table 2.1) but it is not successful otherwise; no correlation is visible between the detailed results for ΔS and $a(0)$ as plotted in fig. 2.4, and the points are all substantially above the theoretical line.

Does the rigid-sphere model tend to overestimate the entropy of the solid phase at the melting point or underestimate that of the

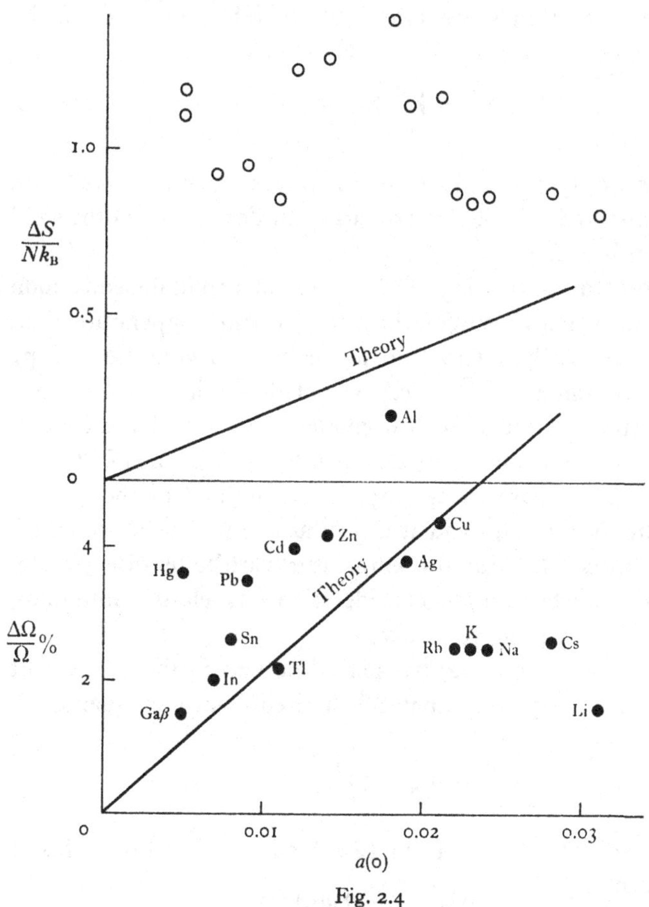

Fig. 2.4

liquid? We can answer this question by calculating what the excess entropy ought to be for a rigid-sphere liquid with $y = 0.46$ and comparing the result with the experimental values shown in table 2.1. If we start with a rarefied system for which the entropy is given correctly by the perfect gas formula and compress it, we arrive at a situation where

$$S_E = \int \left[\left(\frac{\partial S}{\partial \Omega} \right)_T - \left(\frac{\partial S_{PG}}{\partial \Omega} \right)_T \right] d\Omega$$

$$= \int_{\infty}^{\Omega} \left(\frac{p^0}{T} - \frac{Nk_B}{\Omega} \right) d\Omega. \qquad (2.29)$$

Carnahan & Starling's equation (2.20) therefore implies that, by the time $y = 0.46$,

$$-S_E/Nk_B = \int_0^{0.46} \frac{4-2y}{(1-y)^3}\,dy = 4.15. \qquad (2.30)$$

For the majority of liquid metals this agrees remarkably well with what is observed. Hence it is primarily our description of the solid phase that is at fault.

This is hardly surprising. The entropy of a solid depends upon the way in which its specific heat varies with temperature right down to the absolute zero, and no model will give the entropy successfully unless it is also capable of describing the low temperature specific heat. It is clear enough, however, that the rigid-sphere picture is bound to break down for metals when $k_B T$ is no longer large compared with, say, the amplitude of the Friedel oscillations (see p. 69), and it is just as likely to break down for solid Ar. Specific heats at low temperatures are better discussed by treating the substance under consideration as an elastic continuum, in the spirit of the Debye theory.

The volume Ω_0 in (2.22) has played no part in the subsequent analysis but it may be estimated if desired from the equation

$$p_M^0 \simeq \left(\frac{1}{\beta}-\frac{1}{\beta^0}\right)\left(\frac{\Omega-\Omega_0}{\Omega}\right)$$

which is readily derived from (2.21) and (2.23). From this it follows that

$$\frac{\Omega_0}{\Omega} \simeq \frac{1-48a(0)}{1-39a(0)},$$

which implies that for a typical liquid metal, in which $a(0)$ is say 0.01, Ω_0 is about 85% of Ω. It is therefore a good bit bigger than the close-packed volume at the melting temperature, Ω_{CP}. One should not expect to be able to reach Ω_0 by cooling the metal down, however, because then the effective diameter σ increases and so does Ω_{CP}. The repulsive forces between the ions, i.e. the structure-dependent part of the energy which is not included in (2.22), block further contraction when T is comparable with the Debye temperature. The volume of a solid metal at this stage is in practice only about 5% less than its volume at the melting point.

The temperature-dependence of σ is also relevant to the thermal expansion coefficient of the liquid phase, α. Equation (2.21) shows that when the liquid is in equilibrium at zero pressure

$$\frac{p^0\Omega}{Nk_BT} = f(y) = \frac{\Omega}{Nk_BT}\frac{\partial \mathscr{U}}{\partial\Omega}.$$

Differentiation with respect to temperature yields the result

$$\alpha = \frac{Nk_B\beta}{\Omega}\left(f+3yf'T\frac{\mathrm{d}\log\sigma}{\mathrm{d}T}\right). \tag{2.31}$$

Hence for spheres which are completely rigid, with σ independent of T, $(\alpha\Omega/Nk_B\beta)$ should always equal 9.4 just above the melting point. In practice it is rather less than this for liquid metals, as may be seen from table 2.5. On the face of it we may use (2.31) to estimate $(\mathrm{d}\log\sigma/\mathrm{d}T)$ from the experimental data; for liquid Al at its melting point, to take just one example, it suggests that

$$\nu = -3T\frac{\mathrm{d}\log\sigma}{\mathrm{d}T}+\alpha T \simeq 0.08+0.11 = 0.19,$$

where ν is the exponent that describes the temperature dependence of y at zero pressure, i.e. we suppose $y \propto T^{-\nu}$. This is somewhat less than the value anticipated on p. 72. Unfortunately, to the extent that $(\alpha\Omega/Nk_B\beta)$ differs from 9.4 one of the basic assumptions on which our equation of state depends is unreliable. For this reason, (2.31) should be regarded with some suspicion.

2.8. The melting temperature

A number of attempts to explain the magnitude of T_M for simple substances have been made in the past, and the most famous and in many ways most successful is that of Lindemann (1910). Lindemann's picture of the melting process was that a solid shakes itself to pieces when the r.m.s. displacement of each atom due to thermal agitation, say X, becomes a certain fraction of the mean atomic radius R_A defined by (1.68). The quantity X may readily be estimated with the aid of the Debye model in terms of the characteristic Debye temperature Θ_D. The result at the melting point is

$$\left(\frac{X}{R_A}\right)_M = \frac{3\hbar}{\Theta_D}\left(\frac{T_M}{m_A k_B}\right)^{\frac{1}{2}}\left(\frac{4\pi N}{3\Omega}\right)^{\frac{1}{3}}, \tag{2.32}$$

TABLE 2.6. *Lindemann's correlation*

Element	T_M (°K)	Θ_D (°K)	$\left(\frac{X}{R_A}\right)_M$
Ar	84	85	0.18
Li	459	(430)	(0.22)
Na	371	160	0.21
K	336	100	0.22
Cu	1356	310	0.21
Ag	1233	220	0.19
Au	1336	185	0.18
Be	1550	900	0.23
Mg	923	330	0.21
Zn	693	240	0.18
Cd	594	165	0.16
Hg	234	100	0.13
Al	933	410	0.18
Ga	303	240	0.11
In	429	110	0.19
Tl	573	90	0.20
Sn	505	160	0.14
Pb	600	90	0.20
Sb	903	(200)	(0.15)
Bi	545	(1000)	(0.16)

Note. The values quoted for Θ_D, collated from a variety of sources, are ones that enable the Debye theory to fit the observed specific heat with reasonable precision for temperatures in the neighbourhood of Θ_D itself.

and some values calculated from this equation are listed in table 2.6. They are indeed fairly uniform, their average value being close to 0.2. Lindemann offered no explanation of why 0.2 should be the critical value.

From the point of view of the modified rigid-sphere model discussed in the previous section, melting occurs when, as a combined result of the increase of Ω and the decrease of Ω_{CP} which heating brings about, the ratio Ω/Ω_{CP} reaches a critical value of about 1.55. At this stage the mean distance between the centres of spheres which would be in contact at low temperatures is $(1.55)^{\frac{1}{3}}\sigma$ or about 1.16σ, σ being the effective rigid-sphere diameter at the

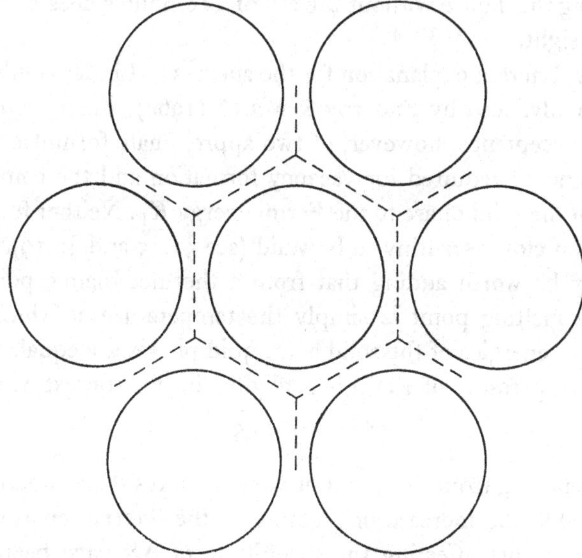

Fig. 2.5. Section through a regular array of spheres with $\Omega/\Omega_{CP} = 1.55$ and through one of the Voronoi polyhedra.

temperature T_M. Fig. 2.5 shows a cross-section through a regular close-packed array of spheres under these conditions. We are now to picture each sphere as endowed with kinetic energy and free to move within limits set by its neighbours, it being understood that the long-range order of the lattice is preserved throughout. What value of the r.m.s. displacement X does this picture suggest? An exact calculation is not straightforward. To obtain a rough estimate, let us suppose that the spheres are unable to break through the surfaces of their own Voronoi polyhedra but move about inside them unhindered. For a cubic close-packed array, we then find

$$\frac{X}{R_A} = \frac{0.16}{1.16}\left(\frac{3}{2}\right)^{\frac{1}{2}}\left(\frac{2\pi}{3}\right)^{\frac{1}{3}} \simeq 0.11,$$

which is not far below 0.2. Since it is clear that the approximate method of calculation is such as to underestimate X/R_A, there is a good chance that the rigid-sphere model may be used to justify Lindemann's empirical rule with some precision. We still have to depend on Alder & Wainwright's computations, of course, to tell us that 1.55 is the critical value of Ω/Ω_{CP}. A reliable method for estimating this figure without the aid of a computer does not seem to be in sight.

A very different explanation for the success of Lindemann's rule has been advanced by Enderby & March (1966). Their argument requires acceptance, however, of two approximate formulae relating the energy required for vacancy formation and the compressibility of the solid phase to the Fermi energy K_F. Neither formula appears on closer scrutiny to be valid (see §1.15 and §2.19).

It may be worth adding that from a thermodynamic point of view the melting point is simply the temperature at which the Gibbs free energies of the solid and liquid phases are equal. Since the normal pressure of 1 atm is negligible in this context, we may write

$$T_M = \Delta\mathscr{U}/\Delta S.$$

To estimate T_M, from this point of view, requires the estimation of $\Delta\mathscr{U}$ and ΔS, the increase on melting of the internal energy and entropy. Factors affecting the magnitude of ΔS have been discussed above. The magnitude of $\Delta\mathscr{U}$ is rather easier to calculate, since it can be expressed in terms of the pair potentials discussed in §1.13. Calculations have been carried through by Hartmann (1971) for a number of metals and the results are in adequate agreement, as they should be, with experimental values for the latent heat of melting.

2.9. Distribution functions

Our discussion of liquid structure in terms of models has proved unhelpful, and we must now consider an alternative approach in which the emphasis is laid entirely on certain *distribution functions* that are likely to determine the macroscopic properties of a liquid in thermal equilibrium. The first and most important of these is the so-called *pair distribution function* $g(R)$, defined in such a way that if there is one molecule with its centre at the point $\mathbf{R} = \mathbf{0}$ the probability of finding another at the same instant with its centre inside a small element of volume $d\mathbf{R}$ is

$$Ng(R)\, d\mathbf{R}/\Omega.$$

This definition in terms of a probability implies that one is concerned only with an ensemble average of the structure, an average, that is, over a large number of replicas of the system of interest, with the same volume and temperature but not necessarily the same microscopic configuration; a time average for a single system comes to the same thing. Since the liquid is on the average isotropic, $g(R)$ depends upon the magnitude but not the direction of $\mathbf{R}$. It is normalised so as to make it tend to unity for large R; for small R it must go to zero, because it is energetically so unfavourable for molecules to overlap. In an ideal gas of point particles, however, g would be unity everywhere.

The pair distribution function cannot by itself express all the subtleties of a liquid structure and one may define in a rather similar fashion a whole series of higher-order distribution functions $g_3(\mathbf{R}, \mathbf{R}')$, $g_4(\mathbf{R}, \mathbf{R}', \mathbf{R}'')$ etc. (For consistency of notation $g(R)$ should be written as $g_2(R)$, but we shall need to refer to this quantity very frequently and it is clumsy to tag a suffix onto it the whole time.) The three-body function, for example, is such that if there is known to be a molecule at the origin the probability of finding two others at the same instant with their centres in the small elements of volume $\mathrm{d}\mathbf{R}$ and $\mathrm{d}\mathbf{R}'$ is $(N/\Omega)^2 g_3\,\mathrm{d}\mathbf{R}\,\mathrm{d}\mathbf{R}'$. It seems likely that the sort of distinction between quasi-crystalline and RCP models which we discussed above emerges primarily in g_3 and the higher functions, rather than in g.

Many of the basic thermodynamic parameters can be expressed directly in terms of g and the effective pair interaction $w(R)$ – if one exists. Thus the internal energy $\mathscr{U}$ is given immediately by

$$\mathscr{U} = \tfrac{3}{2}Nk_{\mathrm{B}}T + \frac{2\pi N^2}{\Omega}\int_0^\infty g(R)\,w(R)\,R^2\,\mathrm{d}R. \tag{2.33}$$

An expression for the pressure p can also be obtained, using the virial theorem of Clausius (see Fisher, 1964 for details). The compressibility, curiously enough, depends only upon g, as can be shown by appealing to the orthodox theory of density fluctuations in statistical mechanics (e.g. Landau & Lifshitz, 1958, p. 350). Let m be the number of molecules which are present at any instant inside some volume ω which is large compared with atomic dimensions but small compared with the volume of the specimen Ω. It is convenient to introduce a continuous density function $\rho(\mathbf{r})$, which is defined in such a way that $\rho(\mathbf{r})\,\mathrm{d}\mathbf{r}$ is unity if the small element of volume $\mathbf{dr}$ contains the centre of a molecule but is zero otherwise, and hence to write

$$m = \int^{\omega} \rho(\mathbf{r})\,\mathrm{d}\mathbf{r}$$

and

$$m^2 = \int^{\omega} \rho(\mathbf{r})\,\mathrm{d}\mathbf{r} \int^{\omega} \rho(\mathbf{r}')\,\mathrm{d}\mathbf{r}'. \tag{2.34}$$

A contribution to the product of integrals in (2.34) arises whenever $\mathbf{dr}$ and $\mathbf{dr}'$ both contain the centre of a molecule and particularly when they contain the centre of the *same* molecule. A little reflection shows that the ensemble average may be written in terms of the pair distribution function as

$$\begin{aligned}\overline{m^2} &= \overline{m}\left(1 + \frac{N}{\Omega}\int^{\omega} g(|\mathbf{r}-\mathbf{r}'|)\,\mathrm{d}\mathbf{r}'\right)\\ &= \overline{m}\left(1 + \frac{N\omega}{\Omega} + \frac{N}{\Omega}\int^{\omega} (g(|\mathbf{r}-\mathbf{r}'|) - 1)\,\mathrm{d}\mathbf{r}'\right)\\ &= \overline{m}\left(1 + \overline{m} + \frac{4\pi N}{\Omega}\int_0^{\infty} (g(R) - 1)\,R^2\,\mathrm{d}R\right).\end{aligned}$$

The final step is justified so long as the region ω has dimensions large compared with the distance over which $(g-1)$ goes to zero, i.e. compared with the range of order in the liquid: otherwise some sort of surface correction would be necessary. Now $(\overline{m^2} - \overline{m}^2)$, which is of course the ensemble average of $(m-\overline{m})^2$, is known from fluctuation theory to equal $(Nk_B T\beta\overline{m}/\Omega)$, where β is the isothermal compressibility. Hence

$$\frac{Nk_B T\beta}{\Omega} = 1 + \frac{4\pi N}{\Omega}\int_0^{\infty} (g-1)\,R^2\,\mathrm{d}R. \tag{2.35}$$

Equation (2.35) is completely valid even for a liquid metal, though (2.31) and the corresponding equation for p require some modification before they can be applied to metals, on account of the volume-dependent terms in the energy which are not incorporated in the effective pair interaction.

If the functions g and w are known not just for the volume of interest but for states of lower density as well, then it is possible to express even the entropy in terms of them with the aid of (2.29). For our present purposes, however, it is of more interest to note that S can be expressed in terms of the higher-order distribution

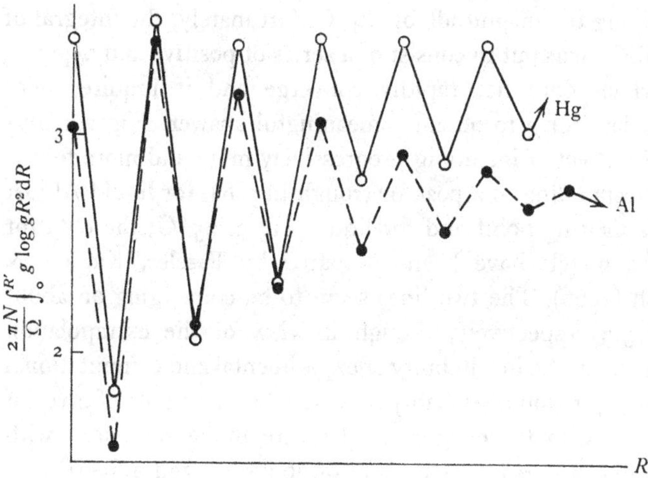

Fig. 2.6

functions without bringing the pair potential into the discussion at all. Fisher (1964) shows how a 'cluster expansion' may be developed for the entropy excess as follows:

$$-\frac{S_E}{Nk_B} = \frac{N}{2\Omega}\int_0^\infty g(R)\log g(R)\,d\mathbf{R}$$
$$+\frac{N^2}{6\Omega^2}\int_0^\infty\int_0^\infty g_3(\mathbf{R},\mathbf{R}')\log\left[\frac{g_3(\mathbf{R},\mathbf{R}')}{g(R)\,g(R')\,g(|\mathbf{R}-\mathbf{R}'|)}\right]d\mathbf{R}\,d\mathbf{R}'$$
$$+\text{etc.} \tag{2.36}$$

Each successive term in the expansion indicates a further reduction in entropy associated with a higher order of correlation between molecular positions. This result confirms the suspicion that the entropy of a liquid contains more information about the subtleties of its structure than is stored in most other properties.

It turns out, as we shall see below, that $g(R)$ can be determined experimentally for all R, but there is no way of determining g_3 directly, let alone any of the higher-order correlation functions; an ingenious method has been suggested by Egelstaff, Page & Heard (1969), but it is unlikely that this will prove of sufficient accuracy to yield useful results for liquid metals near their melting points. It is only by indirect inference that one can learn anything about g_3. One point that is worth consideration straight away is how far the first term in the cluster expansion is capable by itself of explaining the magnitude of S_E. Unfortunately, the integral of ($g \log g\, d\mathbf{R}$) turns out to consist of a series of positive and negative terms which does not rapidly converge and it requires very accurate data for g to obtain a meaningful answer. Fig. 2.6 illustrates the effect of including progressively more and more terms, each corresponding to a peak or trough in $g(R)$, for liquid Al just above its melting point and for liquid Hg at 25 °C; the data for these two metals have been compared by Fessler, Kaplow & Averbach (1966). The two lines seem to be converging on about 2.7 and 3.3 respectively, though in view of the extrapolation required and of the inevitability of experimental and computational inaccuracies, it would probably be wise to attach limits of error of at least ±0.5 to these figures. They are to be compared with experimental values for $-S_E/Nk_B$ of about 3.6 and 4.65 respectively.† The gap is rather bigger than can be dismissed as due to error alone.

2.10. The interference function

The key to the experimental determination of $g(R)$ lies in the realisation that $(g-1)$ is the Fourier transform in three dimensions of $(a-1)$, where $a(q)$ is the *interference function* already introduced in chapter 1. We have seen (§1.9) that the *structure factor*

$$F(\mathbf{q}) = \sum_i \exp(-i\mathbf{q}\cdot\mathbf{r}_i)$$

determines how the amplitude of waves scattered from an assembly of identical molecules is modulated, whether they be electron waves, X-rays or neutrons, by the effects of coherence between wavelets scattered from molecules at different sites. The interference function, defined by the ensemble average in (1.79), determines how the *intensity* of the scattered waves is modulated by coherence effects, and hence is in principle a measurable quantity.

From (1.79) we have

$$\begin{aligned} a(q) &= N^{-1}\overline{\left(\sum_i \exp(-i\mathbf{q}\cdot\mathbf{r}_i) \sum_j \exp(i\mathbf{q}\cdot\mathbf{r}_j)\right)} \\ &= N^{-1}\left(N + \overline{\sum_i \exp(-i\mathbf{q}\cdot\mathbf{r}_i) \sum_{j\neq i} \exp(i\mathbf{q}\cdot\mathbf{r}_j)}\right). \end{aligned} \tag{2.37}$$

† The excess entropy for Hg is $-5.1Nk_B$ at the melting point as quoted in table 2.1, but it changes appreciably on heating to 25°C.

If there were no correlations at all between the positions of different molecules there would also be no correlation between

$$\exp(-i\mathbf{q}\cdot\mathbf{r}_i) \quad \text{and} \quad \sum_{j\neq i}\exp(i\mathbf{q}\cdot\mathbf{r}_j),$$

and the average of the product of the two sums in (2.37) would be the product of their averages. Thus we could write

$$a(q) = N^{-1}\left(N+\frac{N(N-1)}{\Omega^2}\int^{\Omega}\exp(-i\mathbf{q}\cdot\mathbf{r})\,d\mathbf{r}\int^{\Omega}\exp(i\mathbf{q}\cdot\mathbf{r}')\,d\mathbf{r}'\right),$$

both integrals here being taken over the volume of the specimen. If, for the sake of simplicity, we suppose the specimen to be a sphere of radius R_0, we know from equation (1.70) that each integral increases only as fast as R_0, unless q is very small. In general it is true, whatever the shape of the specimen, that for a random, gas-like, structure

$$a(q) = 1+O(R_0^{-1}),$$

where R_0 is some dimension characterising its size, provided that q is larger than R_0^{-1}. For a macroscopic specimen we may replace $a(q)$ by unity over virtually the whole range of q. Physically this means that the molecules in a random structure scatter *incoherently*, the total scattered intensity in any direction being just N times the intensity scattered by a single molecule.

More care is needed if correlations between the positions of adjacent molecules do exist. Using R_{ji} as convenient shorthand for the separation between the jth and ith molecules, we may write (2.37) as

$$\begin{aligned} a(q) &= 1+N^{-1}\overline{\sum_i\sum_{j\neq i}\cos\mathbf{q}\cdot\mathbf{R}_{ji}} \\ &= 1+\Omega^{-1}\overline{\sum_i\int g(R)\cos(\mathbf{q}\cdot\mathbf{R})\,d\mathbf{R}}. \end{aligned}$$

The trouble is that the domain of $\mathbf{R}$ over which the integral has to be taken does not have the same shape for all i. To get round this difficulty we may observe that, since $g(R)$ must be unity for a random structure and since $a(q)$ is then also unity, it follows that

$$\overline{\sum_i\int\cos(\mathbf{q}\cdot\mathbf{R})\,d\mathbf{R}} = 0.$$

Hence, for values of q which are large compared with the inverse of the specimen size, we are entitled to write

$$a(q) = 1+\Omega^{-1}\sum_i \int (g-1)\cos(\mathbf{q}\cdot\mathbf{R})\,d\mathbf{R}.$$

For a liquid the integral now converges rapidly for large R, because $(g-1)$ remains appreciable for a range of only a few intermolecular spacings, and hence the domain of integration is of little importance. Except when i denotes one of the molecules very near to the surface of the specimen, we may indeed assume the domain of integration to be infinite. Since in a macroscopic specimen the surface molecules contribute a negligible amount to the sum over i, it follows that (except for very small q)

$$\begin{aligned} a(q)-1 &= \frac{N}{\Omega}\int (g-1)\cos(\mathbf{q}\cdot\mathbf{R})\,d\mathbf{R} \\ &= \frac{4\pi N}{\Omega}\int_0^\infty (g-1)\frac{\sin qR}{qR}R^2\,dR. \end{aligned} \tag{2.38}$$

Hence $(a-1)$ is a Fourier transform of $(g-1)$ and the converse follows at once: the appropriate inversion of (2.38) is

$$\begin{aligned} g(R)-1 &= N^{-1}\sum_{\mathbf{q}} (a(\mathbf{q})-1)\cos(\mathbf{q}\cdot\mathbf{R}) \\ &= \frac{\Omega}{2\pi^2 N}\int_0^\infty (a-1)\frac{\sin qR}{qR}q^2\,dq. \end{aligned} \tag{2.39}$$

The isotropy of the liquid implies, of course, that $a(\mathbf{q})$ does not depend upon the direction of $\mathbf{q}$.

Certain important properties of the interference function follow from its relation to the pair distribution function and may be enumerated here. As q tends to infinity the factor $\sin qR$ on the right-hand side of (2.38) oscillates more and more rapidly between plus and minus, while the amplitude of the oscillations diminishes because of the factor q. This is enough to show that, in the absence of any singularities in $g(R)$, $a(q)$ must tend to unity for large q.

The limiting value of $a(q)$ when q is very small – small, that is, compared with the inverse of atomic dimensions but still large compared with R_0^{-1} – may be obtained direct from (2.35). Denoting it by $a(0)$ (though when q is strictly zero $a(q)$ by its definition is clearly equal to N) we have

$$a(0) = Nk_B T\beta/\Omega. \tag{2.40}$$

TABLE 2.7. *$a(0)$ at the melting point for simple liquids*

Ar	Li	Na	K	Rb	Cs	Cu	Ag
0.064	0.031	0.024	0.023	0.022	0.028	0.021	0.019

Zn	Cd	Hg	Al	Ga	In	Tl	Sn	Pb	Sb	Bi
0.014	0.012	0.005	0.018	0.005	0.007	0.011	0.008	0.009	0.020	0.011

The figures quoted in table 2.7 are based upon this equation and upon the compressibility data discussed in §2.19 below. Evidently $a(0)$ at the melting point is always much less than unity, but it is significantly larger for the rare gas Ar than for any of the metals. The arguments in §2.7, and in particular equation (2.23), show that this is because $(\partial^2\mathcal{U}/\partial\Omega^2)$ is negative for Ar rather than positive.

In between $q = \infty$ and $q = 0$ the only general constraint laid on $a(q)$, apart from the obvious one that by definition it is positive, is provided by a sum-rule which follows at once from (2.39), as a consequence of the fact that $g(0)$ is bound to be negligible compared with unity for real molecules (as opposed to the point particles that constitute an ideal gas). It may be written in the form

$$\int_0^\infty (1-a)\,q^2\,dq = 2\pi^2 N/\Omega. \tag{2.41}$$

The result is approximate to the extent that (2.38) is not strictly true for values of q comparable with R_0^{-1}, but the error can be made infinitesimal by increasing the size of the specimen.

It is of interest to note that if a liquid is uniformly expanded, so that all its linear dimensions are enlarged by a factor $1+\epsilon$ ($\epsilon \ll 1$), then its interference function scales in such a way that

$$\begin{aligned} a(q) &= a_0(q(1+\epsilon)) \\ &= a_0(q)+\epsilon q\, da_0/dq. \end{aligned} \tag{2.42}$$

This result, which has been used in §1.15, is readily proved from (2.38). It may be verified that if the sum rule is satisfied by a_0 before expansion, then it is satisfied after expansion by a.

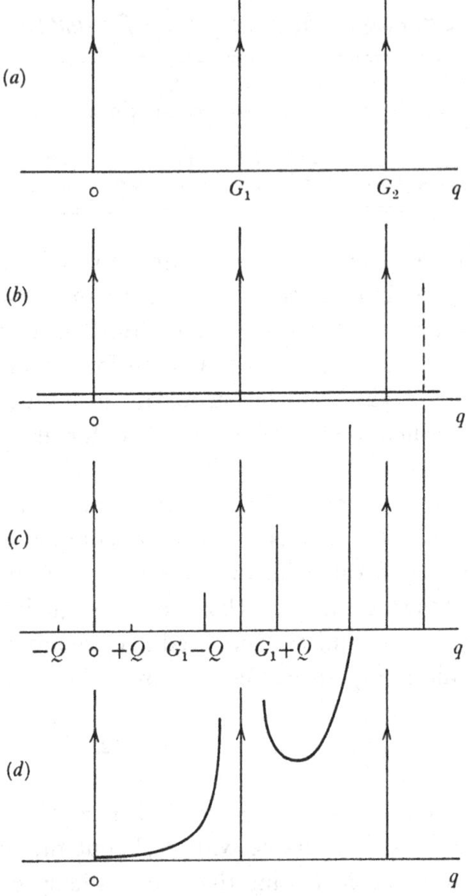

Fig. 2.7. The interference function for a solid crystal (*a*) without imperfections, (*b*) with a vacancy, (*c*) with a single longitudinal mode excited, (*d*) with all modes excited. Arrows indicate δ-functions where $a(q) \simeq N$.

2.11. The interference function for a solid

It is instructive to consider the nature of the interference function for a solid. In a perfectly regular solid, undisturbed by any thermal agitation of the molecules, the structure factor $F(\mathbf{q})$ is of course equal to N whenever $\mathbf{q}$ equals one of the reciprocal lattice vectors $\mathbf{G}$ but is zero elsewhere; this is virtually a definition of what is meant by the reciprocal lattice. Correspondingly, $a(\mathbf{q})$ has sharp spikes at the reciprocal lattice vectors, where it also is equal to N, and is zero elsewhere, as shown in fig. 2.7(*a*). The problem is to determine what the pattern will look like when the crystal is less than perfect and at a finite temperature.

The effect of removing one of the molecules so as to create a vacancy is easily discussed. If the jth molecule is removed we have

$$F(\mathbf{q}) = F_0(\mathbf{q}) - \exp(-i\mathbf{q}\cdot\mathbf{r}_j),$$

whence

$$\begin{aligned}(N-1)\,a(\mathbf{q}) &= \overline{F(\mathbf{q})F(-\mathbf{q})} \\ &= \overline{F_0(\mathbf{q})F_0(-\mathbf{q})} - \overline{\sum_{i\neq j}(\exp(i\mathbf{q}\cdot\mathbf{R}_{ij}) + \exp(i\mathbf{q}\cdot\mathbf{R}_{ji}))} - 1 \\ &= Na_0(\mathbf{q}) - 2(a_0(\mathbf{q}) - 1) - 1;\end{aligned}$$

use has been made here of equation (2.37) and an average taken over all possible values for j. When N is large the result can be written as

$$a(\mathbf{q}) = a_0(\mathbf{q}) + N^{-1}(1 - a_0(\mathbf{q})), \tag{2.43}$$

and in this form it was quoted in §1.15. It tells us that if a vacancy is present the interference function for a solid acquires a uniform background, as indicated in fig. 2.7(*b*), while the height of the original spikes is reduced in such a way that the sum rule is still satisfied.

The consequences of thermal agitation are most easily discussed in terms of a simple Debye-type model. We begin by analysing the effect upon $a(\mathbf{q})$ of exciting a single normal mode of vibration of the crystal as a whole, in which the displacement $\xi(\mathbf{r})$ at some particular instant of time of the molecule whose rest position is $\mathbf{r}$ varies sinusoidally with $\mathbf{r}$ according to the equation

$$\begin{aligned}\boldsymbol{\xi} &= \boldsymbol{\xi}_{\mathbf{Q}} \cos(\mathbf{Q}\cdot\mathbf{r}) \\ &= \tfrac{1}{2}\boldsymbol{\xi}_{\mathbf{Q}}(\exp(i\mathbf{Q}\cdot\mathbf{r}) + \exp(-i\mathbf{Q}\cdot\mathbf{r}));\end{aligned} \tag{2.44}$$

then we add the effects of different modes together. When this model is used in calculations of the specific heat of a solid it is customary to assume that the normal modes are independent of one another – 'loosely-coupled', in the language of statistical mechanics – and we shall adopt this approximation initially, though it turns out to introduce an error into the final answer.

Equation (2.48) below indicates that $\boldsymbol{\xi}_{\mathbf{Q}}$ in thermal equilibrium is an infinitesimal quantity for a large specimen, so that an expansion procedure is appropriate as follows:

$$
\begin{aligned}
F(\mathbf{q}) &= \sum_i \exp\left(-\mathrm{i}\mathbf{q}\cdot(\mathbf{r}_i+\boldsymbol{\xi}_i)\right) \\
&= \sum_i \exp\left(-\mathrm{i}\mathbf{q}\cdot\mathbf{r}_i\right)\left(1-\mathrm{i}\mathbf{q}\cdot\boldsymbol{\xi}_i-\tfrac{1}{2}(\mathbf{q}\cdot\boldsymbol{\xi}_i)^2+\ldots\right) \\
&= \sum_i \exp\left(-\mathrm{i}\mathbf{q}\cdot\mathbf{r}_i\right)\left(1-\tfrac{1}{2}\mathrm{i}\mathbf{q}\cdot\boldsymbol{\xi}_{\mathbf{Q}}\left\{\exp\left(\mathrm{i}\mathbf{Q}\cdot\mathbf{r}_i\right)+\exp\left(-\mathrm{i}\mathbf{Q}\cdot\mathbf{r}_i\right)\right\}\right. \\
&\qquad \left.-\tfrac{1}{2}(\mathbf{q}\cdot\boldsymbol{\xi}_i)^2+\ldots\right) \\
&= F_0(\mathbf{q})\left(1-\tfrac{1}{4}(\mathbf{q}\cdot\boldsymbol{\xi}_{\mathbf{Q}})^2\right)-\tfrac{1}{2}\mathrm{i}\mathbf{q}\cdot\boldsymbol{\xi}_{\mathbf{Q}}\{F_0(\mathbf{q}-\mathbf{Q})+F_0(\mathbf{q}+\mathbf{Q})\}+\ldots
\end{aligned} \tag{2.45}
$$

In this expansion F_0 should strictly represent the structure factor when all the vibrational modes except the $\mathbf{Q}$th are thermally excited. For the sake of simplicity, however, we shall here suppose the temperature to be low enough for it to be replaced by the structure factor of an ideal crystal which is not vibrating at all; i.e. we shall suppose $F_0(\mathbf{q})$ to be finite only where $\mathbf{q} = \mathbf{G}$. This approximation, which involves the neglect of so-called 'multi-phonon' contributions, is not necessarily valid in a solid metal near its melting point, though it can be shown to be perfectly sound for small values of q.

The first term on the right-hand side of (2.45) tells us that the amplitude of the structure factor at $\mathbf{q} = \mathbf{G}$ is somewhat depressed by the excitation of each mode. If *all* the modes are taken into account, $F(\mathbf{G})$ is depressed by a factor known as the *Debye–Waller* factor, which is conventionally written as $\exp(-W)$. We have

$$\exp(-W) = \prod_{\mathbf{Q}}\left(1-\tfrac{1}{4}(\mathbf{G}\cdot\boldsymbol{\xi}_{\mathbf{Q}})^2\right),$$

so that, by taking logarithms of both sides,

$$W = \tfrac{1}{4}\sum_{\mathbf{Q}}(\mathbf{G}\cdot\boldsymbol{\xi}_{\mathbf{Q}})^2 = \tfrac{1}{6}(GX)^2, \tag{2.46}$$

where X, as above, is the *total* r.m.s. displacement of each molecule due to thermal agitation. The spikes in the interference function at $\mathbf{q} = \mathbf{G}$ are of course reduced by the *square* of the Debye–Waller factor, $\exp(-2W)$. From the figures for X in table 2.6 it may be seen that for a reciprocal lattice vector of order π/R_{A} this is typically about 0.9 just below the melting point.

The second term on the right-hand side of (2.45) indicates a finite contribution to $F(q)$ whenever $(\mathbf{q}-\mathbf{Q})$ or $(\mathbf{q}+\mathbf{Q})$ vanishes or is equal to one of the reciprocal lattice vectors. Every peak in the original pattern for $a_0(q)$ therefore acquires two satellites, as shown in fig. 2.7(*c*), whose height is given by

$$\tfrac{1}{4}N\overline{(\mathbf{q}\cdot\boldsymbol{\xi}_{\mathbf{Q}})^2}. \tag{2.47}$$

For the satellites which lie around the origin in $\mathbf{q}$-space, i.e. within the first Brillouin zone, $\mathbf{q}$ is to be set equal to $\mathbf{Q}$ in this formula, which therefore reduces for a *longitudinal* mode to

$$\tfrac{1}{4}N\omega_{\mathbf{Q}}^2\,\overline{\xi_{\mathbf{Q}}^2}/v_{\parallel}^2,$$

where $\omega_{\mathbf{Q}}$ is the angular frequency of the mode and $v_{\parallel}$ the appropriate longitudinal wave velocity; a *transverse* mode contributes nothing to $a(\mathbf{q})$ in this region because $(\mathbf{Q}\cdot\boldsymbol{\xi}_{\mathbf{Q}})$ vanishes. But if the modes are loosely-coupled and the temperature T is well above the characteristic Debye temperature we may deduce from the principle of equipartition that in equilibrium

$$\tfrac{1}{4}Nm_{\mathrm{A}}\,\omega_{\mathbf{Q}}^2\,\overline{\xi_{\mathbf{Q}}^2} = \tfrac{1}{2}k_{\mathrm{B}}T, \tag{2.48}$$

where m_{A} is the molecular mass. Hence within the first Brillouin zone the model predicts

$$a(\mathbf{q}) = k_{\mathrm{B}}T/m_{\mathrm{A}}v_{\parallel}^2 \tag{2.49}$$

in thermal equilibrium. A factor 2 has been introduced at this stage because for every mode for which the displacement varies like cos $(\mathbf{Q}\cdot\mathbf{r})$ there is one for which it varies like sin $(\mathbf{Q}\cdot\mathbf{r})$ to be considered – their contributions to $a(\mathbf{q})$ are additive because although they oscillate with the same frequency there is no phase correlation between them.

The permitted values of $\mathbf{Q}$ are evenly distributed within the first Brillouin zone, and this ensures that the satellite spikes in the interference function are also evenly distributed, not just in the first zone but throughout $\mathbf{q}$-space. The result of thermal agitation, therefore, is that the interference function acquires an essentially continuous background which is represented by a smooth curve in fig. 2.7(*d*). The curve is shown as horizontal for small q, because $v_{\parallel}$ is a constant for long wavelength waves. As q approaches $\tfrac{1}{2}G_1$, i.e. the boundary of the first Brillouin zone for the one-dimensional situation to which this figure applies, the curve tends to rise because, as is well-known from the theory of lattice vibration in crystals, $v_{\parallel}$ tends to fall; the simple Born–von Karman theory, for example, in which only the effects of nearest-neighbour interactions between molecules are included, would suggest

$$a(\tfrac{1}{2}G_1) \simeq (\tfrac{1}{2}\pi)^2 a(0). \qquad (2.50)$$

Outside the first Brillouin zone the curve continues to rise because the wave number $\mathbf{q}$ in (2.47) is now $(\mathbf{G}_1 - \mathbf{Q})$ rather than $\mathbf{Q}$, and hence one should expect 'diffuse peaks' in the interference function centred round every reciprocal lattice vector, as shown schematically in the figure. A detailed analysis is complicated by the fact that $\boldsymbol{\xi}_{\mathbf{Q}}$ for a transverse mode is not necessarily perpendicular to $(\mathbf{G} - \mathbf{Q})$, so that such modes cannot be ignored outside the first zone.† For a full discussion, and a demonstration that the sum rule is always satisfied, the reader is referred to standard treatises on the theory of X-ray diffraction by solids (Born, 1943; James, 1945).

All this is for a single crystal. A number of attempts have been made to discuss the form of $a(q)$ for an isotropic polycrystalline solid (e.g. Gerstenkorn, 1952; Kaplow, Strong & Averbach, 1965*b*) but exact calculations are not easy, especially if the crystallites are only a few molecules across. It is known that an isolated crystallite would show, even in the absence of thermal agitation, an interference function whose sharp spikes would be diffused over a range Δq roughly equal to the inverse of the crystallite diameter (Guinier, 1963). If the crystallites could be treated as randomly disposed with respect to one another, one could obtain the interference function for the whole specimen simply by averaging the curves that would describe their scattering behaviour in isolation; the spikes would be further spread out in the resultant curve because the unit vector of the reciprocal lattice would vary according to the orientation of the crystallites. But the assumption of randomness, though legitimate for a powder, may well be

† To the extent that modes which are classified as 'transverse' may in some crystals involve displacements which have a longitudinal component, the transverse modes cannot necessarily be ignored within the first zone either. The above discussion has been simplified to make it easier to follow.

misleading for a continuous specimen in which the crystallites are stacked together in such a way that a layer of molecules on the surface of one is in direct contact with a layer on the surface of the next.

However, the argument concerning the behaviour of $a(q)$ for small q goes through as above, even for a polycrystalline specimen. In a polycrystalline specimen $v_\parallel$ should be independent of the direction of $\mathbf{q}$ and given by

$$v_\parallel^2 = \Omega(1+\tfrac{4}{3}\beta g)/Nm_A\beta;$$

the shear modulus g enters because a longitudinal vibration in an infinite solid involves an element of shear strain as well as volume strain. If this result is inserted into (2.48) one obtains the result

$$a(0) = Nk_B T\beta/\Omega(1+\tfrac{4}{3}\beta g). \tag{2.51}$$

If the shear modulus is allowed to go to zero it becomes almost identical with the result already established for a liquid in equation (2.40).†

Almost identical but not quite. The difference is that, since vibrations of long wavelengths in any real solid or liquid are known to be *adiabatic*, it seems necessary to use the adiabatic compressibility β_S in (2.51) whereas it is the isothermal one β_T which occurs in (2.40). The latter is always somewhat bigger, since the ratio β_T/β_S is the same as the specific ratio γ described by equation (2.8).

This discrepancy arises because of limitations in our model. If the vibrational modes of a solid were truly harmonic and independent of one another it would have a zero thermal expansion coefficient, in which case the distinction between β_T and β_S would disappear. The fact that a finite distinction does exist in practice is a sign that the modes are not independent and that a much more complicated set of collective coordinates is needed if the thermal agitation is to be described exactly. It is instructive to consider the volume fluctuations, upon which the magnitude of $a(0)$ ultimately

† It has not been sufficiently appreciated, perhaps, that the very general argument upon which (2.40) is based is valid *only* for fluids. A local volume fluctuation cannot occur in a solid without some shear strain arising in the surrounding matrix. This shear strain is ignored in the conventional fluctuation theory from which (2.35) and (2.40) are derived.

depends, as caused partly by fluctuations in local pressure and partly by fluctuations in local entropy – fluctuations in p and S may be shown to be uncorrelated. Egelstaff (1967, p. 163) calculates in this way that for a fluid, in which the complications associated with shear strain do not arise,

$$(\Omega/N)\,a(0) = \frac{\overline{\Delta\Omega^2}}{\Omega} = \frac{\overline{(\Delta\Omega\cdot\Delta p)}^2}{\Omega\overline{\Delta p^2}} + \frac{\overline{(\Delta\Omega\cdot\Delta S)}^2}{\Omega\overline{\Delta S^2}}$$
$$= k_B T\beta_S + k_B T(\beta_T - \beta_S). \tag{2.52}$$

It looks from this as though the Debye-type model is adequate to account for the pressure fluctuations but not for the fluctuations of entropy. The latter may also be described by a set of sinusoidal modes, but these are non-propagating. They are governed by a diffusion equation rather than a wave equation, which means that once excited they decay in amplitude in an exponential fashion.

Non-propagating entropy fluctuations exist in liquids as well as solids and always contribute a fraction $(\gamma-1)/\gamma$ of the total $a(0)$. In most liquid metals, to judge by the figures in table 2.3, this fraction is small enough to be ignored in many contexts; a Debye-type model will give more or less the right answer. For liquid Ar, in which γ is about 2.2, the situation is naturally rather different.

Before leaving this section we may derive a result that will be needed in the discussion of compressibility in §2.19 below. There we will be concerned to evaluate the change in $a(q)$ due to a long-wavelength (small Q) longitudinal compression wave in a liquid. If (2.45) is applicable for a liquid it leads immediately to†

$$a(\mathbf{q}) - a_0(\mathbf{q}) = \tfrac{1}{4}\overline{(\mathbf{q}\cdot\boldsymbol{\xi}_\mathbf{Q})^2}\{a_0(\mathbf{q}-\mathbf{Q}) + a_0(\mathbf{q}+\mathbf{Q}) - 2a_0(\mathbf{q})\} + O(\xi^4). \tag{2.53}$$

This implies that for $\mathbf{q} = \pm\mathbf{Q}$ there are spikes in the interference function, described by (2.47), while elsewhere, in the limit of small Q,

$$a(q) - a_0(q) = \tfrac{1}{4}\overline{(q\xi_\mathbf{Q})^2}\cos^2\theta\left\{\frac{Q^2}{q}\sin^2\theta\frac{\mathrm{d}a_0}{\mathrm{d}q} + Q^2\cos^2\theta\frac{\mathrm{d}^2a_0}{\mathrm{d}q^2}\right\}, \tag{2.54}$$

† Note that the product of say $F_0(\mathbf{q})$ and $F_0(\mathbf{q}-\mathbf{Q})$ disappears on averaging over the ensemble because there can be no phase correlation between these two quantities; the phase difference between them is determined in a quite arbitrary fashion by the choice of origin for $\mathbf{r}$.

where θ is the angle between $\mathbf{Q}$ (or $\boldsymbol{\xi}_{\mathbf{Q}}$) and $\mathbf{q}$. An average over θ yields the result (for $\mathbf{q} \neq \pm\mathbf{Q}$) that

$$a(q) - a_0(q) = \tfrac{1}{4}(Q\xi_{\mathbf{Q}})^2 \left(\frac{2}{15} q \frac{\mathrm{d}a_0}{\mathrm{d}q} + \frac{3}{15} q^2 \frac{\mathrm{d}^2 a_0}{\mathrm{d}q^2}\right). \qquad (2.55)$$

In the neighbourhood of the main peak in $a_0(q)$ the curvature $(\mathrm{d}^2 a_0/\mathrm{d}q^2)$ is large and negative. In this region, therefore, the compression wave reduces $a(q)$, just as it would reduce the height of the spikes at $q = G$ in a regular lattice. It may readily be verified that (2.55) is consistent with the sum rule.

It must be admitted that the application of (2.45) to a liquid disturbed by a compression wave is not rigorously correct. It involves the assumption that the atomic displacements due to the wave are all in the direction of $\mathbf{Q}$ and vary sinusoidally with $\mathbf{r}$ in the manner described by (2.44). But if this were true the average environment of the atoms in a slice of the liquid which was compressed or expanded by the wave would become anisotropic. It is an essential feature of a long-wavelength compression wave in a liquid, however, that the material within any slice perpendicular to $\mathbf{Q}$ is *uniformly* compressed or expanded, isotropy being maintained. This presumably requires that the atomic displacements due to the wave have components perpendicular to $\mathbf{Q}$, and their components parallel to $\mathbf{Q}$ may fluctuate about the mean value which (2.44) describes. It is not possible to take these subtleties fully into account, but there is no reason to suppose that they make any significant difference to the final answer.

2.12. Mixtures

Before discussing the experimental results for $a(q)$ in liquid metals it is necessary to generalise some of the arguments of §2.10 to situations where more than one type of molecule is present.

Suppose that we label the different species of molecule by a set of suffices 0, 1, 2..., the whole set being denoted where necessary by a dummy, α or β. Thus the molecular concentrations might be $c_0, c_1, c_2 \ldots$, with

$$\sum_\alpha c_\alpha = 1. \qquad (2.56)$$

The amplitude of a wave scattered by a mixed assembly may then be expressed as

$$\sum_{\alpha} f_\alpha \sum_{i(\alpha)} \exp(-i\mathbf{q}\cdot\mathbf{r}_{i(\alpha)}), \tag{2.57}$$

where f_0, f_1 etc. are the appropriate form factors and $r_{i(\alpha)}$ denotes the position of the ith molecule, a member of the α species. The mean intensity becomes

$$\sum_{\alpha}\sum_{\beta} f_\alpha f_\beta^* \overline{\sum_{i(\alpha)}\sum_{j(\beta)} \exp(i\mathbf{q}\cdot(\mathbf{r}_{j(\beta)}-\mathbf{r}_{i(\alpha)}))}$$
$$= \sum_{\alpha}\sum_{\beta} f_\alpha f_\beta^* N c_\alpha(\delta_{\alpha\beta}+\overline{\sum_{j\neq i}\cos\mathbf{q}\cdot\mathbf{R}_{ji}}), \tag{2.58}$$

which is reminiscent of equation (2.37) above. The Kronecker $\delta_{\alpha\beta}$ in (2.58), which by convention is unity when α and β stand for the same species but vanishes otherwise, takes care of the terms in the double sum for which $\mathbf{r}_j$ and $\mathbf{r}_i$ refer to the same molecule.

It proves convenient to define a set of *partial* interference functions by an equation analogous to (2.38):

$$a_{\alpha\beta}(q)-1 = \frac{4\pi N}{\Omega}\int_0^\infty (g_{\alpha\beta}(R)-1)\frac{\sin qR}{qR}R^2\,\mathrm{d}R. \tag{2.59}$$

Here $g_{\alpha\beta}(R)$ is the pair distribution function that determines the distribution of β molecules round an α one, normalised in such a way that, like $g(R)$ above, it tends to unity in a liquid for large values of R: i.e. if there is an α molecule at $R = 0$ the probability of finding a β molecule at the same instant with its centre in a small element of volume $\mathrm{d}\mathbf{R}$ is $Nc_\beta g_{\alpha\beta}(R)\,\mathrm{d}\mathbf{R}/\Omega$. Then (2.58) can be written as

$$\sum_{\alpha}\sum_{\beta} f_\alpha f_\beta^* N c_\alpha(\delta_{\alpha\beta}+c_\beta(a_{\alpha\beta}-1))$$
$$= \sum_{\alpha} N c_\alpha f_\alpha f_\alpha^* - \sum_{\alpha}\sum_{\beta} N c_\alpha c_\beta f_\alpha f_\beta^* + \sum_{\alpha}\sum_{\beta} N c_\alpha c_\beta f_\alpha f_\beta^* a_{\alpha\beta}$$
$$= N\{(\overline{ff^*}-\bar{f}\cdot\bar{f}^*)+\sum_{\alpha}\sum_{\beta} c_\alpha c_\beta f_\alpha f_\beta a_{\alpha\beta}\}. \tag{2.60}$$

The result becomes particularly simple for a so-called *substitutional* system in which all the $g_{\alpha\beta}$ and all the partial interference functions are the same: what this requires is that molecules of the various constituent species should be able to replace each other at random and without causing the neighbouring molecules to rearrange themselves in any way – there must be no tendency for molecules of any species to cluster together or to form ordered arrays. In such an ideal system (2.60) reduces to

$$N\{\overline{|f-\bar{f}\,|^2}+|\bar{f}\,|^2 a(q)\}, \tag{2.61}$$

which has a very transparent physical meaning. The first term (which is the same as the first term in (2.60) though written in a different way) arises only when the form factor is a variable quantity; because the variations, for a substitutional system, are entirely random in passing from one molecule to the next the scattering for which they are responsible is essentially *incoherent* and is unaffected by the liquid structure. The structure-dependence is concentrated in the second term, which describes the *coherent* scattering; this depends only on the average form factor and not on its variations.

These results apply to systems with any number of components, but *binary* ones are generally of most interest. In principle it needs three parameters to describe the structure of a binary system, a_{00}, a_{11} and a_{01}; the variation of $g_{01}(R)$, and hence a_{01}, is determined by precisely the same distribution of molecule–molecule distances in the liquid as determines the variation of $g_{10}(R)$, so that a_{01} and a_{10} must be identical. A convenient rearrangement of (2.60) for a binary system is

$$N\{c_0c_1[f_0f_0^*(1-a_{00})-(f_0f_1^*+f_0^*f_1)(1-a_{01})+f_1f_1^*(1-a_{11})] \\ +c_0f_0f_0^*a_{00}+c_1f_1f_1^*a_{11}\}. \tag{2.62}$$

Since c_0c_1 is zero at both ends of the concentration range, the term proportional to this quantity indicates a scattering which arises only as a result of mixing two species together and has a maximum at around $c_0 = c_1 = \frac{1}{2}$. The remaining two terms should vary monotonically with concentration between the limits appropriate for the two constituents in their pure state. It is always necessary to remember, however, that the partial interference functions (and even in some circumstances the form factors f_0 and f_1) may themselves be functions of concentration.

We may note for future reference that the partial interference functions, like $a(q)$ for a pure substance, are obliged to satisfy some general constraints. Thus each $a_{\alpha\beta}$ must tend to unity for large q, as has been pointed out already. The analogues of the condition that $a(q)$ for a pure substance must always be positive are obtained by remarking that (2.62) must always be positive, whatever the individual form factors; they prove to be:

$$\left.\begin{aligned} c_0 a_{00} + c_1 &\geqslant 0, \\ c_1 a_{11} + c_0 &\geqslant 0, \\ (c_0 a_{00} + c_1)(c_1 a_{11} + c_0) &\geqslant c_0 c_1 (a_{01} - 1)^2. \end{aligned}\right\} \qquad (2.63)$$

As for the sum rule expressed by (2.41), the proof of this goes through as above, with a_{00}, a_{11} or a_{01} in place of a.

2.13. Experimental methods for the determination of $a(q)$

The basic method whereby $a(q)$ may be determined experimentally for a pure liquid should be apparent from the discussion on p. 110. The specimen is irradiated by a parallel beam of some monochromatic radiation which has a wavelength λ comparable with atomic dimensions and a frequency which is preferably much greater than the characteristic vibration frequencies of the molecules; if the latter condition is not satisfied it is not legitimate to treat the scattering as *elastic*, and the complications that then arise are to be discussed in chapter 3. X-rays and 'warm' neutrons are both suitable.† Then the intensity scattered through an angle θ is measured with a detector which subtends a small solid angle at the specimen. So long as this solid angle is kept the same when θ is varied the detector always accepts the same number of scattered beams, i.e. the same number of permitted values of the scattering vector $\mathbf{q}$, and its output is therefore directly proportional to the scattered intensity for a single value of $\mathbf{q}$. The magnitude of q is given for elastic scattering (see fig. 1.5 on p. 33) by $4\pi \sin(\theta/2)/\lambda$, so this quantity forms a convenient choice of abscissa when the results are presented graphically.

The corrections that are necessary before $a(q)$ can be extracted are most easily discussed in the case of neutrons. In non-magnetic materials neutrons are effectively scattered only by the nuclei, and these are so small compared with λ that the scattering by an individual nucleus is completely isotropic, i.e. the form factors are always independent of q. This is a great simplification. On the other

† Corrections for inelastic effects *are* sometimes required with the neutrons that are used in practice; see pp. 185–6.

hand the form factor is not necessarily the same from one molecule to the next, either because the nominally pure specimen in fact consists of more than one isotope or because its nuclei have spin: during the scattering process the neutron and the nucleus together form a 'compound' in a definite spin state, and the cross-section depends on whether the spins of the two particles in this compound are parallel or anti-parallel – though if the nucleus has zero spin the distinction cannot arise. Hence even a pure metal tends to look like a mixture to a neutron beam, though it is an ideal substitutional mixture, described by a single interference function. Equation (2.61) is applicable and tells us that there may be appreciable incoherent scattering to be allowed for, though fortunately it must be isotropic. One final consideration to bear in mind with neutrons is that the flux which can be obtained in an adequately monochromatic beam is none too large, and in order to obtain satisfactory counting rates it is essential to use a specimen which is thick enough to scatter a good proportion of the neutrons which are incident upon it; in practice this means specimens of the order of 1 cm across. The inevitable result is that there must be a good deal of multiple scattering. Fortunately this too is likely to be isotropic in many cases (Cocking & Heard, 1965).

The net result is that the raw curve for signal versus $4\pi \sin \frac{1}{2}\theta/\lambda$ may look something like fig. 2.8(a). The problem is to subtract out the uniform background due to incoherent and multiple scattering and then, using the fact that $a(q)$ must go to unity for large q, to divide the remainder by the scale height I_∞ in the figure so as to derive $a(q)$.

If the experimental curve could be continued to high enough and low enough q this would present no problem, especially if a value of β is available so that one knows the limiting value of $a(q)$ for small q (which is in any case very small). But sometimes the experimental range is too narrow for this approach to be accurate. The alternative procedures are (a) to use the sum rule expressed by (2.41) to determine just where the horizontal lines that determine the scale height should be drawn; (b) to calibrate the apparatus using V (vanadium) which has the almost unique advantage of giving no coherent scattering, for the reason that $\bar{f}$ accidentally vanishes, and which is therefore an incoherent,

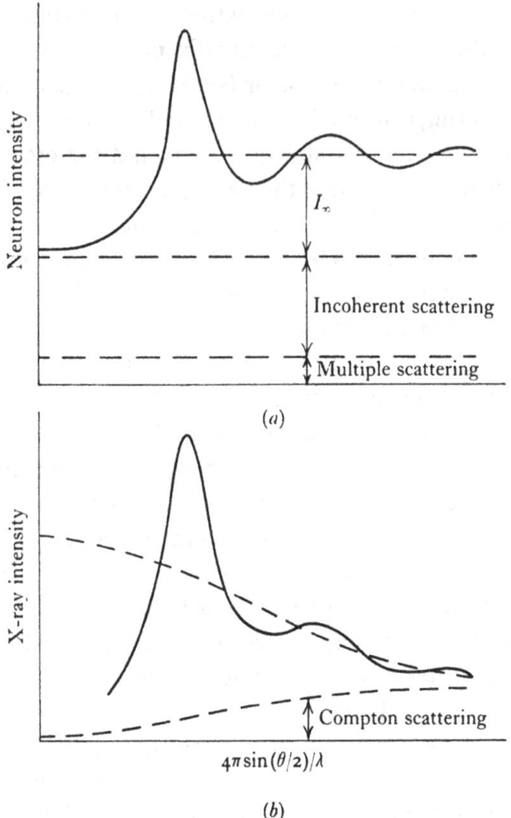

Fig. 2.8. Diffraction patterns for a typical liquid.

isotropic scatterer for neutrons; this method requires a knowledge of the relative values of $\overline{f^2}$ for V and $\bar{f}^2$ for the metal of interest, but neutron scattering cross-sections are now known quite accurately from independent experiments. Best of all, no doubt, is to use all three methods and to check that they give consistent results; recent sets of neutron data have indeed been checked in this way (North, Enderby & Egelstaff, 1968).

X-rays differ in several respects. In the first place they are scattered more strongly, so that an incident beam is stopped by a millimetre or so of metal instead of by tens of centimetres. Consequently, in order to avoid excessive multiple scattering it is desirable to make the specimens rather thin, whereupon radiation scattered by the container becomes a source of error. Of course it is possible to arrange things so that the X-rays strike a free liquid surface and are scattered back without passing through any containing wall, but a correction for multiple scattering and absorption is then required which varies with θ. Unlike neutrons, an X-ray does not perceive any difference between one isotope and the next, nor is it sensitive to nuclear spin; nevertheless, if it happens to be scattered inelastically by the Compton effect, leaving one particular ion in an excited state, then it is scattered incoherently, and this means that there is still a background signal, which unfortunately is *not* isotropic, to be subtracted out. Finally, the form factors for X-rays vary quite strongly with q.

The result is the sort of raw curve shown in fig. 2.8(b). In order to extract $a(q)$ one has to depend on theoretical calculations both of the Compton scattering background and of the form factor curve, shown by broken lines in the figure; these calculations are done for free atoms and may not be completely appropriate for a metal. In scaling them to the experimental curve it is customary to be guided largely by the requirement that $a(q)$ must tend to unity for large q, though the sum rule may also be useful if the measurements have been continued to low enough values of q.

There is little doubt that the neutron method is in general the more accurate,† though a great deal of expert care has been applied to X-ray measurements on liquids over the years and a lot of valuable data (as well as a lot of rather misleading data) have been accumulated.

Once a reliable curve for $a(q)$ over a good range of q is available it may be subjected to Fourier inversion with the aid of a computer so as to yield the pair distribution function $g(R)$. It is now well understood, however, that substantial errors may be introduced during this operation if the curve for $a(q)$ is wrongly extrapolated outside the range of measurement; in particular, quite spurious ripples and bumps may appear in $g(R)$ as a result of too abrupt termination of the oscillations in $a(q)$ (Paalman & Pings, 1963).

† A case has been made by Greenfield, Wellendorf & Wiser (1971) for the superiority of the X-ray method, especially at low values of q, provided that the measurements are done by transmission rather than reflection. Their results, for liquid Na, do seem to be particularly accurate.

2.14. Discussion of results

Three typical experimental curves for $a(q)$ are reproduced in fig. 2.9, one for liquid Ar being included because of the interest that lies in the comparison between metals and rare gases. All of the curves have the same general form, rising from low values near $q = 0$ to a pronounced peak at q_{max} say and subsequently oscillating in an unspectacular way about unity. When subjected to Fourier inversion they yield curves for $g(R)$ like the one illustrated in fig. 2.10. The ripples which are visible there for small values of R are certainly spurious and should be ignored; $g(R)$ must in fact be zero until R reaches the value denoted by σ^* in the diagram, which is the closest distance of approach of neighbouring molecules. Some approximate results for σ^*, taken from published curves for $g(R)$, are listed in table 2.8 and compared with the atomic radius R_A.

Many authors prefer to present their results by plotting the *radial distribution function* (or RDF) $4\pi R^2 g(R)$, as in fig. 2.11. The position of the first peak in the RDF, which is only marginally further out than the first peak in $g(R)$, may be taken to indicate the most probable separation between neighbouring molecules in the liquid, while the area under the first peak, in so far as this can be separated from subsequent peaks (see the broken curves in fig. 2.11), can be interpreted as a mean *coordination number*. It is noticeable that the most probable separation in the liquid often corresponds very closely to the mean separation between neighbouring molecules in the solid phase just below the melting point: for Ar, for example, the two seem to coincide within $\pm 1\%$ though an increase of about 5% might naïvely have been expected on account of the 14.3% increase in volume. It is the coordination number rather than the molecule–molecule spacing which tends to change on melting. Estimates of the coordination number vary considerably according to the source of the data and the precise way in which the enveloping curve for the first peak in the RDF is drawn, but the values most often quoted for simple liquids cluster about 10 or 11, whereas the number of nearest neighbours in a close-packed solid lattice is of course 12. Liquid Te, which may be classified as a semi-metal (see p. 81), forms the only clear

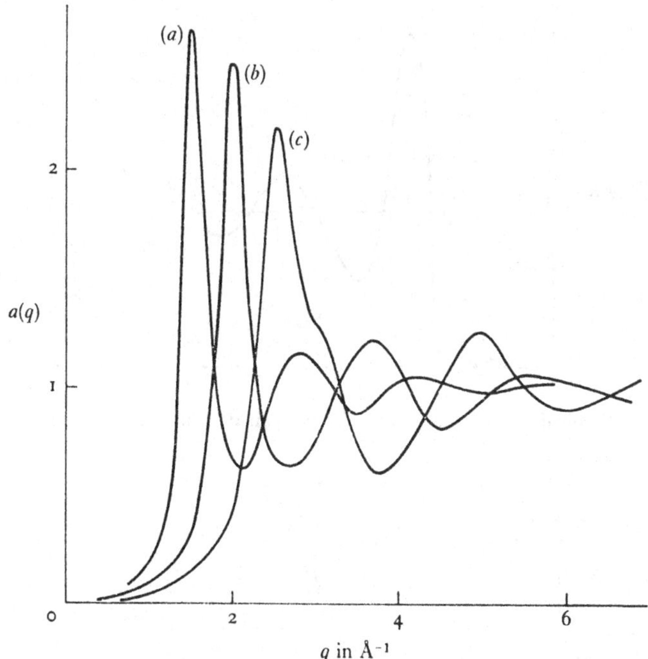

Fig. 2.9. The interference function for (*a*) liquid Rb at 40 °C (Gingrich & Heaton, 1961), (*b*) liquid Ar at 84 °K (Henshaw, 1956), (*c*) liquid Ga at 50 °C (Ascarelli, 1966).

exception; according to Cabane & Friedel (1971) the neutron data for this substance (Tourand & Breuil, 1971) suggest a coordination number of only 3. The way in which these facts have sometimes been quoted in support of quasi-crystalline models has already been mentioned in §2.2.

From the point of view of the rigid-sphere model discussed in §2.7, it is no surprise that the diffraction pattern is much the same for every liquid metal at its melting point. The thing that determines $g(R)$ or $a(q)$ for a rigid-sphere system is the packing fraction y, and we have seen that this should be very little affected by the attractive forces which are responsible for cohesion in metals; at the melting point it should always be close to 0.46, which is the value suggested for non-attracting spheres by the computations of Alder & Wainwright (see §2.6). Thus the $g(R)$ and $a(q)$ curves for different metals should be virtually identical at T_M, provided that

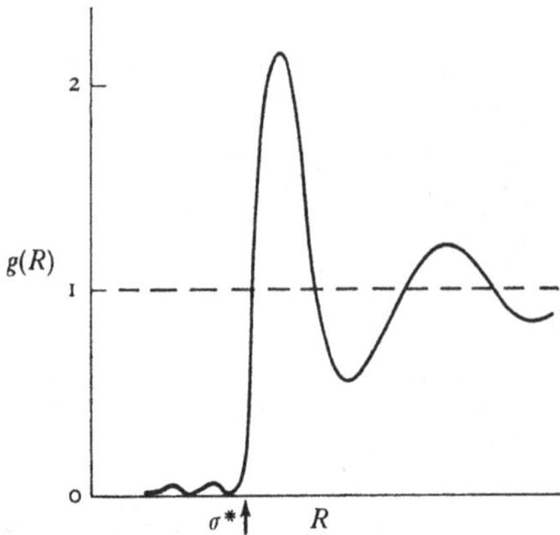

Fig. 2.10. Typical experimental curve for the pair distribution function.

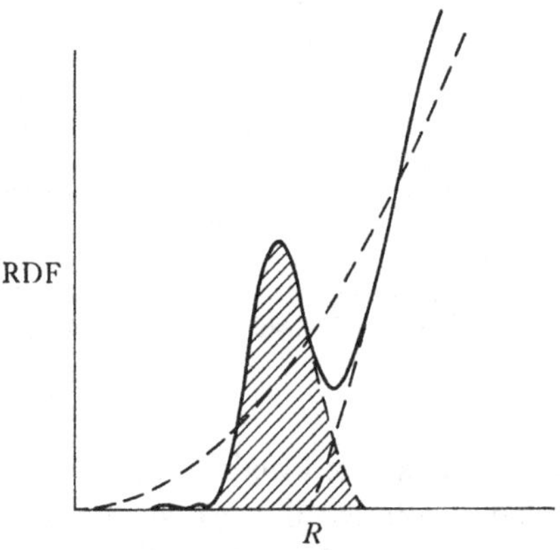

Fig. 2.11. Radial distribution function corresponding to the curve in fig. 2.10. The hatched area corresponds to the first coordination shell.

they are plotted with R/R_A and qR_A respectively as abscissae so as to allow for differences of atomic volume.

A simple check of the rigid-sphere model is to see whether the closest distance of approach, σ^*, agrees with the effective rigid-sphere diameter σ, as it should. For $y = 0.46$ we may expect

$$\frac{\sigma^*}{R_A} = 2(0.46)^{\frac{1}{3}} = 1.54.$$

The values quoted for σ^*/R_A in table 2.8 are mostly rather less than 1.54, but they are subject to considerable error; the behaviour

TABLE 2.8. *Closest distance of approach in simple liquids*

Substance	R_A (Å)	σ^* (Å)	σ^*/R_A
Ar	2.24	3.2	1.43
Li	1.75	2.8	1.60
Na	2.14	3.0	1.40
K	2.65	3.6	1.36
Ag	1.66	2.4	1.45
Zn	1.56	1.9	1.22
Hg	1.79	2.6	1.45
Ga	1.66	2.3	1.39
In	1.86	2.4	1.29
Sn	1.89	2.5	1.32
Pb	1.97	2.7	1.37

Note. These data are for temperatures close to T_M.

of $g(R)$ in the neighbourhood of σ^* is particularly affected by improper termination of the oscillations in $a(q)$. Discrepancies of 15% or so from the expected ratio are no real cause for concern.

To make a more thorough check possible we need a theoretical curve, preferably for $a(q)$ rather than $g(R)$, to compare with the experimental results over the whole range of q. In principle one could compute such a curve by the method of molecular dynamics, but results of sufficient detail for systems of non-attracting rigid spheres are not yet available. We may therefore make use of the Percus–Yevick theory, to be discussed in §2.17, which for rigid spheres has an analytical solution which can readily be plotted out for any value of y one cares to choose. Where it has been tested, this theory seems to match the results of molecular dynamics with reasonable precision. It suggests an equation of state for non-attracting spheres, for example, of the form

$$\frac{p^0\Omega}{Nk_BT} = \frac{1+y+y^2}{(1-y)^3}, \tag{2.64}$$

which for $y \leqslant 0.46$ is very similar to the empirical equation (2.20). It suggests that

$$a^0(0) = \frac{Nk_B\beta^0T}{\Omega} = \frac{(1-y)^4}{(1+2y)^2} \tag{2.65}$$

for non-attracting spheres, which comes out to be 0.023 when $y = 0.46$; the value suggested by Alder & Wainwright's computations seems to be 0.026. For relatively small y it has been tested by Ashcroft & March (1967) against a virial expansion theory for rigid spheres which is exact up to terms of order y^3 and has been found to work extremely well.

The fit between the Percus–Yevick rigid-sphere model and the experimental data for $a(q)$ in liquid metals has been explored in some detail by Ashcroft & Lekner (1966). It is never perfect: the model tends to predict too large a spacing between the first peak and the second, and since the attractive forces in real metals make β less than β^0 (see §2.7) it necessarily overestimates the magnitude of $a(0)$ to a small extent. But with $y = 0.46$, or 0.45, it is able to match the position and shape of the first peak, for temperatures close to T_M, with fair precision. The model has been used for a number of calculations concerning liquid metals which are described in later chapters.

It is no good pretending, however, as some authors have done (e.g. Furukawa, 1960), that significant differences between one metal and another do not exist. The shoulder on the right of the main peak for liquid Ga in fig. 2.9 is a distinctive and well-established feature (compare the neutron data of Ascarelli (1966) with the X-ray data of Rodriguez & Pings (1965)) which is not shown by any of the monovalent metals. A less pronounced shoulder is just detectable in most of the recently published curves for Hg (Black & Cundall, 1965; Kaplow, Strong & Averbach, 1965*a*; Wagner, Ocken & Joshi, 1965; Rivlin, Waghorne & Williams, 1966; Waseda & Suzuki, 1970*a*) and similar features have been reported for other polyvalent liquid metals such as Sn, Sb and Bi (North *et al.*, 1968; Isherwood & Orton, 1968). A shoulder to the left of the main peak has been reported for Zn (Caglioti, Corchia & Rizzi, 1967; North *et al.*, 1968). The reader is referred to review articles for further details (e.g. Frost, 1954; Furukawa, 1962; Kruh, 1962) though the necessity for treating the older data with reserve needs to be emphasised. That there are subtle structural distinctions between Ga, Hg, Sn and so on can also be inferred from the figures for their excess entropy in table 2.1; the variations of S_E at the melting point from one liquid metal to the next are certainly small but they are not completely trivial.

Many attempts have been made to interpret the distinctions in terms of quasi-crystalline models, one example being the work of Fessler, Kaplow & Averbach (1966) on liquid Al. Some authors have tried to link the position of the shoulders which are observed on some $a(q)$ curves with particular 'bond lengths' that occur in the solid phase – though when the $a(q)$ curves are subjected to Fourier inversion it is impossible to pick out any peak in $g(R)$ for which the shoulder is clearly responsible. But the sad truth is that there is insufficient information in the pair distribution function alone to justify this type of approach.

A different one has been adopted by Weaire (1968), who has tried to extend to liquid metals the arguments which Heine & Weaire have applied to solids (see p. 69). Distortions of an $a(q)$ curve from the standard rigid-sphere form are attributed to the effects of the indirect interaction energy; the metal tries to lower this by shifting the weight of $a(q)$ away from the position where $u(q)$ passes through a node. Weaire discussed liquid Hg in particular, but he used a Heine–Animalu pseudo-potential which is now thought to be erroneous (see p. 40). In any case, the two criticisms which were aimed at Heine & Weaire's analysis for solid metals apply with equal force in the liquid phase. If we are to associate the structural distinctions between liquid metals with variations in the form of the pair potential $w(R)$, and hence indirectly with variations in $u(q)$, a more powerful method is required.

In principle, one should be able to make the association with the aid of molecular dynamics. One could choose a plausible form for $w(R)$, compute the corresponding $a(q)$, compare it with experiment, and vary $w(R)$ until exact agreement was obtained. This is scarcely a realistic programme, however. We do not have a sufficiently clear understanding of the nature of $w(R)$ to be confident that our starting point is reasonably close to the truth. The work of Schiff (1969) and the careful comparison which Page *et al.* (1969) have effected between the structural data for liquid Rb and Ar, make it clear that particular care is needed to choose the right form of repulsive core for $w(R)$, and that it is hard to predict this from first principles is fully apparent from the theoretical curves in fig. 1.14. Without inordinate expenditure of machine time we would never achieve perfect agreement, or anything like it. We would be left in a state of uncertainty, not knowing whether to attribute the residual discrepancies to significant errors in $w(R)$, to errors introduced – because of the finite size of the model investigated – during the computation of $a(q)$, or even to errors in the experimental data.

An alternative approach is to use one of several theories of the liquid state to calculate $w(R)$ from $a(q)$, and to use molecular dynamics – since none of the theories is completely trustworthy – simply to verify the result, by trying to regenerate the original data. The progress that has so far been achieved along these lines is discussed in §§2.16 and 2.17 below.

2.15. Temperature dependence of $a(q)$

The general effect on the interference function of heating above the melting point is indicated in fig. 2.12. The peaks tend to become flatter and broader as $a(q)$ moves towards the value of unity which characterises a completely random gas-like structure, while their position stays much the same despite the thermal expansion that occurs (e.g. North *et al.*, 1968); thermal expansion seems to be reflected by a fall in the coordination number rather than by an increase in the mean separation between atoms. Since $a(0)$ is proportional to βT, and since the compressibility tends to increase on heating at constant pressure, the quantity $T(\partial \log a/\partial T)_p$ may exceed unity for very small q. Its behaviour in liquid Na, as indicated by the measurements of Greenfield (1966), is shown in fig. 2.13. The figure includes a curve for $T(\partial \log a/\partial T)_\Omega$, based on the experimental result of Endo (1963) that $T(\partial \log \beta/\partial T)_\Omega$ in liquid Na is approximately zero. But since no measurements have yet been reported on the diffraction properties of liquid metals as a function of temperature at constant volume, the shape of this lower curve is largely hypothetical.

It carries with it the implication that if one could somehow supercool liquid Na at constant volume, to such an extent that thermal agitation of the ions had virtually ceased, the interference function would still not vanish except for the smallest values of q. In this respect liquid Na differs from the Debye solid discussed in

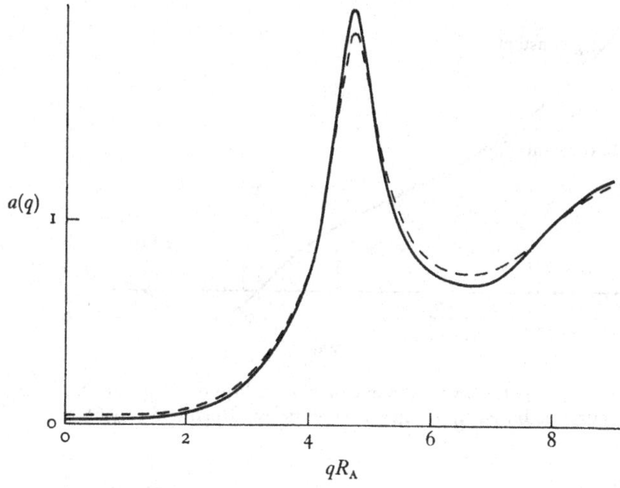

Fig. 2.12. Typical curves for $a(q)$ at two different temperatures.

§2.11, in which, except at the reciprocal lattice vectors, the whole of $a(q)$ is to be attributed to thermal agitation and $T(\partial \log a/\partial T)_\Omega$ does not fall off with q. If one is prepared to make the dubious assumption that the lower curve in fig. 2.13 is typical of any liquid metal, one may use it to estimate very roughly the proportion of $a(q)$ which, being proportional to T, is attributable to thermal agitation rather than to disorder in the rest positions about which the ions instantaneously vibrate. An experimental curve for $a(q)$ in liquid Pb, a metal which has been studied with particular care in the low q region, is reproduced in fig. 2.14 and the broken line shows the effect of multiplying it by $T(\partial \log a/\partial T)_\Omega$, taken from fig. 2.13, to find the temperature-dependent part of $a(q)$. The Debye model suggests that within the first Brillouin zone for a solid, i.e. for qR_A less than about $\frac{1}{2}\pi$, $a(q)$ should be proportional to $(v_\parallel(q))^{-2}$, where $v_\parallel(q)$ is the wave velocity for longitudinal modes of wave vector q. The broken curve in fig. 2.14 could plausibly be interpreted as a curve of $v_\parallel^{-2}$ for liquid Pb; the factor of about 3 by which it rises in going from $qR_A = 0$ to $qR_A = \frac{1}{2}\pi$ is comparable with the factor $(\frac{1}{2}\pi)^2$ in equation (2.50). It is quite consistent with the dispersion curves obtained by slow neutron scattering experiments discussed in §3.14 below.

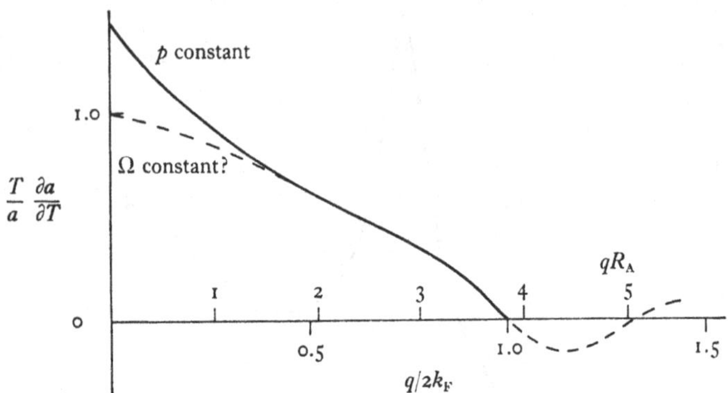

Fig. 2.13. The temperature coefficient of $a(q)$ in liquid Na at 100 °C. The full curve is based upon measurements by Greenfield (1966).

The Debye model is acceptable for liquid Pb, incidentally, because the specific heat ratio γ is only 1.18. It could not be applied in the same way to liquid Ar, for which γ is 2.2 (see table 2.3). Over half of $a(0)$ is contributed by entropy fluctuations in liquid Ar (see p. 120), which the Debye model fails to describe. It is no doubt for this reason that, as the reader will observe from fig. 2.14, the variation of $a(q)$ with q in the small q region is very different for liquid Ar and liquid Pb.

To discuss the temperature-dependence of $a(q)$ near its main peak we may resort once more to the Percus–Yevick rigid-sphere model. The pioneers in this direction were Ashcroft & Lekner (1966), who examined in particular the height of the main peak for liquid Rb; they adjusted the model to explain its fall on heating by allowing the parameter y to vary with temperature like $T^{-0.32}$ at constant pressure. A more detailed comparison has since been effected for liquid Pb by Waseda & Suzuki (1970*b*), whose results suggest a value of about -0.31 for the exponent. Both these estimates are reasonably consistent with the theory outlined in §1.14.

A tedious analysis of equation (2.77) reveals that when y is in the neighbourhood of 0.46 the *position* of the main peak in $a(q)$ should be given, according to the Percus–Yevick rigid-sphere model, by

$$q_{\max}\sigma = 6.83(y/0.46)^{0.294}.$$

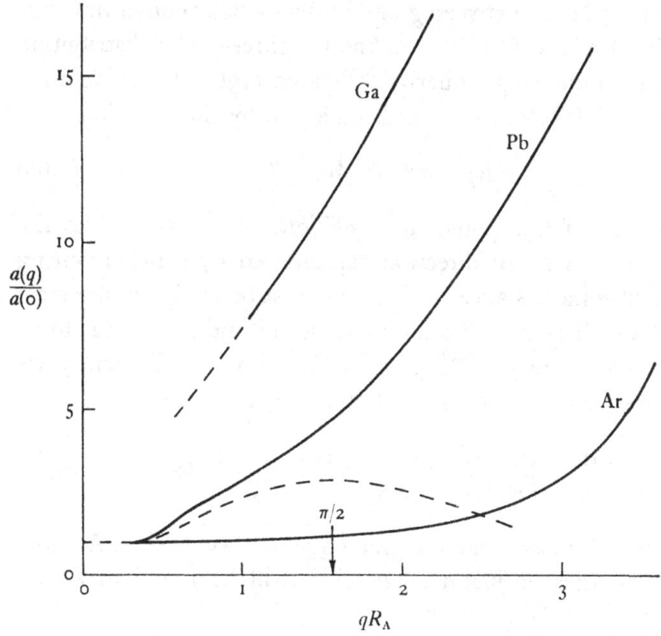

Fig. 2.14. The behaviour of $a(q)$ for small q. (Redrawn from data quoted by Enderby & March, 1965.)

It follows that for liquid Pb, if $y \propto T^{-0.31}$ at constant pressure, we should expect

$$\frac{1}{q_{\max}}\left(\frac{\partial q_{\max}}{\partial T}\right)_p \simeq \frac{0.012}{T_M} - \frac{1}{3}\alpha$$
$$\simeq (2.5-4.2)\times 10^{-5}\ {}^{\circ}\mathrm{K}^{-1}.$$

It is not unexpected, in the light of this result, that the shift of $q_{\max}$ on heating is almost too small to be detected (see p. 134).

2.16. The Born–Green theory

The central problem in the theory of simple liquids is to relate the pair distribution function $g(R)$ to the pair potential $w(R)$ – on the assumption, of course, that a pair potential can be defined. Since thermodynamic properties can be expressed in terms of g and w, as we have seen in §2.9, a formula connecting these two functions would enable us to express them in terms of w alone, and this is the ideal at which most liquid theories are aimed.

An exact relation between g and w does exist, though unfortunately it also involves the unknown three-body distribution function g_3. Following Enderby & March (1965) let us define a new potential $W(R)$ around each molecule by the equation

$$g(R) = \exp(-W/k_B T). \tag{2.66}$$

The gradient of this potential, $-\partial W/\partial R$, presumably gives the mean force in a radial direction experienced by a neighbouring molecule at some distance R. But this must be made up of a term $-\partial w/\partial R$ due directly to the central molecule and a term due to its interaction with all the other molecules close by. Evidently we must introduce g_3 to express the latter, the result being

$$\frac{\partial W}{\partial R} = \frac{\partial w}{\partial R} + \frac{N}{\Omega}\int \frac{g_3(\mathbf{R}, \mathbf{R}')}{g(R)} \frac{\partial w(|\mathbf{R}-\mathbf{R}'|)}{d\mathbf{R}'} \partial \mathbf{R}'. \tag{2.67}$$

This is the fundamental integral equation we desire. A more formal and rigorous justification for it will be found in Fisher (1964) or Rice & Gray (1965).

To proceed further it is necessary to make some assumption about g_3, and a convenient one is that it obeys the famous *superposition approximation*, first proposed by Kirkwood. This may be expressed by the equation

$$g_3(\mathbf{R}, \mathbf{R}') = g(R)g(R')g(|\mathbf{R}-\mathbf{R}'|), \tag{2.68}$$

which has at any rate the symmetry that common sense dictates. An equally symmetrical approximation proposed by Abé is

$$g_3(\mathbf{R}, \mathbf{R}') = \tfrac{1}{3}\{g(R)g(R') + g(R')g(|\mathbf{R}-\mathbf{R}'|) + g(|\mathbf{R}-\mathbf{R}'|)g(R)\} \tag{2.69}$$

and we shall find this to be useful in §4.11 below. Equation (2.68) is usually regarded as superior, however. It enables (2.67) to be reduced after some manipulation to

$$\log g(R) + \frac{w(R)}{k_B T} = \frac{N}{\Omega k_B T}\int (g(R') - 1)\, d\mathbf{R}' \int_{|\mathbf{R}-\mathbf{R}'|}^{\infty} g(R'') \frac{\partial w(R'')}{\partial R''} \partial \mathbf{R}'', \tag{2.70}$$

in which only g and w are involved. This is the starting point of the theory of liquids due to Born and Green (Green, 1960). We may evidently use equation (2.70) to estimate $w(R)$ once we know $g(R)$ from experiment, though the calculation is bound to involve a lengthy and tedious iteration process.

It may be noted that if the superposition approximation were correct the second term in (2.36), our cluster expansion for the excess entropy, would disappear. Yet the first term alone has been shown to be insufficient to account for the magnitude of S_E in the case of liquid Al and Hg. More serious doubts about the validity of the superposition approximation arise from the work of Alder (1964) and Rahman (1964*b*) who have attempted to check it, by the method of molecular dynamics, for systems composed of rigid spheres and of molecules interacting via a Lennard–Jones potential respectively. The computer was used to extract values of $g(R)$ and then of $g_3(\mathbf{R}, \mathbf{R}')$ for the particular case where

$$|\mathbf{R}-\mathbf{R}'| = R' = R.$$

The results of both authors suggest, though the statistical spread is rather large, that for this case $g_3(\mathbf{R}, \mathbf{R}')$ may differ from the value $g(R)^3$ predicted by equation (2.68) by as much as 20% for some values of R. But it is hard to say how significant a discrepancy that represents.

The real test, of course, is to compare the detailed predictions of the Born–Green theory with experimental results. For liquid Ar, moderate agreement is obtained where $g(R)$ is concerned – the match is far from perfect – but the theory is not convincingly successful in predicting thermodynamic properties, at any rate for the sort of densities that prevail in liquid Ar close to its normal melting point. It was for some time possible to dismiss the discrepancies as due to error in the values assumed for $w(R)$; the results for pressure, for example, are extremely sensitive to slight variations in the pair potential. But with the advent of the computer and the methods of molecular dynamics and of Monte Carlo calculation, it became possible to test the theory, not against the behaviour of real Ar, but against that of mock fluids resembling Ar for which the pair potential was known with precision. In such tests the Born–Green theory has proved less successful than theories based on certain rival approximations, which are mentioned in the following section.

2.17. The Percus–Yevick theory

Apart perhaps from equation (2.69), no modification of the superposition approximation suggests itself which is reasonably simple and yet plausible on physical grounds. The rival approximate theories of liquids are therefore not based on any equation for g_3. They involve instead the so-called *direct correlation function* $f(R)$, first introduced by Ornstein & Zernike. This function is defined by the equation

$$g(R)-1 = f(R)+\frac{N}{\Omega}\int(g(R')-1)f(|\mathbf{R}'-\mathbf{R}|)\,d\mathbf{R}'. \qquad (2.71)$$

The extent to which g differs from the value of unity which it would take in a completely random gas-like structure indicates the extent of the correlations which exist in a liquid between the coordinates of neighbouring molecules, and the function $(g-1)$ which occurs on the left-hand side of (2.71) is therefore often known as the *pair correlation function.*† Equation (2.71) is an attempt to split this up into a part, $f(R)$, which depends *directly* on the central molecule and a part which depends on it indirectly, via all the other molecules in the neighbourhood. In a rather similar way we split the mean force acting on a molecule into direct and indirect terms in equation (2.67).

The convenience of (2.71) arises from the fact that $f(R)$ thus defined can be related very simply to the interference function. To see this, multiply both sides of (2.71) by $(N/\Omega)\exp(-i\mathbf{q}\cdot\mathbf{R})$ and integrate over $\mathbf{R}$. The result is

$$\begin{aligned} a(q)-1 &= \frac{N}{\Omega}\int f(R)\exp(-i\mathbf{q}\cdot\mathbf{R})\,d\mathbf{R} \\ &\quad+\left(\frac{N}{\Omega}\right)^2\int(g(R')-1)\exp(i\mathbf{q}\cdot\mathbf{R}')\,d\mathbf{R}' \\ &\quad\times\int f(|\mathbf{R}'-\mathbf{R}|)\exp(i\mathbf{q}\cdot(\mathbf{R}'-\mathbf{R}))\,d(\mathbf{R}'-\mathbf{R}) \\ &= C(q)(1+a(q)-1), \end{aligned}$$

† An alternative name is the *total correlation function*.

or $$C(q) = 1 - 1/a(q). \tag{2.72}$$

Here $$C(q) = \frac{4\pi N}{\Omega} \int f(R) \frac{\sin qR}{qR} R^2 \,\mathrm{d}R \tag{2.73}$$

is a Fourier transform of the direct correlation function. The Fourier inversion of (2.73) allows $f(R)$ to be determined quite simply from experimental data for $a(q)$.

Although convenient from this point of view, there is a weakness in (2.71) which prevents $f(R)$ from having any immediate physical significance: the indirect term is taken to depend upon the mean molecular density at $\mathbf{R}'$ and no proper allowance is made for the effect upon this of the molecule which is by implication situated at $\mathbf{R}$. Were it not for this snag, we might set (cp. equation (2.66))

$$f(R) = \exp(-w(R)/k_B T) - 1. \tag{2.74}$$

The most successful approximate theory of liquids to date seems to be the *Percus–Yevick* theory, which may be summed up by the equation

$$f(R) = g(R)(1 - \exp(w(R)/k_B T)). \tag{2.75}$$

This reduces to (2.74), (as does the result of the rival ***hypernetted chain*** theory), if g can be replaced by $\exp(-w/k_B T)$, i.e. if the difference between w and the effective potential W defined in (2.66) is negligible. Apart from this it is not easy to detect much physical justification for (2.75); the various arguments by which it has been derived (consult e.g. Rice & Gray (1965) or Rowlinson (1965)) are of an abstract and complex nature. Since they are all of quite uncertain validity, the Percus–Yevick theory should be regarded as largely empirical, to be justified or otherwise by comparing its results with experiment. From such a comparison it has emerged more favourably than the Born–Green theory.

The solution of (2.74) for a system of non-attracting rigid spheres is straightforward. If we suppose

$$\begin{aligned} w(R) &= \infty \quad (R < \sigma) \\ &= 0 \quad (R > \sigma), \end{aligned}$$

then (2.74) leads immediately with the aid of (2.73) to

$$C(q) = -\frac{4\pi N}{\Omega} \int_0^\sigma \frac{\sin qR}{qR} R^2 \,\mathrm{d}R,$$

i.e. to
$$\frac{1}{a(q)} = 1+24y\left[\frac{\sin q\sigma - q\sigma\cos q\sigma}{q^3\sigma^3}\right], \tag{2.76}$$

where y is the packing fraction. This simple equation gives qualitatively the right results; it yields a curve for $a(q)$ which starts well below unity at $q = 0$ and oscillates through a series of peaks of diminishing amplitude, which become sharper as y is increased. It exaggerates the magnitude of $a(0)$, however, which should be 0.026; equation (2.76) suggests

$$a(0) = (1+8y)^{-1},$$

and this amounts to 0.21 when $y = 0.46$. The equivalent solution for the Percus–Yevick equation is much more successful, as we have already noted in §2.14, but naturally it is also much more complicated. It may be written in the form

$$(1-y)^4 f(R) = -(1+2y)^2 + 6y(1+\tfrac{1}{2}y)^2(R/\sigma) - \tfrac{1}{2}y(1+2y)^2(R/\sigma)^3 \tag{2.77}$$

for $R < \sigma$; for $R > \sigma$, $f(R)$ evidently vanishes.† A Fourier transform is necessary, of course, to obtain $a(q)$. The equation of state already quoted as (2.64), which agrees so well with the results predicted by molecular dynamics, is derived with the aid of the virial theorem (Rice & Gray, 1965; Rowlinson, 1965).

The rigid-sphere model is convenient for many purposes, but we are interested here in a less restrictive application of the Percus–Yevick theory, to deduce the form of $w(R)$ from experimental results for $a(q)$ in real liquids. The procedure is straightforward in principle: a Fourier transform of $(a-1)$ yields $g(R)$, a Fourier transform of $(a-1)/a$ yields $f(R)$, and $w(R)$ is obtained from the ratio between f and g with the aid of (2.75). Verlet (1968) has tested the method for a system of molecules interacting via a Lennard–Jones potential of the sort appropriate in Ar, using molecular dynamics in more or less the way suggested at the end of §2.14. He has computed $a(q)$ for his model and then analysed the results with the aid of the Percus-Yevick theory, in an attempt to regenerate the $w(R)$ which he initially assumed. The attempt was very successful at low densities corresponding to liquid Ar in

† It is noteworthy that this solution for $f(R)$ is continuous, despite the discontinuous nature of $w(R)$ and $g(R)$.

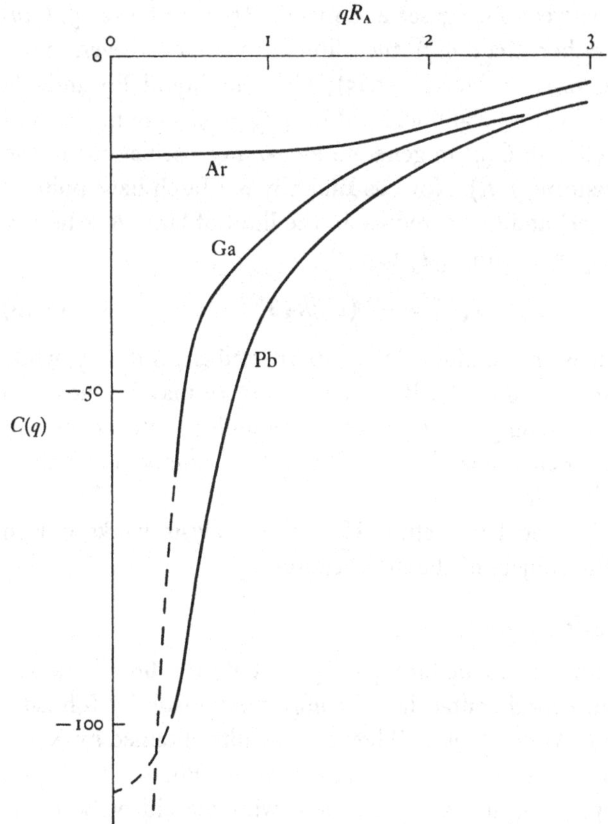

Fig. 2.15. Fourier transform of the direct correlation function. (Redrawn from data quoted by Enderby & March, 1965.)

the neighbourhood of its critical point; at higher densities, corresponding to the melting point, the analysis somewhat underestimated the depth of the attractive well in $w(R)$.

Since the experimental results for liquid Ar and liquid metals are qualitatively very similar (see fig. 2.9) while the differences of $w(R)$ are thought to be profound, it is clear that we must start with accurate data if we are to get meaningful results. In particular, to see the long-range oscillatory tail which is thought to be characteristic of $w(R)$ in liquid metals (see § 1.13) we need accurate data for $a(q)$ in the low q region. The curves for $a(q)$ which are reproduced in fig. 2.14 suggest that there may indeed be a real difference in this region between rare gases and metals. If, as in fig. 2.15, $C(q)$ is plotted rather than $a(q)$ the difference appears even more striking (Enderby & March, 1965); $C(q)$ for liquid Pb and Ga rises sharply in the neighbourhood of $qR_A = \frac{1}{2}$. One might well expect this edge in $C(q)$ to generate long-range oscillations in the Fourier transform, $f(R)$. Now, as Enderby & March have pointed out, both (2.74) and (2.75) reduce in the limit of large R, where w is small compared with $k_B T$, to†

$$f(R) = -w(R)/k_B T; \tag{2.78}$$

so does the master equation of the hypernetted chain theory, while the asymptotic form of the Born–Green theory may be shown to differ from (2.78) only in the presence of a multiplicative constant. Long-range oscillations in $f(R)$ therefore imply long-range oscillations in $w(R)$ as well.

But this is speculative stuff. How the analysis works out in practice is the subject of the next section.

2.18. Results for $w(R)$

The first attempts to deduce pair potentials for liquid metals, starting from experimental data for $a(q)$, were made by Johnson, Hutchinson & March (1964). They took results obtained by X-ray diffraction for alkali metals, for Hg, for Al, for Pb, and for liquid Ar for comparison, and analysed them with the aid of both the Born–Green and Percus–Yevick theories. In every case except that of Ar they landed up with oscillatory potentials. The wavelength did not correlate quite as closely as might have been hoped with π/k_F, the wavelength to be expected for Friedel oscillations (see §1.6), and the amplitude was larger than expected, especially for large R. But these were held to be minor worries; the oscillations were certainly there, whereas the $w(R)$ curve obtained for Ar showed a single attractive minimum, like the curve in fig. 1.15. The Born–Green and Percus–Yevick analyses did not lead to

† If (2.78) could be supposed to be valid for all R it would follow that

$$a(q) = (1 + Nw(q)/\Omega k_B T)^{-1}.$$

Ascarelli, Harrison & Paskin (1967) have proposed this as a useful approximation for liquid metals, especially in the low q region. There is little evidence to support it, however. It certainly fails to describe $a(0)$ correctly (see § 2.19).

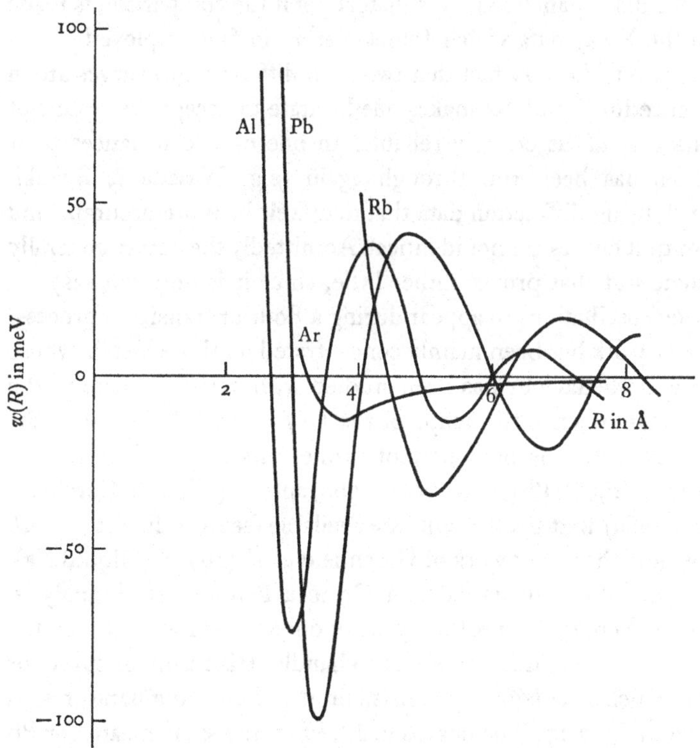

Fig. 2.16. Ion–ion pair potentials according to Johnson, Hutchinson & March (1964).

identical results. Johnson *et al.* favoured the Born–Green theory because it generated more consistent curves for $w(R)$ when they fed into the calculation two different sets of data for a single metal, obtained at different temperatures. Fig. 2.16 shows some of their results.

Paskin & Rahman (1966) later set out to verify the results for liquid Na by trying to regenerate its $a(q)$ curve by the method of molecular dynamics. They used two oscillatory potentials, one of them being very similar to the potential advocated by Johnson *et al.*, though it had to be truncated at large R to make the computation feasible. They claimed to reveal serious discrepancies, though March (1968, p. 44) has since pointed out that they mistakenly compared their computed $a(q)$ curves with *neutron* data for liquid

Na; the discrepancies are less disturbing if the comparison is made with the X-ray data which Johnson *et al.* in fact employed.

However, the very fact that two such different $a(q)$ curves are in existence for liquid Na makes one hesitate to accept the results of Johnson *et al.* as entirely reliable. In one or two instances their analysis has been run through again (e.g. Waseda & Suzuki, 1971*a*), using diffraction data that may well be more accurate, and the output curves are not identical. Admittedly the curves generally oscillate but that proves rather little, since it is only too easy for spurious oscillations to appear during a Fourier transform process.

Later work has been mainly concentrated on those metals whose diffraction behaviour has been studied with particular care in the low q region, since the shape of the $C(q)$ 'edge' discussed in the previous section is obviously of prime importance. It is for Ga (Ascarelli, 1966), Pb (North *et al.*, 1968) and Zn (Caglioti, Corchia & Rizzi, 1969) that good results are available (see also Egelstaff *et al.* (1966) and the recent work of Greenfield *et al.* (1971) on liquid Na). Ascarelli subjected his data for Ga to a Percus–Yevick analysis, since the Percus–Yevick theory is not only better founded than the Born–Green one, it is also easier to handle. His resultant curve for $w(R)$, which is adequately consistent at different temperatures, is shown in fig. 2.17. The curves deduced by the same means for Pb and Zn are similar.

It will be seen by comparing fig. 2.17 with fig. 2.16 that the long-range oscillations found by Johnson *et al.* have now almost disappeared; they have become if anything too small rather than too large. This in itself is not worrying, for Friedel oscillations arise from a singularity in the dielectric constant which scattering may smooth out of existence. It cannot be pretended, however, that the results are in good agreement otherwise with what theory predicts; the repulsive tail extending out to 10 Å and beyond does not appear in any of the theoretical curves in fig. 1.14. Is this a spurious artifact of the Percus–Yevick analysis? Are the experimental data for $a(q)$ to blame? Or is the theory in § 1.13, resting as it does on the assumption that the pseudo-potential is *weak*, misleading?

In the following section we shall concentrate attention upon the magnitude of $a(q)$ as $q \rightarrow 0$, i.e. upon the *compressibility*. This can

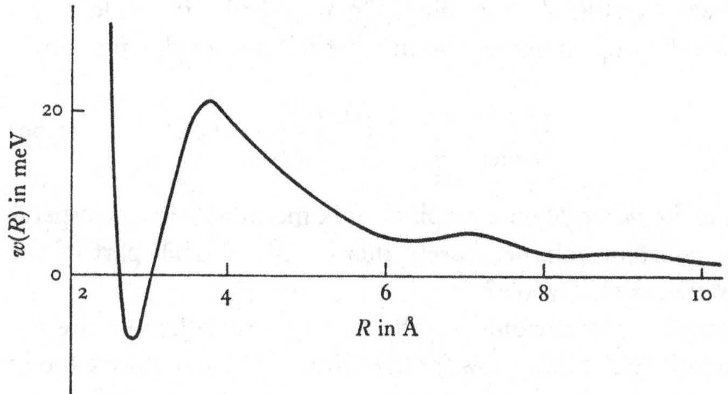

Fig. 2.17. Ion–ion pair potential for Ga. The curve is a smoothed version of ones plotted by Ascarelli (1966).

be related to $w(R)$ without help from the Percus–Yevick theory, and it can be measured directly and accurately, without the need for a diffraction experiment. If we are unable to calculate compressibilities successfully, therefore, we can be almost certain that the theory of $w(R)$ is to blame – and vice versa.

2.19. Compressibility

For the ideal rigid-sphere fluid discussed in §2.7 the compressibility – or rather its inverse, which is the bulk modulus – is composed of two quite distinct terms. One is the bulk modulus, $1/\beta^0$, of a fluid without attractive forces to hold it together, which is due entirely to the kinetic energy of the molecules and is therefore proportional to temperature. The other, according to (2.23), is the quantity $\Omega(\partial^2\mathscr{U}/\partial\Omega^2)_T$, which should depend only upon volume. At the melting point the ratio between the two should be given, according to (2.25), by

$$\beta^0\Omega(\partial^2\mathscr{U}/\partial\Omega^2)_T \simeq 1 - \frac{1}{39a(0)}.$$

Since this amounts for a typical liquid metal to about $\frac{1}{2}$, it looks at first sight as though the kinetic energy term is the major ingredient. What hope have we got of explaining compressibility data from first principles if we insist on discarding at the outset the only theories (i.e. the Percus–Yevick theory and its rivals) which are capable of describing β^0? Very little, it would seem. We shall develop an expression in what follows for the quantity

$$\left(\frac{1}{\beta}\right)_{\text{int}} = \frac{1}{\beta} - T\left(\frac{\partial(1/\beta)}{\partial T}\right)_{\Omega}, \tag{2.79}$$

which is the *intercept* on a graph of bulk modulus versus temperature at constant volume. Surely this is only a small part of the bulk modulus as a whole?

Fortunately, we are aided by the fact that metallic ions are not completely rigid. If we allow for the softness of the repulsive forces between them by supposing σ to decrease on heating, this has a profound effect on the temperature dependence of β^0. It follows from (2.24), in fact, that at the melting point

$$\begin{aligned} T\beta^0 \frac{d(1/\beta^0)}{dT} &= 1 - \nu\left[\frac{0.92f'(0.46)+(0.46)^2 f''(0.46)}{f(0.46)+(0.46)f'(0.46)}\right] \\ &\simeq 1-4\nu, \end{aligned} \tag{2.80}$$

where ν is the exponent that describes the temperature dependence of y. Since ν is thought to be about 0.3 at constant volume (see p. 72) the kinetic energy term in the bulk modulus varies much less rapidly with temperature than might have been supposed, and this term is not wholly lost from $(1/\beta)_{\text{int}}$. The measurements of Endo (1963) suggest that the quantity $T\beta(\partial(1/\beta)/\partial T)_{\Omega}$ is close to zero in liquid Na and only about -0.1 in liquid K, Rb and Cs; those of Davis & Gordon (1967) suggest that in liquid Hg at normal densities† it is $+0.14$. If these results are a fair guide to the behaviour of liquid metals in general, then the difference between the quantity $(1/\beta)_{\text{int}}$ which we are going to calculate and the total bulk modulus $(1/\beta)$ should not amount to more than say 15%.

Before embarking on the calculation it is worth remarking that the bulk modulus of jellium is entirely due to the $\Omega(\partial^2\mathscr{U}/\partial\Omega^2)_T$ term and that it can be calculated with some ease. The simplest answer of all, for which a number of derivations have been

† According to Postill, Ross & Cusack (1968) $T\beta(\partial(1/\beta)/\partial T)$ increases when Hg is expanded by heating; it seems to pass through unity at a density of about 12 gm cm^{-3}.

suggested (e.g. Bohm & Staver, 1951; Bardeen & Pines, 1955), is obtained by neglecting the effects of exchange and correlation. The internal energy of an exchange-free, correlation-free jellium is just the kinetic energy of its electrons, i.e. $\frac{3}{5}n\Omega K_F$, which varies with volume like $\Omega^{-\frac{2}{3}}$. It follows at once that

$$\beta = 3/2nK_F. \tag{2.81}$$

The figures in table 2.9 show that this crude prediction agrees astonishingly well with the experimental data for the heavier alkali metals just above their melting points.† The agreement would seem to be quite fortuitous, for it does not extend to higher temperatures; according to Jarzynski, Smirnow & Davis (1969), the compressibility of Rb increases by 50% on heating from the melting point to 360 °C, whereas $1/nK_F$ increases over the same range of temperature by only about 20%. For polyvalent metals (2.81) fails badly, for it underestimates β by a factor that is close to the valency z.

When the jellium calculation is improved by an allowance for exchange and correlation – this means using (1.16) for the internal energy – the agreement with experiment deteriorates, so we are obliged to take cognisance of those additional terms in the internal energy which arise because metals are made up in reality of discrete ions. To calculate these as a function of volume, however, is not straightforward (see p. 48). Ashcroft & Langreth (1967*a*) have suggested that the most recalcitrant of them may be proportional to $\Omega^{-\frac{1}{3}}$, and by adjusting the constant of proportionality to explain the observed equilibrium volume they have achieved quite impressive agreement, for a number of metals in the solid phase, with the observed compressibility as well. Ascarelli (1968) has used a similar heuristic argument in extending the rigid-sphere model for liquid metals, again with impressive results. But the assumptions made by these authors do not stand up to scrutiny.

Here we shall use a device which relates the compressibility only to those structure-dependent terms in the energy which the arguments of chapter 1 have equipped us to calculate. Let us

† Compressibilities may be measured by ultrasonic techniques; experimental methods and results for liquid metals have been reviewed by Webber & Stephens (1968).

TABLE 2.9. *Isothermal compressibilities of liquid metals*

Metal	z	β (dyne^{-1} cm^2 $\times 10^{-12}$)	$3/2nK_F$ (dyne^{-1} cm^2 $\times 10^{-12}$)
Li	1	(11)	4.6
Na	1	19	13
K	1	38	37
Rb	1	49	52
Cs	1	69	70
Cu	1	1.5	1.9
Ag	1	2.1	3.5
Mg	2	(~4)	1.8
Zn	2	2.5	0.9
Cd	2	3.2	1.5
Hg	2	3.8	1.7
Al	3	2.4	0.55
Ga	3	2.2	0.55
In	3	3.0	1.0
Tl	3	3.8	1.1
Sn	4	2.7	0.7
Pb	4	3.5	0.8
Sb	5	4.9	0.55
Bi	5	4.2	0.65

Note. These data are for temperatures close to T_M. Values in parentheses are rough guesses, based on measurements in the solid phase.

imagine our specimen to be distorted by a longitudinal compression wave of amplitude $\boldsymbol{\xi}_0$ and very small wave number $\mathbf{Q}$, which shifts the ith ion through a distance

$$\boldsymbol{\xi}_i = \boldsymbol{\xi}_0 \cos(\mathbf{Q}\cdot\mathbf{r}_i) \tag{2.82}$$

without altering the total volume. We may suppose if we wish that the temperature remains uniform throughout, though the compression waves that we can most easily excite in practice are, of course, adiabatic. The work required to establish the wave is

$$\Omega(Q\xi_0)^2/4\beta,$$

but some of this is liable to be lost as heat. It is an elementary exercise in thermodynamics to show that the energy retained in the specimen is

$$\Omega(Q\xi_0)^2(1+T(\partial \log \beta/\partial T)_\Omega)/4\beta = \frac{\Omega(Q\xi_0)^2}{4}\left(\frac{1}{\beta}\right)_{\text{int}}. \qquad (2.83)$$

We may derive our answer for $(1/\beta)_{\text{int}}$ by equating (2.83) to the extra structure-dependent energy for which the wave is responsible.

It has been shown in §2.11 that the compression wave described by (2.82) adds two spikes to the interference function of the liquid at $\mathbf{q} = \pm\mathbf{Q}$, of amplitude $\frac{1}{4}N(Q\xi_0)^2$. To satisfy the sum rule, however, it also reduces the magnitude of $a(q)$ elsewhere, and we have a plausible formula to describe the reduction in equation (2.55). From (1.93) it follows that the energy to be equated to (2.83) is

$$\frac{N^2}{4\Omega}(Q\xi_0)^2\left[w(0)+\frac{\Omega}{2\pi^2 N}\int_0^\infty\left(\frac{2}{15}q\frac{\partial a}{\partial q}+\frac{3}{15}q^2\frac{\partial^2 a}{\partial q^2}\right)w(q)\,q^2\,\mathrm{d}q\right]. \qquad (2.84)$$

Let us take $w(0)$ from equation (1.97), which means ignoring the distinction between z and z^* in the interests of simplicity and assuming the variation of $u(q)$ near $q = 0$ to be given by the empty-core model; it requires little effort to adapt the answer if other pseudo-potentials are preferred. As for the $w(q)$ inside the integral in (2.84), it is a fair approximation to replace this by $w''(q)$, which is given by (1.95); a glance at fig. 1.13 will show that in the region of q where $\partial a/\partial q$ and $\partial^2 a/\partial q^2$ are large the indirect part of the interaction, w', is small compared with the direct part. After these simplifications and two integrations by parts, (2.84) becomes

$$\frac{\pi N^2(ze)^2}{\Omega}(Q\xi_0)^2\left[\left(\frac{\epsilon'}{\epsilon q_S^2}\right)_{q\to 0}+R_M^2-\frac{8}{45\pi}R_A^3\int_0^\infty(1-a)\,\mathrm{d}q\right].$$

The integral of $(1-a)$ is readily evaluated from experimental data and amounts to about $2.6R_A^{-1}$ for most liquid metals. Hence the compressibility should be given approximately by

$$\left(\frac{1}{\beta}\right)_{\text{int}} \simeq 4\pi(ne)^2\left[\left(\frac{\epsilon'}{\epsilon q_S^2}\right)_{q\to 0}+R_M^2-0.14R_A^2\right], \qquad (2.85)$$

where ϵ'/ϵ in the limit of low q may be calculated from (1.35) and q_S^2 from (1.24) or, if (m^*/m) is close to unity, from (1.25).

TABLE 2.10. *A test of the theory of compressibility*

Metal	K	Rb	Al	Pb
$(\epsilon'/\epsilon q_s^2)_{q\to 0}$	0.05	0	0.15	0.16
R_M^2	1.32	1.90	0.35	0.35
$-0.14\,R_A^2$	-0.99	-1.14	-0.38	-0.54
Total	0.38	0.76	0.12	-0.03
$(4\pi(n\epsilon)^2\beta)^{-1}$	0.56	0.65	0.06	0.06

Note. The values used for q_s^2 have been calculated from (1.25) on the assumption that $m^*/m = 1$. All entries are expressed in Å².

To get back to the simple jellium formula (2.81) we should have to replace ϵ'/ϵ by unity in (2.85) and omit completely both R_M^2 and $-0.14\,R_A^2$. How little justified such approximations are in practice, especially for a metal like Rb, is shown by the figures in table 2.10. The four metals listed there are ones for which apparently reliable values of R_M have been deduced by adapting the empty-core model pseudo-potential until it explains the shape of the Fermi surface in the solid phase (Ashcroft & Langreth, 1967*a*). For (2.85) to be counted a success the figures in the two lowest rows of the table should agree. In view of the approximations involved, and of the way in which the relatively large figures in the second and third rows cancel one another, the agreement is not unsatisfactory. While it can hardly be said to substantiate the theory of $w(R)$ in detail, there is no evidence that this theory is grossly wrong.

It has been assumed throughout this discussion that the direct interaction between ions in liquid metals is purely coulombic in origin. If the ion cores were to overlap, of course, much stronger repulsive forces would come into play. It was demonstrated in §1.13 that overlap is very unlikely to occur in practice, but the argument may not apply to the noble metals, in which the d electrons are weakly bound and have wave functions which spread out to quite large values of R; direct interaction between the d electrons may be partly responsible for the small compressibilities displayed by liquid Cu and Ag. In one or two cases the compressibility may also be affected by van der Waals attractive forces between the ion cores, though the core polarisability is probably too small and the screening efficiency of the conduction electrons too great for these to be anything like as important as they are in the case of liquid Ar.

The 'long wavelength compression wave' theory of compressibility outlined above can naturally be expressed in terms of real space rather than **q**-space coordinates if preferred. The reader who wishes to see it in a different guise, or to explore in detail the relation between this type of theory and the type that Ashcroft & Langreth, for example, have employed, may consult papers by Price (1971) and by Hasegawa & Watabe (1972). Price, after a more scrupulous calculation than has been attempted here, obtains a value for the compressibility in liquid Na which is in excellent agreement with experiment. Hasegawa & Watabe achieve adequate agreement for all the alkali metals, and although their predicted values for three polyvalent liquid metals are on the high side the discrepancies for Al and Pb amount to only about 30%.

CHAPTER 3

MOLECULAR MOTION IN LIQUIDS

3.1. Prologue

The most characteristic feature of all liquids is their lack of rigidity, their capacity to flow. Solids can flow too, of course, by the process known as *creep*, and in some circumstances, especially at high temperatures and under low stresses, they may behave rather like liquids in that the shear stress and the rate of shear strain are proportional to one another. But the coefficient of viscosity η decreases on melting by a factor of at least 10^{13}, so that for most practical purposes one phase is rigid and the other is not. The distinction is clearly connected with the much higher mobility of individual molecules in the liquid phase, but there must be more to it than just that, for the self-diffusion coefficient D is observed to increase on melting by a factor of only 10^5 or so. It seems that the greater disorder of the liquid phase is also relevant.

Nabarro (1948) has emphasised that a solid specimen is unable to respond to shear stress in a liquid-like fashion if it is a single crystal. If it is composed of crystallites, however, each one about s molecules across, the situation is different; the crystallites may deform as a result of vacancy diffusion, and simple models (see also Herring, 1950) suggest an effective viscosity given by

$$\eta \simeq s^2 k_B T / dD, \tag{3.1}$$

where d is a length of the same order as the molecular diameter. Hence the factors of 10^{13} and 10^5 quoted above imply that the effective value of s in the liquid phase is 10^4 times smaller than in the particular solid specimen under test. As it turns out, equation (3.1) works surprisingly well for liquid metals if s is chosen to be unity.

In this limit the crystallite models of Nabarro and Herring are scarcely convincing, but the reader will discover from the first half of this chapter that there are many other ways in which diffusion and viscosity can be discussed. The multiplicity of theories that have been put forward in the past to describe these phenomena in simple liquids, most of them giving reasonable agreement with experiment, is somewhat confusing, so the treatment here is a highly selective one. Much of the early work was guided by the observation that log D is often – though by no means always – a linear function of $1/T$ at constant pressure. This suggested that diffusion in liquids was an *activated* process, as it is in solids, requiring the presence of 'holes' in the structure analogous to the vacancies that occur in solid crystals. Many attempts have been made to relate the apparent activation energy to the latent heat of evaporation or of melting, or to similar properties of the metals concerned. If this work is dismissed rather brusquely in §3.2, it is because the evidence accumulated by molecular dynamics in recent years is all against it. In chapter 2 we finally abandoned quasi-crystalline models of the structure of liquids in favour of an almost gas-like picture – though the density is necessarily much higher than in a normal gas and the pair distribution function is distinctly different from unity in consequence. Here, too, it is more profitable to draw analogies with gases than with solids.

An expression for the self-diffusion coefficient of a dense gas of rigid spheres can be established by elementary kinetic theory arguments, using the idea of a *friction coefficient* (discussed in §3.3) and some results of chapter 2 concerning $g(R)$. Naturally it involves the packing fraction y, but if this is set equal to 0.46 the theory is found to give an excellent description of the way in which D, measured just above the melting point, varies from one monatomic liquid to another; it works as well for metals as for Ar, say. It also describes the way in which the viscosity at the melting point varies (see §3.4) and hence the success of equation (3.1); only the Li isotopes (see §3.5) are out of line.

However, the simple kinetic theory approach overestimates D by a factor of roughly 1.4 and underestimates η by one of roughly 0.4. The fault seems to lie in the friction coefficient approach rather than in the rigid-sphere approximation. In §3.6 the foundation is laid for an exact theory of D, with the introduction of the auto-correlation function of molecular velocity,

$$\overline{\mathbf{v}(t+s)\cdot\mathbf{v}(t)}.$$

Computations by the method of molecular dynamics, the results of which are summarised in §§3.7 and 3.8, have shown that this function does not decay exponentially with s as was previously supposed, but tends to oscillate. Since it is the area underneath the velocity auto-correlation function curve which determines D, this tendency reduces the rate of diffusion. To calculate an appropriate correction factor from first principles is formidably difficult, but if we are prepared to take it from molecular dynamics we may reconcile the rigid-sphere model with experimental results for D at the melting point with remarkable success. To explain the variation of D with temperature above the melting point we must now appeal to the effects of thermal expansion, and of contraction of the rigid-sphere diameter σ, on the packing fraction y.

Molecular dynamics suggests that if we could determine not just the area under the velocity auto-correlation function curve but also its shape – or else its Fourier transform $Z(\omega)$ – we might begin to see significant differences between one liquid and another, associated with differences in the form of the pair interaction $w(R)$. In principle it is possible to measure $Z(\omega)$ by analysing the energy loss or gain of any form of radiation, especially slow neutrons, when scattered by the liquid. The necessary theory is a generalisation of the theory of elastic scattering in chapter 2 to situations where the molecular sites $\mathbf{r}_i$ are functions of time. Since it is not readily available elsewhere except in rather detailed treatments (e.g. Egelstaff, 1965), its essentials are developed in §§3.9 to 3.12, mainly in classical terms; the corrections required by quantum mechanics are of minor importance at the temperatures at which metals melt. Instead of the single pair distribution function $g(R)$ of chapter 2 we will find it necessary to introduce the two functions $G_s(R,t)$ and $G_d(R,t)$. The former describes the probability that a molecule initially at the origin will be found a distance R away after time t; if this is known for all R and t it is a trivial matter to evaluate quantities such as $\overline{\mathbf{v}(t+s)\cdot\mathbf{v}(t)}$. The latter describes the probability that some other molecule will be found there; $G_d(R,0)$ and $g(R)$ are identical.

The majority of metals are what are called *coherent* scatterers of neutrons; their cross-section for incoherent scattering is relatively small. This is somewhat unfortunate because it turns out that measurements of coherent inelastic scattering yield only the sum of

G_s and G_d. To separate these two functions is simple enough for very short times, less than about 10^{-12} sec say, because then they do not overlap in R. But the results for G_s only begin to be of interest for times longer than this, when the central molecule comes up against the walls of its cage, and overlap of G_s and G_d is then inevitable. A number of ingenious approximations have been suggested, but they are shown in §3.13 to be unreliable. Because of this difficulty of interpretation, and because of the considerable difficulty of making accurate measurements in the first place, the mouse of fact that emerges from the mountain of theory is disappointing.

There is another way in which to treat the results of coherent inelastic scattering experiments. Rather than subject the energy loss or energy gain spectrum to the elaborate analysis required if a curve for $G_s + G_d$ is to be derived, one may attempt to use the peaks in the spectrum to plot out dispersion curves for the normal modes of vibration of which the structure is capable. In a perfectly regular crystal, with perfectly harmonic forces between the molecules, the normal modes are of course sinusoidal waves, with a well-defined wave number q for given frequency Ω, and the spectrum consists entirely of sharp peaks, each of which corresponds to the emission or absorption of a *phonon*. In real crystals the forces are not perfectly harmonic, and normal modes in the strict sense of the term cannot be defined. As is well known, however, it is still useful to talk of phonons, though they must be regarded as scattering one another. As the temperature of the specimen is raised and the phonon density increases, the lifetime of each individual phonon becomes shorter, and the peaks in the energy spectrum for scattered neutrons become correspondingly broader. But even at the melting point these peaks are still distinct enough for phonon dispersion curves to be deduced.

In §3.14 we shall examine the results for liquid metals from this point of view, encouraged by the very marked similarity that has often been noted between the inelastic scattering of liquid specimens and of the corresponding polycrystalline solids before melting. The results obtained are sensible enough. It is surprising at first sight that the apparent dispersion curves repeat themselves in q in a manner that would be associated, for a solid specimen, with the occurrence of *umklapp* scattering, but we will see how a process akin to umklapp scattering is quite conceivable for a liquid, if the peaks in its interference function $a(q)$ are sharp enough. There is some evidence that not only longitudinal phonons may exist in liquids but transverse ones too. The digression in §3.15 concerning visco-elasticity arises out of this.

What difference does the motion of the ions make to the conduction electrons in a liquid metal? Throughout chapter 1 this motion was ignored, on the grounds that the Fermi velocity of the electrons is about 10^8 cm sec^{-1} whereas the r.m.s. velocity of an ion is usually less than 10^5 cm sec^{-1}. In §3.16 two very important formulae are developed, which are needed in chapter 4, for the *lifetime* τ_l and the *relaxation time* τ of the electrons at the Fermi surface. These times are limited by scattering by the ions, and the moment theorems developed in §3.12 may be used to show that the correction needed on account of ionic motion is indeed negligible, except in the case of Li.

The final section of this chapter, §3.17, concerns a different sort of relaxation time, the time for the nuclear spin system to reach equilibrium in the presence of a magnetic field. This time can be measured experimentally, and a comparison of the results obtained with different isotopes has made it clear that spin relaxation is in some cases assisted by an interaction involving the nuclear quadrupole moment. The metals which show this effect are all ones whose crystal structure in the solid phase is anisotropic: it is not detectable in Na, Cu, Al and so on, which form cubic crystals. There is evidence here, though of a tenuous kind, for some persistence in the liquid phase of the local order which characterises the solid. It is difficult to work out a full theory of the effect, since it requires a knowledge not only of the instantaneous three-body distribution function g_3 but of how this function changes with time. Order of magnitude calculations suggest that the three-body correlations persist in these liquid metals for 10^{-11} sec or more, which seems surprisingly long.

3.2. Diffusion as a random walk process

The self-diffusion coefficient is not an easy quantity to measure for a liquid and the results of different investigations on the same

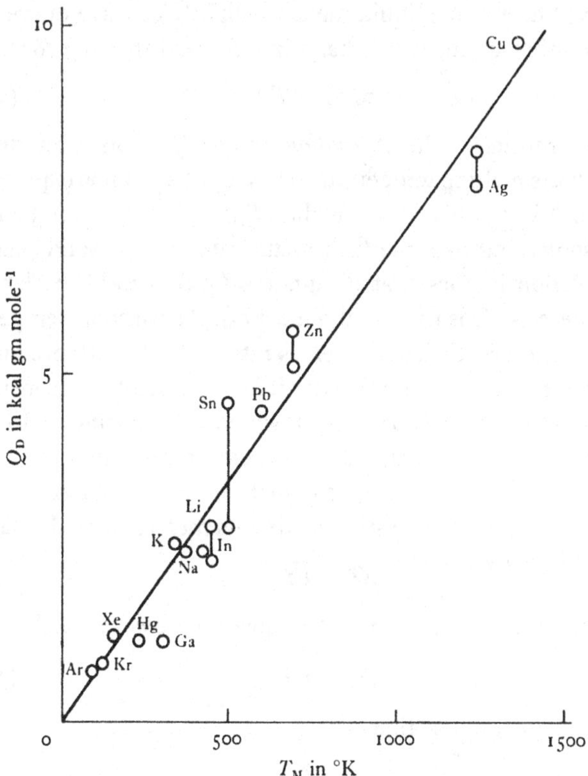

Fig. 3.1. The apparent activation energy for diffusion in liquids. (The data are quoted by Wilson, 1965; Naghizadeh & Rice, 1962; Ott & Lodding, 1965; Murday & Cotts, 1968; and Walls & Upthegrove, 1964. Points linked by a vertical line represent two different experimental values for a single metal.)

material have often been at variance. There seems to be agreement, however, on the following empirical facts (see Wilson, 1965, for references). Just above the melting point for metals and rare gases the magnitude of D lies in the range 10^{-5} to 10^{-4} cm^2 sec^{-1}. At constant pressure it increases rapidly with temperature. Analogy with solids, in which the Arrhenius formula

$$D = D_\infty \exp(-Q_D/Nk_B T) \tag{3.2}$$

normally holds with great precision, has suggested to most experimenters that their results should be presented by plotting $\log D$ versus $1/T$, and quite good straight lines are usually obtained in this way; but in some liquid metals a slightly concave curve fits the points rather better and it has been claimed that the formula

$$D \propto (T - T_0) \tag{3.3}$$

is more appropriate. The *activation energy* Q_D correlates quite closely with the melting temperatures (see fig. 3.1); if it is expressed in the form $Nk_B T_D$ then it seems that T_D is usually about $3.4T_M$, which is something like one fifth of its value in the solid phase. This correlation implies that if equation (3.3) is to be preferred then the intercept T_0 is usually about $0.7T_M$. At constant temperature the diffusion coefficient decreases rather slowly with pressure.

From the theoretical point of view it is often helpful to consider diffusion as a *random walk* process, in which the diffusing molecule takes a series of steps of length $\mathbf{l}_1, \mathbf{l}_2, \ldots$, etc. in random directions, taking a time $\tau_1, \tau_2, \ldots$, etc. to complete each one. After a total time t it has completed $(t/\bar{\tau})$ steps so that its mean square displacement should be given by

$$X^2 = \overline{l^2}t/\bar{\tau} \tag{3.4}$$

A standard solution of the diffusion equation, however, yields the result

$$X^2 = 6Dt, \tag{3.5}$$

so equation (3.4) implies that

$$D = \tfrac{1}{6}\overline{l^2}/\bar{\tau}. \tag{3.6}$$

In the case of a gas, where the molecules travel in straight lines between well-separated collisions, we may set $l = v\tau$ and

$$D = \tfrac{1}{6}\overline{l^2}\bar{v}/\bar{l} = \tfrac{1}{3}\bar{l}\bar{v}; \tag{3.7}$$

the fact that $\overline{l^2}$ is twice $\bar{l}^2$ is an elementary consequence of the exponential distribution of free paths in a gas.† Insertion of $(8k_B T/\pi m_A)^{\frac{1}{2}}$ for $\bar{v}$ in this formula and of experimental results for D in gases leads to values of the mean free path which are much longer than the inter-molecular spacing.

If equation (3.7) is applied to a solid, however, it yields a value of the mean free path which is grotesquely small – 10^{-15} cm or

† Equation (3.7) is not exact, because the direction of travel of a molecule after collision in a gas is not totally uncorrelated with its direction of travel just before. A more exact formula for rigid-sphere molecules is quoted as equation (3.10) below.

less. The reason, of course, is that although a molecule in a solid does possess a mean velocity of $(8k_BT/\pi m_A)^{\frac{1}{2}}$ it spends most of its time travelling to and fro within the same lattice cell. It can only diffuse by jumping to a new cell and this requires (*a*) that there be a vacancy there, and (*b*) that the molecule collect enough energy to surmount the intervening potential barrier. One may treat the diffusion step $\bar{l}$ in (3.6) and (3.7) as the lattice spacing, but the effective mean velocity with which a molecule covers each step is reduced below $(8k_BT/\pi m_A)^{\frac{1}{2}}$ by an exponential Boltzmann factor $\exp(-Q_D/Nk_BT)$; the activation energy per molecule, Q_D/N, is the sum of the free energy required to create the necessary vacancy and the height of the barrier. In a solid rare gas such as Ar the former should be roughly equal to the latent heat of vaporisation per molecule, L_V/N, and since Q_D/L_V in solid Ar is about 2 (Berne, Boato & de Paz, 1962) one may conclude that about half the activation energy is due to the barrier. In a typical solid metal Q_D/L_V is only about 0.3; the difference arises because the pair interaction in metals, in terms of which the energy for vacancy formation can be calculated (see §1.14), expresses only a part of the total energy and hence only a part of L_V.

If equation (3.7) is applied as it stands to a simple liquid just above its melting point – Ar or a typical metal – it yields a value for $\bar{l}$ of about one-tenth of the intermolecular spacing, which is close to the r.m.s. amplitude of thermal vibration in the solid just below the melting point (see table 2.6). It is therefore not implausible to treat diffusion in a liquid as a gas-like process, i.e. to suppose that each molecule executes a series of steps of about one-tenth of a molecular diameter between collisions, the direction of successive steps being more or less uncorrelated. But it is also possible to treat it as a solid-like process, i.e. to suppose that for 99% of the time the molecules step backwards and forwards without making any real progress but are liable to find, at the hundredth step, that a 'hole' has opened in front of them which is sufficiently large to enable them to advance ten times as far as usual, i.e. through a full molecular diameter. Belief in the validity of the Arrhenius equation for simple liquids has encouraged many authors to adopt the latter approach, though the former might turn out to be consistent with equation (3.3).

The fact that the activation energy is lower for the liquid than the corresponding solid is to be understood, according to the solid-like point of view, by supposing that (*a*) the barrier height is smaller, perhaps even negligible, and (*b*) that it costs less energy to open up the necessary hole in a liquid than it does to create a vacancy in a solid, because it need only involve a momentary rearrangement of the molecules, without necessarily any increase in the total volume that they occupy. It is arguable (Furukawa, 1960) that the latent heat of melting provides a better indication of the energy required for this sort of rearrangement than does the latent heat of vaporisation, so that, particularly for metals, it is more meaningful to compare Q_D in the liquid phase with L_M rather than L_V. The data for Q_D/T_M in fig. 3.1 combined with those for ΔS ($= L_M/T_M$) in table 2.1 reveal that a fair correlation between Q_D and L_M does exist; the ratio of these two quantities is in the neighbourhood of 3.

Cohen & Turnbull (1959) have constructed a theory of diffusion in liquids on the assumption that the *energy* of activation is zero; the *free energy* is not, however, because the entropy of a specimen which contains a hole big enough to permit diffusion is less than the entropy of a specimen in which the so-called free volume is uniformly distributed. The Cohen–Turnbull theory is often referred to in the literature and is not unsuccessful, though it cannot account for the smallness of the pressure coefficient of D (Nachtrieb & Petit, 1956). Like several of the ideas outlined in this section, however, it has been supplanted by later work.

3.3. The friction coefficient for rigid spheres

A somewhat different attitude to diffusion is associated with the name of Langevin; it is applied most frequently to the diffusion of macroscopic Brownian particles but its application to molecules is worth consideration.

A spherical particle suspended in a fluid medium is subjected to a fluctuating force, due to bombardment by the molecules of which the fluid is composed, whose mean value must be zero so long as the particle is at rest. But if the particle is moving with a uniform velocity v through the medium it is bombarded more in front than behind and hence experiences a net retarding force.

For small velocities we may take the average value of this force to be proportional to v; the constant of proportionality may be referred to as the *friction coefficient* and denoted by ζ. Now suppose that an assembly of such particles is allowed to sediment in a uniform field of some sort that exerts a force F on each particle in the x direction. Clearly the whole assembly will tend to drift in this direction with a mean velocity F/ζ. On the other hand, once a concentration gradient has been established there will be diffusion in the reverse direction at a rate governed by Fick's law. Equilibrium, which is essentially a kinetic equilibrium, requires

$$nF/\zeta = D\,\mathrm{d}n/\mathrm{d}x,$$

where n is the local concentration of particles. But since we know that in equilibrium

$$n \propto \exp\,(Fx/k_{\mathrm{B}}T)$$

it follows that

$$D = k_{\mathrm{B}}T/\zeta. \tag{3.8}$$

From this point of view a calculation of the diffusion constant hinges on the calculation of ζ.

It is a straightforward exercise in kinetic theory to calculate ζ for a rarefied gas, if we may treat the molecules as smooth rigid spheres of diameter σ and if the diffusing particle is itself a molecule, indistinguishable from its neighbours. Details may be found e.g. in Guggenheim (1960). The answer turns out to be

$$\zeta = \frac{8N\sigma^2}{3\Omega}(\pi m_{\mathrm{A}} k_{\mathrm{B}} T)^{\frac{1}{2}}. \tag{3.9}$$

If the gas is not rarefied, we must remember that the frictional force depends upon the density of neighbours in immediate contact with the diffusing molecule, i.e. we must multiply the right-hand side of (3.9) by $g(\sigma)$, where g is the appropriate pair distribution function. Hence the theory suggests

$$D = \frac{3\Omega}{8N\sigma^2 g(\sigma)}(k_{\mathrm{B}}T/\pi m_{\mathrm{A}})^{\frac{1}{2}}, \tag{3.10}$$

a result which has been derived by Chapman & Enskog (see Guggenheim, 1960, ch. 11), Longuet-Higgins & Pople (1956) and others.

TABLE 3.1. *Self-diffusion coefficients at the melting point*

	D Expt ($\text{cm}^2\ \text{sec}^{-1} \times 10^{-5}$)	D Theory ($\text{cm}^2\ \text{sec}^{-1} \times 10^{-5}$) Eq. (3.12)	Eq. (3.29)
Ar	1.8	2.25	1.6
Li	6.1	9.85	7.1
Na	4.2	5.95	4.3
K	3.7	5.4	3.9
Zn	2.0	3.5	2.55
Hg	0.93	1.33	0.97
Ga	1.6	2.4	1.7
In	1.6	2.5	1.8
Sn	2.0	2.7	1.95
Pb	2.2	2.3	1.7

Now it may be shown (e.g. Rice & Gray, 1965, p. 115) that for non-attracting rigid spheres

$$g(\sigma) = \frac{1}{4y}\left(\frac{p^0\Omega}{Nk_B T} - 1\right), \tag{3.11}$$

where y is the packing-fraction. If we take p^0 from the empirical equation (2.20) which has been found to represent so closely the results of molecular dynamics, we obtain

$$D = \frac{1}{16}\left(\frac{\pi k_B T}{m_A}\right)^{\frac{1}{2}}\left(\frac{6\Omega}{\pi N y^2}\right)^{\frac{1}{3}}\frac{(1-y)^3}{(1-\frac{1}{2}y)}. \tag{3.12}$$

Here is an ostensibly exact result, which should describe the self-diffusion coefficient of any liquid whose structure conforms to the rigid sphere model. With $y = 0.46$ – the reasons for choosing this value or one close to it are given in §§2.7 and 2.14 – it turns out in practice to predict self-diffusion coefficients which are about 40% higher than those observed at the melting point for a number of liquid metals and rare gases. The comparison is effected in table 3.1, for a number of liquids for which recent, and probably reliable, experimental data are available.

That (3.12) is not quite so exact as may appear is indicated by its complete failure in the solid phase; it obviously does not describe the dramatic fall in D which occurs when liquids crystallise. The error of principle in its derivation is exposed in §3.6 below, and a more reliable formula is suggested in §3.7. Further discussion of the temperature dependence of D above the melting point may be deferred till then.

3.4. A rigid-sphere theory of viscosity

When a liquid is subjected to a uniform shear stress a velocity gradient is set up, whose magnitude depends upon the viscosity η. In a dense fluid we may expect the rate at which layers of molecules slip over one another to be limited by the same friction constant that limits the rate at which individual molecules diffuse, so that η is likely to be proportional to ζ and hence to D^{-1}. The constants of proportionality have been worked out by Longuet-Higgins & Pople (1956) for a dense fluid of non-attracting rigid spheres, using elementary kinetic theory arguments. Theirs is the best simple theory that is currently available for comparison with experimental results for liquid metals, though it is only partially successful as we shall see.

It is more convenient to start with a given velocity gradient $\partial V_x/\partial z$ and to calculate the shear stress which it generates across the xy plane, than to start with the stress and calculate the velocity gradient. In a rarefied gas the stress that is exerted directly, by means of collisions between molecules whose centres lie on opposite sides of the plane, is negligible; viscosity is entirely due to the transport of individual molecules across the plane, carrying drift momentum with them. The higher the rate of transport the faster drift momentum is transmitted across the plane, which is why the viscosity increases on heating. In liquids, however, the viscosity decreases on heating. A different mechanism for stress transmission is clearly at work. Longuet-Higgins & Pople make the assumption that near the melting point, where the packing fraction is as high as 0.46, it is only the direct collisions that matter.

The only molecules below the xy plane – or rigid spheres in the model we are discussing – which are in a position at any instant to collide with spheres above it are those whose centres lie in the range $-\sigma < z < 0$. There are $(N\sigma/\Omega)$ of these per unit area. In any one collision the drift velocities of the two spheres involved will differ, on the average, by something of the order of $\sigma(\partial V_x/\partial z)$. In the light of these statements it is easy to accept Longuet-Higgins & Pople's result that the stress per unit area is

$$\frac{1}{10}\left(\frac{N\sigma}{\Omega}\right)\zeta\sigma\left(\frac{\partial V_x}{\partial z}\right),$$

where ζ is given by (3.9), though careful averaging is evidently required to establish the numerical coefficient. Hence

$$\eta = \frac{3y}{5\pi\sigma}\zeta. \tag{3.13}$$

A rather similar answer may be obtained, incidentally, by using Stokes's law, which says that a macroscopic sphere of diameter σ, moving with uniform velocity v through a homogeneous viscous medium, experiences a drag force of magnitude $3\pi\eta v\sigma$. There is no good reason to suppose, *a priori*, that Stokes's law is adequate to describe the drag force experienced by one of the rigid spheres of which our inhomogeneous model fluid is composed. Taken at its face value, however, it suggests

$$\eta = \frac{1}{3\pi\sigma}\zeta,$$

which comes to much the same thing as (3.13) when y is 0.46.

If we take ζ from (3.9) and incorporate the $g(\sigma)$ correction, (3.13) becomes

$$\eta = \eta_{\mathrm{LP}} = 1.2\times 10^{-4}(MT)^{\frac{1}{2}}\Omega^{-\frac{2}{3}}\frac{y^{\frac{4}{3}}(1-\frac{1}{2}y)}{(1-y)^3}\ \text{poise}, \tag{3.14}$$

where M and Ω are the gram atomic weight and gram atomic volume, the latter in cm^3, while T is the temperature in °K. It is of interest to compare this answer with the standard one for a gas, in which the stress is transmitted entirely by individual molecules. As shown e.g. by Guggenheim (1960),

$$\begin{aligned}\eta_{\mathrm{Gas}} &= 0.31m_{\mathrm{A}}(\sqrt{2}\pi\sigma^2)^{-1}(8k_{\mathrm{B}}T/\pi m_{\mathrm{A}})^{\frac{1}{2}}\\ &= 7.8\times 10^{-6}(MT)^{\frac{1}{2}}\Omega^{-\frac{2}{3}}y^{-\frac{2}{3}}\ \text{poise}\end{aligned} \tag{3.15}$$

in the same units as (3.14). The ratio between the two predictions

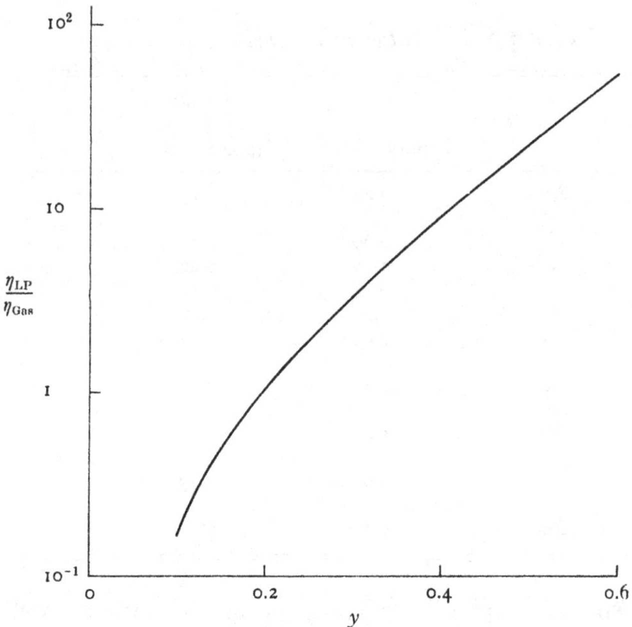

Fig. 3.2. Ratio between the viscosities predicted by (3.14) and (3.15) as a function of the packing fraction.

varies with y in the manner shown in fig. 3.2. For $y = 0.46$ it is about 16, and this looks large enough compared with unity to justify Longuet-Higgins & Pople's initial assumption.

Nevertheless, (3.14) seems to underestimate η at the melting point by a factor of about 0.4. With $y = 0.46$ it becomes

$$\eta = 2.1 \times 10^{-4} (MT_M)^{\frac{1}{2}} \Omega^{-\frac{2}{3}} \text{ poise,}$$

whereas the experimental results are well described by the formula

$$\eta = 5.1 \times 10^{-4} (MT_M)^{\frac{1}{2}} \Omega^{-\frac{2}{3}} \text{ poise}$$

which Andrade (1934) originally proposed, on the basis of arguments that now seem unconvincing. The failure of (3.14) is further exposed in table 3.2.

The rigid-sphere theory of Longuet-Higgins & Pople implies that $(\eta D/T)$ should vary like (y/σ), i.e. it should fall perceptibly on heating above the melting point. In practice it seems to be more or less independent of T for simple liquids, and in some cases it has

TABLE 3.2. *Viscosities at the melting point*

	Expt (poise × 10^{-2})	Theory eq. (3.14) (poise × 10^{-2})
Ar	0.29	0.13
Li	0.60	0.21
Na	0.70	0.22
K	0.54	0.19
Rb	0.67	0.22
Cs	0.69	0.24
Cu	4.1	1.6
Ag	3.9	1.5
Au	5.4	2.2
In	1.9	0.75
Sn	2.1	0.8

been shown to be almost independent of pressure as well (Nachtrieb & Petit, 1956).

Longuet-Higgins & Pople extended their theory, incidentally, to the thermal conductivity κ. Their answer may be expressed in the form

$$\eta/\kappa = 2m_A/5k_B, \tag{3.16}$$

and it agrees rather well with experimental data for liquid Ar; the predicted and observed values for η/κ are 1.9 and 2.1×10^{-7} gm°K erg^{-1} respectively. It cannot be tested for liquid metals because the ionic contribution to the thermal conductivity, which (3.16) should describe, is swamped by a much larger electronic term.

3.5. Isotope effects in lithium

The dependence of D and η on isotopic mass has been studied recently in liquid Li, and the curious results deserve a mention here.

There is no reason to suppose that the forces of interaction between the ions are any different in Li^6 and Li^7, for the lattice parameters in the solid phase are virtually identical for the two isotopes and so are the elastic moduli (Covington & Montgomery, 1957; Robertson & Montgomery, 1960). The difference of ionic mass, however, should affect any property that depends upon the motion of the ions. The Debye temperature in the solid phase, for example, should be proportional to $m_A^{-\frac{1}{2}}$, and this is perfectly consistent with observations of specific heat and electrical resistivity (Martin, 1959; Dugdale, Gugan & Okumara, 1961) and with the fact that the melting temperatures of the two isotopes are almost the same. Classically, the diffusion coefficient should likewise be proportional to $m_A^{-\frac{1}{2}}$; equation (3.12) shows this dependence, but it can be predicted on dimensional grounds, if Planck's constant plays no role in the analysis, without invoking any particular model for the diffusive process (see also Brown & March, 1968). Diffusion coefficients for Li^6 and Li^7 have been measured by McCracken & Love (1960) and Naumov & Ryskin (1965) for diffusion through solid W and Na, and the results are in the expected ratio of $\sqrt{(\frac{7}{6})} = 1.08$. Yet according to Lodding, Mundy & Ott (1970) and Löwenberg & Lodding (1967) the ratio between the diffusion coefficients for Li^6 moving through pure Li^7 and for Li^7 moving through pure Li^6 is significantly greater: it is 1.37 in the solid phase at 0.7 T_M, 1.23 just below T_M, 1.4 in the liquid phase just above it, and about 1.2 at $1.25T_M$. A similar anomaly is shown by the viscosity, which should be proportional to $m_A^{\frac{1}{2}}$; Ban, Randall & Montgomery (1962) report that at T_M the viscosity of Li^7 exceeds that of Li^6 by a factor of 1.44.

To explain these results it is clearly essential to invoke quantum mechanics in some way, and it is suggested that the process whereby Li ions diffuse involves tunnelling through potential barriers and not just thermal excitation over them; quantum-mechanical tunnelling is bound to be much easier for a light element like Li than for a heavy one, of course, and is bound to favour the diffusion of the lighter isotope. If tunelling is important for Li but not for other metals, however, why does it not stand out as anomalous in tables 3.1 and 3.2? This question has not yet been answered.

3.6. The velocity auto-correlation function

In §3.2 the self-diffusion coefficient was expressed in terms of a mean free path between collisions. In §3.3 it was expressed in terms of a friction coefficient. Here we explore yet a third approach, in which the key role is played by the *auto-correlation function* of molecular velocity, defined as

$$\overline{\mathbf{v}(t_1).\mathbf{v}(t_2)} = \overline{v^2}\psi(s), \tag{3.17}$$

where $\mathbf{v}(t_1)$ and $\mathbf{v}(t_2)$ are the velocities of a given molecule at two times such that

$$t_2 - t_1 = s.$$

The bar represents a time average over the history of the molecule, i.e. over all values of t_1. Because it has been subjected to this average the autocorrelation function depends upon the interval s but not upon t_1 and t_2 separately. The function $\psi(s)$ is normalised in such a way that it goes to unity as s goes to zero.

If $\mathbf{R}(t)$ is the displacement of a particular molecule during the interval t, then

$$\mathbf{R}(t)\cdot\mathbf{R}(t) = \int_{t_0}^{t_0+t} \mathbf{v}(t_1)\,\mathrm{d}t_1 \cdot \int_{t_0}^{t_0+t} \mathbf{v}(t_2)\,\mathrm{d}t_2.$$

The domain of integration for this double integral is a square in the $t_1 t_2$ plane, but it is most easily evaluated by considering separately the two halves of the square, for which $t_2 > t_1$ and $t_2 < t_1$ respectively. In terms of the variable s, the integral over the first half may be written as

$$\int_0^t \mathrm{d}s \int_{t_0}^{t_0+t-s} \mathrm{d}t_1\,\mathbf{v}(t_1)\cdot\mathbf{v}(t_2),$$

and the contribution due to the second half is clearly identical for symmetry reasons. Hence an average over t_0 yields

$$\begin{aligned} X^2 = \overline{\mathbf{R}(t)\cdot\mathbf{R}(t)} &= 2\overline{v^2}\int_0^t (t-s)\,\psi(s)\,\mathrm{d}s \\ &= (6k_\mathrm{B}T/m_\mathrm{A})\int_0^t (t-s)\,\psi(s)\,\mathrm{d}s. \end{aligned} \tag{3.18}$$

For very small values of t this reduces to

$$X^2 = 3k_\mathrm{B}Tt^2/m_\mathrm{A}, \tag{3.19}$$

a result which is self-evident from the classical theorem of equipartition, which determines the mean square molecular velocity.

For very long times, in a diffusing system such that no velocity correlations can persist indefinitely, it becomes

$$X^2 = (6k_B T/m_A)t \int_0^\infty \psi(s)\,ds - (6k_B T/m_A)\int_0^\infty s\psi(s)\,ds. \quad (3.20)$$

It is now evident, by comparison of (3.20) with (3.5) that the diffusion coefficient is given in terms of ψ by

$$D = (k_B T/m_A)\int_0^\infty \psi(s)\,ds, \quad (3.21)$$

while the friction coefficient required to satisfy (3.8) is given by

$$\zeta = m_A\left[\int_0^\infty \psi(s)\,ds\right]^{-1}. \quad (3.22)$$

The calculation outlined in §3.3 tells us only the instantaneous value of the mean retarding force experienced by a molecule whose velocity is v, so the friction constant derived by Longuet-Higgins & Pople, for example, is

$$\zeta_{LP} = m_A \operatorname*{Lt}_{s\to 0} \frac{d}{ds}\psi(s). \quad (3.23)$$

It is readily shown that if the decay of ψ with time is *exponential* then these two friction coefficients are identical. Incidentally, the solution of (3.18) in this case is

$$X^2 = 6D(\tau_L \exp(-t/\tau_L) - \tau_L + t), \quad (3.24)$$

where τ_L ($= m_A D/k_B T$) is the time-constant of the decay; this is known as the *Langevin formula* for the mean square displacement. But if the decay is not exponential, then (as Longuet-Higgins & Pople were perfectly aware) ζ_{LP} is not necessarily the same quantity as ζ and their calculation of D is not necessarily correct.

It is not hard to see what conditions must be satisfied, in a system of rigid spheres, for an exponential law to be obeyed: the *average* environment of the diffusing sphere, which we know from statistical mechanics to be isotropic at time t_1, irrespective of the magnitude and direction of $\mathbf{v}(t_1)$, must remain isotropic while the memory of $\mathbf{v}(t_1)$ decays; the average density of spheres in contact with the diffusing one must remain proportional to $g(\sigma)$. But is it possible, in a dense medium, for the neighbouring spheres to relax their configuration at the requisite speed? If they are not able to get out of the way of the moving sphere at the centre it will start to experience an anisotropic environment, on the average, and the retarding force upon it will presumably increase; then $\psi(s)$ will decay more rapidly.

Curiously enough the decay of $\psi(s)$ cannot be exponential even in its early stages for a fluid of real molecules as opposed to rigid spheres. This is because we can no longer treat the interaction between two molecules as an impulse of infinitesimal duration, as Longuet-Higgins & Pople did. Instead we should calculate the retarding force from the gradient of a potential, to be obtained by summing the pair interaction $w(\mathbf{R}-\mathbf{R}_i)$ over the positions $\mathbf{R}_i$ which are instantaneously occupied by neighbouring molecules. If the environment is isotropic on the average at t_1 then the mean potential at this instant must also be isotropic and its gradient is bound to vanish. Hence $\psi(s)$ falls off no faster than s^2 for small s. The fact that it decays at all, from this point of view, depends essentially on the development of some anisotropy in the environment. Presumably an exponential decay law may still be a reasonable approximation in the limit of low densities; this seems to be the implication of the fact that equation (3.10) gives the right answer for a gas when $g(\sigma)$ is replaced by unity.† But when the density is high and relaxation of the surrounding molecules becomes difficult, there will be an increasing tendency for the molecule to oscillate to and fro in a mean potential well whose centre remains anchored to the origin at which the molecule started at time t_1 rather than following it to $\mathbf{R}(t_2)$.

The extreme case is provided by a solid in which no relaxation of the surrounding molecules is possible and the central molecule is condemned to oscillate in the same cell indefinitely. If the oscillations are described with the aid of a Debye model, it is readily shown that

$$\psi(s) = \frac{3}{\omega_D^3}\int_0^{\omega_D} \omega^2 \cos(\omega s)\, d\omega$$

$$= 3\left(\frac{\sin \omega_D s}{\omega_D s} + 2\frac{\cos \omega_D s}{(\omega_D s)^2} - 2\frac{\sin \omega_D s}{(\omega_D s)^2}\right), \qquad (3.25)$$

† At least, it gives the answer derived by Chapman and Enskog's 'first approximation'. Higher approximations in the Chapman–Enskog theory do reveal the existence of small additional terms (see Chapman & Cowling, 1939).

where ω_D is the cut-off frequency ($= k_B\Theta_D/\hbar$). Corresponding to this, the solution to (3.18) is

$$X^2 = \frac{18k_BT}{m_A\omega_D^2}\left(1-\frac{\sin\omega_D t}{\omega_D t}\right), \tag{3.26}$$

which reduces to (3.19) for small t as it must but saturates for large t to twice† the value of X^2 described by equation (2.32), i.e. to about $0.087R_A^2$ in a solid just below its melting point or say 0.2 Å². To incorporate vacancy diffusion into the picture, we need only recognise that there will be moments during the lifetime of a molecule in a solid crystal when the cage of neighbours surrounding it is not complete and when it can continue to travel in a straight line, for a time of order l/v, before it starts once more to oscillate. When (3.25) is modified to take account of this, the integral

$$\int_0^\infty \psi(s)\, ds$$

no longer vanishes but is of order $(l/v)\exp(-Q_D/Nk_BT)$, where Q_D has the significance described in §3.2.

The task of calculating a reliable value of D for any real liquid is obviously one of very great complexity. It is one thing to set up plausible formulae for $\psi(s)$ (see §3.8 below); it is quite another to determine the adjustable parameters in these formulae by reference to other known properties of the liquid concerned. Several ingenious arguments that do start from first principles have been suggested (consult Rice & Gray, 1965), leading to a variety of answers for ζ and D in terms of $w(R)$ and $g(R)$, but none of them seems really convincing. Scarcely any of them, apart from the relaxation model of Gray (1964), allow $\psi(s)$ to oscillate with s, though oscillations seem highly probable on physical grounds and, as we shall see in the following section, there is good evidence that they occur. No model allows for the possibility that a small group of molecules contribute more than their share to D because they can take advantage of temporary 'holes' in the liquid, like the molecules which diffuse in solids. The results obtained, after a

† The factor 2 arises because (2.32) describes the displacement of a molecule from the centre of its cell, whereas (3.26) describes its displacement from the position it occupied at $t = 0$.

good deal of effort represent a rather meagre improvement upon equation (3.12).

Starting from first principles the calculation of the viscosity or thermal conductivity for a real liquid is every bit as difficult as the calculation of D. It may be possible, however, to extract a value for the friction coefficient ζ from a measurement of D, and to use it in conjunction with some theoretical model to predict η or κ. Fair success has been achieved in this way by Rice & Allnatt in the case of liquid Ar (see Rice & Gray, 1965, p. 364).

3.7. The back-scatter correction for rigid spheres

Direct observation of the motion of molecules in simple model liquids, using the methods of molecular dynamics, has illuminated several of the problems discussed above. Given a simple law to describe the pair potential between the molecules, the computer may be programmed to determine how their mean square displacement X^2 varies with time t. For large t the curve extrapolates to a straight line

$$X^2 = 6Dt + \text{constant} \tag{3.27}$$

and the diffusion coefficient for the model may be obtained from the slope. If the Langevin theory were correct the constant in (3.27) would be negative; if it turns out to be positive instead one may deduce immediately from (3.20) that

$$\int_0^\infty s\psi(s)\,\mathrm{d}s \Big/ \int_0^\infty \psi(s)\,\mathrm{d}s$$

is negative, and hence that ψ must somewhere reverse in sign. In fact it is easy to deduce the form of ψ from the curvature of the X^2 plot, since it is apparent from (3.18) that

$$\psi(t) = (m_\mathrm{A}/6k_\mathrm{B}T)\,\mathrm{d}^2X^2/\mathrm{d}t^2. \tag{3.28}$$

The model which has been investigated most thoroughly, by Rahman (1964*a*, 1966) and Levesque & Verlet (1970), is one in which the molecules interact via a Lennard–Jones potential of the sort that is believed to be appropriate in liquid Ar, but a few results have also been reported for rigid spheres (Alder & Wainwright, 1967) and for oscillatory potentials of the sort that may be appropriate for liquid metals (Paskin & Rahman, 1966; Schiff,

1969). It seems to be a common feature of all the calculations that for densities which correspond to real liquids at normal temperatures and pressures the constant in (3.27) *is* positive, i.e. that ψ *does* reverse in sign. This is sometimes described as the *back-scatter effect*. As the density is diminished it becomes less pronounced, until at gaseous densities the decay of ψ is virtually exponential.

It is almost self-evident from (3.21) that back-scatter will reduce the diffusion coefficient of a liquid below the value to be expected from the theory in §3.3. Alder & Wainwright have computed just how severe the reduction is for a rigid sphere fluid, as a function of the packing fraction y. It amounts to a factor of 0.72 for $y = 0.46$ and is roughly proportional to y^{-1} in this neighbourhood. So within the framework of the rigid-sphere model we may allow for back-scatter by adding a correction factor of $(0.33y^{-1})$ to equation (3.12), which then becomes

$$D = 4.9\times10^{-6}M^{-\frac{1}{2}}T^{\frac{1}{2}}\Omega^{\frac{1}{3}}\{(1-y)^3/y^{\frac{5}{3}}(1-\tfrac{1}{2}y)\}\ \text{cm}^2\,\text{sec}^{-1} \qquad (3.29)$$

(with T in °K and Ω in cm³, as in (3.14)). This result, with $y = 0.46$, is compared with experimental data at the melting temperature in table 3.1 above. The agreement is remarkably good, for liquid Ar as well as for liquid metals.

Ascarelli & Paskin (1968) and Vadovic & Colver (1970*a*,*b*) have extended the theory to temperatures above the melting point and have secured an excellent fit with experimental data, over a wide range of T, for an impressive list of metals. Unfortunately, in seeking to eliminate y from (3.29), they make an additional assumption which is impossible to justify, namely that $(\partial\mathscr{U}/\partial\Omega)_T$ is inversely proportional to Ω^2. Granted this assumption one may use the theory outlined in §2.7 to rewrite (3.29) as follows:

$$D \propto \frac{T^{\frac{1}{2}}\Omega^{\frac{1}{3}}}{y^{\frac{2}{3}}}\left(\frac{p^0\Omega}{Nk_BT}-1\right)^{-1} \qquad \text{(see (3.11))}$$

$$\propto \frac{T^{\frac{1}{2}}\Omega^{\frac{1}{3}}}{y^{\frac{2}{3}}}\left(\left(\frac{\partial\mathscr{U}}{\partial\Omega}\right)_T\frac{\Omega}{Nk_BT}-1\right)^{-1}$$

$$\propto \frac{T^{\frac{1}{2}}\Omega^{\frac{1}{3}}}{y^{\frac{2}{3}}}\left(9.4\frac{\Omega_M T_M}{\Omega T}-1\right)^{-1}.$$

It is clear from §2.7, however, that whatever may be the case for liquid Ar the quantity $(\partial\mathscr{U}/\partial\Omega)_T$ *increases* with volume for liquid metals, and quite rapidly too. If the formula is modified accordingly, the agreement with experiment is gravely impaired. One may see this more directly, perhaps, by making the assumption that y varies with temperature like $T^{-\nu}$ and differentiating (3.29) to show that

$$\frac{T_D}{T_M} = -\frac{1}{T_M}\frac{d\log D}{d(1/T)}, = 0.5+3.9\nu+0.33\alpha T_M, \qquad (3.30)$$

where T_D is the activation temperature introduced in §3.2. Experimentally T_D/T_M is always close to 3.4, and one has to choose $\nu \simeq 0.85$ to match this figure. The values suggested by the thermal expansion data which Ascarelli & Paskin have attempted, indirectly, to make use of are in fact only about 0.2 (see p. 103). The arguments in §1.14 and §2.15 favour values in the neighbourhood of 0.3. It seems that, if the crude rigid-sphere model is to be extended to high temperatures by the device of allowing σ and y to fall on heating, there is no single value of ν which will reconcile all the observations.

The reappearance of T_D in (3.30) may suggest that we have somehow reverted to the idea that diffusion is an activated process. This is not the case. If the temperature dependence of D is mainly determined by the variation of y within the function $(1-y)^3/y^{\frac{2}{3}}(1-\frac{1}{2}y)$, then it must be largely accidental that plots of $\log D$ versus $1/T$ are often straight lines. Together with the concept of activation, of course, we have also finally abandoned the idea that 'holes' in the liquid structure are essential before diffusion can occur. Rahman has noted that, according to his computations, molecules whose environment is initially anisotropic tend to move subsequently along the direction in which the Voronoi polyhedron† is elongated. It is conceivable, therefore, that a small fraction of the molecules at any instant are taking advantage of local irregularities in the structure to travel rather further than usual before making collisions and there may be some enhancement of D in consequence; perhaps this explains the interesting result of Alder & Wainwright for rigid-sphere fluids that at low densities (small values of y), where back-scatter is relatively unimportant, equation

† A footnote on p. 78 explains what Voronoi polyhedra are.

(3.12) actually *under*estimates rather than overestimates D. But we are clearly a very long way from the vacancy diffusion mechanism that operates in solids.

It would of course be of great interest to use molecular dynamics to correct the rigid-sphere theory of viscosity (see §3.4), just as we have used it here to correct the theory of diffusion. But the problem of computing viscosities is a hard one and it has not yet been tackled successfully. Vadovic & Colver (1971) have extended their theory of the temperature dependence of D so as to include the temperature dependence of η, but it remains unconvincing.

3.8. The frequency spectrum of molecular motion in liquids

Molecular dynamics has more to tell us about diffusive motion in model liquids than just the back-scatter correction to the diffusion coefficient. To calculate D requires only the *area* under the $\psi(s)$ curve, but its *shape* has also been computed, especially for model 'Ar'. Fig. 3.3 shows some of Rahman's (1964*a*) results.

Many authors prefer to discuss the frequency spectrum of ψ rather than ψ itself. This is defined by

$$Z(\omega) = \frac{2}{\pi}\int_0^\infty \psi(t) \cos \omega t \, dt, \tag{3.31}$$

or else by
$$Z(\omega) = -\left(\frac{m_A\omega^2}{3\pi k_B T}\right)\int_0^\infty X^2(t) \cos \omega t \, dt; \tag{3.32}$$

the equivalence of these two formulae may be demonstrated by integrating the second one twice by parts, making use of (3.28). The fact that $\psi(0)$ is unity means that

$$\int_0^\infty Z \, d\omega = 1,$$

while it follows from (3.21) that

$$Z(0) = 2m_A D/\pi k_B T. \tag{3.33}$$

The spectrum for model 'Ar' near its melting point is plotted in fig. 3.4 and compared with the curves to be expected for (*a*) an equivalent Debye solid and (*b*) a system which obeys the Langevin theory, i.e. for which the decay of ψ is exponential. The oscillatory

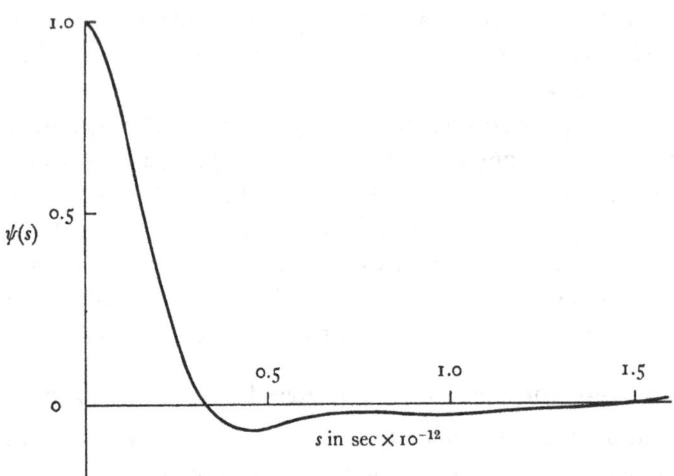

Fig. 3.3. The velocity auto-correlation function for liquid 'Ar' at 94.4 °K according to Rahman.

character of ψ which is apparent in fig. 3.3, for high densities, is naturally reflected by a peak in $Z(\omega)$, but it is a much broader one than in the solid. Efforts have been made to fit the computed spectrum to theoretical models, and the most successful seems to be the *itinerant oscillator* model discussed by Sears (1965) and subsequently elaborated by Nakahara & Takahashi (1966), Damle, Sjölander & Singwi (1968), and Fukui & Morita (1970); the model devised by Berne, Boon & Rice (1966) and Desai & Yip (1968, 1969) has fewer parameters but fits less closely.

Sears's results are based on solution of two coupled equations of motion

$$\ddot{\mathbf{r}}_0 + \mu\dot{\mathbf{r}}_0 + \omega_0^2(\mathbf{r}_0 - \mathbf{r}) = \mathbf{A}(t)$$

$$\ddot{\mathbf{r}} + \nu\dot{\mathbf{r}} = \mathbf{B}(t),$$

where $\mathbf{r}_0$ is the position of the diffusing molecule and $\mathbf{r}$ presumably defines the centre of gravity of a cluster of molecules with the diffusing one at its heart. The two stochastic forces $\mathbf{A}$ and $\mathbf{B}$ represent the random collisions which keep the system in motion, while μ and ν are friction coefficients. It is necessary to define not only the r.m.s. amplitude of $\mathbf{A}$ and $\mathbf{B}$ but also a correlation time for each of them, so that there are seven parameters in the theory altogether, though the list may be reduced to five by invoking the

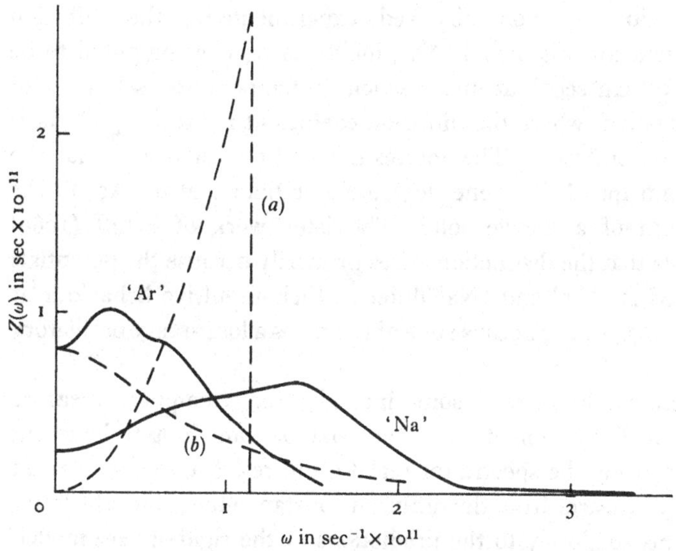

Fig. 3.4. The frequency spectrum for 'Ar' corresponding to the curve for ψ in fig. 3.3, and a spectrum for liquid 'Na' due to Paskin & Rahman. The two broken curves for 'Ar' correspond (a) to the equivalent Debye solid ($\Theta_D = 85$ °K), (b) to Langevin diffusion.

fluctuation–dissipation theorem which links μ to A and ν to B. It is ν which primarily determines the magnitude of D.

The physical content of this model is the idea that each molecule behaves like a damped Einstein oscillator within the cage constituted by its neighbours and only diffuses because the cage is drifting in a Langevin-like fashion. The predicted $Z(\omega)$ naturally consists of a Langevin background with a peak superimposed upon it, centred about the frequency ω_0. Unfortunately, the fact that the theory can be matched with reasonable accuracy to the computed curve for 'Ar' by adjustment of the parameters is no guarantee that the physical picture is correct. To the present author it appears more simple than is altogether plausible.

Included in fig. 3.4 is a frequency spectrum computed by Paskin & Rahman (see Paskin, 1967) for model 'Na', i.e. using an oscillatory potential resembling the one suggested for real Na by the early work of Johnson *et al.* (see p. 145) but adapted so that the pair distribution function generated with its aid, using the method of molecular dynamics, is in reasonable agreement with the pair distribution function observed experimentally; the diffusion coefficient for this model 'Na', incidentally, was computed to be $4.9 \times 10^{-5}\,cm^2 sec^{-1}$ at the particular temperature selected for investigation, where the diffusion coefficient for real liquid Na is $5.8 \times 10^{-5}\,cm^2 sec^{-1}$. The interesting point to notice is that the spectrum for 'Na' is one degree closer than that of 'Ar' to the spectrum of a Debye solid. The later work of Schiff (1969) suggests that the distinction arises primarily because the potentials assumed for 'Ar' and 'Na' differ in their repulsive behaviour at short range, rather because one of them has a long-range oscillatory tail.

These results provide some incentive for attempts to measure X^2, ψ or Z experimentally. There may be something to be learnt by comparing the spectra for real Ar and real Na, say, which can scarcely be learnt from the diffusion constant alone, since the latter conforms so closely to the predictions of the rigid-sphere model. Ideally, the measurements may be made with slow neutrons, and it is the object of the next few sections to expound the theoretical principles involved. It may be admitted from the start, however, that the information obtained to date is rather unhelpful. Values of $X^2(t)$ for times up to 4×10^{-12} sec have been deduced for Ar by Dasannacharya & Rao (1965), and estimates for liquid Na and Pb are available from the work of Randolph (1964) and Brockhouse & Pope (1959) respectively. Unfortunately, the results do not seem to be sufficiently accurate to be useful. The points for Ar lie quite close to a line through the origin with a slope of $6D$; it needs an eye of faith to detect evidence for any curvature at low t, or even to perceive that for large t the points converge onto an asymptote whose intercept on the X^2 axis is positive rather than negative. As for Na and Pb, the data suggest that X^2 starts to fall seriously behind $6Dt$ at times of order 10^{-12} sec, in a manner that is quite unexpected from the computations. This departure is almost certainly a sign of gross error, either in the measurements or in the analysis to which they were subjected; the methods of analysis are described and criticised below. More recent work on Na by Egelstaff (1966) and Cocking (1969) does suggest a rather solid-like spectrum for $Z(\omega)$ but here again the accuracy of the final curves is very uncertain.

3.9. Inelastic scattering

Throughout chapters 1 and 2 the thermal motion of the molecules in liquids was ignored, and the scattering of X-rays, neutrons and conduction electrons was in consequence treated as elastic. In practice the molecular velocities in thermal equilibrium are bound to cause Doppler shifts in the frequency of the scattered radiation such that

$$\Delta\nu/\nu \sim \pm(k_B T/m_A)^{\frac{1}{2}}/c,$$

where c is the wave velocity. Although these shifts are clearly negligible for X-rays, they are not quite so obviously negligible for conduction electrons, while for cold neutrons they may be relatively large: the magnitude of c for a thermal neutron coming from a source at temperature T_N is† of order $(k_B T_N/m_N)^{\frac{1}{2}}$ so that in this case

$$\Delta\nu/\nu \sim \pm(T m_N/T_N m_A)^{\frac{1}{2}}.$$

Let us see how our previous results must be generalised so as to take them into account.

In chapter 1 it was stated, as a standard result of the Born approximation in scattering theory, that the amplitude with which a plane wave $\mathbf{k}_0$ is scattered to $\mathbf{k}_1$ by a perturbation $U(\mathbf{r})$ is proportional to the Fourier component $U(\mathbf{q})$, where $\mathbf{q}$ is $(\mathbf{k}_1 - \mathbf{k}_0)$. For an electron wave U is of course the ionic pseudo-potential, amply discussed above; for a neutron wave it is the Fermi pseudo-potential, centred on the scattering nuclei; and one can without difficulty invent a pseudo-potential to describe the scattering of X-rays. What difference does it make to this result if U is a function of time t as well as of position $\mathbf{r}$? We shall incorporate the time-dependence by writing $U(\mathbf{q}, t)$ as a Fourier sum of harmonic components

$$U(\mathbf{q}, t) = \sum_{\omega} U(\mathbf{q}, \omega) \exp(-i\omega t); \tag{3.34}$$

the sum can be restricted to a set of *discrete* frequencies by the usual device of supposing the motion of the liquid to be periodic over some very long interval of time.

† The definition of frequency, and hence of wave velocity, for a particle such as a neutron is ambiguous. Here $h\nu$ is taken to represent the kinetic energy of the neutron and does *not* include its rest mass energy.

First-order time-dependent perturbation theory tells us that if the specimen is introduced into the path of the incident beam at a time $t = 0$ the amplitude of the scattered wave at time t is proportional to

$$\int d\mathbf{r} \int_0^t \exp i(-\mathbf{k}_1\cdot\mathbf{r}+\omega_1 t')\, U(\mathbf{r}, t')\, \exp i(\mathbf{k}_0\cdot\mathbf{r}-\omega_0 t')\, dt'$$
$$= \sum_{\omega} U(\mathbf{q}, \omega) \int_0^t \exp\,(i(\omega_1-\omega_0-\omega)t')\, dt', \quad (3.35)$$

where ω_0 and ω_1 are the angular frequencies of the neutron before and after scattering. Clearly, the particular component whose angular frequency ω is just equal to $(\omega_1-\omega_0)$ gives rise to a term in the scattered amplitude which rises linearly with t, and hence to a term in the scattered intensity which rises like t^2. The terms due to other components all oscillate in due course, but over the time t we can distinguish a band of frequencies centred around $(\omega_1-\omega_0)$ with a width of order t^{-1} which all contribute to building up the scattered intensity. For this reason the resultant intensity is found to be proportional to $t^2 \cdot t^{-1} = t$, as expected on physical grounds; it is also proportional to the number of permitted values of ω per unit range of angular frequency, say $\mathcal{N}(\omega)$. Thus in the limit of large t we have

$$\begin{matrix}\text{Intensity scattered into } \mathbf{k}_1 \\ \text{per unit time}\end{matrix} \propto |U(\mathbf{q}, \omega_1-\omega_0)|^2 \mathcal{N}(\omega). \quad (3.36)$$

A physical description of the scattering by a *static* perturbation is to say that each of its Fourier components $U(\mathbf{q})$ is capable of giving rise to Bragg reflection of an incident wave; the equation $\mathbf{q} = (\mathbf{k}_1-\mathbf{k}_0)$ is of course equivalent (see fig. 3.5) to Bragg's famous law

$$\lambda = 2d \sin \tfrac{1}{2}\theta.$$

If the perturbation is *not* static we can regard $U(\mathbf{q})$ as made up of a lot of travelling waves. Bragg reflection may occur from each of these but, as with reflection from a moving mirror, a Doppler shift of angular frequency occurs, given by (see fig. 3.5)

$$\frac{2(\omega/q)}{(\lambda/2\pi)} \sin \tfrac{1}{2}\theta = \omega.$$

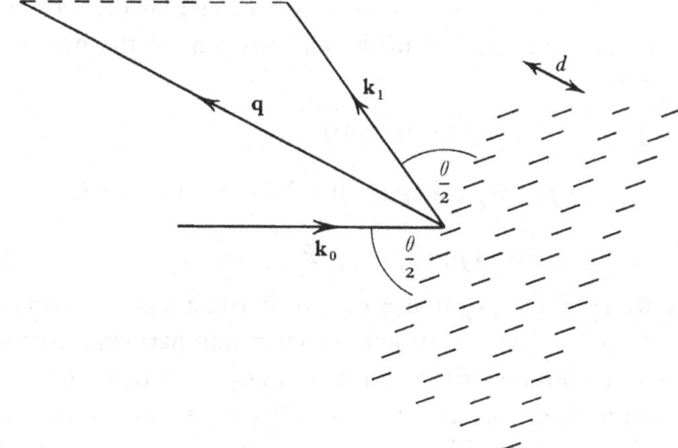

Fig. 3.5. Bragg reflection from a density wave with crests (indicated by bands of hatching) separated by $d = 2\pi/q$. If the crests are moving in the direction of $\mathbf{q}$ a Doppler shift occurs.

Conversely, if we tune our detector to respond only to frequencies which have been shifted upwards by ω, the signal will depend upon the strength of the ω component in $U(\mathbf{q})$. This is just what (3.36) says. A concrete example of its application is the case of a regular crystal with q small, where the magnitude of $U(\mathbf{q})$ is determined by the extent to which the longitudinal vibrational mode of wave vector $\mathbf{q}$ is thermally excited. In §2.11 such modes were treated as standing waves, but to treat them as travelling waves is equally permissible. Hence $U(\mathbf{q}, \omega)$ is finite only for $\omega = \pm(qv_{\parallel})$, the two possible signs corresponding in the present notation to different directions of travel, and the frequency of the scattered wave must be shifted up or down by this amount. In quantum-mechanical terms this corresponds to the absorption or emission of a longitudinal *phonon* on scattering, and the use of inelastic neutron scattering for the determination of phonon dispersion relations in crystals is well known. Of course the phonon may have a finite lifetime, due to interaction with other phonons, in which case a spread of energies may be observed in the scattered beam.

To particularise the argument for a liquid let us suppose, since this is usually the case for neutrons, that the scattering system may be treated as a substitutional mixture (see §2.12) in which the form factor for the ith atom or ion may differ by an amount Δf_i from the mean value $\bar{f}$, but in which there is no correlation between this fluctuation and fluctuations in the environment of the ith atom. Let us write

$$\begin{aligned} U(\mathbf{q}, t) &= \sum_i f_i \exp(-i\mathbf{q}\cdot\mathbf{r}_i(t)) \\ &= \bar{f}\sum_i \exp(-i\mathbf{q}\cdot\mathbf{r}_i(t)) + \Sigma\Delta f_i \exp(-i\mathbf{q}\cdot\mathbf{r}_i(t)) \\ &= \bar{f}F(\mathbf{q}, t) + (\Delta f)_{\text{r.m.s.}} F'(\mathbf{q}, t). \end{aligned} \tag{3.37}$$

Just as we split $U(\mathbf{q}, t)$ into a sum of harmonic components in (3.34), we may split these structure factors into harmonic components with amplitudes $F(\mathbf{q}, \omega)$ and $F'(\mathbf{q}, \omega)$. The randomness of Δf_i makes it clear that the phase of $F'(\mathbf{q}, \omega)$ is random and uncorrelated in any way with the phase of $F(\mathbf{q}, \omega)$, so that on averaging over an ensemble

$$\overline{|U(\mathbf{q}, \omega)|^2} = |\bar{f}|^2\, \overline{F(\mathbf{q}, \omega)F^*(\mathbf{q}, \omega)} + \overline{|\Delta f|^2}\, \overline{F'(\mathbf{q}, \omega)F'^*(\mathbf{q}, \omega)}. \tag{3.38}$$

It is now convenient to introduce two functions describing the spectral distribution of F and F', defined by:

$$\tilde{S}_{\text{coh}}(\mathbf{q}, \omega) = N^{-1}\mathcal{N}(\omega)\, \overline{F(\mathbf{q}, \omega)F^*(\mathbf{q}, \omega)}, \tag{3.39}$$

$$\tilde{S}_{\text{inc}}(\mathbf{q}, \omega) = N^{-1}\mathcal{N}(\omega)\, \overline{F'(\mathbf{q}, \omega)F'^*(\mathbf{q}, \omega)}. \tag{3.40}$$

Consideration of equations (3.36) and (3.38) then shows that by plotting the frequency (or energy) distribution of a monochromatic beam of radiation scattered through a fixed value of $\mathbf{q}$ we obtain a curve with a profile determined by

$$|\bar{f}|^2\tilde{S}_{\text{coh}}(\mathbf{q}, \omega) + \overline{|\Delta f|^2}\tilde{S}_{\text{inc}}(\mathbf{q}, \omega). \tag{3.41}$$

As in the treatment of elastic scattering (§2.12), the two parts of this expression may be distinguished by referring to the *coherent* and *incoherent* scattering respectively.

Naturally, the strength of signal obtained in any particular experiment depends on the strength of the source, the size of the specimen, the bandwidth of the detector and so on, and there is a calibration problem to be overcome before the quantity represented by (3.41) can be deduced. But we shall see in the next section that the total area under the frequency distribution curve, i.e. the signal obtained with a detector of large bandwidth, is not affected by the motion of the atoms and is given correctly by the theory of elastic scattering. Hence, once this total area has been determined, the calibration may be effected by the methods outlined in §2.13. However, although it is possible, if the relative magnitudes of $|\bar{f}|^2$ and $\overline{|\Delta f|^2}$ (the coherent and incoherent scattering cross-sections, to use the language of neutron theory) are known, to determine what proportion of the total area is contributed by coherent rather than incoherent scattering, it is another matter to obtain $\tilde{S}_{\text{coh}}(\mathbf{q}, \omega)$ and $\tilde{S}_{\text{inc}}(\mathbf{q}, \omega)$ separately for each value of ω. The separation of these two functions presents one of the main difficulties when one tries to interpret the results of an inelastic scattering experiment with neutrons. Possible solutions to this difficulty are discussed below. For the moment it suffices to note that there are several metals (e.g. Al, Sn, Pb), for which $|\Delta f|^2$ is either zero or very small compared with $|\bar{f}|^2$ and for these we may clearly determine $\tilde{S}_{\text{coh}}$ but not $\tilde{S}_{\text{inc}}$. The only metal for which $\bar{f}$ happens to be zero, enabling $\tilde{S}_{\text{inc}}$ to be determined but not $\tilde{S}_{\text{coh}}$, is V; but technical difficulties have so far prevented any neutron experiments from being done on V in the liquid state.

One further experimental problem which requires mention here is that unless ω is negligibly small compared with ω_0, in which case of course the resolution of the frequency distribution of the scattered beam is likely to present difficulties, the scattering vector $\mathbf{q}$ must be regarded as changing when ω changes. This means ideally that in order to plot out an accurate curve for (3.41) one should move to a different angle of scattering for each new value of ω, in such a way as to keep the magnitude if not the direction of $\mathbf{q}$ the same; for an isotropic scatterer such as a liquid the direction of $\mathbf{q}$ should of course make no difference to either $\tilde{S}$. In practice it is obviously more convenient to stay at the same angle of scattering, but it complicates the task of analysis.

When 'warm' neutrons are used to measure the static interference function $a(q)$, as described in chapter 2, no effort is made to resolve the energy distribution in the scattered beam; a single measurement of the integrated intensity is made for each scattering angle. Strictly speaking, since the scattering vector is not quite the same for the various components of the scattered beam, this measurement yields a value of $a(q)$ which has been averaged in some way over a small range of q. The corrections which are necessary to obtain the true $a(q)$ have been discussed by Ascarelli & Caglioti (1966) and by North *et al.* (1968).

3.10. Quantum-mechanical corrections

As defined by equations (3.38) and (3.39) the functions $\tilde{S}_{\mathrm{coh}}$ and $\tilde{S}_{\mathrm{inc}}$ are clearly symmetrical in ω, so that for example

$$\tilde{S}_{\mathrm{coh}}(\mathbf{q}, \omega) = \tilde{S}_{\mathrm{coh}}(\mathbf{q}, -\omega),$$

and the argument in §3.9 therefore implies that an incoming neutron is just as likely to gain energy as to lose it when it is scattered through a vector $\mathbf{q}$. But this conflicts with the principle of detailed balance. If one considers a gas of neutrons in thermal equilibrium with the scatterer, inside an equal temperature enclosure, the principle requires that the number of neutrons scattered per unit time from the state $\mathbf{k}_0$ to the state $\mathbf{k}_1$ through the vector $\mathbf{q}$ should be exactly equal to the number scattered from $\mathbf{k}_1$ to $\mathbf{k}_0$ through $-\mathbf{q}$. But since the probability of occupation is less for $\mathbf{k}_1$ than for $\mathbf{k}_0$ by a factor†

$$\exp\left(-\hbar(\omega_1-\omega_0)/k_{\mathrm{B}}T\right)$$

this means that the transition probability for a given neutron to undergo the first process (with gain of energy) must be less than it is for the second (which involves loss of energy) by the same exponential factor. The factor tends to unity if we let $\hbar$ go to zero, so that we are dealing here with an essentially quantum-mechanical correction.

In the language of quantum mechanics, the fact that a neutron is more likely to lose energy than to gain it is due to the possibility of *spontaneous* phonon emission and this is not allowed for by the classical argument upon which equation (3.36) is based.

It is possible to reformulate the whole of §3.9 in quantum-mechanical terms, though naturally this involves the recognition that the molecular position coordinates $\mathbf{r}_i$ are not, strictly speaking, capable of precise definition. But when $\hbar\omega$ is small compared with

† For simplicity it is assumed that the neutron gas is sufficiently rarefied for Boltzmann statistics to apply. The argument can be carried through using Fermi–Dirac statistics for the neutrons and the conclusion is just the same.

$k_B T$ the difference that it makes to the final results is a fairly trivial one. All that we need really modify is (3.41); the energy distribution of the scattered beam is determined instead by

$$|\bar{f}|^2 S_{\mathrm{coh}}(\mathbf{q}, \omega) + \overline{|\Delta f|^2} S_{\mathrm{inc}}(\mathbf{q}, \omega), \tag{3.42}$$

where†

$$S_{\mathrm{coh}}(\mathbf{q}, \omega) = \tilde{S}_{\mathrm{coh}}(\mathbf{q}, \omega) \exp(-\hbar\omega/2k_B T) \tag{3.43}$$

so as to satisfy the detailed balance requirement, and similarly for S_{inc}. Experimental curves for S_{coh} and S_{inc} may be corrected by means of (3.43) to give symmetrical curves for $\tilde{S}_{\mathrm{coh}}$ and $\tilde{S}_{\mathrm{inc}}$, which may then be analysed by essentially classical methods. The maximum value of $\hbar\omega$ for a crystal is roughly $k_B\Theta_D$, where Θ_D is its characteristic (Debye) temperature. Hence $\hbar\omega/2k_B T$ is less than $\frac{1}{5}$ in most solid metals just below the melting point (see table 2.6) and its value in the liquid phase is comparable, so the heuristic correction procedure suggested here is almost certainly an adequate allowance for quantum effects.

For future reference we may note that, whether the scattering is coherent or incoherent, the first moment of S with respect to ω is related to the second moment of $\tilde{S}$. Thus to a good approximation it follows from (3.43) and the symmetry of $\tilde{S}$ that

$$\int_{-\infty}^{\infty} \omega S(\mathbf{q}, \omega)\, d\omega = -\frac{\hbar}{2k_B T}\int_{-\infty}^{\infty} \omega^2 \tilde{S}(\mathbf{q}, \omega)\, d\omega. \tag{3.44}$$

3.11. Time-dependent pair distribution functions

Supposing that an accurate curve for $\tilde{S}_{\mathrm{coh}}$ or $\tilde{S}_{\mathrm{inc}}$ has somehow been obtained, what information can be extracted from it? The answer to this question can best be seen by making use of the Wiener–Khintchine theorem (e.g. Kittel, 1958, p. 133) which describes a connection between the auto-correlation function for any time-varying quantity and the Fourier transform of its power spectrum. Thus from (3.39) and the definition of $F(q, \omega)$ the

† See Aarnodt *et al.* (1962). Actually, these authors find that there should be a second correction factor on the right-hand side of (3.43), $\exp(-\hbar^2 q^2/8m_A k_B T)$. Since we are concerned only with the shape of S or $\tilde{S}$ as a function of ω at constant q, this additional term is of no importance here. In some treatments the correction factor in (3.43) appears as $\exp(+\hbar\omega/2k_B T)$ instead of $\exp(-\hbar\omega/2k_B T)$. This is because $S(\omega)$ is often used to represent the probability of scattering with energy *loss* rather than, as here, with energy *gain*.

Wiener–Khintchine theorem allows one to deduce immediately that

$$\int_{-\infty}^{\infty} \tilde{S}_{\text{coh}}(\mathbf{q}, \omega) \cos \omega t \, \mathrm{d}\omega$$
$$= N^{-1}\overline{F(\mathbf{q}, t_1)F^*(\mathbf{q}, t_2)}$$
$$= N^{-1}\overline{\sum_i \exp(-\mathrm{i}\mathbf{q}\cdot\mathbf{r}_i(t_1)) \sum_i \exp(\mathrm{i}\mathbf{q}\cdot\mathbf{r}_j(t_2))}, \quad (3.45)$$

where
$$t_2 - t_1 = t; \quad (3.46)$$

as in (3.17), the bar may represent an average over t_1 for a single system or an ensemble average for fixed t_1, whichever is conceptually more appealing. By setting t equal to zero we may derive the total area under the curve for $\tilde{S}_{\text{coh}}$: it is just

$$\int_{-\infty}^{\infty} \tilde{S}_{\text{coh}}(\mathbf{q}, \omega) \, \mathrm{d}t = a(\mathbf{q}), \quad (3.47)$$

where $a(\mathbf{q})$ is the interference function which occurs in the elastic scattering theory of chapters 1 and 2. The normalisation of $\tilde{S}_{\text{coh}}$ was deliberately adjusted, of course, so as to lead to this simple result.

Thus if it is impossible to resolve the frequency distribution of the scattered radiation and all one can measure is this total area under the curve, as is actually the case when X-rays are used, then all one can hope to deduce is $a(q)$ or the *instantaneous* pair distribution function $g(R)$. One is provided, as it were, with a snapshot of the liquid's structure of very short exposure. If some resolution is possible, however, one may hope to learn how, on the average, the structure changes with time. It is convenient to introduce a time-dependent pair distribution function $G(R, t)$, such that $NG\,\mathrm{d}\mathbf{R}/\Omega$ gives the probability of finding a molecule in the element of volume $\mathrm{d}\mathbf{R}$ at time t_2 if there was known to be one at $R = 0$ at time t_1. It is usual to regard G as made up of two parts,

$$G(R, t) = G_s(R, t) + G_d(R, t), \quad (3.48)$$

such that $NG_s\,\mathrm{d}\mathbf{R}/\Omega$ is the probability that the atom initially at the origin has moved into the element $\mathrm{d}\mathbf{R}$ during the interval t, while $NG_d\,\mathrm{d}\mathbf{R}/\Omega$ is the probability that some other atom has moved there. The suffices s and d stand for 'same' and 'different', or else 'self' and 'discrete'. As t goes to zero, G_s becomes a δ-function at the origin and

$$G_d(R, 0) = g(R). \tag{3.49}$$

In terms of these functions, then, we may write (cp. the derivation of (2.38))

$$\begin{aligned}\int_{-\infty}^{\infty} \tilde{S}_{\mathrm{coh}}(\mathbf{q}, \omega) \cos \omega t \, d\omega &= \frac{N}{\Omega}\int G(R, t) \cos (\mathbf{q}\cdot\mathbf{R}) \, d\mathbf{R} \\ &= \frac{4\pi N}{\Omega}\int_0^{\infty} (G(R, t)-1)\frac{\sin qR}{qR} R^2 \, dR.\end{aligned} \tag{3.50}$$

In principle, if frequencies separable by $1/\tau$ are resolvable and if curves for $\tilde{S}_{\mathrm{coh}}$ can be obtained over a wide range of q, one may use (3.50) to deduce how $G(R, t)$ develops up to times of order τ.

A similar discussion can be applied to incoherent scattering. The Wiener–Khintchine theorem says that

$$\begin{aligned}&\int_{-\infty}^{\infty} \tilde{S}_{\mathrm{inc}}(\mathbf{q}, \omega) \cos \omega t \, d\omega \\ &= N^{-1}\overline{F'(\mathbf{q}, t_1)F'^*(\mathbf{q}, t_2)} \\ &= N^{-1}(\overline{|\Delta f|^2})^{-1}\overline{\sum_i \Delta f_i \exp(-i\mathbf{q}\cdot\mathbf{r}_i(t_1)) \sum_j \Delta f_j^* \exp(i\mathbf{q}\cdot\mathbf{r}_j(t_2))} \\ &= N^{-1}\overline{\sum_i \exp i\mathbf{q}\cdot(\mathbf{r}_i(t_2)-\mathbf{r}_i(t_1))} \\ &= \frac{4\pi N}{\Omega}\int_0^{\infty} G_s(R, t)\frac{\sin qR}{qR} R^2 \, dR.\end{aligned} \tag{3.51}$$

It is the randomness of Δf_i which ensures that cross-terms for which $i \neq j$ vanish on averaging. By setting $t = 0$ we have

$$\int_{-\infty}^{\infty} \tilde{S}_{\mathrm{inc}}(\mathbf{q}, \omega) \, d\omega = 1. \tag{3.52}$$

It is clear that if one's primary interest is to study the diffusive motion of single molecules it is better to measure $\tilde{S}_{\mathrm{inc}}$ rather than $\tilde{S}_{\mathrm{coh}}$, so as to obtain G_s without confusion by G_d, and from this point of view it is a pity that so many metals are essentially coherent scatterers. In the pioneering work of Brockhouse & Pope on Pb, referred to on p. 180 above, only $\tilde{S}_{\mathrm{coh}}$ could be deduced. For times less than about 10^{-12} sec they could separate G_s from G_d without much difficulty because these functions had scarcely begun to overlap, but beyond 10^{-12} sec their curve for X^2 is probably not to be trusted, especially since a number of rather drastic simplifications were made during the analysis and Fourier inversion of the raw data. The more recent work of Dasannacharya & Rao on liquid Ar is probably more accurate. The incoherent and coherent scattering cross-sections of Ar are comparable, so that the curve obtained by Fourier inversion is a mixture of G_s and (G_s+G_d) in more or less equal proportions. The extra weighting of G_s makes its separation from G_d up to times of order 10^{-12} sec, and the subsequent determination of X^2, somewhat easier. Beyond about 10^{-12} sec, Dasannacharya & Rao assumed that G_s had the Gaussian form

$$G_s(R, t) = (\Omega/N)(2\pi X^2/3)^{-\frac{3}{2}} \exp(-3R^2/2X^2) \qquad (3.53)$$

and deduced the root mean square displacement X from the amplitude of G_s at $R = 0$. They claimed in this way to follow its development up to times of about 4×10^{-12} sec at which stage it has reached 2 Å; for longer times the overlap of G_d becomes serious even at $R = 0$.

Unfortunately, although the Gaussian approximation represented by (3.53) is certainly reliable for short times, because of the Maxwellian distribution of molecular velocities, and becomes reliable again at very long times, the computations of Rahman (1964*a*) show that it is not entirely accurate at intermediate times, especially around 3×10^{-12} sec. The deviation is in the sense that the higher moments of G_s, $\overline{R^4}$ and the like, are larger in comparison with X^4 etc. than (3.53) would suggest, which means that G_s is somewhat flatter than a Gaussian, extending out to larger values of R. Perhaps it is the small group of molecules mentioned on p. 176, which can move relatively unimpeded because the way is open before them, that are responsible for this behaviour.

It may be possible in the future to separate the incoherent from the coherent scattering for some metals, and hence to achieve a more satisfactory determination of G_s, by the use of a polarised neutron source. Spin-flip of the neutron does occur in a small proportion of scattering events, the angular momentum being taken up by re-orientation of one of the scattering nuclei. The fact that only one nucleus is affected is an indication that spin-flip scattering is an essentially incoherent phenomenon. If observations could be confined to those neutrons which have undergone spin-flip, $\tilde{S}_{\text{inc}}$ could be determined without confusion by $\tilde{S}_{\text{coh}}$.

Another idea is to find some way of varying the relative magnitudes of $|\bar{f}|^2$ and $\overline{|\Delta f|^2}$ in (3.41), in which case $\tilde{S}_{\text{coh}}$ and $\tilde{S}_{\text{inc}}$ could be separated by solving two simultaneous equations; this is reminiscent of a method used to separate a_{00}, a_{11} and a_{01} for a binary mixture (see p. 466). The variation of neutron cross-section could be achieved for some metallic elements (Ni is a particularly favourable prospect – see Egelstaff, 1966) by varying the isotopic constitution. Alternatively, one might choose a binary system whose two components (e.g. Cu and Ni) are sufficiently close in atomic volume for an alloy of them to be treated as substitutional, and sufficiently close in atomic mass to diffuse through the alloy at the same rate, and then vary the concentration ratio.

But no applications of these elegant methods have yet been reported. The results of Randolph, referred to on p. 180 above, were obtained by an approximate method to be discussed, along with other approximations, in §3.13.

3.12. Moment theorems

We shall need results for the second moment of $\tilde{S}_{\text{coh}}$ and $\tilde{S}_{\text{inc}}$ which are readily obtained from (3.45) or (3.51) by noting that

$$\langle\omega^2\rangle = \int_{-\infty}^{\infty} \omega^2 \tilde{S}\, d\omega = -\operatorname*{Lt}_{t\to 0} \frac{\partial^2}{\partial t^2} \int_{-\infty}^{\infty} \tilde{S} \cos \omega t\, d\omega. \qquad (3.54)$$

The right-hand side of (3.45) is of course a function only of the interval t given by (3.46) and not of t_2 and t_1 individually, so its second derivative with respect to t may be calculated by differentiating first with respect to t_2 keeping t_1 fixed and subsequently with respect to t_1 keeping t_2 fixed. Thus

$$\langle\omega^2\rangle_{\text{coh}} = N^{-1} \operatorname*{Lt}_{t_2\to t_1} \overline{\sum_i (\mathbf{q}\cdot\dot{\mathbf{r}}_i) \exp(-i\mathbf{q}\cdot\mathbf{r}_i) \sum_j (\mathbf{q}\cdot\dot{\mathbf{r}}_j) \exp(i\mathbf{q}\cdot\mathbf{r}_j)}. \qquad (3.55)$$

Since, according to classical statistical mechanics, there is no correlation between the thermal velocities of different ions in a liquid, it follows, at temperatures such that quantum corrections are negligible, that

$$\langle\omega^2\rangle_{\text{coh}} = N^{-1}\,\overline{\sum_i(\mathbf{q}\cdot\dot{\mathbf{r}}_i)^2} = q^2k_{\text{B}}T/m_{\text{A}}. \qquad (3{\cdot}56)$$

Exactly the same answer may be derived from the second moment of $\tilde{S}_{\text{inc}}$ by starting from (3.51). It follows from (3.44) that the first moments of S_{coh} and S_{inc} are given at high temperatures by

$$\langle\omega\rangle = -\hbar^2q^2/2m_{\text{A}}; \qquad (3{\cdot}57)$$

the negative sign corresponds to the fact that the incident wave is more likely to be scattered with its frequency reduced than enhanced.

Some simple examples of these theorems may be of interest. Consider first the case of scattering by a gas in which $a(q)$ is unity for all q. The mean energy change and the mean square energy change on scattering are respectively

$$\hbar\int_{-\infty}^{\infty}\omega S\,\mathrm{d}\omega\bigg/\!\!\int_{-\infty}^{\infty}S\,\mathrm{d}\omega = -\frac{\hbar^2\langle\omega^2\rangle}{2k_{\text{B}}T}\bigg/\!\!\int_{-\infty}^{\infty}\tilde{S}\,\mathrm{d}\omega$$

and $$\hbar\int_{-\infty}^{\infty}\omega^2S\,\mathrm{d}\omega\bigg/\!\!\int_{-\infty}^{\infty}S\,\mathrm{d}\omega = \hbar^2\langle\omega^2\rangle\bigg/\!\!\int_{-\infty}^{\infty}\tilde{S}\,\mathrm{d}\omega,$$

and from (3.56) together with (3.47) or (3.52) these amount to $-\hbar^2q^2/2m_{\text{A}}$ and $\hbar^2q^2k_{\text{B}}T/m_{\text{A}}$, whether the scattering is coherent or incoherent. These answers are exactly what would have been anticipated classically on the assumption that the recoil momentum $\hbar\mathbf{q}$ is always taken up by a single molecule. If the velocity of this molecule along the direction of $\mathbf{q}$ is v before the collision, its energy increase afterwards is

$$(\hbar^2q^2/2m_{\text{A}})-v\hbar q;$$

the mean value of this is indeed $\hbar^2q^2/2m_{\text{A}}$, while its mean square value tends to $\hbar^2q^2k_{\text{B}}\,T/m_{\text{A}}$ at high temperatures.

Compare this with the case of coherent Bragg reflection through one of the reciprocal lattice vectors $\mathbf{G}$ by a solid crystal containing N molecules. In this instance

$$\int_{-\infty}^{\infty}\tilde{S}_{\text{coh}}\,\mathrm{d}\omega = a(\mathbf{G}) = N.$$

The effect is to replace m_A by Nm_A in the results obtained for the gas, consistent with the idea that it is the crystal as a whole which recoils.

Finally, consider the case of coherent scattering through a small value of q, say Q. The Debye model analysed in §2.11 suggests (see (2.49)) that

$$a(Q) \simeq k_B T/m_A v_\parallel^2,$$

and if this were correct the mean square energy change on scattering would come out to be $(\hbar\Omega)^2$, where $\Omega = Qv_\parallel$ is the angular frequency of a longitudinal phonon. Such a result would of course be consistent with our previous deduction that the low-q scattering by a crystal always involves the emission or absorption of a phonon of this kind.

However, the Debye model is valid only when the difference between the adiabatic and isothermal compressibilities can be neglected; otherwise $a(Q)$ is somewhat more than $k_B T/m_A v_\parallel^2$ and the mean square energy change somewhat less than $(\hbar\Omega)^2$. As noted on p. 120, it is the entropy fluctuations which are responsible. These fluctuations are non-propagating, unlike the pressure fluctuations, and they contribute to $\tilde{S}_{coh}$ a central peak at $\omega = 0$ which increases the total area

$$\int_{-\infty}^{\infty} \tilde{S}_{coh}\, d\omega$$

without affecting the second moment

$$\int_{-\infty}^{\infty} \omega^2 \tilde{S}_{coh}\, d\omega.$$

The frequency-dependence of $\tilde{S}_{coh}(Q, \omega)$ is therefore as shown schematically in fig. 3.6.

The peaks in this figure have been drawn with finite widths to indicate schematically the effects of thermal conduction. This is well-known to cause attenuation of longitudinal pressure waves, so that the frequency Ω undergoes some lifetime broadening. In addition, it causes a sinusoidal temperature oscillation to decay with a lifetime given by $(\mathscr{D}Q^2)^{-1}$, where $\mathscr{D}$ is the thermal diffusivity, so that the central peak in $\tilde{S}_{coh}$ is Lorentzian with a width of

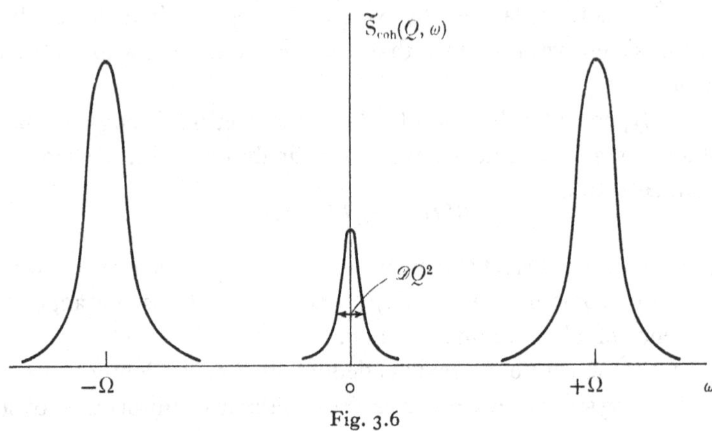

Fig. 3.6

about $\mathscr{D}Q^2$. Of course, if $\mathscr{D}Q^2$ becomes comparable with Ω, the central peak overlaps the two side ones and the situation is then more complicated than fig. 3.6 would suggest; the attenuation of pressure waves becomes large and their contribution cannot be separated from that of the entropy fluctuations. But even in a typical liquid metal, where $\mathscr{D}$ is relatively high, this situation only arises when the wavelength becomes comparable with the interatomic spacing. It does not seem likely that one can ever reach the ideal situation, for very short wavelengths, where $\mathscr{D}Q^2$ is much greater than Ω; in this limit the pressure waves would become isothermal and independent entropy fluctuations would be impossible.

The above illustrations bring out the point that although the second moment of S_{coh} is always given by (3.55) at high temperatures, the mean square value of ω, being inversely proportional to $a(q)$, tends to be small when the coherent scattering is strong and vice versa. It is in fact very noticeable in the experimental results that the line width of coherently scattered radiation becomes small for those values of q that correspond to peaks in $a(q)$. The scattering is then sometimes called *quasi-elastic*.

3.13. Approximate methods for the determination of G_s

An approximation which was fashionable in the early years of the subject of inelastic neutron scattering was the *convolution approximation* first suggested by Vineyard. In mathematical terms this may be expressed by the equation

$$\overline{\sum_i \exp(-\mathrm{i}\mathbf{q}\cdot\mathbf{r}_i(t_1)) \sum_j \exp(\mathrm{i}\mathbf{q}\cdot\mathbf{r}_j(t_2))}$$
$$\simeq \overline{\sum_i \exp(-\mathrm{i}\mathbf{q}\cdot\mathbf{r}_i(t_1)) \sum_j \exp(\mathrm{i}\mathbf{q}\cdot\mathbf{r}_j(t_1))} \times \overline{\exp(\mathrm{i}\mathbf{q}\cdot\mathbf{r}_j(t_2)-\mathbf{r}_j(t_1))}, \tag{3.58}$$

which implies from (3.45) and (3.51) that $\tilde{S}_{\text{coh}}$ may be expressed in terms of $\tilde{S}_{\text{inc}}$ as follows:

$$\int_{-\infty}^{\infty} \tilde{S}_{\text{coh}}(\mathbf{q}, \omega) \cos \omega t \, d\omega = a(q) \int_{-\infty}^{\infty} \tilde{S}_{\text{inc}}(\mathbf{q}, \omega) \cos \omega t \, d\omega. \tag{3.59}$$

The physical assumption behind (3.58) is that the neighbours of the ith molecule, whose average distribution at time t_1 is described by the instantaneous pair distribution function $g(R)$, can each be treated as diffusing subsequently in an *isotropic* fashion at the average rate that G_s describes. This must be true for very short intervals t, since we know from classical statistical mechanics that the instantaneous velocities of these neighbouring molecules are uncorrelated with their positions and unaffected, either in magnitude or direction, by the presence of the ith molecule at the origin. But beyond the smallest values of t the approximation becomes extremely dubious on physical grounds: surely the ith molecule, which must still be near the origin, makes it more likely for the neighbours to move outwards rather than inwards? If the approximation were true for all t then clearly, from (3.59), we should have

$$\tilde{S}_{\text{coh}}(q, \omega) = a(q)\tilde{S}_{\text{inc}}(q, \omega) \tag{3.60}$$

for all ω, which violates the principle demonstrated in the previous section that the second moments of $\tilde{S}_{\text{coh}}$ and $\tilde{S}_{\text{inc}}$ must always be identical. In view of this violation it seems rash to put any faith whatever in the convolution approximation. Several rather complex modifications of it have been suggested (Egelstaff, 1963; Rahman, 1964*a*; Singwi, 1964; Kurkijärvi, 1967; Skold, 1967; Venkataraman, Dasannacharya & Rao, 1967) which are less objectionable on theoretical grounds, and attempts have been made to establish their validity by computer calculations on model fluids. It is, unlikely, however, that any of them will emerge as completely reliable.

It is known that $a(q)$ is much less than unity in liquid metals at small values of q, which suggests that one way to minimise the effects of coherent scattering may be to concentrate upon the low-q region. Naturally one cannot hope to discover the full R-dependence of G_s from measurements of $\tilde{S}_{\text{inc}}$ which are confined to this region. But if one is willing to assume the Gaussian approximation of (3.53), it is easily shown from (3.51) that

$$\int_{-\infty}^{\infty} \tilde{S}_{\text{inc}}(\mathbf{q}, \omega) \cos \omega t \, d\omega = \exp\left(-\tfrac{1}{6}X^2 q^2\right). \qquad (3.61)$$

Thus measurements of $\tilde{S}_{\text{inc}}$ as a function of ω at even one low value of q would be enough to reveal, provided that the Gaussian approximation were sufficient, the time-dependence of the mean square displacement. Alternatively, one might take measurements at a series of low-q values, and obtain X^2 by extrapolating to zero q. Expansion of the right-hand side of (3.51) yields

$$\int_{-\infty}^{\infty} \tilde{S}_{\text{inc}} \cos \omega t \, d\omega = 1 - \tfrac{1}{6}q^2 X^2 + \tfrac{1}{120}q^4 \overline{R^4} + \dots \qquad (3.62)$$

and the higher terms become negligible for very small q, whether or not the Gaussian approximation is correct. It follows from (3.62) as a consequence of (3.32) that

$$Z(\omega) = \frac{m_A \omega^2}{k_B T} \operatorname*{Lt}_{q \to 0} \left(\frac{\tilde{S}_{\text{inc}}}{q^2}\right). \qquad (3.63)$$

So there is the attractive possibility of determining the frequency spectrum $Z(\omega)$, which in many ways is the quantity of most physical interest, direct from the scattering data at low q, without the necessity for any Fourier inversion.

The snag about all this, quite apart from the purely experimental difficulties of carrying out measurements at very low q values, is once more shown up by consideration of the moment relations. At low q we have

$$\int_{-\infty}^{\infty} \tilde{S}_{\text{coh}} \, d\omega = a(q) \int_{-\infty}^{\infty} \tilde{S}_{\text{inc}} \, d\omega \ll \int_{-\infty}^{\infty} \tilde{S}_{\text{inc}} \, d\omega. \qquad (3.64)$$

On the other hand the two second moments are still equal,

$$\langle \omega^2 \rangle_{\text{coh}} = \langle \omega^2 \rangle_{\text{inc}}. \qquad (3.65)$$

Thus although $\tilde{S}_{\text{coh}}$ must be negligible compared with $\tilde{S}_{\text{inc}}$ over much of the frequency range in order to satisfy (3.64), it must actually *exceed* $\tilde{S}_{\text{inc}}$ somewhere in the wings of the distribution if (3.65) is to be satisfied simultaneously. Analogy with the case of scattering by a solid suggests that the frequencies where $\tilde{S}_{\text{coh}}$ is not negligible for small q are centred about $\pm qv_{\parallel}$. Until we know how to calculate the coherent scattering at these frequencies and can subtract it out from the observed total scattering we cannot hope to apply (3.61) or (3.63) with complete confidence. Egelstaff (1966) and Cocking (1969) show curves for $Z(\omega)$ in liquid Na derived by the application of (3.63), but they are unlikely to be correct in detail.

The method adopted by Randolph, in the analysis of his results on Na referred to on p. 180, was to concentrate upon high values of q rather than low values. For sufficiently high q, $a(q)$ becomes effectively unity. In this limit the convolution approximation does not offend any moment theorem and seems plausible enough, i.e. one may set

$$\tilde{S}_{\text{coh}}(q, \omega) = \tilde{S}_{\text{inc}}(q, \omega)$$

without fear of serious error. Hence one can deduce the frequency dependence of $\tilde{S}_{\text{inc}}$ for high q from the frequency dependence of (3.41), whatever the ratio of the coherent and incoherent scattering cross-sections. The time variation of X^2 may then be obtained from (3.61), though this does of course involve the dubious Gaussian approximation.

But the trouble is that for values of X^2q^2 greater than say 20 the Fourier components of $\tilde{S}_{\text{inc}}$ become too small, as is clear from (3.61), to be detectable. Since one ought really to work with values of qR_{A} greater than say 10 in order to satisfy the condition that $a(q)$ is effectively unity, this means that the evolution of X^2 can scarcely be followed beyond the stage at which it amounts to $0.2R_{\text{A}}^2$. Yet it is only for values in excess of this that its behaviour is likely to be of interest.

Randolph did not in fact choose values of qR_{A} quite as big as 10, and this is one thing that makes his published curves for X^2 unreliable. Another is the unsatisfactory nature of the Gaussian approximation. Finally, there is a distinct possibility that Randolph's raw data were erroneous; he found the first moment of S

to be a good deal greater than theory predicts and, to judge by the later work of Randolph & Singwi (1966) on Pb, this is an indication that the results were distorted by multiple scattering. There are of course a great many difficulties in experimental work of this sort, connected with multiple scattering, imperfect resolution of the spectrometer, energy spread in the incident neutron beam and so on, which have been glossed over in the above discussion.

3.14. Longitudinal 'phonons' in liquid metals

So far we have been concentrating on the diffusive motion of single molecules, and hence on G_s to the exclusion of G_d. The fact is that G_d does not appear at the present time to be a function of particular interest. The results of Brockhouse & Pope (1959) make it clear that the peaks and troughs in $G_d(R, 0)$ – which is the same thing, it will be remembered, as the static pair distribution function $g(R)$ – flatten out as time passes and have almost disappeared after say 3×10^{-12} sec. Since the root mean square displacement of each molecule during this time is roughly 2 Å, the loss of memory of the coordination pattern that was present initially was qualitatively to be expected.

The interesting question that *can* be asked about the coherent scattering results for liquids, even when they are too incomplete to allow $G(R, t)$ to be derived in full, is whether they indicate the existence of collective vibrational modes analogous to the normal modes of solid crystals, and whether they allow dispersion curves for these 'phonons' to be determined. The raw data for coherent scattering by liquid metals are often remarkably similar to what are observed with polycrystalline specimens of the solid just below its melting point (consult a survey by Larsson, p. 387 of the volume edited by Egelstaff (1965), for examples) so the analogy with the solid is by no means forced.

The dispersion curve for a longitudinal phonon in a crystal depends, of course, on its direction of propagation relative to the symmetry axes. The curve marked (*a*) in fig. 3.7 shows the general form to be expected for propagation *parallel* to one of these axes, $\mathbf{G}_1$ being the relevant unit vector of the reciprocal lattice. The region of $\mathbf{q}$-space between $-\frac{1}{2}\mathbf{G}_1$ and $+\frac{1}{2}\mathbf{G}_1$ is the first Brillouin zone and is sufficient by itself to define all the longitudinal phonon

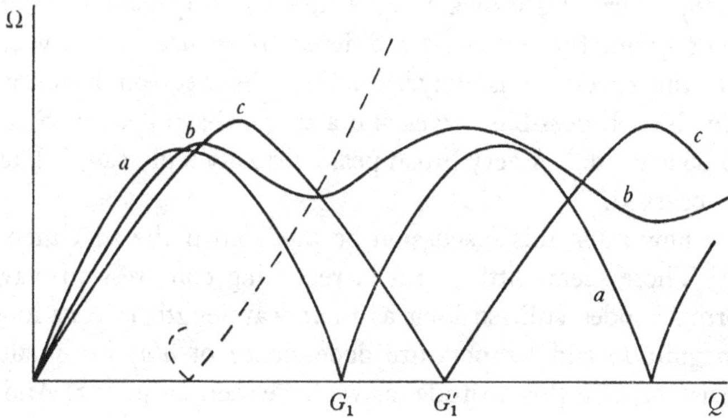

Fig. 3.7. Schematic phonon dispersion curves for a regular crystal. The broken curve suggests the locus on which the laws of conservation of energy and momentum permit phonons to be detected, in an inelastic scattering experiment confined to particular values of k_1 and θ. Its two branches correspond to phonon emission and absorption.

modes that are possible with this direction of propagation; the repetition of the dispersion curve outside this region is merely a reminder of the possibility of *umklapp* scattering processes, in which the scattering vector **q** is not the wave number of the phonon, **Q**, but is $\mathbf{G}_1 \pm \mathbf{Q}$. In order to plot out such a curve experimentally one looks for peaks in $S_{\text{coh}}(q, \omega)$, since the frequencies at which a peak occurs can be identified with the frequency of the phonon which is responsible. At a fixed angle of scattering q is liable to be a somewhat complicated function of ω, as indicated schematically by the broken line in the figure. It is where this line crosses the dispersion curve that a peak in S_{coh} is to be expected for this particular angle. To explore the rest of the curve one must vary the angle, or else the energy of the incident neutrons. It is not difficult to show that to explore the initial rise of the dispersion curve near $q = 0$ requires neutron energies in excess of some critical value which, for a typical liquid metal, is equivalent to a source temperature of say 300 °K. In many scattering experiments colder neutrons are used, to ensure good resolution in the scattered beam.

Of course it is a drastic over-simplification to suppose that **Q** is always parallel to $\mathbf{G}_1$. For propagation in some non-symmetry direction the dispersion curve might have the form suggested by curve (*b*). Curve (*c*) indicates its shape for propagation along a different symmetry axis, with a different reciprocal lattice vector $\mathbf{G}_1'$. If the specimen is polycrystalline, the neutron beam will sample all such possible curves and a single sharp peak in S_{coh} is not to be expected. A very broad peak, virtually a plateau, is likely to be observed.

Now how must this discussion be modified if the specimen is liquid? There seems little harm in regarding compression waves as normal modes still, so long as their wavelength is very long; the magnitude and temperature dependence of $a(q)$ for small q is consistent with this attitude, as we have seen on p. 135. And if we are prepared to admit that the 'phonons' have a very short lifetime and therefore an uncertain frequency, we may well feel justified in retaining the dispersion curve or curves in fig. 3.7 up to $q \sim \frac{1}{2}G_1$, where they reach a maximum. The frequency here should be much the same, presumably, as the frequency at which $Z(\omega)$ is at its largest (see fig. 3.4); it is the mean frequency with which two atoms in contact tend to vibrate against one another in anti-phase. The peak in $Z(\omega)$ appears to be a broad and ill-defined one, so we must not expect to be able to define the frequency of the dispersion curve at its maximum at all precisely. In view of what we know or guess about the diffusive motion of single molecules it is hard to believe that two of them can continue to vibrate against one another undisturbed for more than about 10^{-12} sec, which is after all, only about one period of the oscillation.

But the question we most need to consider is whether it is meaningful to continue the dispersion curve for a liquid into the region of q near G_1. It is well to be clear that if we do we are *not* implying the existence of phonons with $Q \sim G_1$, but merely that a process resembling *umklapp* scattering can occur in liquids as well as solids.†

The key to an understanding of this problem lies in equation

† Schneider & Stoll (1967*a*) have calculated dispersion curves for liquids, using essentially the method applied to long wavelength waves in § 2.19 in order to obtain the compressibility. Near $q = G_1$ and beyond their results are in gross disagreement with the neutron data. They make the mistake of supposing, however, that vibrations of indefinitely high Q can exist as independent modes of motion. In fact the motion of N atoms can be completely specified by specifying the amplitude and phase of $3N$ modes, and all of these can be accommodated within the first Brillouin zone.

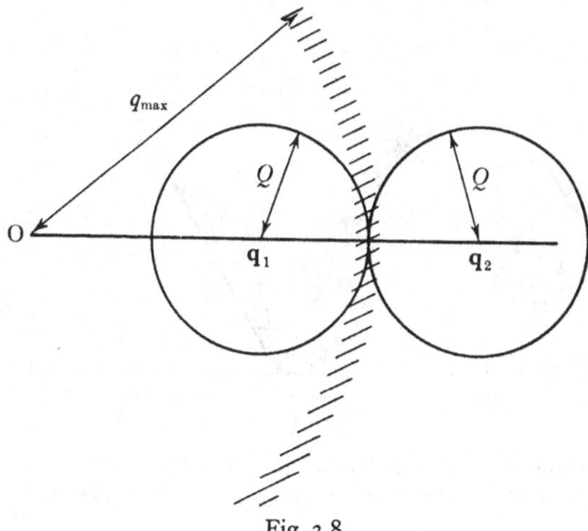

Fig. 3.8

(2.55), which tells us how much a sinusoidal disturbance of wave number $\mathbf{Q}$ and amplitude $\boldsymbol{\xi}_{\mathbf{Q}}$ contributes to the interference function. This equation is readily extended to describe the frequency changes for which the disturbance is responsible. If it has a well-defined frequency Ω, and if we denote by $\Delta\tilde{S}_{coh}$ the contribution to $\tilde{S}_{coh}$ due to the two independent vibrational modes which have opposite directions of propagation, the equation becomes

$$\Delta\tilde{S}_{coh}(q, \omega) = \tfrac{1}{4}(\mathbf{q}\cdot\boldsymbol{\xi}_{\mathbf{Q}})^2\{\tilde{S}_{coh}(\mathbf{q}-\mathbf{Q}, \omega-\Omega)+\tilde{S}_{coh}(\mathbf{q}-\mathbf{Q}, \omega+\Omega) + \tilde{S}_{coh}(\mathbf{q}+\mathbf{Q}, \omega-\Omega)+\tilde{S}_{coh}(\mathbf{q}+\mathbf{Q}, \omega+\Omega)-4\tilde{S}_{coh}(q, \omega)\}. \quad (3.66)$$

This expression takes full account of so-called 'multi-phonon' scattering, because the $\tilde{S}_{coh}$ on the right-hand side includes the effects of all the other vibrational modes which may be excited besides the two of interest.

Except where $\mathbf{q} = \pm\mathbf{Q}$, $\Delta\tilde{S}_{coh}$ is an infinitesimal quantity of order N^{-1}. Nevertheless, the contribution of all the 'phonons' put together is a significant one. At a point in $\mathbf{q}$-space such as $\mathbf{q}_1$ in fig. 3.8, a particularly important contribution to $\tilde{S}_{coh}$ is likely to arise from the group of longitudinal 'phonons' for which Q is close to $(q_{max}-q_1)$, where q_{max} is the radius of the sphere on which $a(q)$ has its main maximum. Not only is $\tilde{S}_{coh}(q, \omega)$ large on this sphere, but, as was noted on p. 194, it tends to be rather sharply

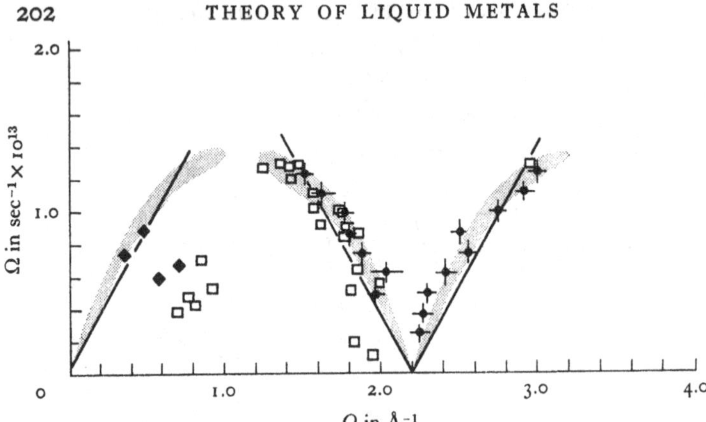

Fig. 3.9. Phonon dispersion in liquid Pb at 352 °C. (Redrawn by permission from Randolph & Singwi (1966), *Phys. Rev.* **152**, 99.)

peaked in terms of frequency about $\omega = 0$. Consideration of equation (3.66) suggests that on this account the group of 'phonons' for which $Q \simeq (q_{max} - q_1)$ will generate two peaks in $\tilde{S}_{coh}(q_1, \omega)$ centred about the frequencies $\omega = \pm\Omega$. Similar considerations apply at the point $\mathbf{q}_2$ in fig. 3.8, where it is the group of 'phonons' for which $Q \simeq (q_2 - q_{max})$ that are liable to be especially important. Measurement of the frequencies at which these peaks in $\tilde{S}_{coh}$ appear, if indeed they are sharp enough to show up against the background due to all the other 'phonons', should enable a dispersion curve to be plotted out for a liquid in the neighbourhood of q_{max}. It should be not dissimilar from curve (a) in fig. 3.7, with q_{max} taking over the role of the reciprocal vector G_1.

A dispersion curve obtained for liquid Pb in this way is shown in fig. 3.9. It includes the results of Cocking & Egelstaff (1965), Dorner, Plesser & Stiller (1965), and Randolph & Singwi (1966); these were obtained at temperatures not far above the melting point, but it has been shown (Wignall & Egelstaff, 1968) that the dispersion curve is little affected by heating the specimen through several hundred degrees celsius. The absence of any experimental points near the origin has been accounted for on p. 199.

The straight lines which are drawn in the figure have slopes which correspond to the velocity of longitudinal adiabatic compression waves in liquid Pb, as measured at low frequencies by ultrasonic techniques, and the points lie reasonably close to these lines. Exact agreement is not to be expected, because most of the points lie in a region where $\mathscr{D}Q^2$ is no longer much less than ω, so that we are dealing with modes which are no longer perfectly adiabatic (see p. 194); in addition, as we shall see in the following section, the stress associated with them is no longer a pure compression. It may therefore be wiser to compare the points with the results which are available over the same frequency range for solid lead. These are indicated, for propagation along three symmetry directions, 100, 110 and 111, by a shaded band in the figure. The agreement is good except for the puzzling group of points at about $Q = 0.8$ Å^{-1} which are anomalously low. These are derived from broad but nevertheless quite well marked maxima in $\tilde{S}_{coh}$. Phonon dispersion curves have now been reported for liquid Sn, Bi, Al and Rb (Cocking, 1967) as well as Pb, and of these the first two at any rate show similar anomalies at low q. Since no such anomalies are visible in the rather sketchy dispersion curves for liquid Ar and Ne published by Chen *et al.* (1965), the impression has gained ground that they are due to some unspecified mode of excitation which is peculiar to liquid metals. It is quite possible, however, that the broad peaks are generated by multiple scattering (Dorner, Plesser & Stiller, 1967; Cocking & Egelstaff, 1968) and in the present context should be ignored.

Attempts have been made in the literature, using rather elaborate solid-like models, to explain the width and shape of the various peaks in $\tilde{S}_{coh}(q, \omega)$ whose position serves to define the dispersion curve. Equation (3.66) may perhaps open the way for a more direct analysis. It suggests that two factors which must contribute to the width of a peak in the *umklapp* region around q_{max} are the appreciable spread in terms of both wave number and frequency of the 'quasi-elastic' peak in $\tilde{S}_{coh}$ which is centred on q_{max} itself and upon $\omega = 0$. If $\tilde{S}_{coh}(q, \omega)$ is available directly from experiment for insertion on the right-hand side of (3.66), it should be possible to estimate the importance of these factors. One must expect, however, to find evidence that an additional broadening mechanism is at work. Equation (3.66) is based on the assumption that the modes of wave number Q have a well-defined frequency Ω. In fact there is some attenuation of compression waves in liquid metals even in the ultrasonic region, for which thermal conduction is mainly responsible, and consequently there is some lifetime broadening of their frequency (see p. 193). At the relatively high frequencies where the neutron measurements lie this attenuation should be considerable, and the lifetime broadening of Ω must add to the width of any peak in $\tilde{S}_{\mathrm{coh}}$ for which 'phonons' are responsible.

3.15. Visco-elasticity and shear waves

In the previous section we concentrated upon longitudinal 'phonons', but it is not inconceivable that transverse ones also exist; indeed, their existence may be inferred from the observed specific heat of liquids (see §2.4). For readers who are not familiar with the relevant theory of visco-elasticity the following brief summary may be helpful.

When a solid is suddenly strained through an angle θ a shear stress $\mathscr{S}$ is set up, given by

$$\mathscr{S} = g\theta$$

where g is the shear modulus. We may assume that instantaneously this relation is obeyed for liquids also, but that the stress subsequently decays as a result of rearrangement of the molecules, even though the strain is kept constant. If we suppose the rate of decay to be described by a function f such that

$$\mathscr{S}(t) = \mathscr{S}(0)f(t), \tag{3.67}$$

it follows that the appropriate stress–strain relation for a liquid is

$$\mathscr{S}(t) = \int_{-\infty}^{t} g\frac{\partial\theta}{\partial t'}\mathrm{d}t' f(t-t'). \tag{3.68}$$

We may now look for a solution which corresponds to a shear wave propagated in the x-direction, such that the displacement of the liquid in the y-direction is given by

$$\xi \propto \exp \mathrm{i}(Qx - \Omega t);$$

the strain θ is then of course just $\partial\xi/\partial x$. Partial differentiation of (3.68) with respect to x and replacement of $\partial\mathscr{S}/\partial x$ by $\rho\,\partial^2\xi/\partial t^2$, ρ being the density, leads to the result

$$\frac{\mathrm{i}\rho\Omega}{gQ^2} = \int_0^{\infty} \exp\,(\mathrm{i}\Omega s)f(s)\,\mathrm{d}s, \tag{3.69}$$

where $$s = t-t'.$$

At very low frequencies it is well known that shear waves in liquids are very rapidly attenuated; the wave number Q is complex, given by

$$Q^2 = i\rho\Omega/\eta. \tag{3.70}$$

This is consistent with (3.69) provided that the stress relaxation function f is linked to the viscosity by the relation

$$\int_0^\infty f(s)\,ds = \eta/g = \tau_M. \tag{3.71}$$

The time τ_M defined in this fashion, which is an indication of the time for which a shear stress can endure, is known as the Maxwell relaxation time. Now an integration of (3.69) by parts yields

$$\frac{i\rho\Omega}{gQ^2} = \frac{1}{i\Omega}\left(-1-\int_0^\infty \exp(i\Omega s) f'(s)\,ds\right)$$

and it is clear that the integral on the right-hand side of this equation must become negligible compared with unity when $\Omega\tau_M \gg 1$. Hence in this high frequency limit the dispersion relation is just

$$Q^2 = \rho\Omega^2/g,$$

which describes the propagation of shear waves, as though through a solid, almost without attenuation.†

No detailed assumption has so far been made about the form of $f(s)$ apart from the obvious ones that it is unity at $s = 0$ and tends subsequently to zero. Maxwell originally proposed that it was exponential, in which case it may be shown from (3.69) that

$$Q^2 = \frac{\rho\Omega^2}{g} + \frac{i\rho\Omega}{\eta}. \tag{3.72}$$

But this equation has been found inadequate to represent the behaviour of the many liquid hydrocarbons etc. which have sufficiently large viscosities, especially when supercooled, for one to be able to investigate the full range of frequencies up to $\Omega \gg \tau_M^{-1}$ using conventional ultrasonic techniques. Presumably the stress relaxation function is *not* exponential. A curve for it may

† Shear waves, like longitudinal ones, may suffer attenuation due to thermal conduction but this is ignored in the simple theory above.

be derived from experimental data using the Fourier inversion of (3.69), i.e.

$$f(s) = \frac{i\rho}{2\pi g}\int_{-\infty}^{\infty} \exp(-i\Omega s)(\Omega/Q^2)\, d\Omega, \qquad (3.73)$$

but it would be inappropriate to describe the results obtained in this way, because there is no reason to suppose that the stress relaxes in the same manner in liquid hydrocarbons as it does in liquid metals and other simple liquids.

For most liquid metals at temperatures not far above the melting point τ_M is about 3×10^{-13} sec, which means that frequencies comparable with τ_M^{-1} are inaccessible by ultrasonic methods. However, τ_M^{-1} is still a good deal less than the maximum value of Ω for a typical dispersion curve like the one plotted for liquid Pb in fig. 3.9. It is therefore not unreasonable to suppose that shear modes can be excited in liquid metals, so long as their wavelengths are only a few interatomic spacings or less. For such short wavelengths it is not necessarily legitimate to treat the liquid as a homogeneous medium, as in the simple theory above, so that (3.72) is hardly to be trusted whether or not the decay of stress is exponential. But undoubtedly a shear wave must be heavily attenuated, and in thinking of dispersion curves one should imagine that the frequency associated with a particular (real) value of Q suffers even more lifetime broadening for the transverse branch than for the longitudinal branch.

Transverse phonons can give rise to *umklapp* scattering of neutrons in solid crystals. May one hope to detect the transverse branch of the dispersion curve for a liquid metal in a neutron scattering equipment? The question requires a detailed analysis before a firm answer can be given. The most likely answer seems to be no, because at a point such as $\mathbf{q}_1$ in fig. 3.8 the important group of 'phonons' for which $\mathbf{Q} \simeq (\mathbf{q}_{max} - \mathbf{q}_1)$ contribute nothing to $a(\mathbf{q}_1)$ unless they are longitudinal; for transverse ones the factor $(\mathbf{q}\cdot\boldsymbol{\xi}_{\mathbf{Q}})^2$ in (3.60) vanishes. However, at least one example has been reported (Larsson, see Egelstaff, 1965, p. 387; Cocking, 1967) of a peak in S_{coh} which is attributed to transverse phonons in the polycrystalline solid (Al, as it happens) and does *not* disappear on melting.

In principle, visco-elastic effects should appear in the longitudinal branch of the dispersion curve for a liquid as well as in the transverse branch. For example, when the frequency increases through τ_M^{-1} and the relaxation of shear stress becomes impossible, the longitudinal velocity for an ideal, isotropic continuum should increase from $(1/\rho\beta)^{\frac{1}{2}}$ to $((1+\frac{4}{3}g\beta)/\rho\beta)^{\frac{1}{2}}$. Here is another reason why the points in fig. 3.9 need not be expected to lie on the straight lines which represent an extrapolation of the low-frequency ultrasonic data. That they lie as close as they do may be because the increase of $v_{\parallel}$ as Ω passes through τ_M^{-1} compensates for a decrease over much the same frequency range due to the change from adiabatic to isothermal propagation.

Below τ_M^{-1}, viscosity is bound to cause some attenuation of the longitudinal modes and hence to add to the lifetime broadening mentioned in the previous section. Measurements of the attenuation coefficient by ultrasonic techniques (reviewed by Stephens (1963) and Webber & Stephens (1968)) suggest that, in a typical liquid metal, only about 85% of it can be attributed to the effects of thermal conduction. The residual 15% is usually too large to be attributed to shear viscosity alone and indicates the existence of a finite *bulk viscosity* b.† The ratio of b to η may be estimated from the data, though the error is liable to be considerable, and some results quoted by Sharma (1968) range from 0.3 for liquid Pb to 2.7 for liquid Cd. Longuet-Higgins & Pople (1956) have estimated that for a rigid-sphere fluid b/η should be $\frac{5}{3}$ and for several liquid metals the experimental value is quite close to this. It is too early to say whether the deviations for Pb, Cd and others are significant or not.

3.16. Relaxation times for conduction electrons

The inelastic scattering of conduction electrons is governed by the same principles as the inelastic scattering of neutrons, and it forms an application of some of the theoretical ideas developed above to calculate two electronic relaxation times which will be needed in later chapters.

† While a fluid is undergoing compression a term in the pressure is liable to arise proportional to the rate of change of the volume. The bulk viscosity is the constant of proportionality, just as η is the constant of proportionality in an equation relating shear stress to rate of change of shear strain.

If the atoms are stationary, so that the perturbing pseudo-potential U is time-independent and the scattering elastic, then the calculation is entirely straightforward. An electron in a state $\mathbf{k}_1$ (see fig. 3.10) may be scattered to any other state $\mathbf{k}_2$ which lies on the same spherical surface in **k**-space of radius k. The density of states in the neighbourhood of $\mathbf{k}_2$ which have the same spin orientation† and which correspond to scattering angles between θ and $\theta+\mathrm{d}\theta$ is

$$\frac{\Omega}{(2\pi)^3} 2\pi k^2 \sin\theta \,\mathrm{d}\theta \frac{\mathrm{d}k}{\mathrm{d}E} = \frac{\Omega m^*}{4\pi^2\hbar^2} k \sin\theta \,\mathrm{d}\theta. \tag{3.74}$$

Time-dependent pertubation theory shows (cp. (3.35) and (3.36) above) that the probability of scattering per unit time through this angular range is

$$\frac{m^*}{2\pi\hbar^3\Omega} k \sin\theta \,\mathrm{d}\theta \,\overline{|U(q)|^2} = \frac{m^*N}{2\pi\hbar^3\Omega} k \sin\theta \,\mathrm{d}\theta \, a(q)|u(q)|^2, \tag{3.75}$$

where u is the pseudo-potential due to a single ion as in (1.73); the bar in (3.75) indicates that an ensemble average has been taken. The resultant mean *lifetime* τ_l of an electron in the state $\mathbf{k}_1$ is given by

$$\frac{1}{\tau_l} = \frac{m^*}{2\pi\hbar^3}\frac{N}{\Omega} k \int_0^\pi a(q)\,|u(q)|^2 \sin\theta \,\mathrm{d}\theta$$

$$= \frac{m^*}{\pi\hbar^3}\frac{N}{\Omega} k \int_0^1 a|u|^2\, 2(q/2k)\,\mathrm{d}(q/2k). \tag{3.76}$$

No allowance has been made in this calculation for the effects of the exclusion principle, which would, of course, prevent a transition to the state $\mathbf{k}_2$ if the latter were already occupied. Curiously enough the exclusion principle is irrelevant, when the scattering is elastic, to any calculation of the overall rate at which the momentum distribution of a degenerate electron gas approaches equilibrium. The reason is that although it reduces the probability of transitions into $\mathbf{k}_2$ by a factor $(1-f_2)$, where f_2 is the probability that this state is occupied, it also reduces the probability of transi-

† The chance of spin-flip occurring during a scattering process is usually so remote as to be negligible for a conduction electron in a metal, because the pseudo-potential does not couple directly with the electron's magnetic moment.

tions occurring in the reverse direction, from $\mathbf{k}_2$ to $\mathbf{k}_1$, by a factor $(1-f_1)$. Hence the net rate of scattering from $\mathbf{k}_1$ to $\mathbf{k}_2$ is

$$f_1(1-f_2)P_{12}-f_2(1-f_1)P_{21}, \tag{3.77}$$

where P_{12} and P_{21} are transition probabilities which there is no need to specify. If the energies corresponding to $\mathbf{k}_1$ and $\mathbf{k}_2$ are the same, then P_{12} and P_{21} are equal and (3.77) can be reduced to

$$f_1 P_{12}-f_2 P_{21}. \tag{3.78}$$

This is just what one would write down were the exclusion principle to be ignored.

Of course, if the electrons are in thermal equilibrium both f_1 and f_2 are described by the Fermi–Dirac function quoted in (1.6) and are equal, so (3.78) vanishes and there can be no change in the distribution with time. Scattering only becomes important when the distribution is disturbed, e.g. by the application of an electric field. It is convenient to write

$$f = f^0+\delta,$$

where f^0 is the equilibrium Fermi–Dirac function and δ is the local deviation from it; this can vary from point to point in $\mathbf{k}$-space but is usually significant only in the neighbourhood of the Fermi surface. Then (3.78) becomes

$$\delta_1 P_{12}-\delta_2 P_{21}$$

for elastic scattering. It is the magnitude of δ which determines the strength of the total current carried by the electrons, if their distribution has been shifted off-centre in $\mathbf{k}$-space. The fact that the effective scattering *out of* any state $\mathbf{k}_1$ is proportional to $\delta(\mathbf{k}_1)$ shows that, in the absence of any electric field to keep it going, the current may be expected to decay in an exponential fashion.

The time constant for current decay is *not* exactly the τ_l of equation (3.76). Consider an electron in the state $\mathbf{k}_1$ initially, contributing a term $\Delta\mathbf{j}_1$ to the total current density $\mathbf{j}$; if the Fermi surface is spherical, $\Delta\mathbf{j}_1$ points in the direction of $\mathbf{k}_1$. When this electron is scattered once through an angle θ its component of current parallel to $\mathbf{k}_1$ changes by a factor of $\cos\theta$; at the same time it acquires a component perpendicular to $\mathbf{k}_1$, but this clearly vanishes on averaging over all possible azimuthal angles and may be ignored. If we denote by Δj_2, Δj_3 etc. the *mean* value of the current density which the electron contributes, always in the direction of $\mathbf{k}_1$, after 1, 2 etc. scattering processes, then we have a set of equations

$$\left.\begin{aligned} \frac{d}{dt}\Delta\mathbf{j}_1 &= -\frac{\Delta\mathbf{j}_1}{\tau_l}, \\ \frac{d}{dt}\Delta\mathbf{j}_2 &= -\frac{\Delta\mathbf{j}_2}{\tau_l}+\frac{\Delta\mathbf{j}_1}{\tau_l}\overline{\cos\theta}, \\ \frac{d}{dt}\Delta\mathbf{j}_3 &= -\frac{\Delta\mathbf{j}_3}{\tau_l}+\frac{\Delta\mathbf{j}_2}{\tau_l}\overline{\cos\theta}, \quad \text{etc.} \end{aligned}\right\} \qquad (3.79)$$

which may be summed to give

$$\frac{d}{dt}\Delta\mathbf{j} = -\frac{\Delta\mathbf{j}}{\tau_l}(1-\overline{\cos\theta}) = -\frac{\Delta\mathbf{j}}{\tau},$$

where $$\Delta\mathbf{j} = \Delta\mathbf{j}_1+\Delta\mathbf{j}_2+\Delta\mathbf{j}_3+\dots$$

is the total mean current density carried by the electron and $\overline{\cos\theta}$ is the mean value of $\cos\theta$ after one scattering. The effective *relaxation time* for $\Delta\mathbf{j}$ is therefore given by

$$\begin{aligned} \frac{1}{\tau} &= \frac{m^*}{2\pi\hbar^3}\frac{N}{\Omega}k_F\int_0^\pi a|u|^2\sin\theta(1-\cos\theta)\,d\theta \\ &= \frac{m^*}{\pi\hbar^3}\frac{N}{\Omega}k_F\int_0^1 a|u|^2\,4(q/2k_F)^3\,d(q/2k_F), \end{aligned} \qquad (3.80)$$

which differs from (3.76) by the inclusion of a $(1-\cos\theta)$ weighting factor. In this formula k has been replaced by k_F because it is the electrons near the Fermi surface that carry the current. Since this expression for τ is the same for all values of $\mathbf{k}_1$ near the Fermi surface, it must also describe the relaxation time with which the total current $\mathbf{j}$ decays.†

Now the problem is how to modify this discussion to take account of inelastic scattering. If the energies of states $\mathbf{k}_1$ and $\mathbf{k}_2$ differ by $\hbar\omega$, then P_{12} and P_{21} are not equal and one cannot pass from (3.77) to (3.78). The net rate of scattering from k_1 to k_0 becomes

$$(f_1^0+\delta_1)(1-f_2^0-\delta_2)P_{12}-(f_2^0+\delta_2)(1-f_1^0-\delta_1)P_{21}. \qquad (3.81)$$

† An alternative justification for the inclusion of the $(1-\cos\theta)$ factor, based upon solution of the Boltzmann equation, will be found in standard texts.

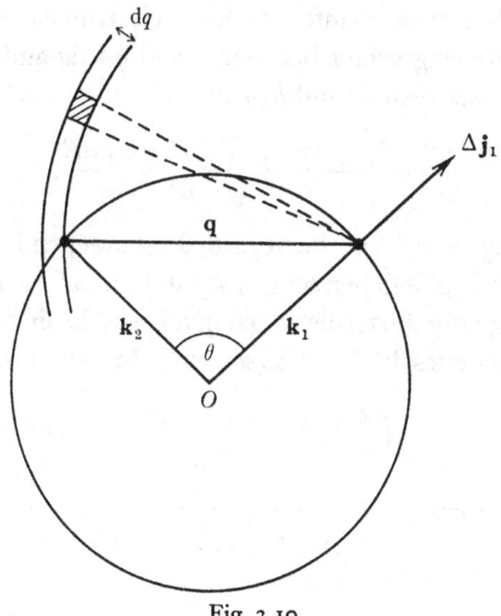

Fig. 3.10

But the principle of detailed balance tells us that

$$f_1^0(1-f_2^0)P_{12} = f_2^0(1-f_1^0)P_{21},$$

and with the aid of this condition (3.81) is readily reduced to

$$\delta_1\left(\frac{1-f_2^0}{1-f_1^0}\right)P_{12}-\delta_2\left(\frac{1-f_1^0}{1-f_2^0}\right)P_{21}, \qquad (3.82)$$

plus terms in $\delta_1\delta_2$ which can be neglected if δ is everywhere small enough to justify a linear theory. Thus the effective rate of scattering out of the state $\mathbf{k}_1$ is altered, compared with the situation for elastic scattering, by a factor $(1-f_2^0)/(1-f_1^0)$.

Since it is a property of the Fermi–Dirac function that

$$\mathrm{d}f^0/\mathrm{d}E = -f^0(1-f^0)/k_\mathrm{B}T$$

and

$$\mathrm{d}^2f^0/\mathrm{d}E^2 = f^0(1-f^0)(1-2f^0)/(k_\mathrm{B}T)^2,$$

we may expand this factor in a Taylor series to read

$$\frac{1-f_2^0}{1-f_1^0} = \left[1+\left(\frac{\hbar\omega}{k_\mathrm{B}T}\right)f_1^0-\tfrac{1}{2}\left(\frac{\hbar\omega}{k_\mathrm{B}T}\right)^2 f_1^0(1-2f_1^0)+\ldots\right].$$

Now P_{12} is proportional (see (3.36)) to $S(\mathbf{q}, \omega)$ and to the density of states near $\mathbf{k}_2$. It is readily shown from the geometry of fig. 3.10

that the number of states into which the electron can be scattered, through a scattering vector between q and $q+\mathrm{d}q$ and which have a wave vector between k_2 and $k_2+\mathrm{d}k_2$ is

$$\frac{\Omega}{4\pi^2}\frac{q\,\mathrm{d}q\,k_2\,\mathrm{d}k_2}{k_1} = \frac{\Omega}{4\pi^2}\frac{m^*}{\hbar^2}\frac{q\,\mathrm{d}q\,\mathrm{d}E}{k_1}.$$

Hence so long as m^* can be regarded as independent of k_2 the density of states is independent of k_2 and therefore independent of ω. Selecting only the terms which involve ω, we find that the rate at which δ_1 decreases due to the first term in (3.82) is proportional to

$$\int_{-\infty}^{\infty} S(q,\omega)\left[1+\left(\frac{\hbar\omega}{k_B T}\right)f_1^0-\frac{1}{2}\left(\frac{\hbar\omega}{k_B T}\right)^2 f_1^0(1-2f_1^0)+\ldots\right]\mathrm{d}\omega$$

for scattering through a vector q. With the aid of the moment theorems proved in §3.12 this becomes

$$a(q)-\frac{\hbar^2 q^2}{m_A k_B T}f_1^0(1-f_1^0)+\ldots \tag{3.83}$$

This quantity should be used to replace $a(q)$ in equations (3.76) and (3.80).

The magnitude of the correction term in (3.82) depends on the value of f_1^0. Now it is well known that provided the scattering is elastic the steady-state solution of the Boltzmann equation in the presence of a field is

$$f(\mathbf{k}) = f^0(\mathbf{k}-\Delta\mathbf{k}), \tag{3.84}$$

corresponding to a uniform displacement of the equilibrium distribution in $\mathbf{k}$-space. If (3.84) is a sufficient approximation to the distribution function when the scattering is inelastic, then as a function of energy in the neighbourhood of $\mathbf{k}_1$ we have

$$\delta_1 \propto \mathrm{d}f_1^0/\mathrm{d}E.$$

Hence the mean value of the correction term in (3.83), weighted by a factor δ_1 which expresses the strength of the contribution to the total current due to the state $\mathbf{k}_1$, is

$$-\frac{\hbar^2 q^2}{m_A k_B T}\int_{-\infty}^{\infty} f_1^0(1-f_1^0)\frac{\mathrm{d}f_1^0}{\mathrm{d}E}\,\mathrm{d}E\bigg/\!\!\int_{-\infty}^{\infty}\frac{\mathrm{d}f_1^0}{\mathrm{d}E}\,\mathrm{d}E = -\frac{1}{6}\frac{\hbar^2 q^2}{m_A k_B T}. \tag{3.85}$$

For the lightest metal, Li, this amounts to about $-0.015(q/2k_F)^2$, which makes it just significant compared with $a(q)$ (see fig. 5.2). In heavier metals, however, the usual elastic scattering theory should be completely adequate.

This conclusion provides some *a posteriori* justification for the neglect of thermal motion of the ions throughout chapter 1. Strictly speaking we should have taken account of their motion when calculating, for example, the extent to which the ionic pseudo-potential is screened by the conductions electrons; this would have meant using for $U(q, \omega)$ a dielectric constant which depended not only on q but on ω as well. But if the effect of the motion is scarcely detectable in τ_l or τ it is also scarcely detectable in ϵ'.

It ought perhaps to be added that the replacement of $\overline{|U(q)|^2}$ by $Na(q)\,|u(q)|^2$ in (3.75) and subsequently in (3.80) requires not only the assumption that the static dielectric constant can be used throughout, but also an assumption that the ions remain in thermal equilibrium when a current is switched on. The customary justification of the latter for solid metals is based on the belief that the time it takes the phonon distribution to reach equilibrium is much shorter than τ, but in liquid metals this condition can hardly be satisfied: τ is generally less than 10^{-14} sec. However, if an electric current were to produce a significant distortion of the 'phonons' in a liquid metal a departure from Ohm's law would presumably be observed. No such departures seem to have been reported.

3.17. Nuclear quadrupole interaction in liquid metals

In many non-metallic liquids the techniques of nuclear magnetic resonances have provided a useful tool for the measurement of self-diffusion coefficients. In liquid metals, NMR is not of much use for this particular purpose (though see Murday & Cotts, 1968), but a study of the nuclear spin relaxation time T_1 has yielded some information about an interaction with the nuclear quadrupole moment which is relevant in the context of the present chapter.

For readers unfamiliar with the subject of NMR a brief explanation of what is meant by T_1 and some related quantities may be necessary (for more detailed information consult Slichter (1963)). When a magnetic field is applied to a metal in the z direction the nuclear spins become preferentially orientated and the specimen acquires a longitudinal nuclear magnetisation M_z parallel to H. The inverse of T_1 defines the rate at which this magnetisation grows exponentially towards its equilibrium value. The rate is limited because the nuclear spin system has to find some way to get rid of an amount of energy $\mathsf{M}_z\mathsf{H}$ either to the 'lattice' (i.e. to the thermal motion of the ions) or to the conduction electrons. In practice, for most metals, the dominant relaxation mechanism is provided by interaction with the conduction electrons; each nuclear dipole is able to reorientate itself by causing an electron to flip its spin in the reverse direction (thereby satisfying the law of conservation of angular momentum) and at the same time to move to a state with a non-magnetic energy which is greater by $\hbar(\gamma_n - \gamma_e)\mathsf{H}$, where γ_n and γ_e are the gyromagnetic ratios of the nucleus and the electron respectively. The probability of this process must clearly be proportional to the density of states available for the electron near the Fermi level, $\mathcal{N}(E)_F$. Moreover, the number of electrons which can be affected must be proportional to $\mathcal{N}(E)_F k_B T$, since those which lie too far below the Fermi level are prevented from undergoing spin reversal by the exclusion principle. Finally, the probability depends upon the 'contact interaction' between the nuclear and electronic magnetic moments, which is responsible for the *Knight shift*, $\mathcal{K} = \Delta\mathsf{H}/\mathsf{H}$, discussed in §4.14 below, though the latter is also determined by the strength of the electronic spin susceptibility χ_p. From these ingredients arises the so-called *modified Korringa relation* for T_1 in a metal:

$$\frac{1}{T_1} = \left(\frac{1}{T_1}\right)_e = \alpha\frac{4\pi k_B T}{\hbar}\left(\frac{\gamma_n}{\gamma_e}\right)^2 \mathcal{K}^2, \tag{3.86}$$

where
$$\alpha = (m^*/m)^2\,(\chi_p/\chi_p(\mathrm{FE}))^{-2}. \tag{3.87}$$

Elementary treatments of electronic spin susceptibility (see §4.15) make it appear that the ratio of χ_p to its free-electron 'Pauli' value, $\chi_p(\mathrm{FE})$, should be just (m^*/m). More careful consideration shows, however, that the correlation and exchange effects between conduction electrons are likely to enhance χ_p by a good deal more than they affect the density of states (see p. 295). Hence α is expected to be rather less than unity.

TABLE 3.3. *Nuclear spin relaxation times in liquid metals*

Isotope	I	$\hbar\gamma_n$ (nuclear magnetons)	Q (barns)	T (°K)	$(T_1)_e$ (msec)	α	$(T_1)_q$ (msec)	τ_q (psec)
Li^6	1	0.82	$8 \cdot 10^{-4}$	460	620	0.62	—	—
Li^7	3/2	2.17	0.036		85	0.65	—	—
Na^{23}	3/2	1.48	0.10	383	11	0.64		
Rb^{85}	5/2	0.54	0.24	315	2.6	0.85	—	—
Rb^{87}	3/2	1.83	0.12		0.23	0.83	—	—
Cu^{63}	3/2	1.48	−0.16	1360	0.66	0.57	—	—
Cu^{65}	3/2	1.59	−0.15		0.57	0.57	—	—
Al^{27}	5/2	1.46	0.15	950	2.1	0.65	—	—
In^{115}	9/2	1.23	1.16	440	0.20	0.93?	1?	> 6.5?
Bi^{209}	9/2	0.90	−0.4	542	0.066	1.43?	0.15?	> 8.7?
Hg^{199}	1/2	1.00	0	300	0.053	0.84	0	—
Hg^{201}	3/2	0.40	0.45		0.33		0.022	> 0.5
Ga^{69}	3/2	1.34	0.18	300	1.0	0.75	1.7	> 0.3
Ga^{71}	3/2	1.70	0.11		0.6		4.3	
Sb^{121}	5/2	1.34	−0.53	925	0.15	0.62	0.29	> 1.4
Sb^{123}	7/2	0.73	−0.68		0.50		0.41	

Once the magnetisation M_z has been established it is possible, by the application of a pulsed auxiliary field, to make it precess very rapidly into the xy plane, where it settles down to precess about H with the nuclear Larmor frequency $\gamma_n H$ ($\sim$ 10 Mc/sec). It can then give rise to an oscillating signal in a pick-up coil set with its axis along the x or y directions. The rate at which this signal decays is determined by the inverse of T_2, where T_2 is the *transverse* as opposed to *longitudinal* relaxation time. There are good reasons to expect T_2 to equal T_1 in any liquid metal, whether or not the relaxation mechanism discussed above is dominant, at any rate in ideal circumstances where H is uniform. In practice it has frequently been found that T_2 is less than T_1 (Holcomb & Norberg, 1955), but the discrepancies can be reduced by taking great care to avoid impurities (Hanabusa & Bloembergen, 1966) or by using a different medium to support the fine metallic particles of which the specimen is composed (Cornell, 1967). They are probably of no theoretical significance. It is best to choose T_1 rather than T_2 for comparison with theory since it should be insensitive to most forms of inhomogeneity.

The theory expressed by equation (3.86) gives an adequate account of the observations for Li, Na, Rb, Cu, Al, Hg^{199} and perhaps for In too. A selection of the best available data is presented in table 3.3, which is taken from Faber (1963) with some modifications and additions on account of recent work. For these seven metals, and for Bi^{209}, the column headed $(T_1)_e$ shows the experimental values of T_1. A comparison of the results for the three pairs of isotopes, Li^6 with Li^7, Rb^{85} with Rb^{87}, and Cu^{63} with Cu^{65}, shows that within the limits of error – often considerable – $(1/T_1)$ is indeed proportional to γ_n^2. It has also been verified that it is proportional to $T\mathscr{K}^2$ within the limits of error. And finally the values of α which are needed to secure quantitative agreement with (3.86), which are included in table 3.3, are acceptable; they are plotted in fig. 3.11 as a function of r_s, which is a measure of electron density, together with a theoretical curve which is believed to represent the behaviour of $(m^*/m)^2\,(\chi_p/\chi_p(\text{FE}))^{-2}$ for an ideal jellium with an accuracy of about $\pm 5\%$ (Hedin & Lundqvist, 1969).† Rossini *et al.* (1967) have suggested that the value of α required for In (0.93) is implausibly high, and the value required for Bi (1.43, according to measurements quoted by Rossini & Knight, 1969) is certainly too high; to bring these metals into line with others of comparable electron density, we should take α to lie in the range 0.7–0.8. It is therefore possible that in In and Bi $(1/T_1)_e$ does not account for the whole of the observed $(1/T_1)$, i.e. that some additional relaxation mechanism is at work. Rossini *et al.* have sought evidence to support this hypothesis in the temperature variation of $(1/T_1)$ but it is hardly conclusive; the more

† The discrepancies could be explained by postulating that $(1/T_1)_e$ contains a small contribution from the non-contact term in the magnetic interaction between the nucleus and the conduction electrons; in principle this may enhance the relaxation rate by an amount proportional to γ_n^2 without significantly enhancing the Knight shift.

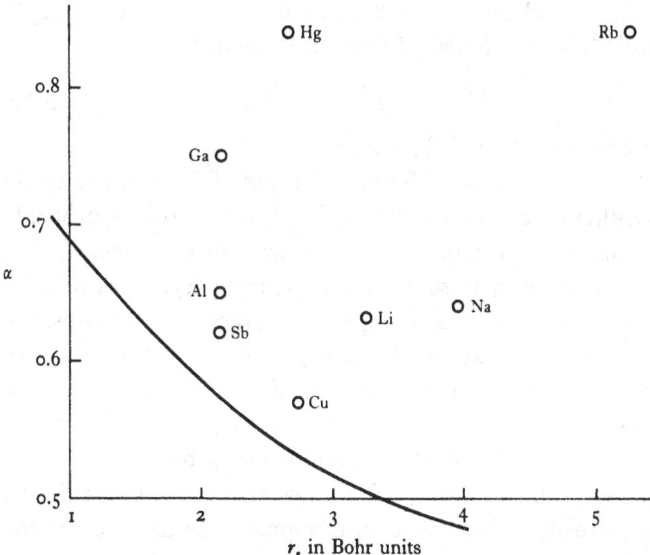

Fig. 3.11. Experimental points and a theoretical curve for the factor α described by (3.87).

extensive measurements of Warren & Clark (1969) reveal that, for In at any rate, $(1/T_1)$ is still quite accurately proportional to $T\mathscr{K}^2$.

It is when the behaviour of Hg^{201} is examined that the existence of an additional relaxation mechanism becomes really apparent. This isotope has a gyromagnetic ratio which is considerably smaller than that of Hg^{199}, so that $(T_1)_e$ should be a lot bigger: 0.33 as opposed to 0.053 msec at 300 °K. Yet the observed value of T_1 for Hg^{201} at this temperature is actually less than for Hg^{199}: only 0.02 msec. Since Hg^{201} has a large quadrupole moment where Hg^{199} has none it seems very likely that its relaxation rate is enhanced by a quadrupole interaction. The nuclear quadrupole moment Q couples with the gradient† q of the electrostatic field in which it finds itself; if q is fluctuating rapidly with time the Fourier components of the interaction spread over a wide band of frequencies, and those with ω close to $\gamma_n\mathsf{H}$ can in principle assist relaxation, by causing transitions from one spin orientation to

† Strictly q is a tensor and the part of it which is of interest is $\mathsf{q}_{zz}(1+(\eta^2/3))^{\frac{1}{2}}$, where η is the 'asymmetry parameter', usually rather less than unity.

another. The quadrupole relaxation time which is listed for Hg^{201} in the table has been obtained from the equation

$$(1/T_1) = (1/T_1)_e + (1/T_1)_q, \tag{3.88}$$

on the assumption that $(T_1)_e$ is 0.33 sec.

Results are now available for two isotopes of Ga and two of Sb and in neither case is $(1/T_1)$ proportional to γ_n^2, so it is probable that the quadrupole interaction is important for these metals also. It is not quite so easy to separate $(1/T_1)_e$ from $(1/T_1)_q$ as it is for Hg, because none of the isotopes concerned has a vanishing quadrupole moment. It can be done, however, by solving two simultaneous equations, if we know the factors by which the two terms should change in passing from one isotope to the other. According to (3.86) the first term should change by the ratio of γ_n^2. As for the second, Abragam (1961) shows that if, for convenience, the auto-correlation function of q is supposed to decay exponentially with a time constant τ_q, then

$$(1/T_1)_q = \frac{3\pi^2(2I+3)}{10I^2(2I-1)} \overline{\left(\frac{e^2\mathsf{Q}\mathsf{q}}{h}\right)^2} \tau_q, \tag{3.89}$$

where I is the nuclear spin quantum number. The result holds only in the 'extreme narrowing' limit, when $\gamma_n \mathsf{H} \tau_q \ll 1$, but since this condition requires only that τ_q should be much less than 10^{-7} sec there is no doubt that it is fulfilled in liquid metals. Equation (3.89) provides the information that is needed, and it has been used in deriving from $(1/T_1)$ the results for $(T_1)_e$ and $(T_1)_q$ which are entered in table 3.3 for Ga and Sb. Within the limits of error $(1/T_1)_e$ comes out to be proportional to T for both metals as it should, while the values of α needed to make it fit (3.86) are sensible. $(1/T_1)_q$ tends if anything to decrease slowly on heating, which one may attribute to a decrease either in q or τ_q.

Hg^{201}, Ga, Sb and for that matter In and Bi are all metals in which the quadrupole interaction is known to be considerable in the solid phase; their crystal structures lack the high degree of symmetry which characterises the cubic metals Li, Na, Rb, Cu and Al. The magnitude of $(e^2\mathsf{Q}\mathsf{q}/h)$ in the solid has been deduced, either by analysis of a low-temperature specific heat anomaly (Hg^{201}, Cornell, 1967) or from the details of the resonance spectrum (Ga,

Knight, Hewitt & Pomeranz, 1956; Sb, Hewitt & Williams, 1963; In, Hewitt & Taylor, 1962; Bi, Hewitt & Williams, 1964). It decreases on heating, presumably due to the displacement of the ions from their regular lattice sites, by only a few parts per cent in the case of Ga but by a factor of 2 or more in the case of In. Just below the melting point this coupling constant appears to reach about 20 Mc/sec in Ga^{69}† and In^{115}, about 50 Mc/sec in Sb^{121}, and something less than 150 and 58.5 Mc/sec in Hg^{201} and Bi^{209} respectively. We may use these figures as upper limits for the liquid phase, since if the coupling constant decreases on heating in the solid it must surely continue to decrease during the further distortion of the lattice that melting involves. In this way lower limits for τ_q in the liquid may be obtained. The results are shown in table 3.3; the ones for In and Bi are based upon highly tentative assignments for $(T_1)_q$, in line with the suggestion of Rossini *et al.*, and may be too big.

How can τ_q be estimated independently? One primitive method which has been used by some authors is to set $(6D\tau_q)^{\frac{1}{2}}$ equal to the interionic spacing, say $2R_A$, which gives an answer of about 7 psec for the first four of the metals under consideration, and about 10 psec for Bi,‡ not inconsistent with the figures in the table. The philosophy behind this method is presumably that each ion can be regarded as diffusing independently and that equation (3.5) is adequate to describe the evolution of its r.m.s. displacement X with time. But the value of q at any one ion is highly sensitive to the configuration of its neighbours, and if indeed they all diffuse independently their r.m.s. displacement need not be anything like as much as one spacing for q to be changed completely; at a guess, one tenth of a spacing should be quite sufficient. For such small displacements equation (3.19) may be a better approximation for X than (3.5), which suggests a relation of the form

$$(3k_B T/m_A)\tau_q \simeq \tfrac{1}{5}R_A. \tag{3.90}$$

† This applies to solid GaI. It is arguable (Ascarelli, 1966) that the structure of liquid Ga is more akin to that of the solid β phase; certainly it crystallises more readily as Gaβ and the entropy of crystallisation is not so anomalous. But no study of the quadrupole resonance spectrum in Gaβ has yet been reported.

‡ The diffusion coefficient has not been measured for liquid Sb and Bi but may be estimated from equation (3.29).

This yields values for τ_q of order 0.1 psec, significantly less than the lower limits shown in the table. The implication may be that the diffusion of neighbouring ions is not independent, and that the local order of a cluster of ions may survive while the cluster as a whole is displaced.

It is tantalising not to be able to reach any more definite conclusion, since this quadrupole interaction is one of very few effects to which the local order in the liquid phase is relevant. An attempt to treat the problem more quantitatively has been made by Sholl (1967) (see also Borsa & Rigamonti, 1967; Yul'met'ev, 1968). He shows how the relaxation rate may be evaluated, without the assumption that the decay of the auto-correlation function of q is necessarily exponential, given a knowledge of (*a*) the potential set up by each screened ion, (*b*) the three-body distribution function g_3 as well as g_2, (*c*) the way in which these distribution functions evolve with time. Unfortunately, in order to obtain numerical answers he is forced to make some rather drastic assumptions, e.g. that the ionic potential behaves like $\cos(2k_F R)/R^3$ (see § 1.6), that g_3 is given by the superposition approximation, and that (3.5) is accurate even for short times. The answers do seem to be about right for Ga and Sb and perhaps for In. To test the theory further it should be applied to Na, Rb, Cu and Al (a start has been made in this direction by Rossini & Knight (1969)) to see whether it predicts correctly that in these liquid metals quadrupole relaxation should be undetectable.

CHAPTER 4

ELECTRON STATES IN DISORDERED STRUCTURES

4.1. Prologue

Now that we understand more about how the ions are arranged in liquid metals we may return to the electrons. The argument in chapter 1 was based throughout upon the NFE model, i.e. upon the assumption that the electrons can be described in terms of a set of plane wave states of well-defined $\mathbf{k}$, the occupied states being enclosed by a Fermi surface of spherical form. We saw that in general the ionic pseudo-potential will perturb the electrons, so that the energy of a plane wave state is not necessarily just $\hbar^2k^2/2m$; the perturbation will also scatter electrons from one value of $\mathbf{k}$ to another. But we assumed that these effects could be allowed for by the introduction of a suitable effective mass m^* and lifetime τ_l, and that we could go on using the NFE model despite them. A theoretical expression for τ_l has been obtained in §3.16.

Yet in a regular crystal the NFE model is not necessarily adequate. The electron states may be significantly modulated by the lattice and should be described by Bloch waves, while the Fermi surface may be significantly distorted from the spherical. It is well known that to calculate the electronic properties of a solid metal correctly, e.g. its transport properties, it is essential to take these subtleties into account. Presumably it is still necessary to do so if the specimen is a fine-grained polycrystalline solid. Can we really be confident that the local order in a liquid metal is sufficiently slight to permit such complications to be ignored? Of course the lifetime τ_l and the electronic mean free path are probably both rather short in liquid metals. This may introduce considerable uncertainty into the definition of $\mathbf{k}$, and one can see intuitively that distortions of the Fermi surface are less likely to matter on that account. But then one comes up against a problem of a different kind, whether it is legitimate to assume, as we did in chapter 1, that the Fermi surface is *sharp*.

We really need an understanding of what the *eigenfunctions* look like in a disordered system such as liquid metal, and it is to this question that most of the present chapter is addressed. It is a fascinating question in its own right, which has received a lot of attention from theorists in recent years; some have been interested more in amorphous solids rather than liquids, or in disordered solid alloys, or even in the spectrum of the vibrational modes for a disordered structure rather than in the spectrum of the conduction electrons, but in many ways these are different aspects of the same basic problem.

For one-dimensional systems the problem may be said to be solved. The electronic eigenfunctions in a disordered metal in one dimension are not travelling waves at all; they have a standing wave character and moreover are *localised* in space. Provided that the short-range disorder is not too extreme, there can be *forbidden gaps* in the energy spectrum, even though long-range order is completely lacking. Simple arguments supporting these conclusions are given in §§4.3 to 4.6 below; §4.2 is devoted to a preliminary survey of the extensive literature in the field for readers who wish to explore it further. In §§4.7 and 4.8 we examine whether it is possible to reach the same results by perturbation theory, starting from the one dimension equivalent of the NFE model. It turns out that if sufficient care is exercised one can get very close. But to substantiate the existence of energy gaps, for example, it is certainly necessary to go beyond the second order.

In three dimensions, unfortunately, the problem is much harder and a complete solution is not yet in sight. It cannot be emphasised too strongly that the topological differences between systems in one dimension and in three are profound; it is quite improper to fly from one to the other on the wings of analogy. No attempt is made here to expound the recent theoretical work on three-dimensional disordered systems: the mathematical arguments are too complex, the reliability of the results is still too uncertain, and their relevance to problems of experimental interest too remote. The reader must be satisfied with some brief references to the original literature, to be found in §4.2.

Since a better theory is out of our reach, we revert once more, in §4.9, to perturbation methods based upon the NFE model; by now we do at least have some experience, based upon the one-dimensional case, by which to gauge the reliability of our conclusions. We arrive at a description of the eigenstates in terms of groups of plane waves; their Fourier components fill a shell in **k**-space, of finite thickness related to the inverse of the mean free path. Up to second order, at any rate, the shells should be spherically symmetric in **k**-space, and the eigenfunctions have a standing wave, non-propagating character; it is most unlikely, however, (see §4.10) that they are localised. Localisation must of course occur if a liquid metal is expanded, perhaps at the stage when the Fermi level falls below the zero of the 'muffin-tin' potential, but that is another story which we have touched on already in §1.12. Forbidden energy gaps and departures from spherical symmetry are certainly conceivable when there is local order, as in a polycrystalline solid, but to generate these features it is clearly necessary, as in one dimension, to go beyond the second order. The difficulties of doing so are examined in §4.11. The two most formidable ones are (*a*) that to carry any calculation even to the third order requires a knowledge of the three-body distribution function, g_3, which we do not possess, and (*b*) that in the third order the linear screening theory of chapter 1 cannot necessarily be trusted (see §4.12). In real liquid metals at normal densities, however, it seems probable that one can allow for the higher-order terms with reasonable accuracy by effecting a relatively slight modification in the mean pseudo-potential for each ion, while retaining the formalism of a second-order calculation.

Attempts to put numbers into this theory are still rather limited. However the density of states has been computed for a number of liquid metals, by a variety of methods based upon the formulae derived below. The best results suggest, as is shown in §4.9, that at normal densities m^*/m is close to unity; at any rate, the structure-dependent effects with which the theory is primarily concerned are small. If the peak in $a(q)$ is not sharp enough to affect m^*/m in a second-order calculation, it is a reasonable inference that higher-order effects are also small, e.g. that the assumption of spherical symmetry in **k**-space is normally legitimate.

The final sections, §§4.13–4.16, contain a critical review of the available data concerning positron annihilation, magnetic susceptibility and Knight shift in liquid metals, and some brief remarks about Compton scattering of X-rays. These are all topics to which the theory developed previously is in principle relevant. With one or two exceptions the results do agree with the NFE model and confirm that m^* is close to m.

Throughout this chapter, encouraged by the results of §3.16, we shall treat the ions as stationary and the scattering as elastic.

4.2. Survey of the literature

One of the simplifications that is legitimate in one dimension but not necessarily in three is to represent the ions by δ-functions of potential, adjusted in strength so as to have the right transmission and reflection coefficients for an incident electron with the energy of interest. It has been shown by Borland (1961*b*) that this involves no loss of generality, at any rate for ions whose potentials do not overlap. Many authors have therefore discussed models which are disordered versions of the familiar Kronig–Penney model for a one-dimensional solid. Perhaps the simplest form of disorder to consider is that in which the δ-functions remain regularly spaced but a certain number of them, selected at random, are made stronger than the rest; this corresponds to a 'solid binary alloy'. Or one may suppose the δ-functions to be displaced at random, but with some symmetrical probability distribution, about their regularly spaced positions; this corresponds to a 'hot solid', in which long-range order is still present. It is also instructive to consider a 'dislocated solid' in which a sequence of δ-functions have all been displaced by the same amount, thus creating an unusually large interval between two successive δ-functions at one end of the sequence and an unusually small one at the other. To represent a one-dimensional 'liquid' one must allow all the intervals between successive δ-functions to vary at random, either between some prescribed upper and lower limits, or according to some convenient distribution function that allows very large or very small intervals to occur from time to time; in either case the long-range order of the solid is lost. Finally, there is the one-dimensional 'gas', in which there is no correlation at all between the position of one δ-function and another.

The binary alloy case need not detain us (consult Schmidt,

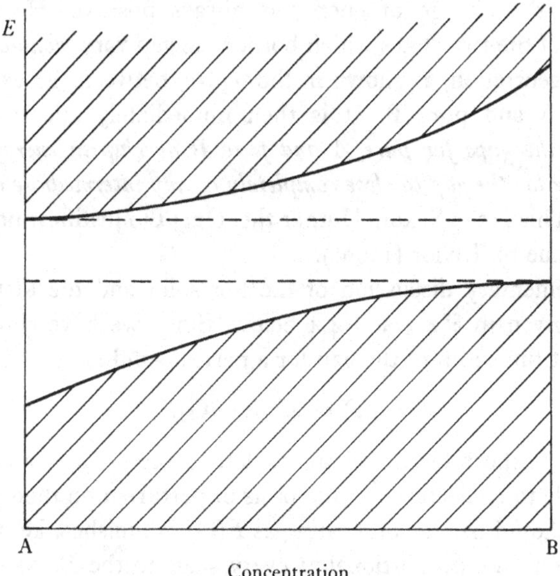

Fig. 4.1. Schematic energy spectrum for the electrons in a one-dimensional alloy; hatching indicates regions where states are allowed. No states are allowed, whatever the concentration, in the energy range between the two broken lines.

1957; Faulkner & Korringa, 1961; Agacy & Borland, 1964; Tong & Tong, 1969) except for one simple theorem, first conjectured by Saxon & Hutner (1949) and proved by Luttinger (1951), which has an analogy in the case of one-dimensional liquids. The theorem may be explained and justified with the aid of fig. 4.1. Let us suppose that a one-dimensional crystal of pure A is converted *continuously* into pure B by slowly intensifying the δ-functions, one after the other in any chosen order. The number of nodes in an eigenfunction cannot suddenly change during this process, so that a particular eigenfunction in A deforms adiabatically into one which has the same number of nodes in B. It follows that the two which represent the states immediately on either side of an energy gap in A must, after deformation, represent the states on either side of the corresponding gap in B. But the change of energy of any state ψ due to some infinitesimal change of potential ΔV is

$$\langle\psi|\Delta V|\psi^*\rangle, \tag{4.1}$$

and if ΔV is always positive, as we assume to be the case in this example, the change of energy is always positive. Hence the energies of the two states which border the gap vary *monotonically* with concentration, as shown in the figure, between the extremes of pure A and pure B. It is then immediately apparent that *provided the gaps for pure A and pure B overlap in energy, it is impossible for the gap to close completely at any intermediate concentration.* This is the Saxon–Hutner theorem; the justification given for it is due to Taylor (1966*b*).

An elementary discussion of the hot solid and the dislocated solid is given in §§4.3 and 4.4 below. Since we have shown on p. 59 that the width of the gap for a perfect solid is

$$2U(\mathbf{G})/\Omega = 2F(\mathbf{G})u(\mathbf{G})/\Omega,$$

where $\mathbf{G}$ is the relevant reciprocal lattice vector, it would seem reasonable to conjecture, on the basis of perturbation theory, that the gap should narrow on heating as $F(\mathbf{G})$ diminishes, i.e. that its width should be proportional at every stage to the *Debye–Waller factor* (see p. 116). The discussion seems to bear out this conjecture.

If, however, one goes on to conclude that the gaps must necessarily disappear when the long-range order is lost, because then $F(\mathbf{q})/\Omega$ is of order N^{-1} for all values of $\mathbf{q}$ other than zero, one is making a severe mistake. It is proved in §4.5 below, by an argument due to Borland (1961*b*) and Roberts & Makinson (1962), that so long as there are upper and lower limits (b_{max} and b_{min} say) to the spacing between successive δ-functions a gap is bound to persist, whether or not there is long-range order, *provided that the gap for a regular one-dimensional solid with the spacing* b_{max} *overlaps in energy the gap for a solid with the spacing* b_{min}. Here is the analogue of the Saxon–Hutner theorem.

The density of states and persistence of gaps in the one-dimensional liquid have been discussed in a variety of ways by Dworin (1965), Hori (1964, 1966, 1968*a*, 1968*b*), Hori & Matsuda (1964), Matsuda (1962, 1966), Matsuda & Okada (1965), Hiroike (1965), De Dycker & Phariseau (1965), Taylor (1966*a*), Blair (1967), Halperin (1967) and Fornazero & Mesnard (1967). Hiroike has concluded that a gap should persist even if the spacings are distributed in a Gaussian fashion, so long as the r.m.s. deviation is not too great; De Dycker & Phariseau argue that it is *only* the r.m.s. deviation which really matters, the nature of the distribution being of minor importance. But it seems probable that these claims, which are justified by arguments that are not rigorously exact, are erroneous, and that some upper and lower limits are essential if states in the energy gap are to be strictly forbidden. If a Gaussian distribution prevails, then, whatever electronic energy one cares to nominate, the ensemble of possible arrangements of the δ-functions must surely include a few freak ones to make this energy allowable. One of the morals that has been emphasised by many workers in this field is that great care has to be exercised in averaging over the ensemble if essential features of the true solution are not to be lost in the process.

It is possible to check some of the theoretical predictions against the results of computer calculations for disordered Kronig–Penney models, though naturally it is impossible to verify by any finite calculation that states in a gap are absolutely forbidden. Monte Carlo computations for chains of up to 50000 δ-functions, with intervals distributed in a parabolic fashion between an upper and lower limit, have been reported by Makinson & Roberts (1960) and Roberts (1963). Their technique was one of *node counting*, initiated by James & Ginsbarg (1953) and Landauer & Helland (1954). If the Schrödinger equation is integrated numerically from one end of the chain to the other it will not in general prove possible to satisfy the boundary conditions at both ends simultaneously, unless a laborious search is undertaken to find an eigenvalue of the energy. It is clear, however, that if the entire energy range could be explored step by step, each successive eigenvalue of the energy would correspond to an eigenfunction with one more node in it, so that the number of nodes in any particular eigenfunction of energy E gives the *integrated* density of states, i.e. the number of states with energy less than E. Hence all one need do to find this integrated density of states is to count nodes, and for this purpose it is clearly *not* essential to hit upon an eigenvalue of the energy in the first place. An energy gap will show up as a range of E in which the number of nodes stays constant.

Computations by a rather different method have since been carried out by Borland & Bird (1964). They looked for approximate solutions to an integral equation which in principle gives

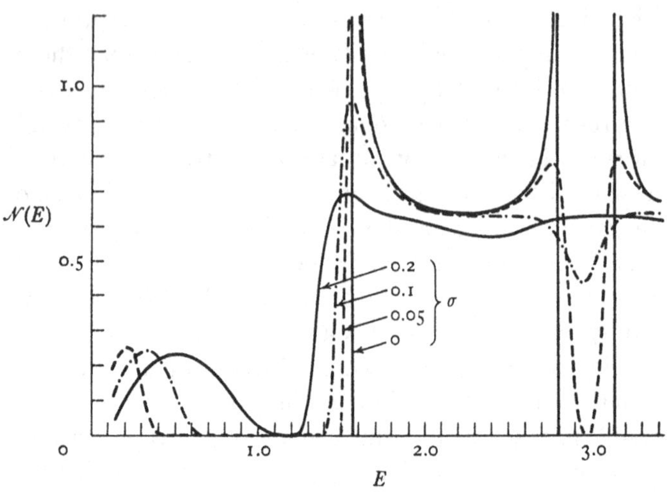

Fig. 4.2. The density of states in a one-dimensional chain for various degrees of disorder. (Redrawn by permission from Borland & Bird (1964), *Proc. Phys. Soc.* **83**, 23. Acknowledgement is also made to the National Physical Laboratory.)

directly the energy spectrum of the electrons, averaged over an ensemble of infinite chains; the Monte Carlo method can only be applied to one finite chain at a time, so that the results of Borland & Bird are probably more accurate. The curves they have obtained for the density of states, assuming a set of *attractive* δ-function potentials of sufficient strength to reduce the energy corresponding to the top of the first band to zero in the case of perfect order, are shown in fig. 4.2; the intervals were assumed to be distributed at random between $(b-\frac{1}{2}\Delta b)$ and $(b+\frac{1}{2}\Delta b)$, and the curves are labelled by the parameter σ, which is the ratio between the r.m.s. deviation $\Delta b/2\sqrt{3}$ and the mean interval b. A value of 0.05 for σ is sufficient to narrow the first gap very considerably and to let states appear right across the second. As Δb and σ are further increased both gaps fill up quite rapidly.

It has been suggested by a number of authors (Mott & Twose, 1961; Roberts & Makinson, 1962; Borland, 1963) that the eigenfunctions for states which appear inside a gap when the chain is disordered are *localised*; that is to say, their amplitude decays fairly steadily to zero on either side of a central maximum. Indeed, it appears that nearly *all* the states in a disordered one-dimensional

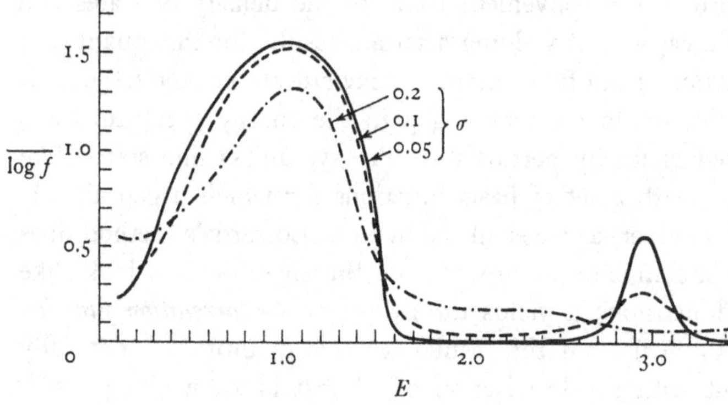

Fig. 4.3. Localisation of states in a disordered chain. (Redrawn by permission from Borland & Bird (1964), *Proc. Phys. Soc.* **83**, 23. Acknowledgement is also made to the National Physical Laboratory.)

chain are localised in this sense, though the decay is most rapid for the ones inside the gap. Borland & Bird have computed the ensemble average of $\log f$ as a function of E, where f is the factor by which the envelope of ψ^2 decays on passing from one δ-function to the next, and their results for the same conditions that yield the curves in fig. 4.2, are shown in fig. 4.3; evidently the localisation of ψ^2 for states within the first gap is very severe. An elementary discussion of this phenomenon is given in §4.6 below. The reader may also wish to consult Mattuck (1962), Faulkner (1964), Lifshitz (1964), Halperin & Lax (1966, 1967), Mott & Allgaier (1967), Mott (1967) and Roberts *et al.* (1968, 1969), all of whom discuss in one way or another the states that appear below a conduction band when disorder is introduced, though some of them are concerned more with the effect of 'alloying' than of 'melting'.

The relatively simple case of the one-dimensional 'gas' is the subject of papers by Lax & Phillips (1958), Frisch & Lloyd (1960), Borland (1961*a*) and Halperin (1965).

Most of the papers quoted so far represent attempts to solve the problem 'exactly', i.e. without the use of perturbation theory. For the application of perturbation methods the reader may consult Sah & Eisenschitz (1960), Klauder (1961), Edwards (1961), Yonezawa (1964) and Gubanov (1965). Edwards's method, which is expounded in elementary language in §4.7 below (see also Faber,

1966), provides a convenient route to the density of states and should be capable of yielding accurate results for this quantity if the deviations from free-electron behaviour are not too large. It is hard, however, to generate a gap in the energy spectrum for a disordered chain by perturbation theory, unless one starts, like Gubanov, with a set of basis functions for which a gap already exists. Moreover, as we shall see in §4.8, Edwards's method does not give a complete picture of what the eigenfunction looks like in one dimension; it yields the average *auto-correlation function* $\overline{\langle \Psi(x+X)|\Psi(x)\rangle}$, but the results for this quantity are not fully consistent with the localisation of Ψ that is known to prevail. Emphasis on the auto-correlation function has also been placed by Taylor & Bambakidis (1967).

Several of the authors quoted above have attempted to extend their arguments to three-dimensional systems, but certain fundamental questions remain unanswered: whether there is an analogue to the Borland–Makinson–Roberts theorem in three dimensions, for example, and just what conditions must be satisfied before the eigenfunctions become localised. Experimental evidence, as we shall see in §4.10, virtually eliminates the possibility of localisation in a real liquid metal at normal densities, for states near the Fermi level at any rate. We have noted on p. 61, however, that expansion may bring the Fermi level below the zero of the 'muffin-tin potential' and that the tight-binding approximation may then become appropriate. Following the extensive work of Anderson (1958, 1970), Mott (1967, 1968, 1969, 1970) and many others, it is now widely accepted that the tight-binding method applied to a three-dimensional disordered array, whether of ions in an expanded liquid metal or of impurity centres in a semiconductor, does predict localised states when the disorder is large enough. The interested reader may consult Eisenschitz & Dean (1957), Beeby & Edwards (1963), Bonch-Bruevich (1965, 1968), Ziman (1969*a*, 1969*b*, 1970), Lloyd (1969) and Brouers (1970), in addition to the papers of Mott and Anderson quoted above. A book by Mott & Davis (1971) discusses the application of the concept of localised states to a wide range of physical problems involving disordered semiconductors.

The perturbation method of Edwards is readily extended to

three dimensions (Edwards, 1962), and, with the modifications suggested below, it seems to provide the most useful way of estimating the density of states in real liquid metals. It has been applied for this purpose by a number of authors, with results discussed in §4.9. One of the advantages of the Edwards formalism is that it is framed in terms of the interference function $a(q)$ which is provided so directly by diffraction experiments. Of course if the pseudo-potential of the individual ions is too strong, so that the Born approximation becomes inadequate, a rigorous calculation is bound to require a knowledge of the higher-order distribution functions. Approximate methods for use in these circumstances have been discussed by Edwards himself, Beeby (1964), Ballentine & Heine (1964), Jalickee, Morika & Tanaka (1965), Ballentine (1965) and Lukes (1965). They appear to be equivalent to the device suggested on p. 48 and further examined in §4.11 below, of replacing the true pseudo-potential by an effective one, to be estimated on the crude assumption that the environment of each ion is spherically symmetric. So do the approximate methods based on the Korringa–Kohn–Rostoker method for solid metals which have been developed by Ziman and his colleagues (Phariseau & Ziman, 1963; Ziman, 1965, 1966; Lloyd, 1967; Lloyd & Berry, 1967; Morgan & Ziman, 1967; De Dycker & Phariseau, 1967), though Fletcher (1967*a*, 1967*b*) and Keller (1971) have attempted to extend the formalism so that angular correlations in the disposition of neighbouring ions may be taken into account.

Finally, there is a paper by Rousseau, Stoddart & March (1970) that fits into none of the categories laid down above. It contains an approximate method for calculating densities of states which the authors apply with interesting results to a model representing liquid Be. One implication of the analysis is that, for the density of states to depend upon structure, the pseudo-potentials of adjacent ions must sometimes overlap. No other argument suggests that overlap is so important – it never occurs, of course, in the Kronig–Penney models with which we began this section – so there is some reason to doubt whether the approximations of Rousseau *et al.* are legitimate.

4.3. Hot solids in one dimension

It is necessary first to establish some elementary results about the way an electronic wave function ψ behaves when it encounters a δ-function of potential. We shall suppose the potential to be attractive, equal to $-V$ within a square well of width t, though it makes no essential difference to the argument or to the results to choose a repulsive potential instead. Inside the well the wave function is proportional to

$$\exp\left(\pm \mathrm{i}(2m(E+V)/\hbar^2)^{\frac{1}{2}}x\right).$$

If t is allowed to go to zero while V goes to infinity at such a rate that (Vt) remains finite, it is apparent that the change of ψ in passing from one side of the well to the other becomes infinitesimal: i.e.

$$\psi_{\mathrm{R}} = \psi_{\mathrm{L}} \tag{4.2}$$

where the suffices stand for 'right' and 'left'. On the other hand there is a finite change in the gradient of ψ: if x is taken to increase from left to right, then

$$\psi'_{\mathrm{R}} - \psi'_{\mathrm{L}} = t\psi'' = -(2mVt/\hbar^2)\psi. \tag{4.3}$$

Let us suppose that outside the well on the left the wave function is represented by

$$\psi = A_{\mathrm{L}} \cos(kx + \zeta_{\mathrm{L}}) \tag{4.4}$$

and similarly on the right, and that the origin of x coincides with the centre of the well. Then it follows from (4.2) and (4.3) and the usual boundary conditions (ψ and ψ' everywhere continuous) that

$$\tan\zeta_{\mathrm{R}} - \tan\zeta_{\mathrm{L}} = 2mVt/\hbar^2 k = p \quad \text{(say)} \tag{4.5}$$

and

$$(A_{\mathrm{R}}/A_{\mathrm{L}})^2 = 1 + p\sin 2\zeta_{\mathrm{L}} + p^2\cos^2\zeta_{\mathrm{L}}. \tag{4.6}$$

The parameter p introduced in (4.5) forms a convenient dimensionless representation of the strength of the δ-function.

We are going to imagine that we have a continuous chain of δ-functions, or that the boundary conditions at the end of the chain are periodic, which comes to the same thing. This means that we are not necessarily restricted to solutions of the Schrödinger equation which, like (4.4), are real; solutions of the form

$$\psi = \tfrac{1}{2}(A^{+}\exp(\mathrm{i}kx) + A^{-}\exp(-\mathrm{i}kx)) \tag{4.7}$$

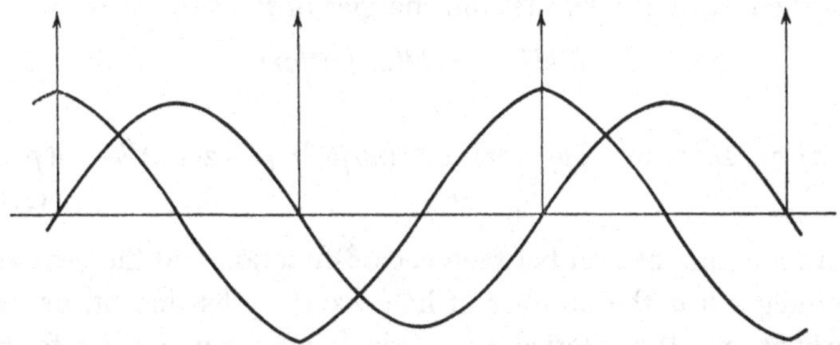

Fig. 4.4. Eigenfunctions on either side of a band gap for the Kronig–Penney model; vertical arrows indicate δ-functions of potential.

which have a travelling-wave character are also conceivable. It is possible, however, to express (4.7) in the form of (4.4) if the phase angle ζ is chosen so that

$$\tan \zeta = -\mathrm{i}(A^+ - A^-)/(A^+ + A^-) \tag{4.8}$$

and the amplitude A so that

$$A = (A^+A^-)^{\frac{1}{2}}, \tag{4.9}$$

and if this is done equations (4.5) and (4.6) remain valid. It is purely a mathematical device, of course, to define a phase angle whose tangent may be imaginary, but it is a convenient one. Real wave functions are easier to visualise and much of the argument that is to follow will be phrased in terms of them. But the mathematical results obtained can be applied equally well to travelling wave functions, if the substitutions expressed by (4.8) and (4.9) are carried through.

For the moment we are interested primarily in the states which border an energy gap in a perfectly regular solid and there is no doubt that the eigenfunctions for these are essentially real. It is hardly necessary to solve the Kronig–Penney model completely in order to see what they look like; they are standing waves of the type shown in fig. 4.4. The condition that $A_{\mathrm{R}}/A_{\mathrm{L}}$ is unity at each δ-function shows that

$$\zeta_{\mathrm{L}} = \pm \tfrac{1}{2}\pi$$

if ψ has a node there, whereas if it has an anti-node

$$\zeta_{\mathrm{L}} = -\tan^{-1} \tfrac{1}{2}p \simeq -\tfrac{1}{2}p \quad \textit{or} \quad \pi - \tfrac{1}{2}p.$$

It is then easily shown that the energies of the two states are

$$\hbar^2k^2/2m = \hbar^2(n\pi/b)^2/2m \tag{4.10}$$

and

$$\hbar^2(n\pi/(b+2k^{-1}\tan^{-1}\tfrac{1}{2}p))^2/2m \simeq \hbar^2(n\pi/b)^2/2m-2(Vt)/b+O(p^2), \tag{4.11}$$

where b is the interval between each δ-function and the next and the integer n is the number of half wavelengths that fit, or very nearly fit, into this interval; fig. 4.4 is drawn for $n = 1$, i.e. for the two states which border the lowest energy gap. Hence if the δ-functions are sufficiently weak for the terms of order p^2 and higher to be negligible the width of the gap is $2(Vt)/b$, which is, of course, the answer given by the one-dimensional analogue of (1.84).

Now suppose that the 'solid' is heated and the δ-functions are displaced slightly from their regularly-spaced positions by amounts Δx_j. If the δ-functions are weak and the displacements small we may calculate the resultant energy shifts with the aid of (4.1). Because the δ-functions are displaced from the nodes in its wave function, the energy of the state at the top of the gap is reduced by

$$2L_0^{-1}\sum_j \sin^2(n\pi\Delta x_j/b)(Vt) \simeq (2/b)(n\pi X/b)^2(Vt),$$

where X is the r.m.s. displacement and L_0 is the total length of the chain. The energy of the state at the bottom of the band is raised by approximately the same amount, so that the gap as a whole is narrowed by the factor

$$1-2(n\pi X/b)^2. \tag{4.12}$$

This is indeed just the form that the Debye–Waller factor takes in one dimension, when X is small.

We have noted previously that the Debye–Waller factor is about 0.9 for the lowest energy gaps in a typical metal at the melting point. This does not represent a very marked narrowing and it therefore seems unlikely that the shape of the Fermi surface and the density of states correction m^*/m should change much on heating from low temperatures to T_M. There is in fact experimental evidence from the Knight shift (discussed in §4.16 below) that m^*/m is insensitive to temperature in the majority of metals, though Cd is certainly an exception. In semiconductors, however, the narrowing of energy gaps on heating is a well-recognised phenomenon, for which much more sophisticated theories have been developed than the one outlined above (e.g. Muto & Oyama, 1950, 1951; Fan, 1951).

Away from an energy gap in a regular one-dimensional solid the electronic wave functions are, of course, Bloch waves, described approximately by

$$\psi = L_0^{-\frac{1}{2}}(1+\alpha^2)^{-\frac{1}{2}} \exp(ikx)(1+\alpha \exp(-iGx)), \quad (4.13)$$

where α is a numerical factor which tends to ± 1 as the gap is approached; it is not difficult to show that if

$$Q = G-2k$$

then

$$\alpha^2 \simeq 1-|\hbar^2 GQb/mu(G)| \quad (4.14)$$

when Q is small. The approximation in (4.13) involves the neglect of all the reciprocal lattice vectors except the one which is closest to $2k$, which reflects the leading component of the wave most strongly.

Although it is not strictly relevant to the theme of this chapter it is of interest to consider the extent to which a Bloch wave near the gap is scattered into the equivalent Bloch wave of opposite k, when the lattice is slightly disordered by heating. A transition from k to $-k$ requires an *umklapp* process to occur, in which a phonon of wave vector Q is involved. The matrix element linking the two Bloch waves, which determines the strength of the scattering, is

$$\begin{aligned}\langle\psi(-k)|U|\psi(k)\rangle &= L_0^{-1}(1+\alpha^2)^{-1}\{U(-2k)+2\alpha U(G-2k)+\alpha^2 U(2G-2k)\}\\ &= L_0^{-1}(1+\alpha^2)^{-1}\{u(Q-G)\,F(Q-G)+2\alpha u(Q)\,F(Q)+\alpha^2 u(Q+G)\\ &\quad\times F(Q+G)\}. \end{aligned} \quad (4.15)$$

Now we have seen in §2.11 that excitation of phonons makes the structure factor of a solid rise to a peak in the neighbourhood of every reciprocal lattice vector, and this is just as true in one dimension as in three. From (2.45) it follows that

$$\left.\begin{aligned} F(Q-G) &= -F(Q)(G-Q)/Q,\\ F(Q+G) &= \quad F(Q)(G+Q)/Q,\end{aligned}\right\} \quad (4.16)$$

and both these quantities diverge to infinity as Q goes to zero. It might be feared that in consequence the matrix element diverges. That this is *not* the case may be seen by expanding (4.15) in powers of Q. With the aid of (4.16) it reduces to

$$L_0^{-1}F(Q)\left\{\frac{2\alpha}{1+\alpha^2}u(Q)+\left(1-\frac{G}{Q}\frac{1-\alpha^2}{1+\alpha^2}\right)\left(u(G)+\tfrac{1}{2}Q^2u''(G)+\ldots\right)\right.$$
$$\left.+\left(\frac{G}{Q}-\frac{1-\alpha^2}{1+\alpha^2}\right)\left(Qu'(G)+\ldots\right)\right\}, \quad (4.17)$$

and it is then clear from (4.14) that all the terms of order Q^{-1} drop out. In fact for small Q the matrix element becomes

$$L_0^{-1}F(Q)\left\{\pm u(Q)+u(G)+Gu'(G)+\frac{\hbar^2G^2b}{2m}\right\}, \quad (4.18)$$

the signs depending upon whether Q is positive or negative (i.e. upon whether the Bloch waves correspond to states a little below or a little above the gap) and also upon the sign of $u(G)$. We shall have occasion to refer back to this result in chapter 5. (For a more detailed discussion, in three dimensions, consult Sham & Ziman (1963), especially pp. 262–8.)

4.4. The dislocated solid

Even for a perfectly regular solid there do exist solutions of the Schrödinger equation for energies inside the gaps. The reason why we have ignored them in the previous section is that their amplitude varies along the chain in an exponential fashion

$$(A_L)_j \propto \exp(\pm\gamma b)(A_L)_{j-1}, \quad (4.19)$$

so they are not capable of satisfying the periodic boundary conditions. They therefore cannot represent eigenfunctions. If one interval in the chain is slightly enlarged or slightly diminished, however, it becomes possible to fit an exponentially growing wave on the left to an exponentially decaying wave on the right. An unimpeachable eigenstate is therefore created with an energy inside the original gap.

An expression for the constant γ in (4.19) may conveniently be obtained from equation (4.6), since A_R/A_L is the same thing as $\exp(\pm\gamma b)$ in this context. To simplify the discussion we may ignore the term in p^2, which makes little difference so long as p is small, and write

$$\exp(\pm 2\gamma b) = 1 + p \sin 2\zeta_L. \qquad (4.20)$$

The two cases $\zeta_L = \pm\frac{1}{2}\pi$ and $\zeta_L \simeq 0$ or π correspond to the top and bottom of the gap respectively, where γ vanishes. As one moves into the 'forbidden' energy region the magnitude of γ increases, and it is clear from (4.20) that somewhere near the middle it reaches a maximum of approximately $(p/2b)$.

Let us consider an energy which lies a small amount

$$\epsilon(2Vt/b)$$

below the top of the gap. It may be shown that for this energy

$$\zeta_L = \pm\tfrac{1}{2}\pi \pm \sqrt{\epsilon},$$

where the negative sign attached to $\sqrt{\epsilon}$ implies a growing wave and the positive sign a decaying one. It is then evident that if one of the intervals between δ-functions is enlarged by

$$\Delta b = 2\sqrt{\epsilon}/k \simeq 2b\sqrt{\epsilon}/n\pi$$

a growing wave to the left of 'dislocation' will turn into a decaying wave on the right, so that an eigenstate with this energy is thereby made possible. Similarly, a 'negative' dislocation, such that one interval is diminished by $(2b\sqrt{\epsilon}/n\pi)$ allows an eigenstate to exist with an energy $\epsilon(2Vt/b)$ above the bottom of the gap. In general, a number of positive and negative dislocations should narrow the gap by a factor

$$1 - \frac{1}{2}\left(\frac{n\pi}{b}\right)^2 (\Delta b)^2_{max}. \qquad (4.21)$$

This result is obviously related very closely to the one expressed by (4.12). If we imagine ourselves starting with a regular chain and moving blocks of eigenfunctions to the left or right, each through a distance X, then the final state would be described by a value of $2X$ for $(\Delta b)_{max}$. Substitution of this value into (4.21) makes it identical with (4.12).

In the situation that we have envisaged, the long-range order of the original chain is not destroyed by the introduction of the dislocations. It is quite possible, however, to imagine a chain containing much the same overall set of dislocations but *no* long-range order, because the positive and negative ones succeed one another in a random fashion. But the narrowing of the gap seems to depend only on $(\Delta b)^2_{\text{max}}$ and not on the exact arrangement of the dislocations. Is it a reasonable inference that the existence of a gap of some sort is not so dependent on long-range order, and therefore not so wholly determined by the magnitude of $U(G)$, as might have been supposed?

4.5. Energy gaps in one-dimensional liquids

The crude argument in the previous section is applicable only when the dislocations are very far apart. We must consider now a more realistic model for a one-dimensional liquid in which all the intervals between successive δ-functions are treated as variables, though none of them is greater than b_{max} or less than b_{min}.

The key to an understanding of the problem lies once more in equations (4.5) and (4.6). These tell us that $|A_{\text{R}}/A_{\text{L}}|$ must exceed unity provided that ζ_{L} lies in either of the ranges

$$-\tan^{-1}\tfrac{1}{2}p < \zeta_{\text{L}} < \tfrac{1}{2}\pi$$
$$\pi-\tan^{-1}\tfrac{1}{2}p < \zeta_{\text{L}} < \tfrac{3}{2}\pi,$$

in which case ζ_{R} lies in the ranges

$$\tan^{-1}\tfrac{1}{2}p < \zeta_{\text{R}} < \tfrac{1}{2}\pi$$
$$\pi+\tan^{-1}\tfrac{1}{2}p < \zeta_{\text{R}} < \tfrac{3}{2}\pi.$$

Borland and Makinson & Roberts refer to these ranges, which are indicated by hatching in fig. 4.5, as *stable*. Now if the interval between the first and second δ-functions in the chain is b_1 and the energy is such that the wave number throughout this interval, where the potential is zero, is k, we have

$$\zeta_{\text{L},2} = \zeta_{\text{R},1}+kb_1.$$

It is then clear by inspection of the figure that if the phase is stable at the first eigenfunction it is bound to be stable at the second, provided

$$n\pi-2\tan^{-1}\tfrac{1}{2}p < kb_1 < n\pi. \qquad (4.22)$$

Hence the phase will remain stable throughout the entire chain, provided that

$$(n\pi-2\tan^{-1}\tfrac{1}{2}p)/b_{\text{min}} < k < n\pi/b_{\text{max}}. \qquad (4.23)$$

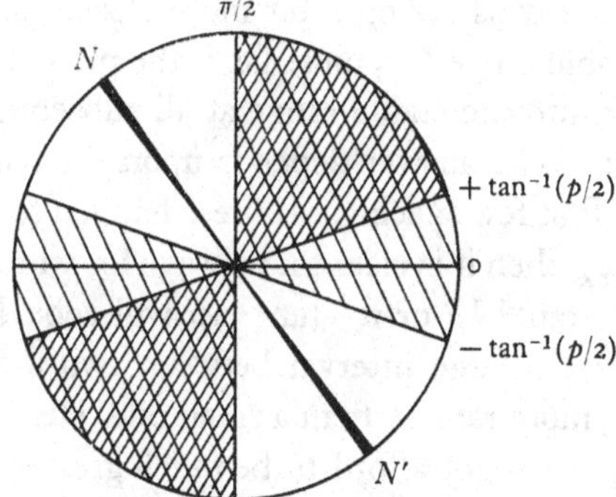

Fig. 4.5. The ranges in which ζ_L (hatching thus \\\\\\) and ζ_R (hatching thus ///) are stable.

Correspondingly, the wave must grow continuously in amplitude from left to right and cannot represent an eigenfunction.

What happens if the phase lies initially in the unhatched region of fig. 4.5? If by chance it remains in this region throughout the entire chain then $|A_R/A_L|$ is always less than unity and the wave must decay continuously; once more it cannot represent an eigenfunction. If (which *a priori* is much more likely) the wave sooner or later arrives at a δ-function with a stable phase, then it is caught. Its phase can never subsequently escape from the stable range, provided that the inequalities in (4.23) are satisfied, and this means that it cannot recover at the end of the chain the non-stable phase with which it started. Since the wave is therefore unable to satisfy the periodic boundary conditions it is no more admissible as an eigenfunction than those waves which grow or decay in a monotonic fashion.

We therefore arrive at the conclusion that eigenstates are impossible for all energies which satisfy (4.23). *A gap is bound to exist in the electronic energy spectrum, despite the lack of long-range order, provided that*

$$(\hbar^2/2m)(n\pi - 2\tan^{-1}\tfrac{1}{2}p)^2/b_{\min} < (\hbar^2/2m)(n\pi/b_{\max})^2. \quad (4.24)$$

It should be apparent from (4.10) and (4.11) that this is just the condition quoted already on p. 226, i.e. that the *top* of the gap for a regular solid with the spacing $b_{\max}$ lies above the *bottom* of the gap for a solid with the spacing $b_{\min}$, so as to allow these two gaps to overlap in energy.

It is worth considering a bit more closely just whereabouts in the non-stable range $\zeta_{L,1}$ must lie, if the phase is to escape the fate of slipping into the stable range at all subsequent eigenfunctions in the chain. The answer depends upon the particular configuration of the first few spacings. If these happen to be all rather large, close to b_{max}, then it is clear that a wave for which $\zeta_{L,1}$ is not much less than $-\tan^{-1}\frac{1}{2}p$ or $\pi-\tan^{-1}\frac{1}{2}p$ will soon be forced into the stable range; a large interval between two δ-functions tends to advance ζ_L more rapidly than a small one. Hence in this particular instance $\zeta_{L,1}$ can not afford to be much greater than $\pm\frac{1}{2}\pi$. On the other hand, if the first few intervals are all rather small, $\zeta_{L,1}$ must be close to $-\tan^{-1}\frac{1}{2}p$ or $\pi-\tan^{-1}\frac{1}{2}p$ if the phase is to remain non-stable indefinitely.

A small arc of angles NN' is blacked in in fig. 4.5 to suggest where $\xi_{L,1}$ may have to lie to ensure permanent non-stability, for a typical disordered chain which starts off neither with a run of large intervals nor with a run of small ones. Since $(\zeta_R-\zeta_L)$ is a monotonically increasing function of ζ_L in the non-stable region, this arc will emerge from the first eigenfunction somewhat expanded and will further expand at every eigenfunction it comes to. The more it expands the more risk there will be of some of the waves whose phases lie within it being carried into a stable range by an encounter with an extra large or extra small interval. Evidently, this blacked-in arc for ζ_L must be *infinitesimal* initially if the chain is a very long one. One might well imagine that *no* value could continue to remain non-stable indefinitely, were it not for the well-established existence of a class of solutions to Schrödinger's equation which are permanently stable and grow monotonically from left to right. Since there is no essential difference between one direction and another for the chain, there must exist an equally large class of solutions which grow monotonically from right to left, and these are just the permanently non-stable waves we have been discussing. The fact is that if one inspects the phase distribution of any class of solutions which have the same general variation of amplitude along the chain, there will be a wide spread in the values of ζ_L wherever the amplitude is small, but where the amplitude is large the values of ζ_L will crowd into a very narrow arc indeed.

4.6. Localisation of states in one dimension

Is the condition expressed by (4.24) a *necessary* as well as *sufficient* one for the existence of a gap? If it is just satisfied for a particular chain and we then make one interval a little larger than the previous b_{max} or a little smaller than the previous b_{min}, do we necessarily create an eigenstate with an energy inside one of the previously forbidden gaps?

In a vast majority of cases the answer would be no. The anomalous interval might jolt the phase of some waves out of the stable region into the edges of the non-stable region, but they would soon slip back to stability again; the result would be a variation of amplitude as shown in fig. 4.6(*a*). For a new eigenstate to be created it is necessary for the phase to be jolted as far as the infinitesimal blacked-in arc of fig. 4.5, so as to allow a growing wave to be transformed into a continuously decaying one as suggested by fig. 4.6(*b*), and in the vast majority of cases this requires an anomalous interval which is *substantially* larger than b_{max} or *substantially* smaller than b_{min}. But in a very small minority the interval we choose to enlarge or diminish will happen to lie near the middle of a sequence of intervals which are already close to b_{max} or b_{min} respectively. If so, the value of ζ_L for a growing wave arriving at the anomalous interval from the left will be close to the edge of the stable region, while the blacked-in arc which describes the permissible range of ζ_L for waves to decay on the right will be close to the edge of the adjacent non-stable region. Only a small jolt is then needed to make ζ_L jump into the blacked-in arc. So if we are to make statements about an ensemble of chains in which all possible configurations are included, the answer to our question is yes: even the smallest excursion outside the range of intervals bounded by b_{max} and b_{min} is liable to create an eigenstate with an energy just inside the erstwhile gap, and (4.24) *is* a necessary condition as well as a sufficient one.†

The eigenfunctions which can be created in this sort of way clearly have a *localised* character, just like the states which appear

† It is possible for the energy gap to persist unaltered if the anomalous intervals are all so much larger than b_{max} (or smaller than b_{min}) that they can accommodate one more (or one less) half wavelength. But this is trivial.

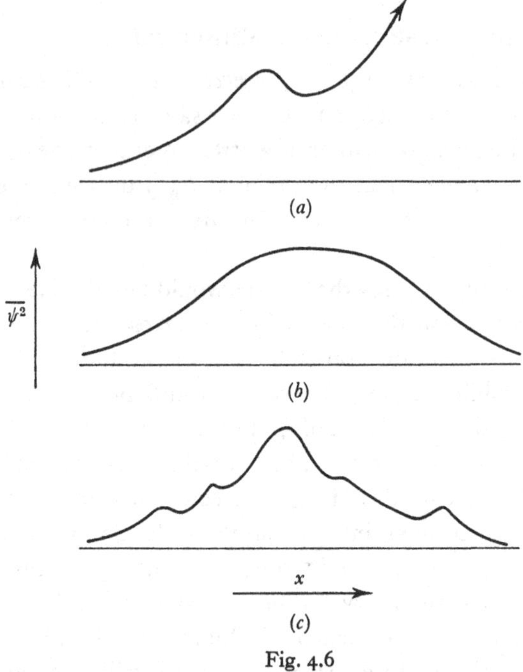

Fig. 4.6

inside the gap in a dislocated solid. Their amplitude decays on either side of a central peak in the way that fig. 4.6(*b*) suggests. The rate of decay near the peak must be small because here, as we have seen, the phase must be close to one of the boundaries between the stable and non-stable regions, and actually at the boundaries the factor

$$f = (A_R/A_L)^2 = 1+p\sin 2\zeta_L + p^2\cos^2\zeta_L \tag{4.25}$$

is unity. For this reason the curve in fig. 4.6(*b*) is drawn with a flat top rather than a sharp cusp. But if one moves away from the peak to the right or left through several δ-functions (just how many depends upon the extent to which b_{max} and b_{min} differ) one may expect to find that ζ_L has migrated away from the boundary and thereafter it should sample all possible values, i.e. all the angles that lie in the stable range if one looks to the left of the peak, or in the non-stable range if one looks to the right. If p is small, it seems likely to sample these angles with equal probability, in which case

$$\overline{\log f} = 2p/\pi + O(p^2); \tag{4.26}$$

it is the arithmetic mean of $\log f$ rather than of f that is of interest, because

$$(x/b)\,\overline{\log f}$$

gives the logarithm of the mean factor by which ψ^2 changes over the distance x. Thus away from its peak, ψ^2 should vary in a more or less exponential fashion on the average, and the distance required for it to change by a factor of e should be roughly $(\pi\bar{b}/2p)$ when p is small.

We have so far considered the situation where only one interval lies outside the range between b_{max} and b_{min}. It makes little difference to the argument, however, if anomalous intervals are quite frequent. Fig. 4.6(*c*) shows the sort of envelope that is then to be expected, for a state inside the original gap.

The results of Borland & Bird, some of which were reproduced in fig. 4.2 and fig. 4.3 above, suggest that although, when the chain is disordered, $\overline{\log f}$ is greatest for states inside one of the original gaps, it is still significant well away from the gaps. Considering the form of equation (4.25) for f this is not surprising. Well away from any gap there is no rational relation between k and $\bar{b}$ so that a wave tends to arrive at each δ-function with more or less random phase; it is rather as though it were propagating through a one-dimensional 'gas'. In the ideal gas successive values of ζ_L would be completely random, and since

$$\log f = p \sin 2\zeta_L + p^2 (\cos^2 \zeta_L - \tfrac{1}{2} \sin^2 2\zeta_L) + O(p^3) \quad (4.27)$$

we should expect† $\overline{\log f} = p^2/4 + O(p^3)$. (4.28)

On the average, therefore, ψ^2 should grow exponentially in a one-dimensional gas, increasing by a factor of e in a distance of order $4\bar{b}/p^2$, and the situation in a one-dimensional liquid far away from any gap should not be very different. There is now, of course, no absolutely stable range of phase, so that ψ^2 for an individual wave must not be expected to grow monotonically; it is bound to have frequent ups and downs, superimposed on a background of exponential growth.

But symmetry still demands that there exists an equally large

† Terms up to p^4 in (4.28) have been evaluated for the one-dimensional gas by Borland (1963), using a more careful argument.

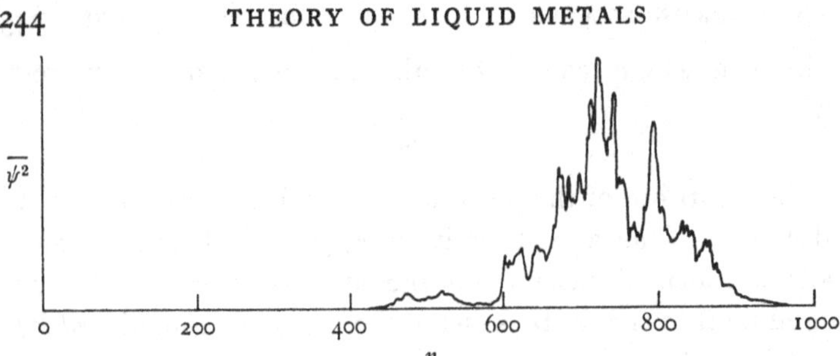

Fig. 4.7. The envelope of an eigenfunction in a disordered one-dimensional chain of 1000 δ-functions. (Redrawn by permission from Borland (1963), *Proc. Roy. Soc.* A **274**, 529. Acknowledgement is also made to the National Physical Laboratory.)

class of solutions for which ψ^2 on the average *decays* exponentially from left to right – or, if you prefer, grows from right to left. The reason why these solutions do not show up when we take the ensemble average of (4.27) and why $\overline{\log f}$ does not vanish in consequence, is a subtle one, indicated by the remarks at the end of §4.5. At each δ-function in the chain the range of ζ_L which is permissible, if ψ^2 is to start decaying exponentially rather than growing, is *infinitesimal*. It needs a freak of chance to make any wave starting from the left enter this range, and the crude averaging process we have used makes no special allowance for such freaks. Nevertheless, we must clearly depend upon their occurring from time to time if there are to be any eigenfunctions which satisfy the periodic boundary conditions.

This is more or less the reasoning which Mott & Twose (1961) in the first instance, followed by Borland (1963), advanced to justify the claim that virtually all electron eigenstates in an ensemble of disordered one-dimensional systems must be localised.† It seems, like all the conclusions that have been drawn above, to be substantiated by the results of numerical investigations. An envelope of ψ^2 for a typical eigenfunction in a one-dimensional gas, computed by Borland for a chain of 1000 δ-functions randomly disposed, is shown in fig. 4.7. We must now see to what extent it is possible to reach the same conclusions by the methods of perturbation theory.

† The word 'virtually' seems a necessary qualification, in case the definition of the ensemble permits regular, solid-like arrangements of eigenfunctions to occur in which non-localised Bloch waves can be propagated without impediment.

4.7. Perturbation theory for wave groups

A normalised wave function for a simple travelling wave in a one-dimensional system of length L_0 is

$$\psi_s = L_0^{-\frac{1}{2}} \exp(\mathrm{i}k_s x) \exp(-\mathrm{i}E_s t/\hbar), \tag{4.29}$$

where the energy E_s may be taken to include the mean potential energy

$$L_0^{-1} \langle \exp(\mathrm{i}k_s x)|U(x)|\exp(\mathrm{i}k_s x)\rangle$$

as well as the kinetic energy $\hbar^2 k_s^2/2m$. Any solution of the time-dependent Schrödinger equation can always be expressed as a *group* of such waves, i.e. in the form

$$\Psi = \sum_s a_s \psi_s, \tag{4.30}$$

but since each component is liable to be attenuated continuously by scattering, and as continuously reinforced by waves scattered from other components, the coefficients a_s must in general vary with time. We may start this discussion by considering the rules which govern the rate of change of a_s. If Ψ is to represent an *eigenfunction* for the system, with energy E, then of course its time-dependence must be completely described by the factor $\exp(-\mathrm{i}Et/\hbar)$, which is only possible if all the coefficients in (4.30) rotate uniformly in phase without change of amplitude, so that

$$a_s(t) = c_s \exp(\mathrm{i}(E_s - E)t/\hbar), \tag{4.31}$$

where c_s is independent of time. A rather subtle dynamic equilibrium is required between the various scattering processes in order to bring this phase rotation about, and by examining the conditions that need to be fulfilled to make it possible we can deduce how the coefficients which describe an eigenfunction have to vary with k and E.

The insertion of (4.30) into the time-dependent Schrödinger equation, followed by multiplication of both sides by ψ_s^* and integration over all x, yields the following standard result:

$$\mathrm{i}\hbar\dot{a}_s = \sum_{t \neq s} a_t U_{ts} \exp(\mathrm{i}(E_s - E_t)/\hbar), \tag{4.32}$$

where

$$U_{ts} = L_0^{-1} \langle \exp(\mathrm{i}k_s x)|U(x)|\exp(\mathrm{i}k_t x)\rangle = L_0^{-1} U(k_s - k_t) \tag{4.33}$$

is a matrix element of the potential. At first sight (4.32) seems to describe only the enhancement of a_s due to scattering *into* the sth component. To see how it also describes the attenuation of a_s due to scattering *out* of this component, it is best to consider a particularly simple group for which at time $t = 0$ one coefficient, say a_0, is unity while all the others are zero. It may appear from (4.32) as though $\dot{a}_0$ must in this case vanish at time $t = 0$, since all the terms on the right-hand side, if we identify a_0 with a_s, are infinitesimal; but this is wrong – there are an infinite number of these infinitesimal terms to consider! In fact by identifying a_0 with some coefficient other than a_s we can see from (4.32) that

$$i\hbar a_s = a_0 \frac{\hbar U_{0s}}{i(E_s - E_0)} \left| \exp\left(i(E_s - E_0)t/\hbar\right) \right|_0^t, \tag{4.34}$$

and if a result of this form is used for all coefficients except a_0 then (4.32) tells us that for small t

$$i\hbar\dot{a}_0 = -a_0 \sum_{s\neq 0} U_{0s} U_{s0} \left\{ \frac{1 - \exp\left(i(E_0 - E_s)t/\hbar\right)}{(E_s - E_0)} \right\}. \tag{4.35}$$

A careful examination shows that the right-hand side does not vanish: at any particular time there is a contribution roughly proportional to t from each state with energy E_s within a band about E_0, and since the width of this band is proportional to t^{-1} the sum does not vanish, even in the limit when t goes to zero. Equation (4.35) may, of course, be used as the basis for a conventional calculation of the lifetime of an electron in the state k_0 (see §3.16).

The moral is that if (4.32) is to be applied to a more general wave group it is essential to look very closely at the infinitesimal part of each a_t which has arisen by scattering from a_s in the first place and is therefore both proportional to a_s and coherent with it in phase. The scattering may have been direct, or it may have occurred via say the uth component, or indeed via any number of intermediate components. If the infinitesimal part which has arisen by scattering direct from a_s is represented by a_t' we have

$$i\hbar\dot{a}_t' = \Sigma_t a_t + a_s U_{st} \exp\left(i(E_t - E_s)t/\hbar\right). \tag{4.36}$$

In this equation an extra term has been inserted, with a coefficient Σ_t which is soon to be determined, to express the continuous decay

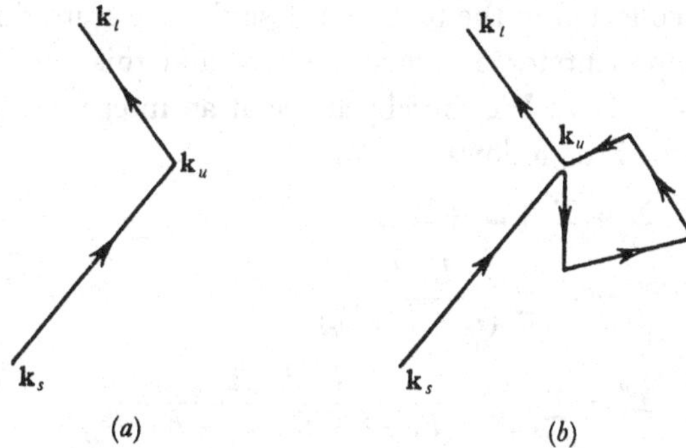

Fig. 4.8. Scattering processes linking states s and t in $\mathbf{k}$-space. (The diagram is drawn for a two-dimensional system though the argument in the text is for one dimension.)

of a_t' due to scattering into other components; such a term ought strictly to have been included in the simple example discussed above, and would have resulted in a slight modification of (4.34). Now if we consider the particular case where Ψ represents an eigenfunction, the time variation of a_s is described by (4.31), and (4.36) can then be integrated to yield

$$a_t' = -\frac{a_s U_{st}}{(E_t - E + \Sigma_t)} \exp(\mathrm{i}(E_t - E_s)t/\hbar). \tag{4.37}$$

It is not difficult to extend the argument to show that the infinitesimal part of a_t which has arisen out of a_s as a result of *two* scattering processes is given by

$$a_t'' = \frac{a_s}{(E_t - E + \Sigma_t)} \sum_{u \neq t,s} \frac{U_{su} U_{ut}}{(E_u - E - \Sigma_u)} \exp(\mathrm{i}(E_t - E_s)t/\hbar), \tag{4.38}$$

and there is an infinite series of higher-order terms, a_t''', a_t'''' etc., for which similar expressions can be written down. Fig. 4.8 illustrates schematically two of the scattering processes which would contribute to a_t''. As well as the simple type of two-stage path in $\mathbf{k}$-space which is shown in fig. 4.8(*a*), an infinite number of more complex re-entrant paths like that in fig. 4.8(*b*) are automatically allowed for by the inclusion of Σ_u in the denominator of (4.38).

Equation (4.32) should now be rewritten in the form

$$\begin{aligned} \mathrm{i}\hbar \dot{a}_s = &\sum_{t \neq s} (a_t' + a_t'' + \ldots) U_{ts} \exp(\mathrm{i}(E_s - E_t)t/\hbar) \\ &+ \sum_{t \neq s} (a_t - a_t' - a_t'' - \ldots) U_{ts} \exp(\mathrm{i}(E_s - E_t)t/\hbar). \end{aligned} \tag{4.39}$$

The terms collected in the first sum describe the scattering out of the sth component for an eigenfunction, and this sum is to be identified with $\Sigma_s a_s$. We thereby arrive at an integral relation for the coefficient Σ as follows:

$$\Sigma = \Sigma' + \Sigma'' + \Sigma''' + \dots$$

where
$$\Sigma'_s = -\sum_{t \neq s} \frac{U_{st} U_{ts}}{(E_t - E + \Sigma_t)} \tag{4.40}$$

$$\Sigma''_s = \sum_{t \neq s} \sum_{u \neq t,s} \frac{U_{ts} U_{su} U_{ut}}{(E_t - E + \Sigma_t)(E_u - E + \Sigma_u)}, \tag{4.41}$$

and so on. Each of these contributions to Σ involves what may be called a *closed ring* of matrix elements, and we shall see below that such rings do not necessarily vanish when an average is taken over an ensemble. The Born approximation amounts in the present context to the neglect of all terms in Σ other than Σ'. For the time being we will suppose this approximation to be justified, returning to consider the higher-order terms in §4.11 below. We shall also suppose U to be *Hermitian*, so that

$$U_{ts} = U^*_{st} \tag{4.42}$$

and $U_{st} U_{ts}$ is real. This is certainly justified for a real potential set up e.g. by a chain of δ-functions. It is not necessarily justified for the pseudo-potential in a metal, as we shall see in §4.11, though in fact the model pseudo-potentials that are available for numerical calculations (see §1.8) do have Hermitian properties.

In general Σ may be a complex quantity and we shall denote by $-A$ and $-i\Gamma$ respectively the real and imaginary parts of its ensemble average. Since Σ_t, say, results from the coupling of the tth component with a very large number of other components of similar energy, it may reasonably be hoped that the fluctuations of Σ_t within the ensemble become negligible when the length of the chain, and hence the density of permitted values of k in $\mathbf{k}$-space, is large. If this is true, then the average of the factor $(E_t - E + \Sigma_t)^{-1}$ which occurs in (4.40) differs insignificantly from the inverse of the average of $(E_t - E + \Sigma_t)$. Within the confines of the Born approximation we may therefore write

$$A_s = \sum_{t \neq s} \frac{(E_t - E - A_t)}{(E_t - E - A_t)^2 + \Gamma_t^2} \overline{|U_{st}|^2} \tag{4.43}$$

and
$$\Gamma_s = \sum_{t \neq s} \frac{\Gamma_t}{(E_t - E - A_t)^2 + \Gamma_t^2} \overline{|U_{st}|^2}. \tag{4.44}$$

By itself A_s describes a continuous phase rotation of a_s and is equivalent to a second-order shift in the energy E_s; it is Γ_s which describes the decay of the amplitude of a_s.

These results differ from those of conventional second-order perturbation theory in two respects. First, it is the energy of the whole eigenstate E which figures in the denominators, rather than E_s. Secondly, the inclusion of the damping term $\Sigma_t a_t'$ in (4.36) makes Σ_t crop up again on the right of (4.40). However, if E is replaced by E_s and A_t and Γ_t are allowed to go to zero one may obtain from (4.43) the conventional result for the second-order energy shift of an electron in the state k_s (quoted above as equation (1.78)) while (4.44) leads to the conventional expression for its lifetime; the factor $\Gamma_t/((E_t - E - A_t)^2 + \Gamma_t^2)$ becomes a δ-function centred about the energy E_s. We may note for future reference that the connection between Γ and τ_l is always such that

$$\Gamma = \hbar/2\tau_l. \tag{4.45}$$

The factor 2 arises because Γ determines the rate at which the amplitude of a wave decays, whereas $1/\tau_l$ describes the decay of the square of the amplitude.

Let us now return to equation (4.39), which may be integrated to read

$$a_s(E_s - E + \Sigma_s) = -\sum_{t \neq s} (a_t - a_t' - \ldots) U_{ts} \exp(\mathrm{i}(E_s - E_t)t/\hbar),$$

or, in terms of the time-independent coefficients defined by (4.31),

$$c_s(E_s - E + \Sigma_s) = -\sum_{t \neq s} (c_t - c_t' - \ldots) U_{ts}. \tag{4.46}$$

This is an exact result. The problem is to extract useful information from it. It is not so much a single equation, of course, as a set of simultaneous equations for the coefficients c_s, c_t, c_u etc., and the eigenvalues of the energy of the system are the values of E for which all these equations have a solution. In principle the eigenvalues are the roots of a secular equation, and for a regular solid, in which U_{ts} is non-zero only when $(k_s - k_t)$ takes one of a limited number of discrete values, so that only a limited number of values for k are represented in each eigenfunction, it may be feasible to write down the secular equation and to solve it directly. But this approach is quite impracticable for a liquid, in which U_{ts} is non-zero for all possible values of $(k_s - k_t)$ and the components of each eigenfunction spread over the whole range of k.

To make progress one may multiply (4.46) by its complex conjugate and take an ensemble average of the result. The most important terms to survive the averaging process on the right-hand side are clearly those of the form

$$(c_t - c_t' - \ldots)(c_t^* - c_t'^* - \ldots)|U_{st}|^2 \tag{4.47}$$

and these are straightforward enough. By subtracting c_t' we have removed the only part of c_t which arises by scattering via U_{st} and there is no reason for the remainder to be correlated with U_{st} in any way, so the ensemble average of (4.47) should be

$$\overline{(c_t - c_t' - \ldots)(c_t^* - c_t'^* - \ldots)} \cdot \overline{|U_{st}|^2} = \overline{c_t c_t^*} \cdot \overline{|U_{st}|^2};$$

the terms $c_t', c_t'' \ldots$ can be dropped at this stage because by comparison with c_t they are of order N^{-1}. The cross-product terms of the form

$$c_t U_{ts} c_u^* U_{su} \tag{4.48}$$

are rather more tricky. Each c_u contains an infinitesimal part which has arisen by scattering from c_t, and vice versa. Hence terms arise in (4.48) which are proportional to $c_t c_t^*$ or $c_u c_u^*$ and to some closed ring of matrix elements. In §4.11 it will appear that these should be negligible so long as the Born approximation is justified, so for the present we may suppose

$$\overline{c_s c_s^*}((E_s - E - A_s)^2 + \Gamma_s^2) = \sum_{t \neq s} \overline{c_t c_t^*}\, \overline{|U_{st}|^2}. \tag{4.49}$$

Write this with the aid of (4.44) in the form

$$B_s \sum_{t \neq s} \frac{\Gamma_t \overline{|U_{st}|^2}}{(E_t - E - A_t)^2 + \Gamma_t^2} = \sum_{t \neq s} B_t \frac{\Gamma_t \overline{|U_{st}|^2}}{(E_t - E - A_t)^2 + \Gamma_t^2}, \tag{4.50}$$

where $$B_s = \overline{c_s c_s^*}((E_s - E - A_s)^2 + \Gamma_s^2)/\Gamma_s$$

and similarly for B_t. It is not possible to satisfy (4.50) for all values of s simultaneously unless all the B_t are identical. Hence for the ensemble average we deduce that†

$$\overline{c_k c_k^*} = B\frac{\Gamma_k}{(E_k - E - A_k)^2 + \Gamma_k^2}, \tag{4.51}$$

where B is a constant, independent of k.

The magnitude of B is fixed by the normalisation condition that

$$\sum_k \overline{c_k c_k^*} = 1. \tag{4.52}$$

This sum may be converted into an integral, if we introduce a dimensionless variable

$$z = (E_k - E - A_k)/\Gamma_k \tag{4.53}$$

and use $\mathcal{N}(E_k)$ to represent the density of states per unit range of E_k, i.e. the density of states in the absence of any second- or higher-order perturbation by the potential. It becomes

$$B\int_{-(E+A)/\Gamma}^{\infty} \mathcal{N}(E_k)\frac{1}{\Gamma_k}\left(\frac{\partial E_k}{\partial z}\right)_E \frac{dz}{1+z^2} = 1, \tag{4.54}$$

and if Γ is small, so that $\mathcal{N}(E_k)(\partial E_k/\partial z)/\Gamma_k$ can be treated as constant over the region where the integrand is significant and the lower limit of the integral can be replaced by $-\infty$, then

$$B^{-1} = \pi\mathcal{N}(E_k)\left(\frac{1}{\Gamma_k}\left(\frac{\partial E_k}{\partial z}\right)_E\right)_0. \tag{4.55}$$

The suffix zero is used to indicate a value taken at $z = 0$.

There is, however, a second normalisation condition. It is an elementary consequence of the fact that both the eigenfunctions $\Psi(E)$ and the basis functions $\psi(k)$ must form complete sets of orthonormal functions for the system, that

$$\sum_E \overline{c_k c_k^*} = 1, \tag{4.56}$$

where the sum is now taken for a fixed value of k rather than a fixed value of E. It then follows as above that, provided Γ is small,

$$B^{-1} = -\pi\mathcal{N}(E)\left(\frac{1}{\Gamma_k}\left(\frac{\partial E}{\partial z}\right)_k\right)_0, \tag{4.57}$$

† Here a general suffix k is introduced in place of s, t etc. Thus E_k takes the value E_s when $k = k_s$; the suffix distinguishes it from E, which remains the energy of the whole wave group.

where $\mathcal{N}(E)$ is the density of states in E, i.e. with the perturbing effects of the potential fully switched on. For (4.55) and (4.57) to be consistent we must have

$$\begin{aligned}\mathcal{N}(E_k)/\mathcal{N}(E_k) &= -(\partial E_k/\partial z)_0/(\partial E/\partial z)_0\\ &= (\partial E_k/\partial E)_{z=0}.\end{aligned} \tag{4.58}$$

In this way we arrive at last at an expression for the perturbed density of states.

Some conclusions which have emerged from this lengthy discussion may be summarised as follows:

(*a*) Corresponding to each eigenvalue of the energy, E, there is no one value of k for an electron in a disordered one-dimensional system. Instead its eigenfunction may be pictured as a wave group, containing as many components of positive as of negative k.

(*b*) A plot of $\overline{c_k c_k^*}$ versus k for a given eigenfunction should in general display a Lorentzian profile as indicated by equation (4.51), though since A_k and Γ_k can both be functions of k the Lorentzian need not be perfect.

(*c*) Unless the k-dependence of Γ_k is very marked (and in most circumstances it is not likely to be so) $\overline{c_k c_k^*}$ has its maximum where (E_k-E-A_k) vanishes. The locus of this maximum on a plot of E versus k as in fig. 4.9, replaces the sharp E–k curve for a free electron.

(*d*) When Γ is small the density of states at any particular value of E is inversely proportional to the slope of this locus, as one might naively expect. Thus since

$$\mathcal{N}(E_k) = (L_0/\pi)(\mathrm{d}k/\mathrm{d}E_k)$$

for a one-dimensional system, it follows from (4.58) that

$$\mathcal{N}(E) = (L_0/\pi)(\partial k/\partial E)_{z=0}. \tag{4.59}$$

(*e*) It is readily shown that

$$\left(\Gamma_k\frac{\partial z}{\partial E_k}\right)_0 = 1-\left(\frac{\partial A}{\partial E_k}\right)_0$$

and
$$\left(\Gamma_k\frac{\partial z}{\partial E}\right)_0 = -1-\left(\frac{\partial A}{\partial E}\right)_0$$

so that
$$\mathcal{N}(E)/\mathcal{N}(E_k) = (1+\partial A/\partial E)_0/(1-\partial A/\partial E_k)_0. \tag{4.60}$$

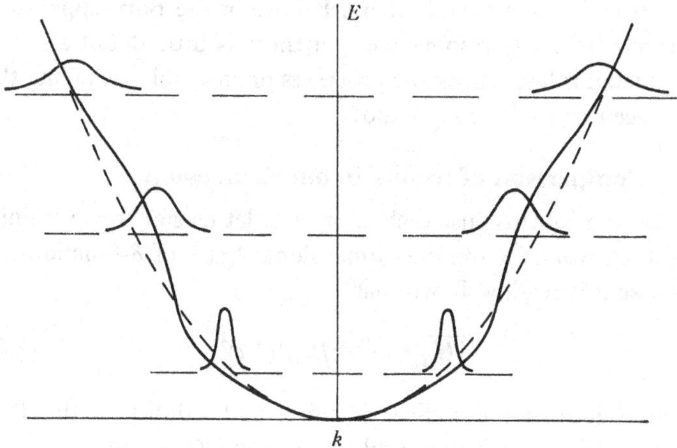

Fig. 4.9. Hypothetical profiles of $\overline{c_k c_k^*}$ for three different energies. The full curve is the locus of its maximum.

This equation may serve as a basis for estimating the correction to the density of states due to the potential, if the level shift A can be computed self-consistently from (4.43). Like (4.55) and (4.57) it is strictly valid only when Γ is small.

The mode of presentation adopted in this section is taken from Faber (1966), but it corresponds closely to the original argument of Edwards (1961, 1962, 1965), though not framed in the language of Green functions. Edwards uses a density matrix $\rho(E, E_k)$ which is the same thing as $\mathcal{N}(E)\overline{c_k c_k^*}$ in the present notation. The only point at which the results of the two treatments disagree concerns the density of states. The Green function approach appears to fix the magnitude of B^{-1} unambiguously as $\pi\mathcal{N}(E)$, with no freedom to adjust the normalisation. If this is correct the density of states can be determined by the application of (4.52) alone, and the answer when Γ is small is then

$$\mathcal{N}(E)/\mathcal{N}(E_k) = (1 - \partial A/\partial E_k)_0^{-1} \tag{4.61}$$

rather than (4.60). But unless $\partial A/\partial E$ is zero, which is not necessarily the case, one is then in difficulties because (4.56) is not satisfied. This problem is examined by Edwards in his 1965 paper.

We shall see in the next section that the results are not wholly in agreement with those obtained by the methods of Borland,

Makinson & Roberts and others. Failure of the Born approximation may be partly responsible, but there is little doubt also that some detail is lost during the processes of ensemble averaging that have been carried through above.

4.8. Comparison of results in one dimension

To test the conclusions reached in §4.7, let us start by examining how they work for the one-dimensional 'gas' of δ-functions. In this case it is readily shown that

$$\overline{|U_{st}|^2} = (N/L_0)^2(Vt)^2 \tag{4.62}$$

for all values of the suffices s and t, and validity of the Born approximation may be ensured by making (Vt) small.

It is not difficult to see that when $\overline{|U|^2}$ is a constant a self-consistent solution is possible for equations (4.43) and (4.44) for which A_k and Γ_k are also constants, independent of k for a given E. A more careful scrutiny shows that

$$A = -(\hbar^2 k_0/8m\bar{b})p^2 + O(p^3) \tag{4.63}$$

$$\Gamma = (\hbar^2 k_0/4m\bar{b})p^2 + O(p^3), \tag{4.64}$$

where p describes the strength of the δ-functions as in (4.5), $\bar{b}$ is once again the mean spacing between them, and k_0 is such that

$$E_{k_0} = E. \tag{4.65}$$

The energy shift described by (4.63) is of no consequence. It is true that since $(\partial A/\partial E)$ does not vanish, even though $(\partial A/\partial E_k)$ does so, there should according to (4.60) be a slight distortion of the free-electron density of states due to the perturbing δ-functions, but we did not pursue the argument far enough in §4.6 to be able to check this particular prediction. We are more concerned with the magnitude of Γ and with what this implies about the variation of the electronic eigenfunctions with x.

It is of course impossible to reconstruct Ψ exactly, given only the magnitude of $\overline{c_k c_k^*}$ as a function of k; for this we would need to know the phase of all the components as well as their amplitude. What we can deduce, however, (by another application of the

Wiener–Khintchine theorem which was used in §3.11) is the *auto-correlation* function of Ψ. Given

$$\Psi = L_0^{-\frac{1}{2}} \sum_k c_k \exp(ikx)$$

then
$$\langle \Psi(x+X) | \Psi(x) \rangle = \sum_k c_k c_k^* \exp(-ikX). \qquad (4.66)$$

Hence if (4.51) is correct the ensemble average of this auto-correlation function for the one-dimensional gas should be given when p and Γ are small by

$$B \sum_k \frac{\Gamma \exp(-ikX)}{(E_k - E - A_k)^2 + \Gamma^2}$$

$$\simeq \frac{1}{\pi} \int_{-\infty}^{\infty} \frac{dz}{1+z^2} \cos kX = \cos k_0 X \exp(-m\Gamma X/\hbar^2 k_0),$$

which becomes
$$\cos k_0 X \exp(-p^2 X/4\bar{b}) \qquad (4.67)$$

with the aid of (4.64). This answer is correctly normalised because it goes to unity as X goes to zero. If one likes to define a *coherence length* l by the relation (see (4.45))

$$l = \hbar k_0 \tau_l / m = \hbar^2 k_0 / 2m\Gamma, \qquad (4.68)$$

then the average auto-correlation function of Ψ can also be expressed by
$$\cos k_0 X \exp(-X/2l). \qquad (4.69)$$

Now the argument in §4.6, and equation (4.28) in particular, suggests that Ψ should have an oscillatory character with an amplitude that on the average varies exponentially with x, i.e. that

$$\Psi \propto {\cos \atop \sin} (k_0 x) \exp(\pm p^2 x / 4\bar{b}).$$

If we take the negative sign in the exponential, i.e. the case of a wave that decays from left to right, and evaluate

$$\int_0^{L_0} \Psi^*(x+X) \Psi(x) \, dx,$$

averaging over the cos and sin solutions which are equally probable, we do indeed arrive at (4.67). Equally, we may arrive at it for a solution which grows from left to right by evaluating

$$\int_{-L_0}^{0} \Psi(x-X) \Psi(x) \, dx.$$

We have seen, however, that an *eigen*function is only possible if at some point in the chain (say at $x = 0$) a growing wave is converted into a decaying wave. If we suppose that

$$\Psi \propto \begin{matrix}\cos\\ \cdot\\ \sin\end{matrix}(k_0 x)\exp(-p^2|x|/4b) \tag{4.70}$$

and evaluate $$\int_{-\frac{1}{2}L_0}^{\frac{1}{2}L_0} \Psi^*(x+X)\Psi(x)\,dx,$$

we find an answer that is *not* exactly (4.67) but

$$\cos k_0 x \exp(-p^2X/4\bar{b})(1+p^2X/4\bar{b}) \tag{4.71}$$

instead. To generate this sort of auto-correlation function from (4.66) one requires c_k to vary with E_k in a Lorentzian fashion, rather than $c_k c_k^*$.

The moral seems to be that, although the perturbation theory of §4.7 yields an answer which is consistent with the expected exponential envelope for Ψ, it is *not* sufficiently powerful to reveal that the eigenfunctions in a one-dimensional gas have to be localised, i.e. that their envelope has a cusp at some point in the chain. Perhaps this failure is not surprising; in §4.6 the necessity for this cusp was inferred in a very indirect fashion, there being no evidence for it from equation (4.28) alone.

Equations (4.43) and (4.44) are integral equations which are coupled to one another, and the solutions are not transparent in cases where $\overline{|U_{st}|^2}$ is not a constant but varies with the length of the scattering vector, $q = k_s - k_t$. Perhaps the most instructive example to consider is one in which it has a fairly broad and symmetrical peak centred about $q_{\max}$ but is zero elsewhere, so that it tends to couple the kth component with components around $-(q_{\max}-k)$, the coupling being especially strong when k is close to $\frac{1}{2}q_{\max}$. The broken curve to the left of the origin in fig. 4.10(*a*) shows schematically how

$$(E_k - E - A_k)/[(E_k - E - A_k)^2 + \Gamma_k^2]$$

should behave in this case as a function of k, passing through zero at $\pm k_0$ where

$$z = (E_k - E - A_k)/\Gamma_k = 0.$$

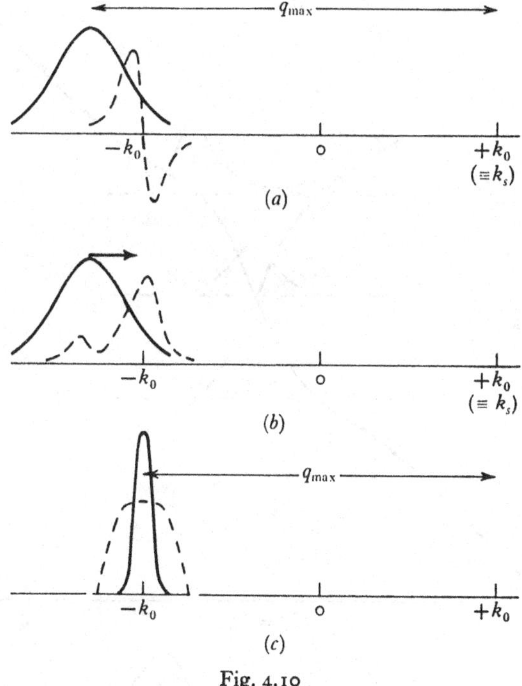

Fig. 4.10

The full curve shows the suggested form of $\overline{|U|^2}$, centred about $-(q_{max}-k_0)$. According to (4.43) it is the convolution of these two functions that determines the magnitude of A_k at $k = +k_0$, on the other side of the origin. Evidently in the case illustrated, for which k_0 is less than $\frac{1}{2}q_{max}$, A_0 is positive,† and this means that E is *less* than E_0. When k_0 is greater than $\frac{1}{2}q_{max}$ the signs are reversed and E is *greater* than E_0. When k_0 is just equal to $\frac{1}{2}q_{max}$, A_0 vanishes (the broken curve being almost exactly anti-symmetric about k_0 in this case) and then E is *equal* to E_0. As a result the locus of $z = 0$ on a plot of E versus E_k has a wiggle in it, as shown schematically in fig. 4.11.

According to (4.58) this wiggle distorts the free-electron density of states; $\mathcal{N}(E)$ acquires a minimum at $k = \frac{1}{2}q_{max}$, flanked by two maxima. Because the width of the wiggle, i.e. the separation between the values of E_0 at which $|E-E_0|$ is greatest, is

† By A_0 is meant the value of A_k for $k = k_0$. Similarly E_0 means the value of E_k when $k = k_0$.

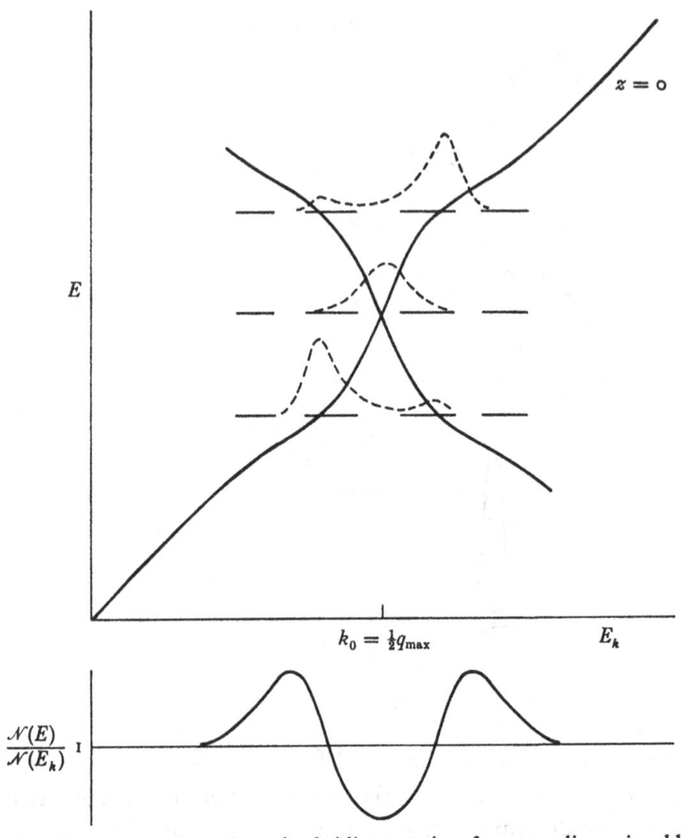

Fig. 4.11. Loci of the main and subsidiary maxima for a one-dimensional liquid, where the corresponding solid would have an energy gap. The lower curve indicates the effect on the density of states.

determined by the width of the peak in $\overline{|U|^2}$, while the amplitude of the wiggle is proportional to the height of the peak, the minimum in $\mathscr{N}(E)$ becomes more pronounced as the peak is made narrower and higher.

The magnitude of Γ_0 is determined by the convolution of the two functions plotted schematically in fig. 4.10(*b*). This time the broken curve, whose shape is explained below, represents the behaviour of

$$\Gamma_k/[(E_k - E - A_k)^2 + \Gamma_k^2]. \tag{4.72}$$

It is to be noted that according to (4.51) and (4.52) the area under the broken curve is proportional to B^{-1}. Thus if the peak in $\overline{|U|^2}$

is sufficiently broad in comparison with Γ for $\overline{|U|^2}$ to be treated as a constant over the region where (4.72) is appreciable, it follows that Γ_0 is also proportional to B^{-1} and hence, from (4.57), to the density of states in the *perturbed* system $\mathcal{N}(E)$ and to the factor $(1+\partial A/\partial E)_0^{-1}$. In fact, if $\mathcal{N}(E)$ is written as m^*/m times the free-electron density of states, the solution for Γ_0 in this case turns out to be

$$\Gamma_0 = \frac{\hbar}{2\tau_{l,0}} = \frac{m^*}{2\hbar^2 k_0}\frac{N}{L_0}a(2k_0)|u(2k_0)|^2(1+\partial A/\partial E)_0^{-1}. \quad (4.73)$$

Apart from the factor $(1+\partial A/\partial E)_0^{-1}$ the time $\tau_{l,0}$ defined by this result is exactly the same as the lifetime of an electron in the state k_0, as calculated from the one-dimensional analogue of equation (3.76).

To deduce how Γ_k varies as k is increased above k_0 for a fixed E, we need to imagine the full curve in fig. 4.10(*b*) to be moved to the right while the broken curve is kept fixed. It is evident that for the case illustrated Γ_k increases with k to begin with and reaches its largest value when

$$k \simeq q_{max} - k_0. \quad (4.74)$$

This k-dependence of Γ_k has to be taken into account, of course, in the estimation of how $\Gamma_k/[(E_k-E-A_k)^2+\Gamma_k^2]$ varies with k. It may be expected to generate the sort of asymmetry in this quantity which is suggested by the broken curve in the figure. The curve has been drawn with a small subsidiary maximum at the value of k (or strictly at minus the value of k) for which (4.74) is satisfied, though if the peak in $\overline{|U|^2}$ is a broad and low one the asymmetry should not be very noticeable and the subsidiary maximum could well disappear. Symmetry is restored, even if the peak in $\overline{|U|^2}$ is a strong one, for the particular energy for which $k_0 = \frac{1}{2}q_{max}$.

The reason for emphasising the behaviour of (4.72) is that it is this quantity, according to (4.51), which determines the magnitude of $\overline{c_k c_k^*}$. The broken curve in fig. 4.10(*b*) therefore, and the various similar curves which are sketched for different energies in fig. 4.11, serve to indicate the expected 'profile' of the electronic eigenfunctions in **k**-space. It is to be noted that the subsidiary maxima lie near a curve of negative slope in fig. 4.11, which is the locus on which (4.74) is satisfied; Edwards's suggestion that the *main* maxima in $\overline{c_k c_k^*}$ (or in his related density matrix ρ) lie near this backwards sloping curve would seem to be erroneous.

From the profile of an eigenfunction in **k**-space one may deduce, by the use of (4.66), its auto-correlation function in real space. If we neglect, for the moment, the variation of Γ_k with k for a fixed E and suppose

$$\overline{c_k c_k^*} \propto \frac{\Gamma_0}{(E_k - E - A_k)^2 + \Gamma_0^2}$$

$$\propto \frac{\Gamma_0}{(E_k - E_0)^2 (1 - \partial A/\partial E_k)_0^2 + \Gamma_0^2},$$

this Lorentzian profile ensures that the mean auto-correlation function decays exponentially with X; it may still be described by (4.69), in fact, but the 'coherence length' is now given by

$$\begin{aligned} l &= (\partial E_k/\partial k)_0 (1 - \partial A/\partial E_k)_0 / 2\Gamma_0 \\ &= \hbar^2 k_0 (1 + \partial A/\partial E)_0 / 2m^* \Gamma_0. \end{aligned} \tag{4.75}$$

This result, taken in conjunction with (4.73), suggests that where there are anomalies in the density of states l should be proportional to $(m/m^*)^2$ and to $(1 + \partial A/\partial E)_0^2$.

But (4.73) and (4.75) are valid only when the peak in $\overline{|U|^2}$ is broad – what happens when its width becomes comparable with or less than Γ? The situation is then complex, because the asymmetry of $\Gamma_k/[(E_k - E - A_k)^2 + \Gamma_k^2]$ and the subsidiary maxima in this quantity may become pronounced, but a semi-quantitative discussion is possible for the particular case for which $k_0 = \frac{1}{2} q_{max}$. Fig. 4.10(*b*) should then be redrawn as in fig. 4.10(*c*). The convolution of the two curves in this figure clearly produces an answer for Γ_0 which is *less* than the one predicted by (4.73) and in the limit when the peak in $\overline{|U|^2}$ is very narrow we have

$$\Gamma_0 = \sum_{t \neq s} \frac{\Gamma_0}{0^2 + \Gamma_0^2} \overline{|U_{st}|^2},$$

or

$$\Gamma_0^2 = \sum_{t \neq s} \overline{|U_{st}|^2}. \tag{4.76}$$

If the pseudo-potential is due to a chain of δ-functions one may sharpen the peak in $\overline{|U|^2}$ by permitting less and less variation in the spacings between them. A one-dimensional analogue of the sum rule discussed on p. 113 ensures that the right-hand side of (4.76) is independent of just how much disorder is left in the chain, and the answer in this limit comes out to be

$$\Gamma_0 \simeq u(q_{\max})/\bar{b} = \hbar^2 k_0 p/2m\bar{b}. \tag{4.77}$$

For values of k other than k_0 (but still for $k_0 = \frac{1}{2}q_{\max}$) the solution is

$$\begin{aligned} \Gamma_k^2 &\simeq \Gamma_0^2 - (E_k - E - A_k)^2 \quad (\text{for } |E_k - E - A_k| < \Gamma_0) \\ &= 0 \quad (\text{for } |E_k - E - A_k| > \Gamma_0). \end{aligned} \tag{4.78}$$

Thus the profile for $\overline{c_k c_k^*}$ should be non-Lorentzian, with a cut-off at

$$(E_k - E - A_k) = \pm \Gamma_0.$$

These predictions agree nicely with some conclusions reached in §4.6 about the eigenfunctions which form inside an energy gap when a one-dimensional chain is sufficiently disordered. The fact that the width of the profile in **k**-space comes out to be proportional to p rather than p^2 is consistent with equation (4.26) (compare (4.28) for states well away from any gap), while the cut-off in the profile is consistent with the conjecture that the envelope of Ψ in real space must have a rounded top (see fig. 4.6(*b*)) rather than a sharp cusp.

It has been taken for granted in the above discussion that a self-consistent solution of equations (4.43) and (4.44) does exist in the middle of the erstwhile energy gap when the chain is slightly disordered. It is by no means obvious, however, that this is true. Certainly when there is *no* disorder, i.e. when the width of the peak in $\overline{|U|^2}$ is infinitesimal, the only self-consistent solution turns out on closer analysis to be one for which Γ_k vanishes for all k. To see this one may note that, if U_{st} is finite for only one value of $(k_s - k_t)$,

$$-\Sigma_s = A_s + \mathrm{i}\Gamma_s = \frac{U_{st}U_{ts}}{E_t - E + \Sigma_t}$$

while

$$-\Sigma_t = \frac{U_{st}U_{ts}}{E_s - E + \Sigma_s}.$$

For an energy E that corresponds to the middle of the gap

$$(E_t - E) = -(E_s - E),$$

in which case $$\Sigma_s = -\Sigma_t.$$

The solution for Σ must then be entirely real, with

$$-\Sigma_k = A_k = \tfrac{1}{2}(E_k - E) \pm \left[\left(\frac{E_k - E}{2}\right)^2 + \overline{|U|^2}\right]^{\frac{1}{2}}; \quad (4.79)$$

the positive sign applies when $E_k < E$ and the negative sign when $E_k > E$. Since Γ_k vanishes for all k so does $\overline{c_k c_k^*}$. It is gratifying to verify in this roundabout fashion that no eigenstates exist in the middle of an energy gap so long as the chain is completely ordered!†

An imperfect analysis of the complex mathematical problem involved suggests that it is *only* when the width of the peak in $\overline{|U|^2}$ is infinitesimal that Γ has to be zero in the middle of the gap; the non-zero solution described by (4.77) and (4.78) seems to be allowed by the slightest broadening of it. It is very unlikely, therefore, that the perturbation method of §4.7 is sufficiently powerful to yield the exact condition for the closure of a gap in one dimension which is expressed by equation (4.24), at any rate so long as the Born approximation is adopted.

It was emphasised in chapter 1 that the Born approximation is valid only when the modulation of an electron wave, as it travels through the perturbing potential, is slight enough to be ignored. We have seen in §4.6, however, that states in the gap region tend to be associated with stretches of the chain where there are several intervals in succession which are all of much the same length – either close to b_{max} or to b_{min}. A wave travelling through such a stretch is liable to be significantly modulated by coherent reflection from successive eigenfunctions, even though each of them is weak. Perhaps this is why, if the Borland–Makinson–Roberts theorem is ever to be substantiated by perturbation methods, an improvement on the Born approximation seems to be essential.

† The reader may like to convince himself that for states which border the energy gap the subsidiary maxima in $\overline{c_k c_k^*}$, illustrated in fig. 4.11, serve in the limit of almost complete order to describe the reflected components of the Bloch waves in terms of which the eigenfunctions are most appropriately expressed. See §5.18.

4.9. Extension to real metals in three dimensions

The extension of the perturbation method of §4.7 is entirely straightforward. Virtually all the results obtained in that section may be taken over as they stand, with the trivial modification that since $\mathbf{k}$ is a vector in three dimensions it deserves bold-faced type in places. The method suggests that so long as the perturbation is isotropic, i.e. so long as $U(\mathbf{q})$ is independent of the direction of $\mathbf{q}$, the eigenfunctions are themselves isotropic. At any rate the ensemble average $\overline{c_k c_k^*}$ should be pictured as independent of the direction of $\mathbf{k}$; generally this quantity should be significant within a spherical shell in $\mathbf{k}$-space of radius k_0, its intensity varying across the thickness of the shell in a more or less Lorentzian fashion, as described by (4.51).

No problem arises over the evaluation of Γ in the three-dimensional case, provided that it is small enough compared with K (which is the *kinetic* component of E, $\hbar^2k_0^2/2m$) to be treated as a linear function of k throughout the spherical shell where $\overline{c_k c_k^*}$ is significant and for the variation of Γ_k with k to be ignored. Equation (4.44) may then be reduced, with the aid of equation (4.75) for the coherence length, to the integral form

$$\Gamma_s = \frac{m^*}{(2\pi)^3\Omega\hbar^2k_0(1+\partial A/\partial E)_0}\int \overline{|U(q)|^2}\,\frac{(1/2l)}{(k_t-k_0)^2+(1/2l)^2}\,d\mathbf{q}; \tag{4.80}$$

fig. 4.12 illustrates once again the relation of the quantities k_s, $\mathbf{q}$, $\mathbf{k}_t$ and k_0. In the limit when Γ is vanishingly small, i.e. when the mean free path is very long, an integration of (4.80) over the angle ϕ reduces it to

$$\Gamma_s = \frac{m^*}{4\pi\hbar^2k_s(1+\partial A/\partial E)_0}\frac{N}{\Omega}\int_{|k_s-k_0|}^{k_s+k_0} a(q)\,|u(q)|^2\,q\,dq. \tag{4.81}$$

An expression for Γ_0 may be obtained from this by setting k_s equal to k_0, and it is readily seen to be equivalent to (3.76) apart from the factor $(1+\partial A/\partial E)_0^{-1}$. Contrary to the situation in one dimension, Γ is always proportional to u^2 rather than u, whether or not there is a sharp peak in $a|u|^2$.

Manipulation of (4.81) shows that

$$(\partial\Gamma_k/\partial E_k)_0 = \beta\Gamma_0/2K_0,$$

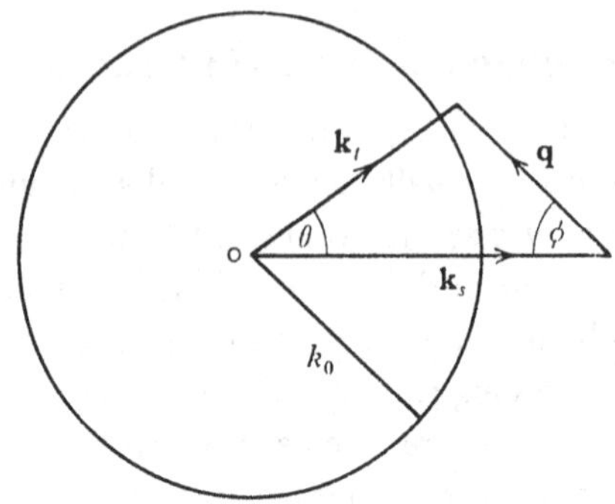

Fig. 4.12

where $$(1+\beta)\int_0^1 \overline{|U(q)|^2}\, 2(q/2k_0)\, \mathrm{d}(q/2k_0) = \overline{|U(2k_0)|^2}. \qquad (4.82)$$

In the majority of liquid metals the dependence of $U(q)$ on q is such that the numerical factor β defined by (4.82) is a good deal less than unity,† and our neglect of the variation of Γ_k with k (for fixed E) is therefore probably justified. On the other hand

$$\partial\Gamma_k/\partial E = (1+\beta)\Gamma_0/2K_0,$$

so we may take Γ to be proportional to $K^{\frac{1}{2}}$.

The estimates of $\hbar/2\tau K_F$ which are listed for liquid metals in table 1.1 suggest that this ratio (which is almost the same thing as $1/lk_F$) may be as large as 0.1 in some cases. It is doubtful whether this is quite small enough to render equation (4.81) entirely accurate. A little reflection shows that the most important effect of shortening the coherence length should be a blurring of the sharp upper and lower limits to the integral in that equation, in the manner indicated schematically by fig. 4.13. Edwards has shown, however, that in some cases (4.81) may be retained as it stands provided that an appropriate modification is made to the interference function $a(q)$. His argument is of sufficient physical appeal to deserve a brief explanation.

The first step is to note that, since

$$\mathbf{k}_t = \mathbf{k}_s + \mathbf{q},$$

† For a three-dimensional 'gas' of δ-functions, for which $\overline{|U(q)|^2}$ is in dependent of q, β vanishes.

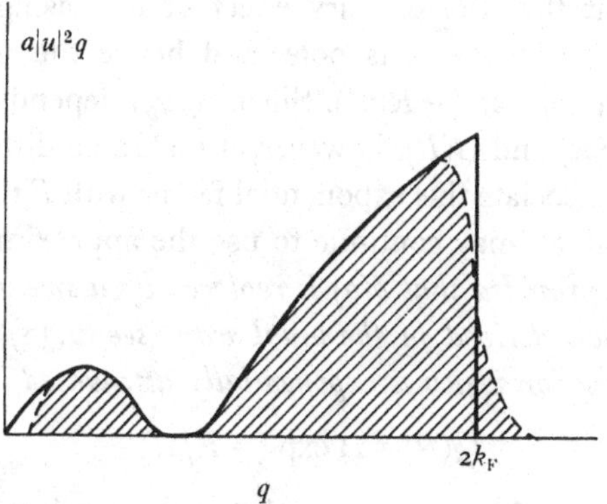

Fig. 4.13. The full curve shows how the integrand of (4.81) varies with q in a typical case; the broken curve suggests the sort of correction which is needed because the coherence length is not infinite.

the integral in (4.80) has the general form,

$$\int F(\mathbf{q})\, G(\mathbf{k}_s+\mathbf{q})\, d\mathbf{q}, \tag{4.83}$$

and if $F(\mathbf{q})$ and $G(\mathbf{k}_t)$ are the Fourier transforms of two functions in real space, $F(\mathbf{R})$ and $G(\mathbf{R})$, one may recognise (4.83) as the Fourier component of their product, $F(\mathbf{R})\,G(\mathbf{R})$. Since $F(\mathbf{q})$ and $G(\mathbf{k}_t)$ depend only on the magnitudes of $\mathbf{q}$ and $\mathbf{k}_t$ respectively, and not on their directions, $F(\mathbf{R})$ and $G(\mathbf{R})$ depend only on the magnitude of $\mathbf{R}$. *If* (and this is a severe limitation in practice) the variation of the ionic pseudo-potential $u(q)$ is sufficiently slow for $|u(q)|^2$ to be treated as a constant, then the whole variation of $F(q)$ arises from the interference function and it follows that

$$F(R) \propto \delta(R)+g(R)-1,$$

$(g-1)$ being the pair correlation function. On the other hand

$$G(R) \propto \int_0^\infty \frac{(1/2l)}{(k_t-k_0)^2+(1/2l)^2}\,\frac{\sin k_t R}{k_t R}\,k_t^2\, dk_t$$

$$\propto \text{Im}\,\frac{1}{2R}\int_{-\infty}^{\infty} \frac{(1/2l)}{(k_t-k_0)^2+(1/2l)^2}\exp{(ik_t R)}\,k_t\, dk_t. \tag{4.84}$$

The integral in (4.84) may readily be evaluated by contour methods and arises entirely from the pole at

$$k_t = k_0 + i/2l.$$

It is apparent that the primary effect of increasing $(1/2l)$ is to decrease the residue at this pole, and hence the magnitude of $G(R)$, by a factor $\exp(-R/2l)$. Since (4.83) depends only on the product of $F(R)$ and $G(R)$, however, it makes no difference to the answer if we associate this exponential factor with F rather than G. It follows that we may continue to use the approximate equation (4.81) for Γ, *provided that $a(q)$ is replaced by a smeared out interference function derived in the usual way (see (2.38)) from a pair correlation function which is exponentially attenuated,*

$$(g(R)-1)\exp(-R/2l). \tag{4.85}$$

It is, of course, only to be expected that the conduction electrons in a liquid metal are unable to appreciate any correlation that may exist between the positions of two molecules which are separated by a distance much greater than the electronic coherence length l.

Although it is Γ that determines such properties as the resistivity, it is A that determines the density of states. Unfortunately, the accurate evaluation of A presents severe problems, mainly because the factor $(E_k-E-A_k)/[(E_k-E-A_k)^2+\Gamma_k^2]$ which replaces $\Gamma_k/[(E_k-E-A_k)^2+\Gamma_k]$ – compare (4.43) with (4.44) – is not necessarily important only in the neighbourhood of $k = k_0$. Shaw & Smith (1969) have emphasised how essential it is on this account to remember that the pseudo-potential may be non-local, i.e. that $U(\mathbf{q})$ may depend not only on k but on the angle ϕ. The angular dependence is likely to be particularly severe if virtual bound states exist within the ions; our method may in principle be used in such a case to improve upon the 'spherical approximation' of §1.11, but an elaborate calculation is likely to be required. Another snag is that one cannot necessarily treat (E_k-E-A_k) as a linear function of k or ignore the variation of Γ_k. Only if these difficulties are set aside can an integral expression for A, equivalent to (4.81) for Γ, be written down; it is

$$A_k = \frac{m^*}{(2\pi)^2\hbar^2 k(1+\partial A/\partial E)_0} \times \frac{N}{\Omega}\int_0^\infty a(q)\,|u(q)|^2 q\,dq\,\tfrac{1}{2}\log\left|\frac{((k+q)^2-k_0^2)^2+(k_0/2l)^2}{((k-q)^2-k_0^2)^2+(k_0/2l)^2}\right|. \tag{4.86}$$

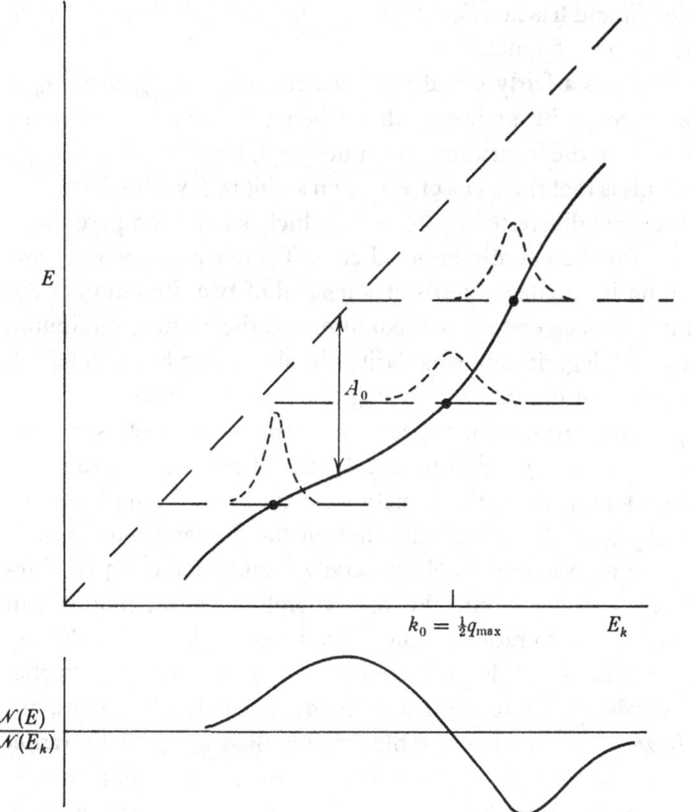

Fig. 4.14. The equivalent in three dimensions of fig. 4.11.

If Γ is small enough this may be further simplified by omitting $(k_0/2l)$ inside the logarithm. For $k = k_0$ it becomes

$$A_0 = \frac{m^*}{(2\pi)^2 \hbar^2 k_0(1+\partial A/\partial E)_0} \frac{N}{\Omega} \int_0^\infty a|u|^2 q \,dq \log \left| \frac{2k_0+q}{2k_0-q} \right|. \quad (4.87)$$

It is worth noting that A_0, according to (4.87), must always be positive, which was not true of the solution in one dimension.

For a three-dimensional 'gas' of δ-functions the integral in (4.87) diverges, unless it is supposed that $u(q)$ cuts off to zero at some very large value of q. It may be shown, however, that $\partial A/\partial E_k$ tends to zero like $k_0/q_{\text{cut-off}}$ and can therefore be neglected, while $\partial A/\partial E$ is also negligible. Hence the second-order energy shift is in this case a constant, which can be absorbed by a redefinition of the zero for E_k and has no effect on properties of physical interest such as the density of states.

If $a|u|^2$ has a fairly broad peak centred about q_{max}, then A_0 is liable to reach its greatest value when this coincides with the singularity in the logarithmic term in (4.87), i.e. when $2k_0 \simeq q_{max}$. The result is that the locus of $z = 0$ on a plot of E versus E_k should take the form illustrated by fig. 4.14, which is to be compared with fig. 4.11 for the one-dimensional case. There should be only one maximum in the density of states instead of two. Equation (4.87) is bound to exaggerate this maximum, and the ensuing minimum, because the logarithmic singularity should in fact be rounded off by inclusion of the terms involving the mean free path.

A peak in $a|u|^2$ is to be expected for many liquid metals because of the way that $a(q)$ behaves, but it will not necessarily produce a significant anomaly in the density of states at the Fermi level. To make $2k_F$ coincide with the position of the first and most prominent maximum in $a(q)$ would generally require about 1.3 conduction electrons per atom. We may therefore expect monovalent liquid metals, with rather smaller Fermi radii, to have their density of states at the Fermi level *increased* by the second-order perturbation; divalent ones to have it *reduced*; and polyvalent ones, for which $2k_F$ may be considerably greater than q_{max}, to be rather little affected. But if, as is not uncommon, the maximum in $a(q)$ happens to lie near the first node in $u(q)$, then the structure in $a|u|^2$ may well be insufficient to give rise to any marked anomaly, whatever the valency.

So much for qualitative predictions. Detailed computations based on various model pseudo-potentials and on various experimental and theoretical curves for $a(q)$ have been reported by Watabe & Tanaka (1964), Ballentine (1966*b*), Schneider & Stoll (1967*c*), Shaw & Smith (1969), Srivastava & Sharma (1969*a*), Jena & Halder (1971), Stoll, Szabo & Schneider (1971) and Chan & Ballentine (1971*a*,*b*). In each case approximations are involved (the reader should consult the original papers for details) which make it difficult to assess the accuracy of the results. Shaw & Smith have emphasised that they can be extremely sensitive to the details of the model employed, especially the position of the first node in $u(q)$, unless care is taken to allow for the non-local character of the

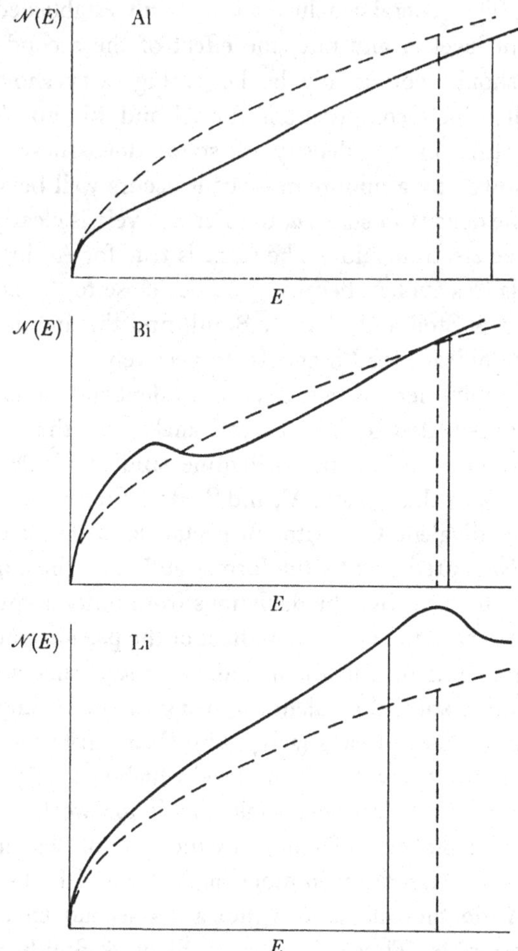

Fig. 4.15. The calculated density of states for three liquid metals, with free-electron parabolae for comparison. Vertical lines indicate the position of the Fermi level relative to the bottom of the conduction band. (After Ballentine, 1966*b* and Shaw & Smith, 1969.)

pseudo-potential in full, i.e. for the variation of $u(q)$ with ϕ as well as k. Not all the authors quoted have been sufficiently scrupulous in this respect.

To refine the calculations would require a great deal of labour and scarcely seems worth while, in view of the uncertainty that would remain as to whether the model pseudo-potential was correct or not. The general conclusion is certainly established that, near the Fermi level at any rate, the effect of the second-order perturbation should not usually be large. Fig. 4.15 shows the results of Ballentine's computations for Al and Bi. For Bi the expected structure in the density of states does show up – a maximum followed by a minimum – but it occurs well below the Fermi level; the density of states *at* the Fermi level† is clearly very close to the free electron value. The same is true for Al, in which the structure is less evident because q_{max} lies close to the node in $u(q)$. Schneider & Stoll and Shaw & Smith find the anomalies to be equally insignificant for Pb and In respectively.

These are all polyvalent metals. Several divalent and monovalent metals were investigated by Watabe & Tanaka, but their results are now regarded as erroneous. Ballentine studied divalent Zn, though less carefully than Bi and Al, and Shaw & Smith and Jena & Halder studied divalent Cd, with unspectacular results: e.g. in liquid Cd m^*/m is predicted by the former authors to be 0.94 and by the latter to be 0.83, but the deviations from unity seem to be due almost entirely to the first-order effect of the pseudo-potential upon $\mathcal{N}(E_k)$ (see §1.10), which is included in these calculations.‡ Chan & Ballentine studied divalent Hg using the promising non-local pseudo-potential of Evans (see p. 40); their curve for $\mathcal{N}(E)$ displays a minimum below the Fermi level which is clearly due to the second-order effect, but once again this is too weak to affect m^*/m to any marked degree. To judge by the work of Schneider & Stoll it is equally insignificant in monovalent Na and K. Only for monovalent Li do the calculations indicate a second effect that might be measurable. Thus according to Shaw & Smith, whose curve for Li is also reproduced in fig. 4.15, m^*/m for this metal is 1.25 though m_1^*/m is only 1.09 (see table 1.2). Stoll *et al.* confirm that m^*/m is significantly greater than unity for liquid Li, but the

† The position of the Fermi level relative to the bottom of the conduction band has to be found by integrating the density of states and is naturally not quite the same for the full curves in fig. 4.15 as for the corresponding free electron parabolae.

‡ Care has been taken in §4.7 and throughout the subsequent argument not to imply that $\mathcal{N}(E_k)$ is necessarily free-electron-like, so that the results obtained are equally valid whether we have to deal with a set of weak δ-functions, in which the potential is a straightforward local one, or with a set of ions in a real metal, for which $u(q)$ may be k-dependent.

three different pseudo-potentials they use lead to results that vary between 1.15 and 1.52.

Shaw & Smith obtained a curve for the density of states in solid Li as well, using a simple OPW method and the same pseudo-potential and atomic volume as for the liquid phase. It has a much higher and sharper maximum above the Fermi level than the curve in fig. 4.15, but at and below the Fermi level, curiously enough, it is almost exactly the same.

4.10. Localisation and band gaps in three dimensions

If the density of states in a typical liquid metal is almost free-electron-like, what can be said about the electronic eigenfunctions? Since they are made up of a group of plane waves of different **k** their amplitude and phase are bound to vary from point to point in real space. But do they, like the eigenfunctions of the free-electron model, extend throughout the specimen, or are they localised? In one dimension localisation is the answer, but it is now well recognised (e.g. Ziman, 1968, 1969*b*) that the topological differences between one-dimensional and three-dimensional systems are far too profound for us to argue by analogy in such a subtle matter as this.

The method of §4.7 tells us only the ensemble average of the auto-correlation function for Ψ. If we assume for simplicity that the profile of $\overline{c_k c_k^*}$ is accurately Lorentzian, then the answer in three dimensions turns out to be

$$\langle \Psi(\mathbf{r}+\mathbf{X})|\Psi(\mathbf{r})\rangle = \frac{\sin k_0 X}{k_0 X} \exp(-X/2l), \tag{4.88}$$

where l is still given by (4.75). This is to be contrasted with (4.69) in one dimension. It is interesting to note that whereas (4.69) is not altogether consistent with the localisation that is known to occur in the one-dimensional case, which caused us indeed to have some doubts about the reliability of the perturbation method, (4.88) *would* be consistent with localisation in three dimensions: it is satisfied by eigenfunctions whose envelope decays like

$$R^{-1} \exp(-R/2l) \tag{4.89}$$

about the point $\mathbf{R} = 0$. It is also satisfied, however, by a wide variety of non-localised functions.

The experimental evidence that virtually eliminates the possibility of localisation in liquid metals at normal densities, at any rate for the states near the Fermi level that chiefly matter, is simply that they remain electrically conducting, with conductivities not very much less than in the solid phase. To understand this point we must anticipate some of the results of the next chapter, §§5.10 and 5.14 in particular. There it is shown that the conductivity as a function of frequency, $\sigma(\omega)$, depends upon the mean square value of a matrix element

$$\left\langle \Psi_n \left| \frac{\partial}{\partial x} \right| \Psi_m \right\rangle$$

between eigenstates whose energies differ by $\hbar\omega$; the average has to be taken over all values of E_m (this being the lower of the two energies) that lie within $\hbar\omega$ of the Fermi level and over an ensemble of possible configurations for the ion. If the states are localised (not necessarily otherwise, depending upon the boundary conditions adopted) this matrix element may be rewritten with the aid of (5.37) as

$$-(m\omega/\hbar)\,\langle\Psi_n|x|\Psi_m\rangle,$$

which strongly suggests, since† $\langle\Psi_n|x|\Psi_m\rangle$ is surely a finite length of the order of l, that $\sigma(\omega)$ should tend to zero at least as fast as ω^2. The argument has been formalised by Halperin (1967) (but see Mott (1967) for a fuller exposition) as follows.

We are interested in the average of

$$|\langle\Psi_n|x|\Psi_m\rangle|^2 = \iint \Psi_n^*(\mathbf{r}_1)\Psi_n(\mathbf{r}_2)\Psi_m(\mathbf{r}_1)\Psi_m^*(\mathbf{r}_2)\,x_1x_2\,d\mathbf{r}_1\,d\mathbf{r}_2$$

with $n \neq m$. Because the Ψ_n form a complete set of orthonormal functions it follows that

$$\sum_n \Psi_n^*(\mathbf{r}_1)\Psi_n(\mathbf{r}_2) = \delta(\mathbf{r}_1-\mathbf{r}_2),$$

and hence that

$$\sum_n |\langle\Psi_n|x|\Psi_m\rangle|^2 = \int \Psi_m(\mathbf{r})\Psi_m^*(\mathbf{r})\,x^2\,d\mathbf{r} = \langle\Psi_m|x^2|\Psi_m\rangle, \quad (4.90)$$

† Note that the choice of origin for x is immaterial because Ψ_n and Ψ_m are orthogonal; this ensures that, for any fixed length a, $\langle\Psi_n|a|\Psi_m\rangle$ is always zero.

the sum here being taken over all n, including m. Let us define the x coordinate of the centre of Ψ_m by a length

$$x_m = \langle\Psi_m|x|\Psi_m\rangle$$

and its spread by

$$\langle\Delta x^2\rangle_m = \langle\Psi_m|(x-x_m)^2|\Psi_m\rangle; \tag{4.91}$$

if (4.89) were correct this would be just $\frac{2}{3}l^2$. Expanding (4.91) and making use of (4.90) we find, for a specimen of infinite volume such that any sum over states can be expressed as a continuous integral, that

$$\begin{aligned}\langle\Delta x^2\rangle_m &= \sum_n |\langle\Psi_n|x|\Psi_m\rangle|^2 - |\langle\Psi_m|x|\Psi_m\rangle|^2 \\ &= \sum_{n\neq m} |\langle\Psi_n|x|\Psi_m\rangle|^2 \\ &= \frac{h^4}{m^2}\int_{-\infty}^{\infty} \frac{\left|\left\langle\Psi_n\left|\frac{\partial}{\partial x}\right|\Psi_m\right\rangle\right|^2}{(E_n-E_m)^2}\,\mathcal{N}(E_n)\,\mathrm{d}(E_n-E_m).\end{aligned} \tag{4.92}$$

Now take an average over the ensemble. So long as $\overline{\langle\Delta x^2\rangle_m}$ remains finite, i.e. so long as the states are localised for all but an infinitesimal proportion of the ionic configurations which can in principle occur,

$$\overline{\left|\left\langle\Psi_n\left|\frac{\partial}{\partial x}\right|\Psi_m\right\rangle\right|^2}$$

must tend to zero at least as fast as $(E_n-E_m)^2$ to prevent the integral on the right of (4.92) from diverging. The conclusion that $\sigma(\omega)$ tends to zero at least as fast as ω^2 follows at once from this result.

The argument applies, of course, to the idealised model discussed throughout this chapter, in which the ions are stationary and the scattering entirely elastic. To allow for ionic motion it may well be necessary to average the theoretical expression for $\sigma(\omega)$ over a range of frequencies (the band width to be determined by the rate of change) and a finite value for $\sigma(0)$ may then appear, even though the eigenfunctions are still localised at each instant of time. In quantum-mechanical language the conduction process becomes one of 'phonon-assisted hopping', the sort of process that is well known to be important in heavily doped semiconductors (Miller &

Abrahams, 1960; Mott & Twose, 1961). It seems very unlikely, however, that conduction in liquid metals at normal densities occurs only by virtue of phonon-assisted hopping. If this were the case, a liquid metal supercooled to well below its Debye temperature, where all but the zero point motion of the ions is frozen out, would become an insulator. While it is clearly impossible to achieve this degree of supercooling experimentally, one *can* perform measurements at very low temperatures on amorphous metallic films whose structure, to judge by diffraction experiments, is very similar to that of the liquid phase. Far from becoming insulators, these films show a liquid-like conductivity which is essentially independent of temperature.

So, at densities where the perturbation treatment that led us to (4.88) is likely to be valid for a liquid metal, the instantaneous eigenfunctions near the Fermi level almost certainly extend throughout the specimen. At lower densities, however, when the Fermi level falls below the zero of the 'muffin-tin potential' and 'pseudo-gaps' begin to open up (see §1.12) the situation may well change. The tight-binding approximation then becomes more appropriate, i.e. one should picture the eigenstates as built up by the coupling together of overlapping atomic orbitals, each of which is highly localised in real space, rather than by the coupling together of plane waves which are localised only in **k**-space. Formally there is some analogy between the tight-binding method and the method employed in §4.7 above. But whereas in **k**-space each plane wave is coupled by the pseudo-potential to a very large number of others, the number of neighbours to which each atomic orbital is coupled in real space may be rather small. It is largely this fact that makes the analysis of the tight-binding model so difficult for a disordered system; even greater care is needed in averaging over ensembles. It is generally agreed that the states within a pseudo-gap are likely to be localised if it is deep enough, whether or not the electron density exceeds the critical value given by (1.86) at which the Mott transition is expected to occur. But for the arguments in support of this contention the reader must refer to the papers quoted in §4.2.

Reference to the pseudo-gaps created by volume expansion reminds us that it might be possible to create one in a rather different way, by rearranging the ions at constant volume so as to form a polycrystalline solid; as the size of the crystallites is increased, and the peaks in $a(q)$ are narrowed, there must surely be a continuous transition to the spectrum characteristic of a single crystal. Equations (4.81) and (4.86) must be inadequate to describe this, however. They imply that when the first energy gap begins to open up it will enclose a *spherical* zone in **k**-space with a radius close to $\frac{1}{2}q_{max}$, where q_{max} denotes the position of the main peak in $a(q)$, i.e. q_{max} is the smallest reciprocal lattice vector. For a metal with the f.c.c. structure

$$q_{max} = 2\pi\sqrt{3}/(4\Omega/N)^{\frac{1}{3}},$$

which means that this spherical zone would be large enough to accommodate only about

$$\frac{8\pi\Omega}{3(2\pi)^3}\left(\frac{q_{max}}{2}\right)^3 = 1.36N$$

electrons. This is clearly nonsense. The first zone must surely accommodate two electrons per atom, as it does in a perfect crystal.

The truth is that the concept of a spherical zone must be quite inappropriate for a polycrystalline solid in which the crystallites are of any size. In the interior of any one large crystallite the electronic eigenstates must be virtually identical with those of an infinite crystal, and the eigenstates for the polycrystalline array as a whole must surely be constructed by matching these Bloch waves together at the crystallite boundaries. Thus we ought to think of the first zone in space as being polyhedral and large enough for two electrons per atom, as for a single crystal, though the orientation of its axes may change as one goes from one crystallite to the next. But to build up any such picture as this by perturbation theory it is clearly essential to go beyond the Born approximation, upon which (4.81) and (4.86) are based. It is only when one does this that the higher-order distribution functions g_3, g_4 etc., which are of vital importance in distinguishing between quasi-crystalline structures and close-packed amorphous ones, enter into the problem.

It is tempting to speculate that for polycrystalline solids one can state another theorem analogous to that of Saxon & Hutner for disordered alloys, i.e. that if there is a range of energy for which no Bloch wave eigenstates are possible in a single crystal, whatever the orientation of relative to the crystal axes, then this range of energy is still forbidden in a polycrystalline aggregate. But there is clearly no hope of proving this by analysis of equations (4.81) and (4.86), if we were unable to prove the one-dimensional equivalent at the end of §4.8.

4.11. Beyond the Born approximation

At an early stage in the argument of §4.7 some second- and higher-order terms, a''_t, a'''_t etc., were neglected by comparison with a'_t and this simplification has coloured many of the subsequent conclusions. Our first object in the present section is to see whether the higher terms may formally be included by the use of an *effective potential* U' as suggested in chapter 1.

The quantities a'_t, a''_t are all functions of the potential U, and we may define an effective potential by the requirement that

$$a'_t(U)+a''_t(U)+\ldots = a'_t(U'). \tag{4.93}$$

Equations (4.37) and (4.38) then show that

$$U'_{st} = U_{st}-\sum_{u\neq t,s}\frac{U_{su}U_{ut}}{(E_u-E-\Sigma_u)}+\ldots \tag{4.94}$$

Each of the higher terms which have not been written down here corresponds to the summation of a lengthier set of paths whereby (in the sense indicated by fig. 4.8) a wave initially in the sth state may be scattered to the uth and thence direct to the tth. A little reflection should convince the reader that these may all be included by writing

$$U'_{st} = U_{st}-\sum_{u\neq t,s}\frac{U'_{su}U_{ut}}{(E_u-E+\Sigma_u)}. \tag{4.95}$$

The effective potential advocated in chapter 1 was defined in real space by the equation

$$U'(\mathbf{r}) = (1+\gamma(\mathbf{r}))\,U(\mathbf{r}), \tag{4.96}$$

where $\gamma(\mathbf{r})$ describes the modulation in the amplitude of an incident wave due to the potential that it experiences, but the equivalence of (4.95) to (4.96) is easily demonstrated. It requires that

$$-\frac{1}{\Omega}\sum_{u\neq t,s}\frac{U'(\mathbf{q}_{su})U(\mathbf{q}_{ut})}{(E_u-E+\Sigma_u)} = \int\gamma(\mathbf{r})\,U(\mathbf{r})\exp(-i\mathbf{q}_{st}\cdot\mathbf{r})\,d\mathbf{r},$$

or, by an elementary Fourier transform, that

$$\gamma(\mathbf{r})U(\mathbf{r}) = -\frac{1}{(2\pi)^3\Omega}\int d\mathbf{q}_{st}\exp(i\mathbf{q}_{su}\cdot\mathbf{r}+i\mathbf{q}_{ut}\cdot\mathbf{r})\sum_u \frac{U'(\mathbf{q}_{su})U(\mathbf{q}_{ut})}{(E_u-E+\Sigma_u)}.$$

The right-hand side may be integrated over $\mathbf{q}_{ut}$ rather than $\mathbf{q}_{st}$ without affecting the result, and this yields a factor of $U(\mathbf{r})$ which cancels, leaving

$$\gamma(\mathbf{r}) = -\frac{1}{\Omega}\sum_u \exp(i\mathbf{q}_{su}\cdot\mathbf{r})\frac{U'(\mathbf{q}_{su})}{(E_u-E+\Sigma_u)}. \tag{4.97}$$

This is indeed just the answer that one would expect; apart from the factor Σ_u in the denominator and the rather subtle point that E appears in place of E_s, it may be obtained directly by elementary first-order perturbation theory – compare, for example, equation (1.26). Note that the extent to which an electronic wave function is modulated in real space is determined by U' rather than by U.

Now the quantity Σ which plays an essential role in the theory may be expressed (see (4.40) and (4.41)) as

$$\Sigma_s = -\sum_{t\neq s}\left[\frac{U_{st}U_{ts}}{(E_t-E+\Sigma_t)} - \frac{1}{2}\sum_{u\neq t,s}\frac{(U_{su}U_{ut}U_{ts}+U_{st}U_{tu}U_{us})}{(E_t-E+\Sigma_t)(E_u-E+\Sigma_u)}+\dots\right]. \tag{4.98}$$

Let us assume, as before, that U is Hermitian, so that

$$(U_{su}U_{ut}U_{ts})^* = U_{st}U_{tu}U_{us}.$$

Then it follows from (4.98) that

$$\Gamma_s = \overline{\mathrm{Im}\,\Sigma_s} = \sum_{t\neq s}\left[\overline{U_{st}U_{ts}}\frac{\Gamma_t}{(E_t-E-A_t)^2+\Gamma_t^2} - \sum_{u\neq t,s}\mathrm{Re}\,(U_{st}U_{ut}U_{ts}) \times\left(\frac{\Gamma_t(E_u-E-A_u)+\Gamma_u(E_t-E-A_t)}{((E_t-E-A_t)^2+\Gamma_t^2)((E_u-E-A_u)^2+\Gamma_u^2)}\right)\right].$$

A slight rearrangement of terms shows that to third order in U this is equivalent to

$$\Gamma_s = \sum_{t\neq s}\frac{\Gamma_t}{(E_s-E-A_t)^2+\Gamma_t^2}\overline{|U'_{st}|^2},$$

i.e. that *the only modification required to our previous formula for* Γ *is the replacement of* U *by* U'. The same principal holds good for much of the discussion in §4.7. Thus third-order terms appear in each cross-product such as (4.48), making

$$\overline{c_t U_{ts} c_u^* U_{su}} \simeq -\overline{c_t c_t^*}\frac{\overline{U_{su}U_{ut}U_{ts}}}{(E_u - E + \Sigma_u^*)} - \overline{c_u c_u^*}\frac{\overline{U_{su}U_{ut}U_{ts}}}{(E_t - E + \Sigma_t^*)},$$

so that the right-hand side of (4.49) becomes

$$\sum_{t\neq s}\overline{c_t c_t^*}\left(\overline{U_{st}U_{ts}} - 2\sum_{u\neq t,s}\frac{\overline{U_{su}U_{ut}U_{ts}}(E_u - E - A_u)}{(E_u - E - A_u)^2 + \Gamma_u^2} + \ldots\right).$$

Since this is equivalent to

$$\sum_{t\neq s}\overline{c_t c_t^*}\,\overline{|U'_{st}|^2}$$

up to third order, the proof of (4.51) with U' in place of U, goes through unchanged. This much is encouraging for it means that even if the difference between U' and U should turn out to be appreciable a good many of our previous conclusions should remain intact. The device of replacing U by U' is not always sufficient, however. It fails even in the third order when one sets out to calculate the energy shift A, as may be seen by extracting the real part of (4.98), and beyond the third order it fails for Γ and for $\overline{c_k c_k^*}$ as well.†

If we are to estimate the importance of third-order corrections in practice, e.g. in the calculation of Γ, we must try to express $\overline{|U'|^2}$ in terms of functions that we know, i.e. the ionic pseudo-potential $u(q)$ and the interference function $a(q)$. It is easy to get as far as

$$\overline{|U'(\mathbf{q}_{st})|^2} = Nu(\mathbf{q}_{st})u(\mathbf{q}_{ts})a(\mathbf{q}_{ts}) - \frac{2}{\Omega}\sum_{u\neq t,s}\frac{\operatorname{Re}\{u(\mathbf{q}_{su})u(\mathbf{q}_{ut})u(\mathbf{q}_{ts})\overline{F(\mathbf{q}_{su})F(\mathbf{q}_{ut})F(\mathbf{q}_{ts})}\}}{(E_u - E + \Sigma_u)} + \ldots \quad (4.99)$$

where $F(q)$ is the structure factor defined by (1.72), but the average of the product of three structure factors is bound to involve the *three-body distribution function* g_3. Some approximations are therefore required.

† The reader who attempts to extend the argument beyond the third order will find himself hampered by the fact that even if U has Hermitian properties, U' does not.

The simplest approximation which suggests itself† is to write

$$\overline{F(\mathbf{q}_{su})F(\mathbf{q}_{ut})F(\mathbf{q}_{ts})} = \overline{\sum_i \exp(-i\mathbf{q}_{su}\cdot\mathbf{r}_i)\sum_j \exp(-i\mathbf{q}_{ut}\cdot\mathbf{r}_j)}$$
$$\overline{\times\sum_k \exp(-i\mathbf{q}_{ts}\cdot\mathbf{r}_k)}$$
$$= \overline{\sum_j\sum_i \exp(-i\mathbf{q}_{su}\cdot\mathbf{R}_{ij})\sum_k \exp(-i\mathbf{q}_{ts}\cdot\mathbf{R}_{kj})}$$
$$= Na(q_{su})\,a(q_{ts}). \tag{4.100}$$

If this is sufficient, we have

$$\overline{|U'(\mathbf{q}_{st})|^2} = Nu'(\mathbf{q}_{st})\,u'(\mathbf{q}_{ts})\,a(q_{ts})$$

to third order, where the effective pseudo-potential for an individual ion is defined by

$$u'(\mathbf{q}_{st}) = u(\mathbf{q}_{st}) - \frac{1}{\Omega}\sum_{u\neq t,s}\frac{u(\mathbf{q}_{su})\,u(\mathbf{q}_{ut})\,a(q_{su})}{(E_u - E + \Sigma_u)}; \tag{4.101}$$

it is not difficult to show, by the same line of argument that was used to demonstrate the equivalence of (4.95) to (4.96), that $u'(q)$ defined in this way is the Fourier transform of

$$\overline{u'(\mathbf{R})} = (1 + \overline{\gamma(\mathbf{R})})\,u(\mathbf{R}). \tag{4.102}$$

Thus if we use (4.100) we are taking account of the way in which an electronic wave function is *on the average* modulated in the neighbourhood of an ion. We are not recognising, however, that $\gamma(\mathbf{R})$ and $u'(\mathbf{R})$ may vary from one ion to another, according to the particular way in which the neighbouring ions are arranged.

Equation (4.100) is inexact, because in the final line we have replaced an average of a product by the product of two averages. Such a replacement is legitimate only if the fluctuations in $\sum_j \exp(-i\mathbf{q}_{su}\cdot\mathbf{R}_{ij})$ and $\sum_k \exp(-i\mathbf{q}_{ts}\cdot\mathbf{R}_{kj})$ from one value of j to the next are entirely uncorrelated, and in practice some degree of correlation is inevitable. It may be noticed that by a slightly different manipulation of the calculation in (4.100) we could have arrived at

$$\overline{F(\mathbf{q}_{su})F(\mathbf{q}_{ut})F(\mathbf{q}_{ts})} = Na(q_{ut})\,a(q_{ts}) \quad or \quad Na(q_{su})\,a(q_{ut}).$$

† This is sometimes referred to as the 'geometrical' approximation.

The second of these alternatives, though not the first, could lead to quite different results for the third-order correction to Γ. It has been suggested by Greenwood (1966) and others that

$$\overline{F(\mathbf{q}_{su})F(\mathbf{q}_{ut})F(\mathbf{q}_{ts})} = \tfrac{1}{3}N\{a(q_{su})\,a(q_{ut})+a(q_{ut})\,a(q_{ts})+a(q_{ts})\,a(q_{su})\} \tag{4.103}$$

is more accurate than (4.100) but no physical justification for this formula is apparent.

In terms of the three-body distribution function we have

$$\begin{aligned}\overline{F(\mathbf{q}_{su})F(\mathbf{q}_{ut})F(\mathbf{q}_{ts})} &= N[1+(a(q_{su})-1)+(a(q_{ut})-1)+(a(q_{ts})-1)]\\ &\quad+\left(\frac{N}{\Omega}\right)^3\iiint g_3(\mathbf{r},\mathbf{r}',\mathbf{r}'')\exp(-i\mathbf{q}_{su}\cdot\mathbf{r})\exp(-i\mathbf{q}_{tu}\cdot\mathbf{r}')\\ &\quad\times\exp(-i\mathbf{q}_{ts}\cdot\mathbf{r}'')\,d\mathbf{r}\,d\mathbf{r}'\,d\mathbf{r}''; \end{aligned}\tag{4.104}$$

the four terms in the first line of (4.104) correspond to $i = j = k$, $i \neq j = k$, $j \neq k = i$ and $k \neq i = j$ respectively. Now it turns out that *Abé's approximation* for g_3, expressed by (2.69), enables this exact result to be reduced to a relatively simple form; the superposition approximation, in the present context, is clumsy by comparison. By an extension of the argument in §2.10 one arrives at

$$\begin{aligned}\overline{F(\mathbf{q}_{su})F(\mathbf{q}_{ut})F(\mathbf{q}_{ts})} &= N\{-1+\tfrac{1}{3}[a(q_{su})+a(q_{ut})+a(q_{ts})\\ &\quad+a(q_{su})\,a(q_{ut})+a(q_{ut})\,a(q_{ts})+a(q_{ts})\,a(q_{su})]\}.\end{aligned}\tag{4.105}$$

This is one degree more complicated for the purposes of computation than (4.103), but in the absence of any firm knowledge about g_3 it is probably the best approximation that we can use.

To obtain a quick idea of whether it is worth bothering at all about these third-order corrections in a real liquid metal, let us go back to (4.100) and (4.101). Just how big compared with unity is $\overline{\gamma(\mathbf{R})}$ likely to be? How big is it, for example, at the centre of an ion, for $R = 0$? This is easily computed from (4.97), since if the origin for $\mathbf{r}$ is chosen to coincide with the centre of an ion the mean value of $U(\mathbf{q})$ may be shown from (2.37) to equal $u(q)\,a(q)$. Hence

$$\overline{\gamma(0)} = -\frac{1}{\Omega}\sum_{u\neq s}\frac{u(\mathbf{q}_{su})\,a(q_{su})}{(E_u-E+\Sigma_u)}.$$

If we make the dubious assumption that $u(\mathbf{R})$ is spherically symmetric in real space and localised, so that $u(\mathbf{q})$ is real and independent of the direction of $\mathbf{q}$, then we may integrate over all directions for $\mathbf{q}_{su}$, to obtain (cp. (4.86))

$$\operatorname{Re}\overline{\gamma(0)} \simeq -\frac{m^*}{(2\pi)^2\hbar^2 k}\int_0^\infty u(q)\,a(q)\,\tfrac{1}{2}\log\left|\frac{(2k+q)^2+(k/2lq)^2}{(2k-q)^2+(k/2lq)^2}\right| q\,dq. \tag{4.106}$$

Such calculations as have been attempted on the basis of this formula are not initially encouraging. Thus if the pseudo-potentials of Heine & Animalu are used, $\operatorname{Re}\overline{\gamma(0)}$ for an electron at the Fermi level comes out to be about -0.25 for liquid Ga, which is by no means small compared with unity. Much the same answer is obtained for liquid Na, despite its longer mean free path.

Further cause for gloom is provided by the work of Springer (1964), who attempted to assess the importance of third-order corrections to resistivity of liquid Zn; the resistivity, as we shall see in chapter 5, is largely determined by Γ. Springer used the equivalent of (4.100) and came up with corrections that were quite unpleasantly large.

Two rays of light illuminate the gloom, however. First, it is not uncommon in scattering calculations to find that, even though the third-order corrections seem to be large, inclusion of all the corrections of higher order still brings the true answer much closer to what was given in the first place by the Born approximation. Ashcroft & Schaich (1970) have developed an iterative method, based upon (4.100) and a number of other rather drastic approximations, for summing higher-order corrections and they find that although the predicted resistivity of liquid Zn is increased by about 70% after one iteration, the net increase after nine iterations is only about 20%. They have done calculations for liquid Na and Al as well, with comparable results.

Secondly, there is a weakness in equation (4.106) which Greenwood (1966) was the first to point out. It is based on the assumption that in calculating the extent to which a pseudo wave function is modulated in the neighbourhood of a particular ion it is necessary to include the pseudo-potential due to that ion and not just the pseudo-potential due to its neighbours. There is no doubt that this procedure is correct for a pseudo-potential derived by the Heine–Animalu method, but it would *not* be correct if we were dealing with well-separated ions whose potentials did not overlap and could use the proper T-matrix (see p. 42) for $u(q)$. Multiple scattering within a single ion would then be allowed for by the choice of u, and to include the effect of the central ion in our calculation of γ would be to allow for it twice over. In these circumstances we should replace $a(q)$ in (4.106) by $(a(q)-1)$.

Calculations suggest that this substitution is not sufficient by itself to make $\gamma(0)_F$ small. We need to remember, however, that the screened coulomb potentials around each ion in a liquid metal are bound to overlap to some extent. This makes it doubtful whether a satisfactory T-matrix can ever be defined, for one cannot necessarily assume that an electron wave scattered by one ion reaches its asymptotic form before encountering another. Most model pseudo-potentials which set out, unlike those of Heine & Animalu, to represent the T-matrix do so only as far as the core potential is concerned; outside the core they are made to converge rapidly onto the true potential. In calculating $\gamma(0)_F$ with a pseudo-potential of this sort one should certainly omit the modulating effect due to the core of the central ion, but there is no good reason to omit the effect of its screened coulomb field outside the core. Hence the right procedure may be to replace a by $(a-1)$ for large values of q inside the integral in (4.106), but to continue to use a where q is small. Since $(a-1)$ and a are much less than unity for large and small values of q respectively, this procedure should lead to a substantially smaller answer for $\gamma(0)_F$ than one obtains by using either a or $(a-1)$ over the whole range of q.

If our ionic pseudo-potential is a T-matrix, then some of the details of the argument leading to equation (4.105) require amendment. In evaluating

$$\overline{F(\mathbf{q}_{su})F(\mathbf{q}_{ut})F(\mathbf{q}_{ts})}$$

we must exclude all terms for which $i=j$, since these describe multiple scattering within a single ion, and (4.105) then becomes

$$\overline{F(\mathbf{q}_{su})F(\mathbf{q}_{ut})F(\mathbf{q}_{ts})} = N\{-1+\tfrac{1}{3}[a(q_{su})+a(q_{ut})-2a(q_{ts}) \\ +a(q_{su})a(q_{ut})+a(q_{ut})a(q_{ts})+a(q_{ts})a(q_{su})]\}. \quad (4.107)$$

Perhaps, with the pseudo-potentials that are available in practice, we ought to rely on (4.107) for large values of q but revert to (4.104) when q_{su} or q_{ut} is small.

There are grounds for hope, therefore, and the empirical fact that calculations of resistivity which stop at the second order are often surprisingly successful, especially if the pseudo-potential has not been computed entirely from first principles (e.g. by the Heine–Animalu method) but has been adjusted to fit some other observed property of the metal in question, is distinctly sustaining. It suggests that the pseudo-potential employed is a quite good approximation to the $u'(q)$ of (4.101), and that the additional corrections implied by (4.104) or (4.107), to which the precise form of g_3 is relevant, are rather trivial.

If one's interest lies in the density of states rather than the resistivity, then it is necessary to discuss A rather than Γ, and the concept of an effective pseudo-potential is less successful. However, a calculation of A can readily be based upon (4.98), the third-order term in A_s amounting to

$$-\frac{1}{\Omega}\sum_{t\neq s}\sum_{u\neq t,s}\mathrm{Re}\,[u(\mathbf{q}_{su})u(\mathbf{q}_{ut})u(\mathbf{q}_{ts})]$$
$$\times\frac{[(E_t-E-A_t)(E_u-E-A_u)-\Gamma_t\Gamma_u]}{[(E_t-E-A_t)^2+\Gamma_t^2][(E_u-E-A_u)^2+\Gamma_u^2]}$$
$$\times\overline{F(\mathbf{q}_{su})F(\mathbf{q}_{ut})F(\mathbf{q}_{ts})}. \tag{4.108}$$

Since $\Gamma_t\Gamma_u$ is always positive whereas $(E_t-E-A_t)(E_u-E-A_u)$ may have either sign, it may be a reasonable approximation in many circumstances to ignore the latter term in the numerator of (4.108). Now $\Gamma_t/((E_t-E-A_t)^2+\Gamma_u^2)$ and $\Gamma_u/((E_u-E-A_u)^2+\Gamma_u^2)$ are large when $\mathbf{k}_t$ and $\mathbf{k}_u$ lies near the spherical surface of radius k_0. Moreover,

$$\overline{F(\mathbf{q}_{su})F(\mathbf{q}_{ut})F(\mathbf{q}_{ts})}$$

according to (4.107), is large when all three of the scattering vectors concerned are close in magnitude to $q_{\max}$, where $a(q)$ has its main peak. It would seem, therefore, that the situation in which the third-order term in A is most likely to be significant is the one illustrated by fig. 4.16(*a*). This evidently requires $k_0 = q_{\max}/\sqrt{3}$, whereas the second-order term, it will be remembered, tends to be

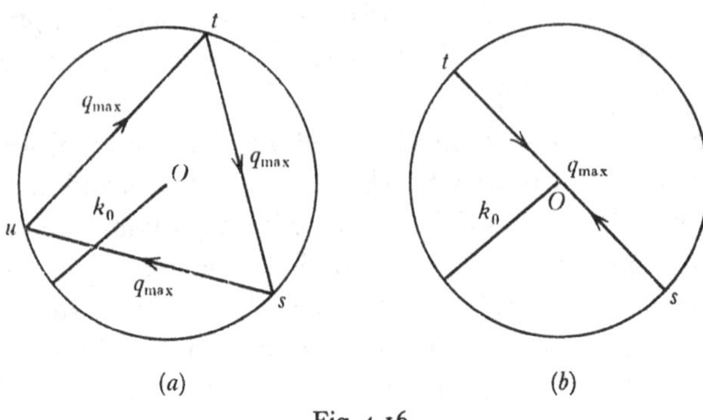

Fig. 4.16

large when $k_0 = \frac{1}{2}q_{max}$ as in fig. 4.16(*b*). If $u(q_{max})$ is positive, as is normally the case, the third-order term will also be positive, and it will therefore tend to accentuate any pseudo-gap produced by the second-order term alone. What is of interest in this argument is the suggestion that when the third-order effects are properly allowed for the number of electrons per atom with energies below the pseudo-gap may rise from 1.36 (see p. 275) to something nearer $1.36 \times (2/\sqrt{3})^3 = 2.1$. The latter figure is gratifyingly close to 2.

The argument is a very speculative one which requires testing by realistic computations. Rough calculations suggest that in the majority of liquid metals with which we are concerned the influence of the pseudo-potential on m^*/m, which seems to be small even in the second order as we have seen above, is entirely negligible in the third.

One last problem deserves a brief mention here, though it has not yet been seriously explored. How, when the scattering of the electrons is strong, do higher-order terms affect the energy, not just of a single electron but of the whole metal? One may speculate that just as one can describe some of the higher-order terms in Γ by using an effective pseudo-potential u' in place of u, so some of the higher-order terms in the energy can be included by modifying the pair potential $w(R)$. Conceivably, when we understand better the nature of the modification that is required, 'experimental' curves for $w(R)$ like the one for Ga that is reproduced in fig. 2.17 will seem more reasonable. But since an accurate expression for the energy is bound to involve g_3, g_4 etc., it is clear that at some stage the whole concept of a pair potential must be abandoned. Liquid Te is presumably a case in point.

4.12. Second thoughts on screening

The theory of the dielectric constant which was developed in chapter 1 was based on the picture that the electrons occupy states of well-defined **k**, and that the Fermi surface dividing the occupied from the unoccupied states in **k**-space is perfectly sharp. It now appears, however, that we should think of the electron states as wave groups, spread over a range of k. Even at the absolute zero (if one could supercool a liquid metal so far!), when the Fermi level is perfectly sharp in the sense that the occupation probability $f(E)$ changes abruptly from 1 to 0 at a definite value of E, it is impossible to define a sharp Fermi surface. If we define an occupation probability in **k**-space by

$$f(\mathbf{k}) = \sum_{E<E_F} \overline{c_k c_k^*(E)} \tag{4.109}$$

it may readily be shown from (4.51) that, when A_k is small enough to be ignored and Γ_k is independent of k but proportional to $E^{\frac{1}{2}}$,

$$f(k) = \tfrac{1}{2} - \frac{1}{\pi}\tan^{-1}(K_F(k^2 - k_F^2)/\Gamma_F k_F^2). \tag{4.110}$$

According to (4.110), $f(k)$ changes from 1 to 0 over a range of order $k_F(\Gamma/K_F)$. What difference does this make to the dielectric constant?

One obvious difference, mentioned already in §1.6, is that it is likely to remove the singularity which is responsible for the Friedel oscillations in the limit of large R. The singularity is evident in equation (1.28) for q_s^2 and arises when the factor

$$\frac{m}{\hbar^2 qk}\log\left|\frac{2k+q}{2k-q}\right|$$

in (1.27) is averaged over all values of k which are less than k_F. A plausible procedure when the Fermi surface is blurred is to weight the average by $f(k)$, i.e. to use

$$\frac{3m}{\hbar^2 qk}\int_0^\infty \frac{f(k)\,k^2\,dk}{qk}\log\left|\frac{2k+q}{2k-q}\right|. \tag{4.111}$$

Gaskell & March (1963) have adopted this procedure, though they use a form for $f(k)$ which is different from (4.110). Their results have been discussed briefly on p. 29.

But is it sufficient merely to insert the weighting factor $f(k)$? Equation (1.27) derives from (1.26), which is a result of simple first-order perturbation theory. In the light of the theory outlined in §4.7 above, ought we not to modify the denominators $(E(\mathbf{k})-E(\mathbf{k}\pm\mathbf{q}))$ in (1.26) by changing $E(\mathbf{k})$ to E and by introducing $\Sigma(\mathbf{k}\pm\mathbf{q})$ to allow for scattering?

In fact the best procedure for correcting our previous calculation of q_s^2 should be as follows. Consider the $\mathbf{k}$th component of the eigenfunction of energy E and find how much its amplitude is modulated by a particular Fourier component of the pseudopotential, $U(\mathbf{q})$; the modulation is the result of the continuous scattering that takes place into the $(\mathbf{k}+\mathbf{q})$th component. Average this modulation factor over all possible values of $\mathbf{k}$, using a weighting factor $\overline{c_k c_k^*}(E)$. Finally, average it over all possible values of E, using as a weighting factor the ordinary Fermi–Dirac function $f(E)$. A modified form of (1.28) should result.

This is a somewhat elaborate recipe and no attempt has yet been made to apply it to calculations of $\epsilon(q)$ in the neighbourhood of $q = 2k_F$. In the limit when q is very small, however, it is not difficult to carry through the calculation exactly, granted the approximations suggested above, i.e. that A_k can be ignored and that Γ_k is proportional to $E^{\frac{1}{2}}$ but independent of k. The analysis is too lengthy to reproduce here, but the result is simple: in the limit when q is very small the previous estimate for q_s^2 should be reduced by a factor

$$\frac{1}{2}\int_0^\infty \frac{f(K)\,\mathrm{d}K}{(KK_F+\Gamma^2)^{\frac{1}{2}}} \simeq 1-\frac{\Gamma_F}{K_F}, \qquad (4.112)$$

where $f(K)$ is the Fermi–Dirac distribution function. This result has been quoted on p. 22.† It can *not* be reproduced by the simpler procedure which Gaskell & March have adopted, which

† For a further discussion of the effects of electron scattering on the dielectric constant, consult Kliewer & Fuchs (1969). The formula which they suggest, however, does not contain the correction factor which (4.112) describes.

may be shown in the limit of small q to yield a correction factor $1+O(\Gamma_F/K_F)^2$.

The ratio Γ_F/K_F seems to be small enough in most liquid metals (see table 1.1) for the correction represented by (4.112) to be a fairly trivial one. A very much more important question to reconsider, in the light of arguments developed in this chapter, is whether the whole idea of a dielectric constant is sound. Is the perturbing pseudo-potential really weak enough for the screening to be linear?

Some of the difficulties of this problem may be illustrated by discussing once again how a sinusoidal perturbation ΔV_b would be. screened in an ideal jellium. When ΔV_b is weak one supposes that each electronic wave function is modulated by a factor $(1+\gamma(\mathbf{r}))$, where γ varies sinusoidally also and is much less than unity, and that a sinusoidal screening cloud is thereby built up with a density proportional to $2\bar{\gamma}$; the bar here represents an average over all the electron states inside the Fermi surface, which are not all modulated to the same degree. The first correction that becomes necessary as ΔV_b is made stronger arises from the fact that the modulation is effected by $\Delta U'$ rather than ΔU. This magnifies the screening efficiency of the electrons by a factor $(1+\bar{\gamma})$ so that for an electron at the Fermi level the effective value of the dielectric constant becomes

$$\Delta U'_F/\Delta V_b = (1+\gamma_F)\,q^2/(q^2+(1+\bar{\gamma})\,q_s^2). \qquad (4.113)$$

In so far as $\bar{\gamma}$ may be close to γ_F, though there is no reason to expect the two to be equal, the factor $(1+\gamma)$ may be cancelled from the right-hand side of (4.113) when q is much less than q_s. But for values of q which are comparable with q_s we find an effective dielectric constant which does depend upon γ and hence upon the strength of the perturbation.

This is not the whole story, however. When ΔV_b is too strong for γ to be small compared with unity, it is no longer a reasonable approximation to suppose that the screening charge varies sinusoidally in density. Harmonics in the screening potential, and hence in $\Delta U'$ must then appear, with wave-numbers $2\mathbf{q}$, $3\mathbf{q}$ etc. And if ΔV_b contains more than one sinusoidal component, with wave numbers $\mathbf{q}_1$, $\mathbf{q}_2$ etc., $\Delta U'$ will contain components with wave numbers $(\mathbf{q}_1+\mathbf{q}_2)$ etc. At this stage the concept of a dielectric constant becomes completely untenable. If it is reached in a liquid metal, then the screened pseudo-potential for an individual ion is not to be found without an elaborate iterative calculation of the sort envisaged in §1.5, while a rigorous treatment of screening for an assembly of ions is bound to involve g_3 etc.

4.13. Positron annihilation

When a positron with several MeV of energy is fired into a metal it suffers inelastic collisions and is brought virtually to rest in a time believed to be of order 10^{-12} sec, at a depth of say 10^{-2} cm. It may survive for a further 10^{-10} sec, which is a long time by comparison with the period of oscillation of the ions or the relaxation time of the conduction electrons, but in due course it is bound to annihilate with an electron. Either two or three high-energy photons are then produced, depending on the angular momenta about their common centre of gravity of the annihilating pair; the two-photon mode of decay is much the more probable and is all we need consider here. If momentum is to be conserved the two photons must leave the metal in almost opposite directions, as shown in fig. 4.17. A small angle θ between them indicates that the pair had a linear momentum $\hbar\mathbf{q}$ before annihilation, with a component

$$\hbar q_z = 2mc \sin \tfrac{1}{2}\theta \simeq mc\theta$$

in the z direction OP, and most of this is presumably to be attributed to the electron. Is it possible to measure the momentum distribution of the conduction electrons, i.e. the function $f(k)$ introduced in the previous section, by studying the relative probability of different values of θ, using coincidence counters and a suitable arrangement of slits?

The simplest quantitative theory of annihilation (see e.g. Wallace, 1960) is encouraging. Based upon the NFE model and second-order perturbation theory, it says that the probability of a particular process, involving a positron and an electron with unperturbed wave functions ψ_+ and ψ_- respectively, should be proportional to the square of a matrix element

$$\int \psi_+ \psi_- \exp(-i\mathbf{q}\cdot\mathbf{r})\, d\mathbf{r}.$$

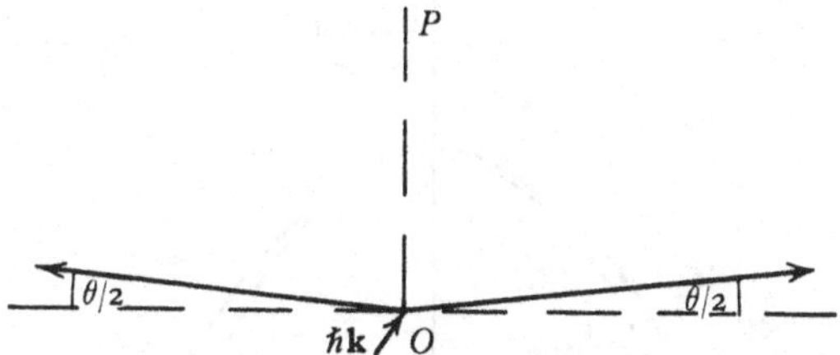

Fig. 4.17. Two photons produced by annihilation at O of a positron and an electron with momentum $\hbar\mathbf{k}$.

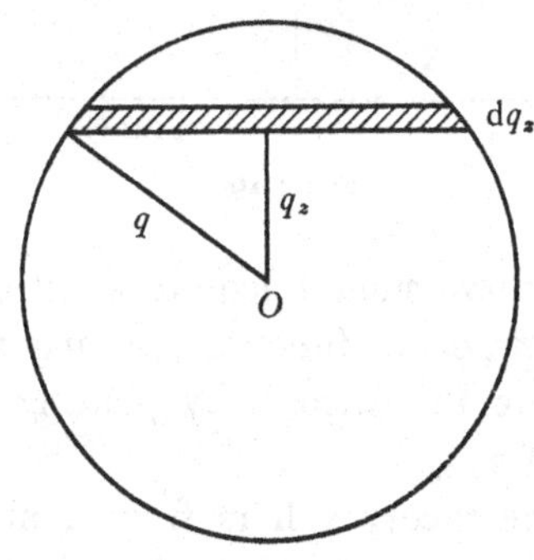

Fig. 4.18

If the kinetic energy of the positron is indeed zero, or at any rate not much greater than $k_B T$ and therefore negligible compared with K_F, we may reasonably treat ψ_+ as a function which is uniform in space. Then the matrix element vanishes unless the wave number $\mathbf{k}$ associated with ψ_- is equal to $\mathbf{q}$. Since the magnitude of the matrix element when $\mathbf{k} = \mathbf{q}$ is independent of q it follows that the probability of observing a coincidence for which q_z (or θ) lies in a particular range dq_z (or $d\theta$) is simply proportional to the number of occupied electronic states which lie within the slice of $\mathbf{k}$-space which is indicated by shading in fig. 4.18, i.e. it should be proportional to

$$dq_z \int_{q_z}^{\infty} f(k)\, k \, dk.$$

If $f(k)$ were unity up to a sharp cut-off at k_F, then a plot of the number of coincidence observed per unit time, n_{coin}, against q_z or θ should have the parabolic form indicated by the full curve in fig. 4.19; if the cut-off is blurred, as equation (4.110) for example

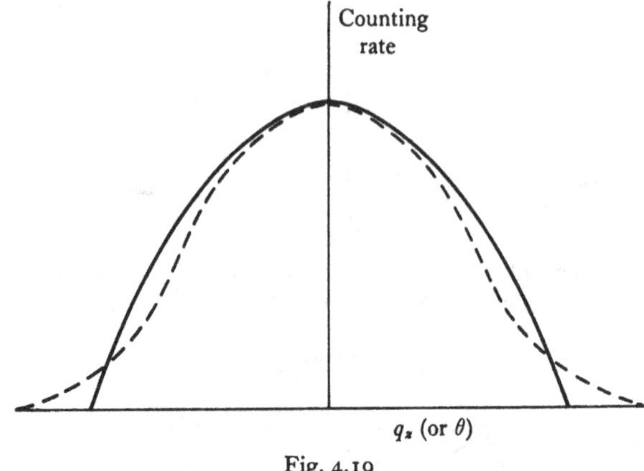

Fig. 4.19

suggests, then the curve should acquire a tail as indicated by a broken line. The form of the function $f(k)$ may be exhibited very directly, according to this theory, by plotting $-q_z^{-1}(\mathrm{d}n_{\mathrm{coin}}/\mathrm{d}q_z)$ against q_z, instead of n_{coin}.

Unfortunately, the theory suffers from a number of serious limitations. First, one ought to remember that even though the potential set up by the positron is screened it nevertheless constitutes a substantial perturbation so far as the conduction electrons are concerned. Is it fair to assume that the momentum distribution of the electrons in the immediate neighbourhood of a positron is the same as when the positron is absent? This question has been examined by Carbotte & Kahana (1965) and Majumdar (1965), who use the apparatus of many-body theory since this is a problem where the independent quasi-particle model for electrons is liable to be misleading. They come to the conclusion that even in the absence of any scattering by the ions the $f(k)$ measured in an annihilation experiment should have a small tail above k_F, though a sharp fall *at* k_F is still to be expected.†

Secondly, there is the complication that, even within the framework of the independent quasi-particle model and in the absence of any scattering of the electrons, their wave functions ψ_- contain

† The perturbation due to the positron also has a very marked effect in heaping up the charge density of the conduction electron gas, and hence increases the annihilation rate. The lifetimes observed for positrons in metals are still not understood in detail.

a range of high-k Fourier components because they oscillate inside each ion core; it is only a pseudo wave function ϕ which can be described by a single value of $\mathbf{k}$. Fortunately, the positron experiences a strongly repulsive potential inside the cores, so that ψ_+ is necessarily small there, and this helps to ensure that the smooth part of ψ_- contributes more to the matrix element than the oscillatory part. Nevertheless, the oscillatory part should not be ignored; it is likely to contribute a tail to the observed momentum distribution, extending out to quite large values of k, which has nothing to do with the lifetime broadening of the Fermi cut-off. Come to that, unless ψ_+ is vanishingly small inside the cores there will be a finite chance that the positron will annihilate with one of the core electrons and not with a conduction electron at all, and this may much enhance the tail. A background of coincidences, extending to values of q_z two or three times larger than k_F, is indeed observed in practice and is customarily attributed to annihilation processes inside the cores. The measured curves for n_{coin} cannot be split into a core part and a conduction-electron pseudo wave function part with any certainty; curves have been computed for the former (West, Borland, Cooper & Cusack, 1967) but since they do not fit the observations at large q_z they can scarcely be trusted in the region of interest near k_F.

Thirdly, it is really very dubious to suppose that ψ_+ is independent of $\mathbf{r}$. Not only are the positrons repelled strongly from the ion cores, they also experience the not insignificant screened coulomb potential that surrounds each ion. Surely the mean free path of a positron is liable to be just as short as that of a conduction electron, i.e. a few interatomic spacings or less? If so, even its ground state wave function in the space between the ion cores must contain a spread of Fourier components, which is bound to broaden the curve for n_{coin}.

Add to these limitations the fact that the instrumental resolution in positron annihilation experiments still leaves something to be desired, and it is not surprising that the results so far obtained have been difficult to interpret in detail. The central part of the curve for n_{coin} can usually be fitted quite closely by a parabola, using the free-electron value for k_F, and the cut-off at k_F does at first sight seem to be smeared out in the expected manner. But the attempts which have been made to correlate the degree of smearing with the magnitude of the electronic mean free path as deduced from the resistivity of the liquid metal have enjoyed only moderate success. Nor does the smearing always increase on melting to the extent to be expected from the increase in the resistivity. Kusmiss & Stewart (1967) divide the low melting point metals into three groups: Li, Na and Tl yield curves for n_{coin} which, once the core part is subtracted out, do not change their shape at all, either on heating up to the melting point nor during the liquefaction process; Zn, Cd, Al, In and Pb yield curves which broaden as the solid is heated but show little further change on melting; it is only the curves for Hg, Ga, Sn, Sb and Bi, which are insensitive to temperature in the solid phase, that broaden significantly during melting. While there is no explanation for such purely qualitative distinctions, it seems premature to attempt any comparison of the apparent results for $f(k)$ with equation (4.110), let alone with the more accurate curves which Ballentine (1966*b*), for example, has computed.

There is one qualitative effect, first noted by Gustafson, Mackintosh & Zaffarano (1963) in their work on Hg, which seems to be observed for all metals; the background due to core annihilation becomes relatively less important in the liquid state. At the same time, the positron lifetime tends to rise (McGervey, 1967; Brandt & Waung, 1968). West *et al.* estimate that 69% of the area under the n_{coin} curve can be attributed to core annihilation in solid Hg but only 57% in liquid Hg; the corresponding figures for Bi are 26% and 23%. Combining these estimates with results for the mean positron lifetime in solid and liquid Hg, West *et al.* deduce that the core annihilation rate decreases from 5.0×10^9 to 3.5×10^9 sec^{-1} on melting, whereas the rate for annihilation outside a core rises from 2.3×10^9 to 2.6×10^9 sec^{-1}. The volume change of Hg on melting is quite insufficient by itself to explain these observations.

The customary explanation of this effect is that the more open structure of the liquid phase allows the positrons to find 'voids' where they can settle down in states of relatively low energy, relatively far from any ion. The picture is not entirely satisfactory; there is no evidence that voids of appreciable size occur in liquids with any greater frequency than vacancies occur in solids just below the melting point, and if a void does open up from time to time it should last for only about 10^{-12} sec, which is short compared with the lifetime of a positron. Perhaps a more accurate picture is that the positrons, rather than take advantage of voids which are already present in the liquid before their arrival, are able to *create* voids by repelling the surrounding ions.

4.14. Compton scattering

The theory of inelastic, Compton scattering of X-ray photons by a metal deserves a brief exposition, since this is another phenomenon that depends upon the momentum distribution of the electrons.† Consider a process in which a photon is scattered through a vector $\mathbf{Q}$ with a decrease in energy of ϵ, while an electron is excited from a state $\mathbf{k}_1$ to a state $\mathbf{k}_2$, (see fig. 4.20), such that

$$k_2^2 - k_1^2 = Q^2 - 2k_1 Q \cos\theta = 2m\epsilon/\hbar^2.$$

Such a process is possible only if $k_1 \geqslant q$, where

$$q = (Q^2 - 2m\epsilon/\hbar^2)/2Q.$$

Its probability depends upon the probability $f(\mathbf{k}_1)$ that the initial state for the electron is indeed occupied, and also on the probability that the final state is empty, but for the values of ϵ which are of importance in practice k_2 is so large compared with k_F that $(1-f(\mathbf{k}_2))$ can be replaced by unity. Hence the probability of scattering through a particular value of Q into an energy range $d\epsilon$ should be proportional, when the electronic momentum distribution is isotropic, to

$$\int_q^\infty f(k_1)\, 2\pi k_1^2\, dk_1 \sin\theta\, d\theta = \frac{2\pi m\, d\epsilon}{\hbar^2 Q}\int_q^\infty f(k_1)\, k_1\, dk_1.$$

Exactly the same integral is therefore involved as in the theory of positron annihilation, so that the 'Compton line shape' of the scattered photons (intensity versus loss of energy for given Q) should be similar in many respects to the curve of annihilation rate n_{coin} versus q_z. The background due to the core electrons should be marked since photons can penetrate the core without difficulty.

† Relativistic corrections are ignored in this discussion. They do not affect the conclusions reached.

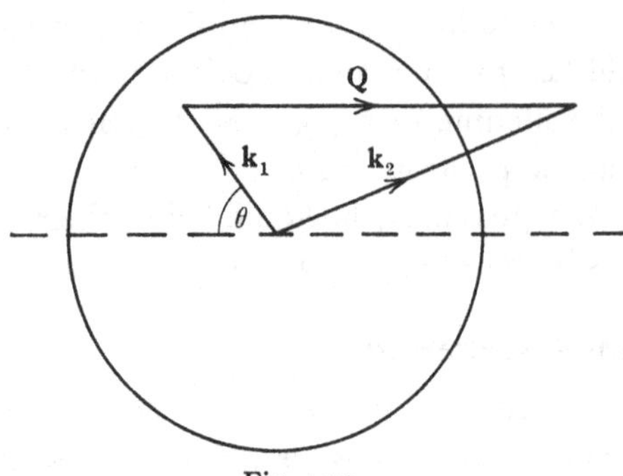

Fig. 4.20

On the other hand, the complications that arise in the positron case due to the spatial variation of ψ_+ and to the perturbation of the electrons by the positron do not arise in the theory of Compton scattering.

Experimental difficulties have so far prevented X-ray crystallographers from obtaining useful results for metals from measurements of Compton scattering; the cut-off in $f(k)$ at the Fermi radius has been detected for solid Be and that is all (e.g. Weiss, 1966). But techniques are being improved and measurements on some of the lighter liquid metals may shortly become worth while.

4.15. Magnetic susceptibilities

There remain two topics to be discussed which are relevant to the theory developed earlier in this chapter. The first concerns the paramagnetic spin susceptibility χ_p of the conduction electrons in liquid metals. Elementary theory suggests that this should be proportional to the density of states at the Fermi level, i.e. to m^*/m. We must consider whether it is possible to measure χ_p and whether the results can be used to check theoretical predictions such as those reported in §4.9 above, about what m^*/m ought to be.

The elementary theory, based on the free-electron model, is due to Pauli. In the presence of a magnetic field H the energy of an electron is shifted up or down by $\mu_B\mathsf{H}$ according to its spin orientation, and the Fermi–Dirac function (1.6) becomes

$$f(E) = [\exp((E \pm \mu_B\mathsf{H} - E_F)/k_BT)+1]^{-1},$$

where E is the energy in the absence of a field. The magnetic moment is then

$$\mu_B \int (f_-(E) - f_+(E))\mathcal{N}(E)\,dE \simeq -2\mu_B^2 H\mathcal{N}(E)_F \int_0^\infty \frac{df}{dE} dE$$
$$\simeq 2\mu_B^2 H\mathcal{N}(E)_F,$$

and the spin susceptibility *per unit mass*,† from (1.5) and (1.7), is

$$\chi_p(FE) = \frac{m\mu_B^2}{\pi^2\hbar^2 m_A}\left(\frac{3\pi^2\Omega^2 z}{N^2}\right)^{\frac{1}{3}}, \tag{4.114}$$

independent of temperature. The free-electron model has been invoked in this calculation only in order to obtain an explicit expression for the density of states. The answer would seem to be generally valid so long as the electron mass m is replaced by m^*.

Unfortunately, the calculation does not allow properly for the effects of exchange. It was explained in chapter 1 that every conduction electron is accompanied by an exchange+correlation hole from which other electrons are excluded, and that the existence of this hole lowers the electron's potential; without it, indeed, the electrons in a jellium would not be bound. Now the exchange repulsion acts only between electrons of like spin. When a metal is magnetised by an applied field, an electron whose moment points along H will have more neighbours of like spin than an electron whose moment points in the opposite direction. This means that, in so far as the hole is caused by exchange repulsion rather than coulomb repulsion, it is as it were deeper for the first electron than for the second. The energy difference between the two electrons is therefore greater than $2\mu_B H$ and the spin susceptibility is *enhanced*. The enhancement is thought to be substantial. Fig. 4.21 shows two theoretical curves for the ideal jellium – the reader should consult Hedin & Lundqvist (1969) for a summary of the calculations upon which these curves are based – and the one for $\chi_p/\chi_p(FE)$ is well above the one for m^*/m.

The only methods available for the measurement of χ_p depend

† The normal definition of susceptibility is, of course, magnetic moment *per unit volume* divided by field, but in this particular subject it seems to be more conventional to use the magnetic moment per unit mass; sometimes magnetic moment *per mole* is used. Readers who wish to consult the original literature are warned that the same symbol χ is used for three different quantities.

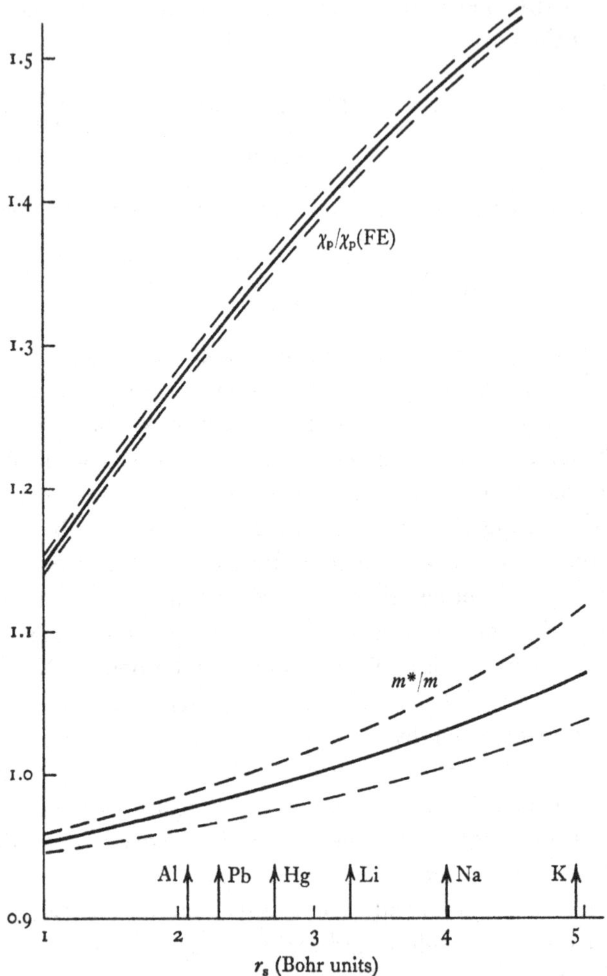

Fig. 4.21. The density of states and spin susceptibility enhancement in an ideal jellium. Broken lines indicate the range of uncertainty in these theoretical results.

upon the techniques of electron spin resonance, and they cannot be applied unless the resonance line is reasonably narrow; this requires in effect that the spin-orbit coupling be weak and restricts one to the lighter metals. Fairly reliable results are now available for Li, Na and K in the *solid* phase, though values deduced from the area under the resonance line are not in complete agreement

TABLE 4.1. *Effective mass ratios for solid alkali metals*

	$\chi_p/\chi_p(FE)$	Exchange enhancement	m^*/m
ESR data			
Li	2.5	1.41	1.77
Na	1.58	1.44	1.10
K	1.61	1.45	1.11
	m_{opt}/m	(m^*_{xc}/m)	m^*/m
Optical data			
Li	1.57	1.01	1.59
Na	1.00	1.03	1.03
K	1.08	1.07	1.15
	$\gamma/\gamma(FE)$	Phonon enhancement	m^*/m
Specific heat data			
Li	2.3	(1.5)	1.5
Na	1.26	1.18	1.07
K	1.20	1.14	1.05

with values inferred from studies of spin wave propagation (Hedin & Lundqvist, 1969). Average values for the ratio $\chi_p/\chi_p(FE)$ are shown in the first section of table 4.1. After division by a theoretical exchange enhancement factor, taken from fig. 4.21, they yield the values for m^*/m shown in the right-hand column.

We shall find in the next chapter that it is possible to deduce an effective mass ratio of sorts by optical measurements on solid metals in the infrared and visible, and the results obtained by Hodgson (1966) and Palmer & Schnatterly (1971) for solid Li, Na and K are shown in the second section of table 4.1. The theory suggests (see §5.18) that so long as the Fermi surface is almost spherical, which is thought to be the case for these three metals, m_{opt} should differ from m^* only because it is not influenced by the effects of exchange and correlation. We are therefore entitled to estimate m^*/m by multiplying m_{opt}/m by a factor m^*_{xc}/m taken from the lower curve in fig. 4.21. The results are as shown in the table. In the third section there is yet another set of results for m^*/m, derived from the low-temperature specific heat data already presented in table 1.2. The three sets are probably consistent within the limits of error. They confirm the suggestion of the theoretical calculations reported in §1.10 that the effective mass ratio is close to unity in solid Na and K but significantly greater than unity in solid Li. It looks as though m_1^*/m must be at least 1.3 for Li, and therefore larger than Weaire (1967) and Shaw & Smith have suggested (see table 1.2); furthermore, the ratio $\mathcal{N}(E)/\mathcal{N}(E_k)$ at the Fermi level, which describes the second-order effects of the pseudo-potential on the effective mass, would seem to be at least as large as the factor 1.17 which Shaw & Smith have calculated for the solid phase (see p. 271). The product of 1.3 and 1.17 is only 1.5, which is if anything on the low side to match the experimental data.

It is easier to study the temperature dependence of χ_p by ESR techniques than it is to make absolute measurements and the temperature dependence for Li has now been determined right up to the melting point and into the liquid phase (Enderby, Titman & Wignall, 1964). It is very slight. The ratio $(\chi_p)_L/(\chi_p)_S$ at the melting point itself is only 1.04 ± 0.05, so $\mathcal{N}(E)/\mathcal{N}(E_k)$ must be rather insensitive to the difference of structure between solid and liquid. This somewhat surprising result is in agreement with the calculations of Shaw & Smith, whose value for the liquid phase is 1.15. If we are to believe Shaw & Smith, $\mathcal{N}(E)/\mathcal{N}(E_k)$ falls by about 1.5% on melting, but since $\mathcal{N}(E_k)$ should rise by about 2% due to thermal expansion the net change in χ_p should be extremely small.†

The width of the ESR line is more affected by melting than its area (McMillan, King, Miller & Carlson, 1962; Taupin, Lambert & Mazieres, 1971). From the width a spin relaxation time for the conduction electrons may be deduced. This is very much longer than the lifetime τ_l because the majority of scattering events change an electron's **k** value without affecting its spin, but simple theories suggest that the two might be proportional to one another. If so, and if the distinction between τ_l and τ (see p. 209) were unimportant in this context, the line width would be proportional to the resistivity ρ. Recent measurements by Devine & Dupree (1970*a*,*b*) show that the two do vary with temperature in the same manner for solid Na. On melting, however, the line width for Na increases

† According to Seymour (private communication) χ_p is insensitive to melting in Na also, but this is less surprising.

TABLE 4.2. *Magnetic susceptibilities at the melting point*

Metal	χ_p(FE)	$(\chi_L - \chi_i)$	$(\chi_L - \chi_S)$
Li	1.5	3.9	0.035
Na	0.69	0.9	0.015
K	0.62	0.85	0.005
Rb	0.33	0.5	−0.006
Cs	0.24	0.5	−0.002
Cu	0.11	0.13	0.015
Ag	0.08	0.08	−0.1
Au	0.04	0.03	−0.02
Zn	0.17	0.05	0.015
Cd	0.12	0.02	0.03
Hg	0.07	0.02	−0.05
Al	0.50	0.5	−0.1
Ga	0.20	0.15	0.34
In	0.15	0.11	0.01
Tl	0.09	0.05	0.025
Ge	0.23	0.16	0.18
Sn	0.17	0.14	−0.06
Pb	0.10	0.06	0.04
Sb	0.18	0.13	0.4
Bi	0.14	0.08	0.7

Note. The values quoted are all *mass* susceptibilities, in units of $gm^{-1} \times 10^{-6}$. The sources of the data and the meaning of the symbols are given in the text.

by only 40% whereas there is a 45% rise in ρ, and the latter is also more sensitive to further heating above the melting point.

For other metals besides Na, Li and K only the total susceptibility has been measured, which includes a diamagnetic term due to the ion cores χ_i and the Landau diamagnetism of the conduction electrons χ_d. Some figures are listed in table 4.2 for the measured total susceptibility of the liquid phase just above the melting point with the ionic contribution removed; the latter can be estimated theoretically or from measurements on non-metallic compounds (Klemm, 1950). The figures are taken from a compilation by Busch & Yuan (1963) and from papers by Takeuchi & Endo (1961), Nachtrieb (1962) and Collings (1965). In many cases there are conflicting results available for χ_L and there is room for dispute as to what is the best value to choose for χ_i (Dupree & Seymour, 1970); the latter is a large correction for the heavier elements. The present author, no less than his predecessors, may have succumbed to the temptation to select the values that fit best with his theoretical preconceptions.

According to the free-electron model the Landau diamagnetism χ_d should amount to $-\frac{1}{3}\chi_p(\text{FE})$. It is sometimes stated that whereas χ_p is proportional to m^*/m, χ_d is proportional to m/m^*, but the recent calculations of Timbie & White (1970), specifically for liquid metals, do not seem to support this generalisation. The diamagnetic term is little affected by exchange and correlation (Pines & Nozieres, 1966), but in a real metal it may well be influenced by the distortion of the conduction electron wave functions inside the cores, in a way that χ_p is not (Kjeldaas & Kohn, 1957). The values of about -0.2 and -0.15 which are indicated for χ_d in liquid Na and K respectively – these are the differences between the figures quoted for $(\chi_L-\chi_i)$ and χ_p as measured directly – seem to be just about what should be expected. The value of $+0.1$ which is indicated for χ_d in liquid Li is of course nonsense; one or more of the experimental results for Li must be erroneous.

For the other monovalent metals in table 4.2 and for the polyvalent metals from Al downwards, we may suppose that the Landau diamagnetism is compensating rather neatly for the enhancement of χ_p due to exchange, since the figures for $(\chi_L-\chi_i)$ and $\chi_p(\text{FE})$ are remarkably close. For the divalent metals Zn, Cd and Hg, $(\chi_L-\chi_i)$ is relatively small, and it is tempting to cite this as evidence that m^*/m is significantly less than unity, which would accord with one of the qualitative predictions in §4.9. It is scarcely conclusive evidence, however. The quantitative work of Ballentine (1966*b*), Shaw & Smith (1969) and Jena & Halder (1971), whose results have been outlined in §4.9, suggests that m^*/m is really very little less in liquid Zn and Cd than it is in many other liquid metals – only a few parts per cent below unity. We cannot rule this conclusion out of court on the strength of the data for χ_L.

The final column in table 4.2 shows the measured change in the susceptibility at the melting point, $(\chi_L-\chi_S)$. As expected from the spin-resonance data, this represents an increase of only 1% in $(\chi_L-\chi_i)$ in the case of Li, and it is relatively very small for all the other alkali metals. For a few of the polyvalent metals, e.g. Ga, Ge, Sb, Bi, the changes are relatively very large, but this is not surprising; they can be understood in terms of anomalies in the solid phase. Of more interest is the behaviour of metals such as Al, In, Sn, Pb etc. which are supposed to be relatively 'good' metals when solid, with Fermi surfaces which are not far from spherical and effective mass ratios which are probably not far from unity (see table 1.2). For none of these does the susceptibility change so dramatically as it does for Bi, for example, but nevertheless, if the experimental data are to be trusted, $(\chi_L - \chi_S)$ is by no means trivial compared with $(\chi_L - \chi_i)$. Changes in the Landau diamagnetism are probably responsible, but we cannot pretend to explain the figures in detail.

Transition metals are largely outside the scope of this book, but the results of Nakagawa (1956) on Mn, Fe, Co and Ni are of sufficient interest to deserve a brief mention. All of these are, of course, strongly paramagnetic because their d-shells are incomplete, but they are well above their Curie points by the time they melt. The susceptibility decreases on melting for δ-Fe and Co, in which the exchange forces between adjacent ions are such as to favour parallel (ferromagnetic) alignment of their spins. For Mn and γ-Fe (extrapolated to the melting point), in which the exchange forces favour anti-ferromagnetism, the change is in the other direction. For Ni the change is too small to be detectable. The general conclusion is that the exchange coupling between paramagnetic ions is weaker in the liquid phase than in the solid.

The temperature coefficient of χ_L above the melting point is small and not easy to measure accurately. Values of $(\chi_L - \chi_i)^{-1} \times (d\chi_L/dT)$ for Hg, In, Ge, Sn and Sb have been collected by Busch & Yuan (1963) and compared with $\frac{2}{3}\Omega^{-1}(d\Omega/dT)$; the two quantities should be almost identical according to equation (4.114), if the effect of volume change, and hence of temperature, on the enhancement of χ_p due to exchange is small and if the Landau diamagnetism is proportional to $\Omega^{\frac{2}{3}}$ as predicted by elementary theory. In fact the first quantity is always the bigger, by a factor of 2 or 3 for In, Ge and Sn, and by a factor of 9 or 10 for Hg and Sb. The discrepancies seem rather large to be dismissed as due to experimental errors in the choice of χ_i. They are not fully understood (but see p. 496).

4.16. The Knight shift

This effect (briefly referred to in §3.17) is a shift $\Delta\nu$ in the frequency ν_0 at which nuclear magnetic resonance is observed for a given field, when the nuclei form part of a metal. It is measured relative to the resonance frequency for an isolated nucleus, which is obtained from NMR experiments on non-metallic compounds; a correction for the so-called 'chemical shift' in the compound is needed in principle and sometimes applied in practice, but in most cases it is assumed small enough to be neglected. The Knight shift is always *positive*; the nuclei in a metal appear to see a field which is always larger, by some amount $\Delta\mathsf{H}$, than the applied field H_0. The ratio $\Delta\mathsf{H}/\mathsf{H}_0$ or $\Delta\nu/\nu_0$ is represented here by the symbol $\mathscr{K}$.

If h is the field that acts upon the nuclear moment in a particular environment, then

$$\nu = \gamma_n \mathsf{h}/2\pi \tag{4.115}$$

is undoubtedly the resonance condition. But just what is h? Is it the applied field H_0? Or, if we imagine the nucleus as immersed in an ideal, homogeneous medium with magnetisation I, is it the field H, which differs from H_0 by a demagnetising correction that would be $-\frac{4}{3}\pi\mathsf{I}$, for example, in the case of a spherical specimen? Or is it the so-called *local* field H_{loc}, which exceeds H by the Lorentz correction $+\frac{4}{3}\pi\mathsf{I}$ due to distant parts of the medium, and which is therefore the same as H_0 for a spherical specimen? The answer is none of these. The nucleus is acted on by the field

$$\mathsf{B} = \mathsf{H}+4\pi\mathsf{I} = \mathsf{H}_{loc}+\tfrac{8}{3}\pi\mathsf{I},$$

and in the case of a spherical specimen the Knight shift should therefore be

$$\mathscr{K} = 8\pi\chi N m_A/3\Omega. \tag{4.116}$$

It is due to the so-called 'contact interaction' between the medium and the resonating nucleus.

However, this simple theory predicts values of $\mathscr{K}$ which are only about 10^{-5} in metals, barely large enough to be detectable. In practice values of order 10^{-2} are observed and can be measured with high precision. The theory is wrong because the metal of which the nucleus is a part does *not* constitute a homogeneous medium. The conduction electrons spread throughout it, but their wave functions are much greater in the vicinity of the nucleus than elsewhere. Consequently, the contact interaction associated with their spin paramagnetism is accentuated by a factor

$$\Omega\langle|\psi(R=0)|^2\rangle_{\mathrm{F}} = \Omega P_{\mathrm{F}}/N \tag{4.117}$$

which can be at least 10^{+3}; here $\langle\ \rangle_{\mathrm{F}}$ implies an average over the states at the Fermi level which are responsible for the paramagnetism. The Landau diamagnetism, which is essentially generated by currents running around the outside of the specimen, is not accentuated in the same way and except in one or two special cases (e.g. solid Be) both this and the ion core diamagnetism may be ignored. Hence we arrive at the standard formula (Knight, 1956)

$$\mathcal{K} = (8\pi/3)m_{\mathrm{A}}\chi_{\mathrm{p}}P_{\mathrm{F}}. \tag{4.118}$$

It is sometimes convenient to express this with the aid of a dimensionless parameter

$$\xi = P_{\mathrm{F}}/|\psi_{\mathrm{A}}(0)|^2 \tag{4.119}$$

in terms of the wave function ψ_{A} of a valence electron in an s state in the free atom. If the atom is monovalent and has a $^2S_{\frac{1}{2}}$ configuration the appropriate value of $|\psi_{\mathrm{A}}(0)|^2$ may be obtained directly from spectroscopic data. If the atom is polyvalent it is necessary to use spectroscopic data for a free *ion* with the $^2S_{\frac{1}{2}}$ configuration and then to apply a crude correction for the effect of ionisation (Knight, 1956).

A reliable calculation of P_{F} from first principles is not easy. It has been estimated with sufficient precision (Heighway & Seymour 1971*a*) to show that equation (4.118) is approximately correct but not well enough to enable accurate values of χ_{p} to be deduced. If our interest in the Knight shift arises from the information it may provide about the magnitude of χ_{p}, and hence of the density of states, the best we can do is to compare the behaviour of different metals, looking for anomalous cases. Table 4.3 lists the values available for $\mathcal{K}_{\mathrm{L}}$, the Knight shift in the liquid phase just above the melting point,† and for ξ deduced from (4.118) and

† These data are collected from Knight, Berger & Heine (1959) (Li, Na, Rb, Cs, Hg, Al, Ga, In, Sn, Bi); Flynn & Seymour (1959) (Bi); Flynn & Seymour (1960) (In); Jones, Graham & Barnes (1960) (In); Seymour & Styles (1964) (Cd); Odle & Flynn (1965) (Cu, Sb); Van der Lugt & Knol (1967) (K); Warren & Clark (1969) (In, Sb); Cabane & Froidevaux (1969) (Te); El-Hanany & Zamir (1969) (Cu); Rigney & Blodgett (1969) (As).

TABLE 4.3. *The Knight shift*

Metal	$10^2\mathscr{K}_L$	ξ	$\frac{\Omega}{\mathscr{K}_L}\left(\frac{\partial \mathscr{K}_L}{\partial \Omega}\right)_T$	$\frac{10^4}{\mathscr{K}_L}\left(\frac{\partial \mathscr{K}_L}{\partial T}\right)_p$	$\frac{10^4}{\mathscr{K}_L}\left(\frac{\partial \mathscr{K}_L}{\partial T}\right)_\Omega$	$\frac{10^2(\mathscr{K}_L-\mathscr{K}_S)}{\mathscr{K}_L}$
				$(°C)^{-1}$	$(°C)^{-1}$	
Li	0.026	1.15	—	—	—	0
Na	0.116	1.05	+0.3	+1.8	+1.09	+2
K	0.253	1.03	—	+1.1	—	+2
Rb	0.662	1.07	—	+1.7	—	+1
Cs	1.46	1.27	—	−3.0	—	−2
Cu	0.27	0.8	—	+0.77	—	+5
Cd	0.80	2.4(?)	—	0	—	+24
Hg	2.45	0.5	—	—	—	0
Al	0.164	0.5	—	—	—	0
Ga	0.44	0.5	−0.1	−0.67	−0.55	—
In	0.79	0.7	—	−0.77	—	−1
Sn	0.73	0.5	—	—	—	−3
As	0.32	—	—	—	—	—
Sb	0.72	—	—	0±0.7	—	—
Bi	1.40	0.14	—	0±0.7	—	80
Te	0.38	0.13	—	14	—	100

(4.119) with χ_p(FE) for χ_p. Of course if χ_p is systematically greater than χ_p(FE) on account of exchange, then these values of ξ are systematically greater than the true values. This is of no concern to us, however; the point is simply to consider whether any value of ξ looks so anomalous as to require the hypothesis that m^*/m is significantly different from unity.

On the whole the results for ξ look very reasonable as they stand and quite consistent with a free-electron density of states. The apparent tendency for ξ to diminish as the valency increases may result from the dubious ionisation correction mentioned above and in any case is no cause for alarm, since we should not expect ξ to stay constant except within groups of similar metals. Its constancy within the group of alkali metals is striking.† In fact the only case which is clearly anomalous is that of Cd. Ziman (1967) has pointed out that there is probably an error in the identification of $|\psi_A(0)|^2$ for Cd and he suggests that the correct answer for ξ probably lies in the range 0.6–0.85. If so, it is entirely reasonable to suppose that

† This constancy must be partly fortuitous. If we use the *measured* χ_p for Li, Na and K (see table 4.1) we arrive at values for ξ of 0.46, 0.66 and 0.64 respectively, which are not quite so harmonious as 1.15, 1.05 and 1.03.

m^*/m is not far from unity for liquid Cd; it is certainly not likely to be much *less* than unity, as the magnitude of $(\chi_L - \mathcal{X}_i)$ might suggest.

Before discussing the change at $\mathcal{K}$ at the melting point we need to consider how it varies with pressure and temperature. The pressure dependence has been studied for only two metals in the liquid state, Na (Watabe, Tanaka, Endo & Jones, 1965) and Ga (Cornell, 1967); rough values of $(\Omega/\mathcal{K}_L)(\partial\mathcal{K}_L/\partial\Omega)_T$ for these two are shown in table 4.3. If we can trust the theoretical curve in fig. 4.21 the exchange enhancement of χ_p is roughly proportional to $\Omega^{0.07}$ while $\chi_p(\text{FE})$ is proportional to $\Omega^{0.67}$. From the entries in table 4.3, therefore, it appears that P_F must vary roughly like $\Omega^{-0.45}$ in Na and like $\Omega^{-0.85}$ in Ga. Similar results have been reported for solid metals and have never been satisfactorily explained. The amplitude of the true wave function ψ should be proportional to that of the pseudo wave function ϕ, which, in the first approximation, is just $\Omega^{-\frac{1}{2}}$, and on this account a variation of P_F like Ω^{-1} is to be expected. We need to assume (*a*) that the ratio between $\psi(0)$ and $\phi(0)$ falls when the metal is compressed, perhaps due to the change in k_F, or (*b*) that the sort of modulation of ϕ which was discussed in §4.11 above causes $\phi(0)$ to differ significantly from $\Omega^{-\frac{1}{2}}$ and to vary with volume less rapidly. Attempts to follow up these assumptions have not produce encouraging results.

The temperature dependence at constant pressure has been more widely studied; table 4.3 shows a number of values for $\mathcal{K}_L^{-1}(\partial\mathcal{K}_L/\partial T)_p$ taken from papers quoted above.† Since $\mathcal{K}_L$ should be a single-valued function of temperature and volume, it follows that

$$\frac{1}{\mathcal{K}_L}\left(\frac{\partial\mathcal{K}_L}{\partial T}\right)_p = \frac{1}{\mathcal{K}_L}\left(\frac{\partial\mathcal{K}_L}{\partial T}\right)_\Omega + \frac{\alpha\Omega}{\mathcal{K}_L}\left(\frac{\partial\mathcal{K}_L}{\partial\Omega}\right)_T, \qquad (4.120)$$

where α is the volume expansion coefficient, and this relation has been used to deduce values of the temperature coefficient at constant volume in the case of Na and Ga.‡ We should be able to

† See also Suzuki & Uemura (1971) for some curious results for supercooled liquid Ga.

‡ To be accurate, it has been used in the opposite way in the case of Na; Watabe *et al.* measured $(\partial\mathcal{K}_L/\partial T)_\Omega$ directly and equation (4.120) enables $(\partial\mathcal{K}_L/\partial\Omega)_T$ to be deduced from their results.

discuss $\mathscr{K}_L^{-1}(\partial \mathscr{K}_L/\partial T)_\Omega$ solely in terms of the modulation of $\phi(0)$ by the surrounding ions; so long as no change in volume is concerned we may expect

$$P_F \propto 1+2\,\mathrm{Re}\,\overline{\gamma(0)_F}, \tag{4.121}$$

the fluctuations of γ about its mean value being irrelevant because, during the time required for an NMR experiment, each nucleus samples the full range of environments that characterise the ensemble. An approximate formula for $\mathrm{Re}\,\overline{\gamma(0)_F}$ has been given above in equation (4.106), which involves the interference function $a(q)$. It has been pointed out that it may be proper in some circumstances to replace a by $(a-1)$ over part of the range of integration, and that only if this replacement is effected is γ likely to be small enough to justify the approximations upon which the formula is based. But this is irrelevant here, because at constant volume the temperature dependence of γ should be due almost entirely to the temperature dependence of a. We may expect

$$\frac{1}{\mathscr{K}_L}\left(\frac{\partial \mathscr{K}_L}{\partial T}\right)_\Omega \simeq -\frac{2m^*}{(2\pi)^2\hbar^2 k_F} \times \int_0^\infty u(q)\left(\frac{\partial a}{\partial T}\right)_\Omega \tfrac{1}{2}\log\left|\frac{(2k_F+q)^2+(k_F/2lq)^2}{(2k_F-q)^2+(k_F/2lq)^2}\right| q\,dq, \tag{4.122}$$

with the same proviso that applies to (4.106), that the pseudo-potential may in this context be treated as local.

One has to use a lot of guesswork to evaluate (4.122); contrary to the assumption made by Watabe *et al.*, the most important contributions to the integral seem to arise from the region $q \gtrsim 2k_F$ where $(\partial a/\partial T)_\Omega$ has not been measured and where the tabulated pseudo-potentials may not be trustworthy. It seems to be capable of giving the observed result in the case of Na if $(T/a)(\partial a/\partial T)_\Omega$ is about -0.1 in the neighbourhood of the first main peak of the interference function, which is quite a plausible assumption (see fig. 2.13); this statement is based on an integration using the Heine–Animalu pseudo-potential. The negative coefficient for liquid Ga can be understood qualitatively, because the region $q \gtrsim 2k_F$ coincides with the first trough in the interference function and here both $(\partial a/\partial T)_F$ and $u(q)$ should be positive; a mean value for the former quantity of about $+0.04$ in this region seems to be required, which is again not unreasonable. It is difficult to go further than this at present.

When a metal solidifies there is of course a small change of volume and this is liable to produce a small change in the Knight shift for reasons mentioned above. It would not, however, be fair to expect

$$\frac{\mathscr{K}_L - \mathscr{K}_S}{\mathscr{K}_L} = \frac{\Omega}{\mathscr{K}_L}\left(\frac{\partial \mathscr{K}_L}{\partial \Omega}\right)_T \left(\frac{\Delta\Omega}{\Omega}\right), \tag{4.123}$$

because the change of volume that occurs on solidification may involve a quite different rearrangement of the neighbouring ions from the change that occurs on compression. And indeed (4.123) does not seem to hold for Na, the only metal for which it can be checked.

The question of how the change of environment on passing into the solid phase may be expected to affect P_F and $\mathscr{K}$ has been examined in the case of Na by Lackmann-Cyrot (1964). She computed $1+2\,\mathrm{Re}\,\overline{\gamma(0)_F}$ using an equation similar to (4.106) for the liquid phase and the equivalent sum over the first few reciprocal lattice vectors for the solid. Her answers were 0.71 and 0.69 respectively. They should not be taken too literally because a very crude approximation for $u(q)$ was adopted in the calculations, but they do suggest that the change in P_F may be of the right sign and about the right magnitude to explain the observed value of $(\mathscr{K}_L - \mathscr{K}_S)/\mathscr{K}_L$ in Na. The smallness of the change in this factor may occasion some surprise; perhaps the sum rule on $a(\mathbf{q})$, which holds in the solid phase as much as in the liquid, is significant in this connection.

The final column in table 4.3 shows a number of results for $(\mathscr{K}_L - \mathscr{K}_S)/\mathscr{K}_L$;† in every case except that of Cd the Knight shift is very little affected by the change of phase. While we are unable to explain the volume dependence of $\mathscr{K}$ there is little hope of explaining these results in detail, but they point to the general conclusion that neither P_F nor χ_p normally changes by more than a few parts per cent. The idea which was at one time current that they might be changing substantially but in opposite directions, their product remaining almost constant, no longer seems

† These data are taken from references already cited in the footnote on p. 303. See also Adams, Berry & Hewitt (1966) for results on solid In.

plausible; no warrant for a compensation of this sort is provided by the theory. The constancy of χ_p through the melting point, now demonstrated for other metals besides Li, implies that whether or not m^*/m is unity it cannot be greatly distorted by the second-order, structure-sensitive effects; the evidence cannot be said to rule out distortions of 20% or so, but factors of 2 seem unlikely. This is quite in accord with tentative conclusions reached in §4.15 and with the theoretical predictions discussed in §4.9. The case of Cd is an interesting one because the Knight shift is unusually sensitive to temperature in the solid phase; it increases by a factor of about 1.65 on heating from 0 °K up to the melting point. The experimental data available are collected in a paper by Kasowski (1969), who has provided a quantitative explanation for the phenomenon, using an empirical pseudo-potential with a highly non-local character which was devised in the first instance to fit the Fermi surface of solid Cd. According to Kasowski, P_F and χ_p both increase with temperature in the solid phase and at almost exactly the same rate; the Debye–Waller factor (see p. 116) is mainly to blame. Kasowski's work has been extended by Jena & Halder (1971), who calculate that when melting occurs P_F should increase by a further 8% and χ_p by a further 27%, which is just about right to explain the observed increase in $\mathscr{K}$. The calculations are reasonably consistent both with the value of 0.54 quoted by Allen *et al.* (1968) for m^*/m in solid Cd at low temperatures (see table 1.2) and with the values of 0.94 and 0.83 quoted for liquid Cd by Shaw & Smith (1969) and Jena & Halder (1971) respectively.

To judge by the behaviour of the total susceptibility, a substantial change in the Knight shift on melting is also to be expected for Ga, Sb and Bi. Large anisotropy and quadrupole effects complicate the measurement of $\mathscr{K}_S$ for these metals. According to Hewitt & Williams (1964), however, the isotropic component of $\mathscr{K}_S$ for Bi is 0.3 ± 0.2%, compared with a figure of 1.4% for $\mathscr{K}_L$. The increase on melting seems not inconsistent with the increase of χ.

CHAPTER 5

ELECTRONIC TRANSPORT PROCESSES

5.1. Prologue

Chapters 1 to 4 have prepared the ground for a discussion of the transport properties of liquid metals. A very simple theory of these can be constructed on the basis of the NFE model, especially if the distinction between m^* and m can be ignored. All that is needed is a formula for the electronic relaxation time τ, and this we have already derived in §3.16. The theory, essentially due to Ziman, is elaborated in the first third of this chapter (§§5.2 to 5.8) and its predictions compared with observation. It turns out to provide an elegant and satisfactory explanation of a semi-quantitative kind for many of the data concerning resistivity, thermoelectric power and thermal conductivity. It explains, for example, why the resistivity of Hg is ten times that of Na; why the temperature coefficient of the resistivity at constant volume is so very small for liquid metals, unless they are monovalent; why this coefficient is actually negative for divalent ones; why the thermoelectric power is negative for liquid Na and positive for liquid Cu; and so on. Discrepancies of detail do appear if the comparison is pushed too far, but these are frequently dismissed as the result of error in the raw data assumed for the interference function $a(q)$ or the ionic pseudo-potential $u(q)$. We have seen in chapter 4 that we cannot necessarily trust the Born approximation in a scattering calculation, nor the linear dielectric screening theory of chapter 1; on both these counts the effective pseudo-potential may differ from the value that one would calculate from first principles. Here is one factor upon which discrepancies of detail could be blamed. Others include the effective mass correction and the blurring of the Fermi surface by scattering.

The Ziman theory is bound to fail at very low densities, of course, for gaps or pseudo-gaps must open in the energy spectrum of the conduction electrons and the eigenstates may become localised (see §§1.12 and 4.10). A quite different approach to transport properties is then required, with the tight-binding model (see p. 274) rather than the NFE model as its starting point (e.g. Mott & Davis, 1971). A brief account is given in §5.10 of the behaviour displayed by Cs, Hg and other metals, when they are expanded continuously into the vapour phase by heating them through the critical point but no attempt is made to interpret the results except at a qualitative level. Liquid Cs has been studied over a greater range of density than any other liquid metal, for not only is its critical point relatively accessible, its compressibility is also exceptionally high. The interesting properties of liquid Cs when compressed are outlined in §5.9.

Ziman's theory is based upon the semi-classical Boltzmann approach, which is not entirely satisfying when the mean free path of the electrons is as short as it appears to be in many liquid metals. In §5.11, therefore, we embark on an attempt to calculate the resistivity, or rather the conductivity σ, in a more fundamental manner. The idea is to represent the valence electrons in terms of the eigenfunctions discussed in chapter 4. The fact that these eigenfunctions, unlike the plane waves of the NFE model, are not necessarily propagating does not prevent conduction; current flows when electrons are excited from one eigenstate to another, and such transitions are induced whenever a field is applied. It is a fairly simple exercise in Kramers–Heisenberg dispersion theory to reproduce Ziman's formula with the aid of this picture, in the limit when $m^*/m = 1$ and when the mean free path is long; the only tricky step, relegated to an appendix, is to demonstrate the necessity for the $(1-\overline{\cos\theta})$ factor which distinguishes the relaxation time τ from the lifetime τ_l. One can go on to show (§5.12) that the effective mass is of no concern, except in so far as m_1^*/m enters through the normalisation corrections discussed at an early stage in §5.3. As for the blurring of the Fermi surface, there is no doubt that corrections to Ziman's formula of order $(1/Lk_F)^2$ – where L is the mean free path – are really needed, but an analysis reported in §5.13 suggests that they are usually small enough to be neglected.

The remainder of the chapter is mainly devoted to the optical properties of liquid metals. The theory of the DC conductivity is extended to finite frequencies in §5.14, where the familiar *Drude equations* for $\sigma(\omega)$, and for the dielectric constant $\epsilon(\omega)$ are derived. The derivation involves a number of approximations which are scrutinised in §§5.15–5.18, in detail that may be wearisome to the non-specialist reader. These sections concern the anomalous skin effect, dynamic screening and other many-body effects, and the possibility that liquid metals may display some memory of the inter-band absorption peaks that are characteristic of solids. The conclusions reached are checked against a sum rule of great generality in §5.19, which is also useful (§5.20) for estimating the importance of the ion core polarisability. The deviations from Drude-like behaviour are expected to be rather slight.

Experimental determination of $\sigma(\omega)$ and $\epsilon(\omega)$ requires optical experiments on liquid metal surfaces, and it is emphasised in §5.21 that a number of errors may creep in during analysis of the measurements, though they should be smaller for liquid metals than for solid ones. In view of all these tiresome reservations it is remarkable how closely the actual results, which are presented in §5.22, agree with the simplest of all models for the specimen surface and with Drude's equations in their classical form. Only Hg is significantly out of line (§5.23). None of the liquid metals so far investigated shows any trace of inter-band absorption peaks associated with the excitation of valence electrons into empty states above the Fermi level. This is one of the strongest bits of evidence available, already invoked in chapter 1, for the validity of the NFE model. Inter-band absorption peaks associated with excitation of core electrons up to the Fermi level, however, are rather little affected by melting (§5.24).

Some remarks about photo-emission and about plasma oscillations are contained in §§5.25 and 5.26. Experiments on these phenomena could supplement the information available from optical experiments of a more traditional kind, though there are few results for liquid metals to report.

Finally, in §§5.27 and 5.28, the effect of a magnetic field is examined and data for the Hall coefficient in liquid metals are presented. Unless one is content with an elementary treatment, based upon the Boltzmann equation, the theory offers considerable difficulties which are not yet solved. But, as in the field of optical properties, the simplest approach is remarkably successful.

5.2. Ziman's theory

Information about the electrical resistivity of liquid metals has been available for many years and there have been a number of attempts to explain the observations. In particular, since theoreticians are easily fascinated by dimensionless ratios, there have been attempts to explain the magnitude of ρ_L/ρ_S at the melting point (see table 5.1)† or at least to correlate it with other measurable properties. Mott's suggestion (1934) that it may be correlated with the latent heat of melting has already been discussed in §2.3. None of the early theories is any more convincing than Mott's and there is little point in describing them here (for references, consult Cusack, 1963).

The weakness of this early work is that it relies too heavily on ideas that have proved successful for solid metals, in particular on the idea that the resistivity can be divided into two parts (*a*) a thermal term, proportional to the mean square amplitude of vibration of the ions and hence to the temperature, T, and (*b*) a residual term, independent of T, due to vacancies, dislocations and such-like. In Mott's theory the increase of resistivity that normally occurs on melting was associated entirely with the thermal term; in an alternative theory advocated by MacDonald (1959) and others it was associated entirely with the residual term; and it is clearly possible to adopt an attitude somewhere between these two extremes. The trouble is, however, that no approach of this sort seems to be consistent with the way that the resistivity varies with temperature above the melting point. Mott's theory suggests

$$(\partial \rho_L/\partial T)_\Omega \simeq \rho_L/T,$$

and MacDonald's theory suggests

$$(\partial \rho_L/\partial T)_\Omega \simeq (\partial \rho_S/\partial T)_\Omega \simeq \rho_S/T.$$

In practice, as the typical curves in fig. 5.1 serve to illustrate, the

† The results listed in this table for Ce and U are due to Busch, Güntherodt, Künzi & Schapbach (1970) and Busch, Güntherodt & Künzi (1970). Other sources may be traced through a review article by Wilson (1965), which has aided the compilation of several other tables in this chapter. References which are already accessible in Wilson's extensive bibliography are quoted sparingly here.

TABLE 5.1. *The change of resistivity on melting*

Metal	ρ_L/ρ_S	Metal	ρ_L/ρ_S	Metal	ρ_L/ρ_S
Li	1.59	Al	2.2	Mn	0.6
Na	1.45	Ga	0.45–3.1	Fe	1.05
K	1.58	In	2.2	Co	1.09
Rb	1.60	Tl	2.1	Ni	1.33
Cs	1.67				
Cu	2.1	Sn	2.1	Ce	1.06
Ag	2.1	Pb	1.9		
Au	2.3				
Mg	1.8	Sb	0.6	U	1.05
Ba	1.6	Bi	0.35–0.47		
Zn	2.2				
Cd	2.0				
		Te	0.05–0.09		
Hg	3.7–4.9				

Note. Where a range of values is quoted, this corresponds to different orientations for the solid phase.

temperature coefficient of ρ_L is remarkably small for all except the monovalent metals, much less then either of these predictions. Is it perhaps more fruitful to compare the liquid with a gas rather than a solid, i.e. to suppose that the arrangement of the ions looks so disordered to the conduction electrons, even just above the melting point, that no amount of heating can make it more disordered? If this is the case, then the distinction between thermal and residual resistivity is not likely to be meaningful.

It seems to have been Bhatia & Krishnan (1948) who first appreciated clearly that the interference function $a(q)$ derived from X-ray or neutron diffraction experiments contains just the information that is required to make a calculation of electron scattering possible, without its being necessary to adopt any specific model for the ionic structure of the liquid metal. Using the free-electron model they obtained a formula for ρ_L, involving $a(q)$, which is basically correct, and they pointed out how it might serve to explain the surprising fact that in liquid divalent metals $(\partial\rho_L/\partial T)_\Omega$ is actually slightly negative. Subsequently Gerstenkorn (1952) adopted a similar approach in trying to calculate ρ_L/ρ_S.

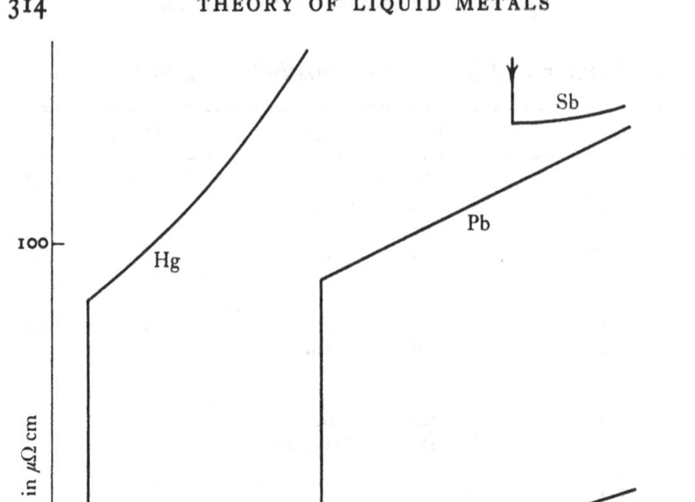

Fig. 5.1. The resistivity of a number of metals as a function of temperature at constant volume.

However, these authors did not know how to express the form factor for electron scattering by a single ion, and perhaps on this account their work was neglected. It was left to Ziman (1961, see also Bradley, Faber, Wilson & Ziman, 1962) to rediscover the basic formula, by which time the concept of the ionic pseudo-potential had been developed.

The essence of what is now commonly known as Ziman's formula has been derived in §3.16 above; it lies in equation

(3.80) for the current relaxation time τ. The effect of an electric field $\mathbf{E}$ on an assembly of free electrons should be to shift this distribution uniformly through $\mathbf{k}$-space at a rate given by

$$\hbar\,\mathrm{d}\mathbf{k}/\mathrm{d}t = e\mathbf{E}$$

and thus to generate current at a rate

$$\mathrm{d}\mathbf{j}/\mathrm{d}t = (ne^2/m)\,\mathbf{E}. \tag{5.1}$$

Although it is the total number of free electrons per unit volume, n, that features in this formula, the current is effectively carried by the electrons near the Fermi surface, where the displaced distribution function $f(\mathbf{k})$ differs from the original equilibrium value $f^0(\mathbf{k})$. The scattering processes discussed in §3.16 cause the contribution to $\mathbf{j}$ of any particular electron, wherever it is situated around the Fermi sphere, to decay with a relaxation time τ, so that when these are included the equation for $\mathrm{d}\mathbf{j}/\mathrm{d}t$ becomes

$$\mathrm{d}\mathbf{j}/\mathrm{d}t+\mathbf{j}/\tau = (ne^2/m)\,\mathbf{E}. \tag{5.2}$$

A steady state is reached in which $\mathrm{d}\mathbf{j}/\mathrm{d}t$ vanishes and from the definition of resistivity $\mathbf{j}$ is then given by $\mathbf{E}/\rho$. Hence† if we ignore for the moment the distinction between m^* and m, it follows from (5.2), (1.8) and (3.80), that

$$\begin{aligned}\rho &= m/ne^2\tau \\ &= \frac{3\pi m^2}{\hbar^3 e^2}\frac{N}{\Omega k_{\mathrm{F}}^2}\langle |u(q)|^2 a(q)\rangle;\end{aligned} \tag{5.3}$$

throughout this chapter and the next we shall be using brackets thus $\langle\ \rangle$ to indicate an average of some function of q over the range between zero and $2k_{\mathrm{F}}$, weighted towards the upper end of this range by a factor q^3, i.e.

$$\langle f(q)\rangle = \int_0^1 f(q)\,4(q/2k_{\mathrm{F}})^3\,\mathrm{d}(q/2k_{\mathrm{F}}). \tag{5.4}$$

Equation (5.3) includes no correction for inelastic scattering effects; according to (3.83) and (3.85) we ought really to replace $a(q)$ by

$$a(q)-\tfrac{1}{6}\hbar^2q^2/m_{\mathrm{A}}k_{\mathrm{B}}T, \tag{5.5}$$

but except for Li the difference that this makes is negligible.

† For a more careful discussion based upon the Boltzmann equation consult any standard text such as Ziman (1960).

In the first part of this chapter we shall explore the success of equation (5.3) in explaining the observed resistivities and thermo-electric properties of liquid metals, and the way they vary with temperature and volume. But we should note at the start that it involves a number of approximations and assumptions in which we have no right to feel complete confidence. Some of them arise from the use of a pseudo-potential formalism and have been aired in chapter 1: is it completely safe, for example, to treat the pseudo-potentials of adjacent ions as additive? What normalisation corrections are needed in recognition of the fact that only the *pseudo* wave functions are free-electron-like? And so on. Then there is the question of whether the pseudo-potential is weak enough to justify the Born approximation; if it is not, then presumably $u(q)$ needs to be replaced by an effective pseudo-potential $u'(\mathbf{q})$ as discussed in §4.11 and there is the possibility that the answer for ρ involves the higher distribution functions, g_3, g_4, etc. Thirdly, can the formula remain valid when $\hbar/\tau_l$ or $\hbar/\tau$ is as much as a quarter of the kinetic energy at the Fermi level, K_F, i.e. when the electronic mean free path is only one or two interatomic spacings? The sharp upper limit to the integral in (3.80), such that scattering processes for which q is in excess of $2k_F$ are completely ignored, implies that the cut-off in the electronic distribution function at the Fermi surface is completely sharp; this is scarcely consistent with the arguments of chapter 4. Finally, although the theories and results discussed in the previous chapter go a long way towards justifying Ziman's use of the free-electron model, we must recognise that strictly speaking the electrons are only nearly free. Ought we perhaps to replace m by m^* in equation (5.3)?

In §5.11 below an alternative approach to the whole calculation is described, which enables many of these queries to be answered and which leads on naturally to a discussion of the optical properties of liquid metals. The effective mass problem, however, is so liable to cause confusion that we had better try to dispose of it straight away.†

† The present author has added to the confusion in the past. Several of the statements made by Faber (1966) will need to be corrected in what follows.

5.3. The role of the effective mass

Several modifications are required to the simple argument in § 5.2 if the electrons are not completely free. In the first place one has to recognise that $e\mathbf{E}$ is no longer the total force on an electron, since it does not include the force of interaction with the ions, and therefore it cannot be equated with the rate of change of momentum. On the other hand the momentum of an electron is not necessarily given by $\hbar\mathbf{k}$; it is better described by $m\mathbf{v}_F$, where

$$\mathbf{v}_F = \hbar^{-1}(\nabla_k E)_F = \hbar k_F/m^* \tag{5.6}$$

is the *group velocity* of an electron at the Fermi surface. If $e\mathbf{E}\cdot\mathbf{v}_F$ describes the rate at which the field does work upon the electron and if we may equate this with the rate of increase of its energy, then

$$e\mathbf{E}\cdot\mathbf{v}_F = \mathrm{d}E/\mathrm{d}t = \nabla_k E\cdot(\mathrm{d}\mathbf{k}/\mathrm{d}t),$$

in which case the relation $\hbar\dot{\mathbf{k}} = e\mathbf{E}$ is still correct; the field displaces the electrons through $\mathbf{k}$-space at the same rate as if the electrons were free. But because the current carried by each electron is $e\mathbf{v}_F$, equation (5.1) now becomes

$$\mathrm{d}\mathbf{j}/\mathrm{d}t = (ne^2/m^*)\,\mathbf{E}. \tag{5.7}$$

Since the probability of scattering depends upon the density of states there is an additional m^* in equation (3.80) for $1/\tau$, so we arrive at

$$\rho = m^*/ne^2\tau = \frac{3\pi m^{*2}}{\hbar^3 e^2}\frac{N}{\Omega k_F^2}\langle|u|^2 a\rangle \tag{5.8}$$

in place of (5.3).

The above analysis is incomplete when the effects exchange of and correlation are important, but Landau's theory of interacting quasi-particles (Pines & Nozieres, 1966) suggests that the end result is nevertheless correct. Imagine an assembly of completely free electrons whose distribution is off-centre in $\mathbf{k}$-space by an amount $\Delta\mathbf{k}$, so that the total current density is $ne\hbar\Delta\mathbf{k}/m$. Now suppose that some interactions are slowly turned on. If they are interactions with an ionic lattice they will change the total current by the factor m/m^* which (5.7) includes, but it is clear that interactions between the electrons themselves can have no such effect; the momentum of the whole electron assembly must be unaffected by exchange and correlation. Hence we should multiply the right-hand side of (5.7) by a factor m^*_{XC}/m so as to remove whatever contribution is made by exchange and correlation to the total effective mass m^*. According to Landau, however, it is a consequence of the interaction between electrons that when their distribution is displaced through $\Delta\mathbf{k}$ the Fermi surface – i.e. the contour upon which the energy of an individual quasi-particle is equal to the chemical potential – is also displaced, by an amount

$$\Delta\mathbf{k}' = \Delta\mathbf{k}(1 - m^*_{XC}/m).$$

Scattering causes the electrons to relax with a time constant of τ towards the displaced Fermi surface, which reverts to its usual equilibrium position as they do so. Evidently

$$\frac{\mathrm{d}}{\mathrm{d}t}(\Delta\mathbf{k}) = -\frac{\Delta\mathbf{k} - \Delta\mathbf{k}'}{\tau} = -\frac{m^*_{XC}}{m}\frac{\Delta\mathbf{k}}{\tau}.$$

Hence (5.2) becomes

$$\mathrm{d}\mathbf{j}/\mathrm{d}t + (m^*_{XC}/m)\mathbf{j}/\tau = (ne^2/m^*)(m^*_{XC}/m)\mathbf{E}. \tag{5.9}$$

The steady state solution is the same as before, for the two factors of m^*_{XC}/m cancel.

All this is orthodox stuff for a solid metal, except, of course, that when the Fermi surface is distorted by band structure effects the Ziman formula for τ is liable to require modification; the probability of scattering through any particular $\mathbf{q}$ depends on a matrix element between Bloch waves which is not necessarily proportional to $|u(q)|^2 a(q)$. It is instructive to consider a one-dimensional solid, for which we may use the results of §4.3. If $u(q)$ is taken to be more or less independent of q, then when the solid is slightly disordered by heating the quantity $|u|^2 a$ varies with q as indicated by the full curve in fig. 5.2(*a*); it has an infinite singularity at the first reciprocal lattice vector G_1. Fig. 5.2(*b*) shows the relation between energy and wave number for the Bloch wave eigenstates; the disorder narrows the gap at $\frac{1}{2}G_1$ but only slightly. Now suppose that the Fermi level lies just below the gap, as indicated by the horizontal line in fig. 5.2(*b*). The

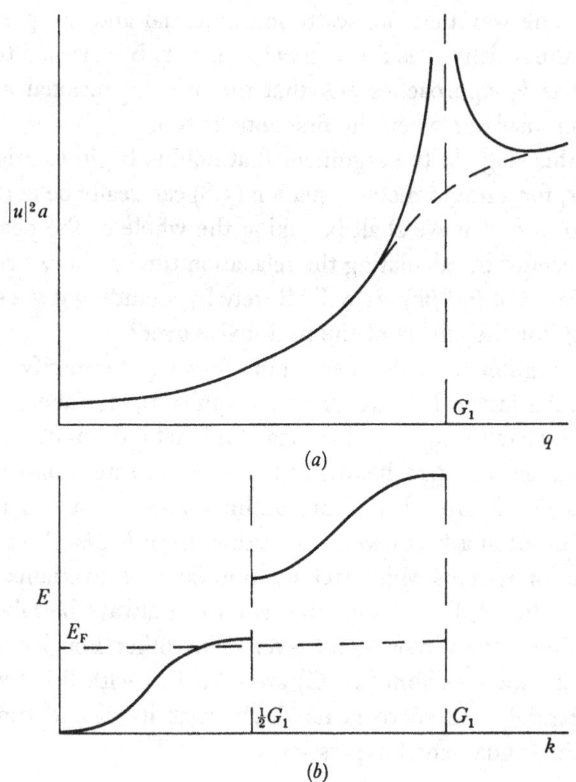

Fig. 5.2. (*a*) The quantity $|u|^2a$, and (*b*) the electronic energy spectrum, for a one-dimensional solid metal.

one-dimensional equivalent of $\langle|u|^2a\rangle$, which is $|u(2k_F)|^2a(2k_F)$, rises towards infinity as k_F approaches $\frac{1}{2}G_1$. According to (4.18), however, the square of the matrix element for scattering from k_F to $-k_F$ remains quite finite; if we wish to apply (5.8) to this problem we must not use the true value of $|u|^2a$ in the neighbourhood of G_1, but rather some smooth curve, indicated schematically by a broken line in fig. 5.2(*a*), from which the infinite peak has been removed. In fact the square of the matrix element for $k_F = \frac{1}{2}G_1$ is given (when $u(q)$ is small) by

$$N^{-1}(\hbar^2G_1^2/2m)^2a(0),$$

so that it does not even depend on the strength of the scattering potential! The way that the scattering potential and the peak in $a(q)$ affect the resistivity is through m^*; it is only because m^* tends to infinity as k_F approaches $\frac{1}{2}G_1$ that the one-dimensional metal becomes an insulator when the first zone is full.

It is at this stage in the argument that doubts begin to arise as to whether, for a liquid metal, equation (5.8) can really be correct. If we make use of it we shall be taking the whole of the peak in $a(q)$ into account in calculating the relaxation time τ. Do we really need the factor of $(m^*/m)^2$ as well? Surely by including it we shall be allowing for the effects of the peak twice over?

Various arguments have been put forward to justify suppression of this factor (Edwards, 1962; Toombs, 1965; Faber, 1966; Ziman, 1967; Szabo, 1971) and we shall find that it drops out of the answer obtained in §5.12 below. It is not easy to pin-point what is wrong with the orthodox theory outlined above, but it may be that the concept of a Fermi velocity distinct from $\hbar k_F/m$ should be abandoned for systems which lack the translational invariance of a crystalline solid. A Bloch wave in a solid can always be labelled by its leading wave vector $\mathbf{k}$, however many other Fourier components with wave vectors $(\mathbf{k}+\mathbf{G})$ are mixed in with this by the lattice potential, and there is no doubt that its momentum as given by the fundamental expression

$$(\hbar/2mi)\int(\Psi^*\nabla\Psi-\Psi\nabla\Psi^*)\,d\mathbf{r}$$

is equal to $\hbar^{-1}\nabla_k E$ and not to $\hbar\mathbf{k}$ (for a proof, see Mott & Jones, 1936, p. 92). As soon as one switches on the potential in a disordered system like a liquid, however, the zeroth-order state of wave number $\mathbf{k}$ becomes coupled to a set of degenerate states which extend right round the Fermi sphere. The momentum of the electron is certainly *not* $\hbar^{-1}\nabla_k E$; in so far as it differs from $\hbar\mathbf{k}$ it is presumably zero.

It is generally agreed, therefore, that the $(m^*/m)^2$ factor in (5.8) should be suppressed, at any rate if our calculation refers, like that of Edwards (1962), to a model in which it is only because of the structure-dependent, second-order term that m^*/m differs from unity. In a real liquid metal, however, we are dealing always with pseudo wave functions and pseudo-potentials and the final answer for the resistivity requires a normalisation correction

$$k^2 \left| \int \psi_k^* \nabla \psi_k \, d\mathbf{r} \right|^{-2} \tag{5.10}$$

which is frequently ignored (Faber, 1966). Because the pseudo-plane wave ψ_k oscillates like the wave function of a free atom inside each ion core, this correction may differ appreciably from unity. To estimate it, let us suppose that we can find some way to switch off the matrix elements of the pseudo-potential $U_{\mathbf{kk}'}$ which are responsible for scattering, while leaving the diagonal terms $U_{\mathbf{kk}}$ which shift the energy E_k from its free-electron value.† In these artificial circumstances ψ_k becomes an eigenfunction satisfying the Schrödinger equation

$$\nabla^2 \psi_k + \frac{2m}{\hbar^2} (E_k - V) \psi_k = 0,$$

V being the true potential associated with the pseudo-potential U. It then requires a trivial modification of the standard proof in Mott & Jones (1936) to show that

$$\int \psi_k^* \nabla \psi_k \, d\mathbf{r} = \frac{im}{\hbar^2} \int \nabla_k (E_k - V) \psi_k^* \psi_k \, d\mathbf{r} = (m_1^*/m)^{-1} k. \tag{5.11}$$

Our expression for the resistivity now becomes

$$\rho = \frac{3\pi m_1^{*2}}{\hbar^3 e^2} \frac{N}{\Omega k_F^2} \langle |u|^2 a \rangle. \tag{5.12}$$

By adding the normalisation correction to (5.8) we have re-introduced with one hand a part of the effective mass dependence that we have just deleted with the other.

The reader will note that the derivation of (5.11) involves an assumption that $\nabla_k V$ is zero everywhere, and he may query whether this is justified unless the complications associated with exchange and correlation may be set aside. The truth is that once these many-body effects become important a rigorous calculation of the resistivity by the method developed in § 5.11 below becomes

† The distinction between E_k and E is explained on p. 251.

extremely difficult.† Here we shall follow Landau in assuming that, under DC conditions, they do not affect the end result.

From the resistivity, incidentally, it is traditional to deduce an effective *mean free path* L, by means of the formula

$$L = v_F \tau = \hbar k_F \tau / m^* = \hbar k_F / n e^2 \rho, \tag{5.13}$$

which is based on (5.8); the figures that are quoted in table 1.1 for $\hbar / 2\tau K_F$ apply more accurately (if $m^* \neq m$) to $(Lk_F)^{-1}$ as given by this formula. Of more interest to us than L, however, is the *range of coherence* l which was introduced in §§4.8 and 4.9. This differs from L by the same $(1-\overline{\cos\theta})$ factor that distinguishes τ_l from τ (see §3.16). Moreover, we noted in connection with (4.75) that for a given scattering potential l is proportional to $(m/m^*)^2$. It follows, as we shall verify in greater detail below, that if (5.12) is correct then

$$l = L(1-\overline{\cos\theta})/(m_I^*/m^*)^2. \tag{5.14}$$

For the majority of liquid metals at normal densities the above discussion is of rather academic interest, since we have come to the conclusion that, except for one or two such as Li, m_I^*/m and m^*/m are probably close to unity. Moreover, the $\overline{\cos\theta}$ that figures in (5.14) is probably rather small; calculations based upon Heine–Animalu pseudo-potentials suggest that $(1-\overline{\cos\theta})$ may be as much as 1.3 in Li and as little as 0.55 in Na, but for a typical polyvalent metal a value close to 1.0 is more probable. We shall ignore the distinction between m^* and m and between L and l throughout most of the next few sections.

5.4. Comparison with experiment: the magnitude of ρ_L

The calculation of ρ_L from Ziman's formula requires a knowledge of how $|u|^2$ and a vary with q in the range between 0 and $2k_F$. The former quantity is at its largest near the origin (see fig. 1.7

† The method hinges on the evaluation of matrix elements between states that differ by the excitation of a single electron from an energy E_1, say, to E_2, and we shall represent these states by one-electron eigenfunctions Ψ_1 and Ψ_2. To be rigorous, however, one ought to use many-body eigenfunctions and integrate over the coordinates for *all* the electrons in the system. In so far as m_{XC}^*/m differs from unity, one may infer that the excitation of one electron has some effect upon the configuration of the others. Integration over their coordinates is then a far from trivial matter.

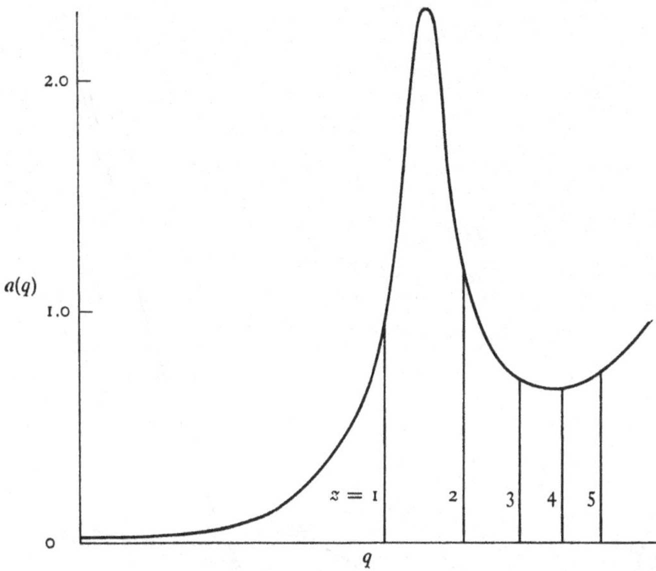

Fig. 5.3. Vertical lines mark the limit of integration in Ziman's formula corresponding to $q = 2k_F$. Its position relative to the main peak in $a(q)$ depends on the valency of the metal.

on p. 41) and $|u(2k_F)|^2$ may be only about 10^{-2} times $|u(0)|^2$. On the other hand $a(2k_F)$ is of order unity whereas $a(0)$ (see table 2.7 on p. 113) is usually about 10^{-2}. Hence $|u|^2a$ is much the same at both ends of the range of integration. But a lot depends upon how it varies in between. In particular, a lot depends upon where the main peak in the interference function lies.

It turns out that because the Fermi radius increases with valency like $z^{\frac{1}{3}}$ the limit of integration $2k_F$ lies just to the *left* of the main peak in a for monovalent liquid metals, just to its *right* for divalent ones, and progressively further to the right for metals which have a valency greater than two. The situation is illustrated schematically by fig. 5.3. Remembering that the average $\langle |u|^2a \rangle$ is heavily weighted towards the upper end of the integration range, one may hazard a guess that it will not make much difference to the answer for ρ_L to replace a by unity in the case of a polyvalent liquid, which is equivalent to using a gas-like model. But a similar replacement is not likely to work well for monovalent liquids. In fact $\langle a \rangle$ turns out to be close to 1.0 for metals such as Al, Sn,

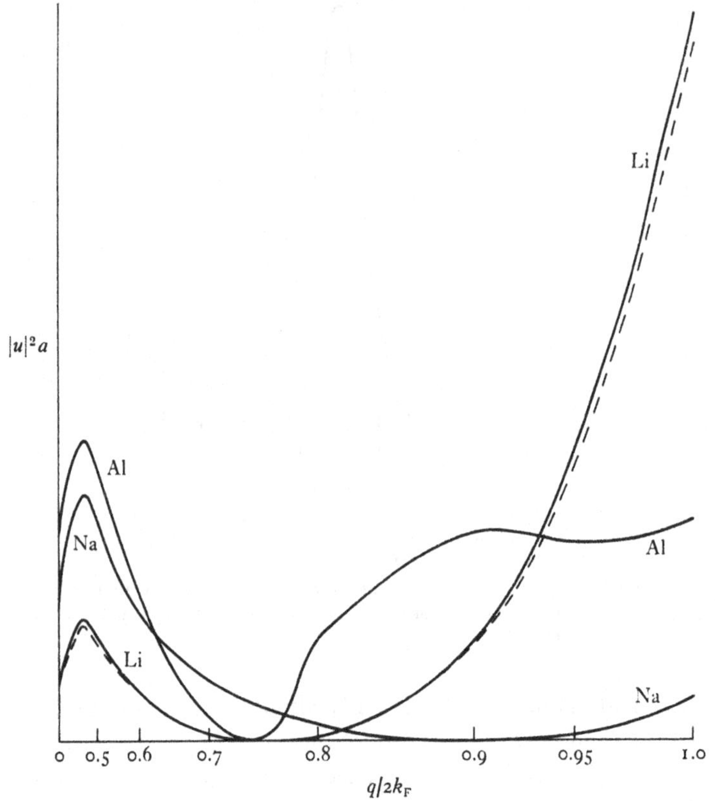

$q/2k_F$

Fig. 5.4. $|u|^2a$ as a function of $(q/2k_F)^4$ for three liquid metals. The broken curve for Li includes a correction for inelastic scattering.

Sb etc., whereas it is only about 0.4 for monovalent Na, Cu etc. For divalent metals it seems to average about 1.15, being rather larger than unity because the peak in a lies in a position where it is heavily weighted. These figures indicate that the short-range order which still exists in the liquid phase may have a significant effect in pulling the resistivity down below the value it would take for a gas-like array of ions in a monovalent specimen, but may actually enhance it in a divalent one. Since a is likely to tend towards unity as the liquid is heated, one can see already in a qualitative fashion how Ziman's theory may explain the slopes of the curves for ρ_L in fig. 5.1, even the slightly negative slope observed for Zn.

Fig. 5.4 shows curves of $|u|^2a$ for monovalent Na and Li and trivalent Al, calculated using the Heine–Animalu pseudo-potentials. They are plotted against $(q/2k_F)^4$ – though the abscissa axis is still labelled in terms of $(q/2k_F)$ – so the area underneath them is what determines the resistivity. All the curves go to zero round about $q/2k_F = 0.8$, because the pseudo-potentials pass through zero here; this is a typical feature of nearly all theoretical pseudo-potentials, except for metals such as Cs and Hg (see fig. 1.7). The area underneath the curves for $|u|^2a$ is divided rather clearly by this zero into two parts: in physical terms the *small-q* part is mainly due to the screened coulomb potential outside each ion core and to the long-range density fluctuations which ensure that a remains finite in the limit $q \to 0$, while the *large-q* part arises from effectively incoherent scattering by the cores themselves. In Ziman's 1961 paper a distinction was drawn between the so-called *plasma* term in the resistivity and the *pseudo-potential* term, and although this distinction is now seen to be somewhat artificial the small-q and large-q contributions do correspond more or less to the two terms which Ziman had in mind. Note how in Na, for which the core pseudo-potential is believed to be unusually small, nearly all ρ_L is due to the small-q or plasma part, whereas in Al, which is typical of the polyvalent metals, the large-q part is more important. Note also how the correction for inelastic scattering, as expressed by (5.5), reduces the area under the curve for Li by about 2 %.

The value of ρ_L obtained from Ziman's formula can vary quite a bit according to whose pseudo-potential one prefers and whose data for $a(q)$. The variation seems to be greatest for the monovalent metals, where a small shift in the zero of the potential (say 0.2 eV or so) is liable to change the answer by a factor of 2 or more, and where it is also sensitive to the precise shape of the $a(q)$ curve to the left of the main peak, which is not often determined accurately in diffraction experiments. Ziman (1967) has summarised the results of early attempts to apply the theory, and some results of later work are shown in table 5.2.

The predictions in column (a) are due to Animalu (1965), Ocken & Wagner (1966), Halder & Wagner (1966), Isherwood & Orton (1968), North *et al.* (1968), North & Wagner (1969); they

TABLE 5.2. *Resistivities at the melting point, in μΩ cm*

Metal	ρ_L (observed)	ρ_L (predicted) (*a*)	(*b*)	(*c*)	(*d*)	(*e*)	(*f*)
Li	24.7	25	16.0	93.3	65	29.6	23.7(CGR)
Na	9.6	7.9	4.9	17.0	—	11.4	11.6(SS), 9.3(CGR)
K	13.0	23	7.0	27.5	19.3	16.6	13.8(CGR)
Rb	22.5	10	7.5	25.6	32	14.1	13.9(SS), 22.5(CGR)
Cs	36.0	10	8.5	39.9	53	—	12.7(SS), 36.2(CGR)
Cu	21	—	—	8.35	16	—	22(M)
Ag	17	—	—	14.0	21	—	28(M)
Au	31	—	—	17.6	25	—	52(M)
Be	—	—	—	—	—	54.9	—
Mg	26	—	27	—	—	17.3	—
Ca	33	—	16	—	—	—	11(E)
Sr	85	—	7	—	—	—	—
Ba	306	—	15	—	—	—	—
Zn	37	37	15	—	—	19.3	28(E), 47(M)
Cd	34	23	17	—	—	17.2	17(E)
Hg (25 °C)	91	30	28	—	—	—	83(E)
Al	24	27	16	—	—	16.8	27(SS)
Ga	26	—	23	—	—	—	—
In	33	24	25	—	—	34.8	29(E)
Tl	73	60	55	—	—	—	79(E)
Si	(80)	—	25	—	—	—	—
Ge	(70)	—	39	—	—	—	—
Sn	48	—	32	—	—	51.4	—
Pb	95	64, 77	69	—	—	—	58(SS), 84(E)
Sb	114	—	57	—	—	—	—
Bi	128	87	87	—	—	—	—
Te	(500)	—	—	—	—	—	—

are based upon the Heine–Animalu pseudo-potentials (see p. 34) and on curves derived in some cases by neutron and in others by X-ray diffraction. Ashcroft & Lekner (1966) have argued that certain features in these experimental curves look decidedly implausible – e.g. the curve used for K is much higher to the left of the main peak than for any of the other alkali metals, which accounts for the relatively large value predicted for K in column (*a*) – and they therefore prefer to use for all metals a theoretical $a(q)$ curve, obtained by applying the Percus–Yevick approximation to a rigid-sphere model (see §2.17). Their theoretical formula for $a(q)$, derived from (2.77), contains an adjustable parameter y, which is the packing fraction. Ashcroft & Lekner quote values for ρ_L corresponding to $y = 0.45$ and 0.46, differing by not more than 10% in most cases. Either choice of y gives an adequate fit to the experimental $a(q)$ curves, though the theoretical expression does not always match the spacing of the peaks properly, nor does it match such features as the 'shoulder' to the right of the main peak observed for Hg and one or two other metals (see §2.14); in the small-q region to the left of the main peak the theoretical curve lies distinctly below the experimental curves used for all the alkali metals, especially that for K. The predictions in column (*b*) are Ashcroft & Lekner's results for $y = 0.46$, again using the Heine–Animalu pseudo-potentials.

The figures in columns (*c*) and (*d*) are taken from papers by a group of authors (Meyer *et al.*, 1967; Dickey, Meyer & Young, 1967; Young, Meyer & Kilby, 1967) who derive their pseudo-potential by a phase shift calculation, as outlined in §1.5. These two columns bear much the same relation to each other as (*a*) and (*b*), in that the experimental $a(q)$ curves were used for the first, and theoretical ones based on the Percus–Yevick rigid-sphere model for the second. However, in these calculations the parameter y was allowed to vary slightly from one metal to another; the effective rigid-sphere diameter, and hence y, was related, using a heuristic argument which is not worth discussion here, to the Fermi wave number k_F. The theoretical $a(q)$ curves obtained in this way agree rather more closely with experiment than Ashcroft & Lekner's universal curve, and for this reason the differences between columns (*c*) and (*d*) are less marked than between (*a*) and (*b*). The figures in column (*e*), due to Paasch & Trepte (1971), are similarly based upon Percus–Yevick curves for $a(q)$ with y modified so as to improve the fit with experimental data, but in this case the pseudo-potential employed was that of Shaw (see p. 38). Finally, the figures in column (*f*) are a mixed bag due to Schneider & Stoll (1966), Moriarty (1970) and Fvans (1970, 1971,) whose pseudo-potentials have been mentioned in §1.8, and Cubiotti, Giuliano & Ruggeri (1971*a*,*b*).

The spread that exists amongst these theoretical values is evidence of the considerable uncertainty that still prevails as to what is the best way to calculate pseudo-potentials and to determine $a(q)$. In view of this uncertainty, the overall agreement between theory and experiment must be regarded as distinctly satisfactory; it is certainly more impressive than has yet been achieved in attempts to explain the resistivity of solid metals. The fears which were aired in §4.11 above (see also Springer, 1964) that the Born approximation might prove seriously inadequate do not seem to be substantiated; perhaps for the rather subtle reason suggested on p. 282, the effective pseudo-potentials which are needed to make the Ziman formula work exactly do not seem to be grotesquely different from the theoretical ones so far available.

One or two cases call for special mention. It has sometimes been doubted whether the Ziman theory is applicable to the liquid noble metals, Cu, Ag and Au. The d bands are not far below the Fermi level for these and the d wave functions of adjacent ions overlap to some extent, and in these circumstances the pseudo-potential approach might not be very useful; at any rate the dependence of $u(\mathbf{q})$ on the magnitude and direction of $\mathbf{q}$ might be unusually rapid. Heine & Animalu did not attempt to calculate pseudo-potentials for the noble metals, but some preliminary estimates based on the size of the band gaps in the solid phase led to estimates of ρ_L which were 5 to 10 times too big (Ziman, 1967). The success achieved for Cu, Ag and Au by Meyer *et al.* and by Moriarty is therefore particularly notable.

Amongst the polyvalent metals Hg stood out for some years as one for which the theoretical predictions seemed to be seriously too low. It is also anomalous, as we shall see later, in its thermoelectric power and in the way that its conductivity varies with volume, frequency and impurity content. Mott (1966) attributed all the anomalies to an incipient pseudo-gap near the Fermi level (see §1.12), arguing that m^*/m might be only about 0.7 in Hg, even at normal densities just above the melting point, but this hypothesis proved hard to reconcile with theoretical work on solid Hg (Brandt & Rayne, 1966; Keeton & Loucks, 1966) and was withdrawn (Mott, 1967) well before the results of Chan & Ballentine (1971*a*), reported on p. 270, became available. It will be apparent, from the entry for Hg in column (2) of table 5.2, however, that the recent calculations of Evans have disposed of at least one of the difficulties with which Mott was concerned. The d band in Hg does not overlap the conduction band, as it does in the noble metals, but according to Evans it is close enough to the Fermi level to make the phase shift η_2 unusually large. The effect of this large phase shift on $u(q)$ may be seen from fig. 1.7. Evidently, by preventing $u(q)$ from passing through a node in the range between 0 and $2k_F$, it enhances ρ_L by a factor of 3 or so compared with what the model of Heine & Animalu would suggest. The possibility that the resistivity of Hg is significantly *diminished* by correction terms to the Ziman formula which are usually ignored is discussed on p. 360 below.

Experimental work on the alkaline-earth metals, Ca, Sr and Ba, is still in an early stage (Van Zytveld *et al.*, 1972) but their resistivities are clearly large. Whereas Zn, Cd and Hg have full bands below the Fermi level, and quite close to it in the case of Hg, the alkaline-earth metals have empty d bands above the Fermi level, quite close to it in the case of Ba. Their high resistivities may be attributed in general terms to strong scattering by virtual bound d states, but such preliminary attempts to develop model pseudo-potentials for them as have so far been made have not produced encouraging results.

As stated above, there have been no very successful calculations of ρ_S so far. Consequently there are few theoretical estimates of the ratio ρ_L/ρ_S at the melting point which are worth quoting for comparison with the experimental values in table 5.1. Ziman (1961) has argued that since Na is a good free-electron metal even in the solid phase and since its resistivity, to judge from fig. 5.4, comes largely from the small-q region, one might reasonably expect

$$\frac{\rho_L}{\rho_S} \simeq \frac{a(0)_L}{a(0)_S} \simeq \frac{\beta_L}{\beta_S} \tag{5.15}$$

in this case; he interprets the steady increase of ρ_L/ρ_S within the sequence Na, K, Rb, Cs in terms of an increasing core contribution to the resistivity. In fact the ratio of the compressibilities in liquid and solid Na seems to be only about 1.2 rather than 1.45, so that (5.15) is not particularly successful, but the argument leading to it is much oversimplified; the values of $q/2k_F$ which matter most as far as Na is concerned lie in a range about 0.5, which is not all that small.

Attention should be drawn to the data for one or two transition and rare earth metals which have been included in table 5.1, though it is probably beyond the scope of Ziman's theory to explain them (but see Evans, Greenwood, Lloyd & Ziman, 1971). For most of these metals the resistivity of the solid phase is unusually insensitive to temperature just below the melting point, so it is not altogether surprising that ρ_L/ρ_S is unusually close to unity. In Fe, for example, $(T_M/\rho_S)(\partial\rho_S/\partial T)_p$ is only about 0.2, while ρ_L/ρ_S is only about 1.05 (Powell, 1953). It is as though four-fifths of the resistivity of Fe at high temperatures were contributed by some mechanism independent of structural disorder in the lattice, the other fifth being proportional to T in the solid phase in the usual way and increasing by perhaps 25% rather than 5% on melting. One theory (Coles, 1958; Mott, 1964) attributes the larger part of the resistivity at high temperatures – the melting point is well above the ferromagnetic Curie point – to scattering by disordered spins.

5.5. Temperature dependence of ρ_L

The temperature coefficient of ρ_L at *constant pressure* has of course been measured for practically every liquid metal, and a number of authors have set out to explain the results in terms of Ziman's formula, using experimental $a(q)$ curves obtained at different temperatures above the melting point (e.g. North *et al.*, 1968; Wingfield & Enderby, 1968; Srivastava & Sharma, 1969*b*; Ruppersberg & Winterberg, 1971; Cubiotti *et al.*, 1971*b*; Evans, 1971; Waseda & Suzuki, 1971*b*). Most of these authors have made the dubious assumption, however, that $u(q)$ stays the same if the specimen is allowed to undergo thermal expansion. Here we shall concentrate upon the coefficient *at constant volume*, leaving the effects of volume changes to be discussed in §5.7. Direct measurements of $(\partial\rho_L/\partial T)_\Omega$ have been made for all the alkali metals except Li, and for Ga and Hg. Where it has not been measured directly it can be inferred from $(\partial\rho_L/\partial T)_p$ if the thermal expansion coefficient is known and also $(\partial\rho_L/\partial\Omega)_T$; the latter quantity can sometimes be estimated from the thermo-electric

TABLE 5.3. *The temperature coefficient of resistivity at the melting point*

Metal	$\frac{T_M}{\rho_L}\left(\frac{\partial \rho_L}{\partial T}\right)_\Omega$	Metal	$\frac{T_M}{\rho_L}\left(\frac{\partial \rho_L}{\partial T}\right)_\Omega$
Li	(0.6)	Zn	(−0.24)
Na	0.82	Cd	(−0.22)
K	0.77	Hg	−0.12
Rb	0.65	Ga	0.14
Cs	0.69	In	(0.16)
Cu }		Sn	(0.13)
Ag }	(0.4)	Pb	(−0.02)
Au }		Bi	0.02

power, with the aid of a semi-empirical correlation discussed on p. 341. The best available results (for recent work, see Lien & Sivertsen, 1969) are listed in table 5.3, figures in parentheses involving an element of guesswork. Note how Zn and Hg fall into line in this table despite the fact (see fig. 5.1) that at constant pressure their temperature coefficients have opposite signs; it is only the unusual sensitivity of ρ_L to thermal expansion that makes $(\partial \rho_L/\partial T)_p$ positive in the case of Hg.

The theoretical prediction based upon Ziman's formula is that

$$\frac{T_M}{\rho_L}\left(\frac{\partial \rho_L}{\partial T}\right)_\Omega = \frac{T_M\langle|u|^2 a(\partial \log a/\partial T)_\Omega\rangle}{\langle|u|^2 a\rangle}, \tag{5.16}$$

and to evaluate the right-hand side requires a knowledge of $(\partial \log a/\partial T)_\Omega$. Unfortunately, this quantity has not yet been measured and we have little to go on but the broken curve in fig. 2.13, based partly on Greenfield's measurements on liquid Na at constant pressure and partly on conjecture. We could of course, like Ashcroft (1966*b*), appeal once again to the Percus–Yevick rigid-sphere model but there is no experimental evidence to support its application in this context. The rigid-sphere model allows no change of $a(q)$ with temperature at constant volume so long as σ, and hence y, are kept constant. Ashcroft adjusts the rate of variation of σ to ensure that $(\partial \log a/\partial T)_\Omega$ takes the right value near $q = 0$, so far as this is known from compressibility measurements.

The curve in fig. 2.13 undoubtedly possesses the right qualitative features to explain the data in table 5.3. For the monovalent metals the limit $q = 2k_F$ more or less coincides with the point at which $(\partial \log a/\partial T)_\Omega$ changes sign; $(\partial \log a/\partial T)_\Omega$ is positive over the whole of the range between 0 and $2k_F$. Hence we should expect $(T_M/\rho_L)(\partial \rho_L/\partial T)_\Omega$ to be positive, but to decrease in magnitude as the large-q part of the resistivity becomes more important, which is just what is observed. For divalent metals the range of q over which $(\partial \log a/\partial T)_\Omega$ is negative is heavily weighted, so that negative values are to be expected for $(T_M/\rho_L)(\partial \rho_L/\partial T)_\Omega$. And for metals with more than 2 valence electrons per atom it should become rather small.

There is still some doubt, however, as to whether quantitative agreement can be obtained. In the case of Na, for example, it appears from fig. 5.4 that $|u|^2 a$ is greatest round about $q/2k_F = 0.5$, and it is the value of $(\partial \log a/\partial T)_\Omega$ in this region that should primarily determine $(T_M/\rho_L)(\partial \rho_L/\partial T)_\Omega$. The broken curve in fig. 2.13 does not suggest so large a value as 0.85. Nor, for that matter, does it suggest an answer for Li quite as big as 0.6. It is still in dispute whether there is a real discrepancy here or whether Greenfield's measurements are erroneous; Ashcroft's theoretical curve for $(\partial \log a/\partial T)_\Omega$ would certainly fall off less rapidly than the one drawn in fig. 2.13 and would be more easy to reconcile with the observations.

If there is a discrepancy is it due, as Greenfield has claimed, to a break-down of the Born approximation? Since the effective pseudo-potential U', which should replace U in the theory when corrections to the Born approximation are significant, is determined by $a(q)$ it may indeed be temperature-dependent, even at constant volume, and in principle this may add to the temperature dependence of ρ_L. However, for reasons given on p. 287, the difference between U' and U is probably negligible near $q = 0$. Although detailed calculations have yet to be done, it seems unlikely that the difference between them near $q/2k_F = 0.5$ is large enough, or its temperature dependence sufficiently rapid, to remove the apparent discrepancy.

It is perhaps worth pointing out that the correction due to thermal broadening of the Fermi distribution is likely to be

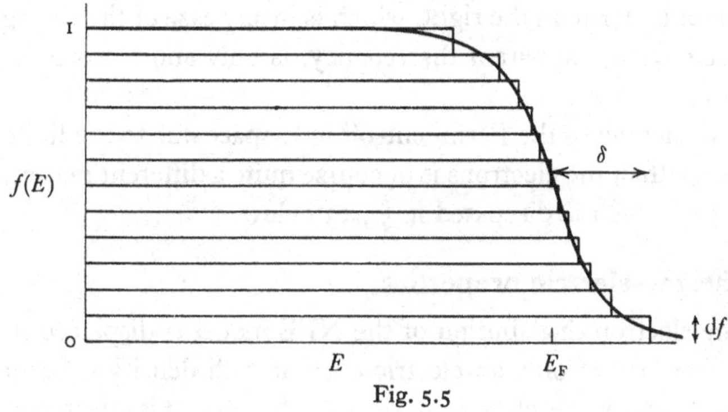

Fig. 5.5

negligible in this context as in nearly every other. The simplest way to evaluate the correction is to think of the Fermi distribution at a finite temperature as made up by the superposition of a number of elementary distributions, each with the sharp cut-off characteristic of the absolute zero but with Fermi energies which differ from E_F by amounts δ. This device is illustrated by fig. 5.5. The conductivity for each element can be calculated by (5.3) but we must recognise that it is a function of δ. The total conductivity is then given by

$$\begin{aligned}\sigma &= \int_0^1 \sigma(\delta)\,\mathrm{d}f \\ &= \int_{\infty}^{-\infty} (\sigma(0)+\delta\sigma'(0)+\tfrac{1}{2}\delta^2\sigma''(0)+\ldots)\frac{\mathrm{d}f}{\mathrm{d}E}\,\mathrm{d}E,\end{aligned} \tag{5.17}$$

where σ', σ'' denote derivatives of $\sigma(\delta)$ with respect to δ in the neighbourhood of $\delta = 0$. Since $\mathrm{d}f/\mathrm{d}E$ is an even function of δ the term in σ' contributes nothing to the integral and the result is

$$\sigma = \sigma(0)+\tfrac{1}{6}\pi^2(k_B T)^2\sigma''+\ldots \tag{5.18}$$

In the case of liquid Na, as we shall see in the next section, $\sigma(0)$ should be more or less proportional to k_F^6 and this means that

$$\sigma'' \simeq 6\sigma(0)/K_F^2. \tag{5.19}$$

From (5.18) and (5.19) it follows that

$$\frac{T_M}{\rho_L}\left(\frac{\partial\rho_L}{\partial T}\right)_\Omega \simeq \frac{T_M}{\rho_L}\left(\frac{\partial\rho_L(0)}{\partial T}\right)_\Omega - 2\pi^2\left(\frac{k_B T_M}{K_F}\right)^2+\ldots \tag{5.20}$$

The correction term on the right, which is in any case of the wrong sign to remove the apparent discrepancy, is only about 0.002 for liquid Na.

The broadening of the Fermi cut-off in **k**-space due to the finite mean free path of the electrons is of course quite a different matter, the effect of which is discussed in §5.13 below.

5.6. Thermo-electric properties

When the electron distribution of the NFE model is displaced in **k**-space there is not only an electric current with density **j**, but a flux of energy **Q**. At absolute zero, since all the current is effectively carried by electrons with the total Fermi energy E_F, it is clear that

$$\mathbf{Q} = (E_F/e)\mathbf{j} = (E_F/e)\sigma\mathbf{E}.$$

At finite temperatures the argument at the end of the previous section is readily extended to show that

$$\begin{aligned}\mathbf{Q} &= \frac{\mathbf{E}}{e}\int_{\infty}^{-\infty}(E_F+\delta)(\sigma(0)+\delta\sigma'+\ldots)\frac{df}{dE}\,dE \\ &= \mathbf{j}\left(\frac{E_F}{e}+\frac{\pi^2(k_BT)^2}{3e}\frac{\sigma'}{\sigma(0)}+\ldots\right).\end{aligned} \tag{5.21}$$

This formula is the key to an understanding of the thermo-electric properties of metals at ordinary temperatures.† When two different metals A and B are placed in contact, charge flows from one to another until a potential difference is set up such that

$$E_F^A = E_F^B.$$

If a current is then driven across the interface from A to B it follows from (5.21) that heat must be released at a rate per unit area given by

$$(Q^A - Q^B) = j\,\frac{\pi^2(k_BT)^2}{3e}\left(\left(\frac{\sigma'}{\sigma(0)}\right)^A - \left(\frac{\sigma'}{\sigma(0)}\right)^B\right). \tag{5.22}$$

This is the origin of the *Peltier heat*, and the absolute Peltier coefficient for each metal is given by

$$\Pi = \frac{\pi^2(k_BT)^2}{3e}\frac{\sigma'}{\sigma(0)}. \tag{5.23}$$

† At low temperatures the heat current is augmented by a phonon drag effect which there is no need to consider here.

TABLE 5.4. *The thermo-electric parameter ξ at the melting point*

Metal	ξ (observed)	ξ (predicted) (*a*)	(*b*)	(*c*)
Li	−9.3	−7.2	−5.0	−5.2 (CGR)
Na	2.8	2.8	0.6	3.2 (CGR)
K	3.5	2.9	2.8	3.6 (CGR)
Rb	1.7	1.3	0.1	1.6 (CGR)
Cs	−1.4	1.1	−7.8	−1.3 (CGR)
Cu	−3.6	—	−3.4	−5.3 (M)
Ag	−1.8	—	−3.2	−4.5 (M)
Au	−0.8	—	−3.6	−6.2 (M)
Zn	−0.3	2.1	—	1.0 (M), 0.4 (E)
Cd	−0.2	—	—	1.0 (E)
Hg	4.1	−1.3	—	5.3 (E)
Al	1.0	1.7	—	—
Ga	0.55	—	—	—
In	0.8	—	—	1.2 (E)
Tl	0.3	2.0	—	1.4 (E)
Sn	0.4	—	—	—
Pb	2.1	2.7	—	2.7 (E)
Bi	0.55	2.5	—	—

Note. The 'observed' values of ξ in this table have been derived from P with the assumption that m^*/m is unity. If m^*/m is really about 1.5 for Li (see p. 298), then the true value of ξ may be only about −6 for this metal.

The other thermo-electric properties may be deduced from the Peltier coefficient using the standard Thomson relations. In particular, the *absolute thermo-electric power P* is given by

$$P = \Pi/T. \tag{5.24}$$

It is customary to define a dimensionless parameter ξ such that

$$\xi = \frac{k_F}{2\sigma}\left(\frac{\partial\sigma}{\partial k}\right)_F; \tag{5.25}$$

the partial differential here expresses the rate of change that would be observed for the conductivity σ if the Fermi sphere could be expanded, but without altering in any way the arrangement of the ions or the screening of their pseudo-potentials. In terms of this quantity it follows from (5.23) and (5.24) that

$$P = \frac{\pi^2 k_B^2 T}{3eK_F}\left(\frac{m^*}{m}\right)\xi. \tag{5.26}$$

This equation predicts a negative thermo-electric power so long as ξ is positive, because the charge on the electron is negative. It may be used to deduce values of ξ from measurements of P; the free-electron formula (1.7) is available for K_F and it is a customary simplification to set m^*/m equal to unity. Some results obtained in this way are listed in table 5.4 (for recent work see Randsalu & Lundén, 1965; Ricker & Schaumann, 1966; Ubbelohde, 1966; Marwaha, 1967; Howe & Enderby, 1967; Kendall, 1968).

Differentiation of Ziman's formula with respect to k_F yields a theoretical prediction for ξ in the form

$$\xi = 3-2\frac{|u(2k_F)|^2 a(2k_F)}{\langle |u|^2 a\rangle}-\frac{\langle k_F(\partial |u|^2/\partial k)_F a\rangle}{\langle |u|^2 a\rangle}. \tag{5.27}$$

The final term is necessary because, even though the screening is not supposed to alter when the Fermi sphere suffers its notional expansion and the real potential round each ion remains quite unchanged, the pseudo-potential may be sensitive to the energy of the electron on which it acts and to the angle of scattering (for given q), both of which depend on the Fermi radius. Most of the results for ξ which are quoted in table 5.4 may be understood in a qualitative fashion on the assumption that this final term is negligible. In a polyvalent metal we do not expect $|u|^2 a$ to vary rapidly over the important region close to $q = 2k_F$, in which case $|u(2k_F)|^2 a(2k_F)$ should not be very different from $\langle |u|^2 a\rangle$. Hence ξ should be roughly unity, as observed. In a monovalent metal such as Li it is evident from fig. 5.4 that $|u|^2 a$ is rising sufficiently fast in the neighbourhood of $2k_F$ to make $|u(2k_F)|^2 a(2k_F)$ a good deal larger than $\langle |u|^2 a\rangle$ and hence to make ξ negative. In Na $u(2k_F)$ should be small enough to let ξ approach 3, as it does.

Early attempts to verify this line of explanation by quantitative calculations were only moderately successful (Marwaha, 1967). Some predictions due to Sundström (1965), which are based upon a pseudo-potential due to Heine & Abarenkov (1964), are shown in column (*a*) of table 5.4. If one remembers that the calculation really concerns the quantity $(3-\xi)$ rather than ξ itself, then the agreement does not seem particularly impressive.

It is not improved if the later Heine–Animalu potential is used instead.

If one is inclined to blame one's model pseudo-potential for any discrepancy rather than the theoretical expression for ξ, it is tempting to turn the handle of Ziman's theory backwards and to use the experimental data to determine an empirical $u(q)$ that will give the right answers. The magnitude of ξ serves to fix $u(2k_F)$, $u(0)$ is known from the general considerations advanced in chapter 1, and the shape of $u(q)$ between these limits may be adjusted without much latitude to yield the observed resistivity. An example of this procedure is provided by the discussion of Ga by Cusack, Kendall & Marwaha (1962). If the Born approximation is not valid then one may hope that it yields the effective pseudo-potential u'. The trouble is, however, that it depends a great deal on the neglect of the final term in (5.27). There is no guarantee that this term is really negligible. In fact Young *et al.* (1967) find that it contributes a large amount to the values of ξ they predict for monovalent metals, which are listed in column (*b*) of table 5.3. The values in column (*c*) are due to Moriarty, Cubiotti *et al.* (1971 *a*,*b*) and Evans (1970, 1971). In these too the final term in (5.27) is important. It must obviously be important for Hg, since without it it is impossible to generate a value for ξ which is greater than 3. According to Evans the first two terms of (5.27) amount to only -1 for Hg.

The parameter ξ is more or less independent of temperature above the melting point except for Zn, Hg and Tl. In Hg at any rate its variation seems to be the result of thermal expansion, for Bradley (1963) has shown that if the metal is heated at constant volume ξ remains unaffected.

5.7. Thermal conductivity

It is well known that if a temperature gradient is applied to a metal the conduction electrons will carry a heat current along it even though an electric current is prevented from flowing, and that indeed they are responsible for the major part of the thermal conductivity. To obtain a formal expression for the electronic term in the thermal conductivity, κ_e, requires solution of the Boltzmann equation. The theory is worked out in many standard

TABLE 5.5. *Values of* $10^8(\kappa\rho_L/T)$ *at the melting point, in* $W\Omega\,°K^{-2}$

Li	2.6	Al	2.4
Na	2.2	Ga	2.07†
K	2.1	Tl	3.2
Cs	2.4	Sn	2.9
Zn	3.2	Pb	2.4
Cd	2.5	Sb	2.6
Hg	2.75	Bi	2.5

† The figure for the Ga (Duggin, 1969) has been corrected for the effects of ionic conductivity, but it would be less than 2.45 even without this correction.

texts and only the answer need be quoted here; it can be written as

$$\frac{\kappa_e \rho}{T} = \frac{\pi^2 k_B^2}{3e^2} = 2.45 \times 10^{-8} \mathrm{W}\Omega\,°\mathrm{K}^{-2}, \tag{5.28}$$

the right-hand side constituting the so-called *Lorenz number*. Equation (5.28) does not depend upon specific assumptions about the shape of the Fermi surface, the density of electron states, the nature of the scattering potential, etc., but it does involve the general assumption that the scattering is elastic; in liquid metals, as we have seen above, this should be sound enough (Rice, 1970).

The total thermal conductivity κ includes a contribution due to ionic motion, but estimates based upon the viscosity using equation (3.16) suggest that this should amount to only a few parts per cent at temperatures just above the melting point. And the experimental estimates of $\kappa\rho/T$ shown for a number of liquid metals in table 5.5 are indeed quite close to the theoretical Lorenz number.† The agreement is perhaps not quite so close has been found for solid metals at high temperatures (well above the Debye temperature) but the measurement of thermal conductivity in a liquid is not easy and the data for κ may well be erroneous; there is considerable disagreement between the results of different investigators for some of the metals in the table. For several of them it has been verified that $\kappa\rho/T$ is essentially independent of temperature over a very substantial range, e.g. from the melting point up to 1200 °C in the case of liquid K at its saturated vapour pressure (Grosse, 1966).

† Recent references are Shpil'rain & Krainova (1967), Filippov (1968), Duggin (1969).

If the discrepancies that do appear in table 5.5 are not due to experimental error, nor to the ionic conductivity – and they sometimes have the wrong sign for this – it is just possible that they are associated with short mean free paths. No theory of electronic transport properties which is based upon the Boltzmann equation is clearly valid unless $\hbar/\tau$ is very small compared with K_F. Corrections to the electrical resistivity of order $(\hbar/\tau K_F)^2$ are discussed in §5.13 below, but no way of evaluating similar corrections to the thermal conductivity has yet been worked out.

5.8. The effects of compression

The effect of isothermal compression upon the resistivity has been studied for a rather limited range of liquid metals, with results that are summarised in table 5.6. The result for K was at one time in doubt, since the work of Endo (1963) indicated a much larger value of $(\partial\rho_L/\partial\Omega)_T$ than that found by Bridgman (1958); Bridgman's figure seems to be supported by the measurements of Lien & Sivertsen (1969) and is therefore preferred here. The fact that the volume coefficient for Hg is unusually large has already been noted.

To estimate $(\partial\rho_L/\partial\Omega)_T$ theoretically is a complex task. A change of volume implies, of course, a change in the size of the Fermi sphere and one must therefore face to begin with all the problems that arise in the estimation of ξ. In addition, one needs to allow for changes in the pseudo-potential due e.g. to changes in the screening properties of the conduction electron gas, which are irrelevant to the thermo-electric power. Finally, one needs to know how $a(q)$ depends upon volume and this has not been studied experimentally. If one chooses to rely on the Percus–Yevick rigid-sphere model for $a(q)$, there is the problem of deciding how σ and y are likely to depend upon volume. At first sight one may be inclined to regard the effective rigid-sphere diameter as a function of temperature alone, in which case y would be proportional to Ω^{-1}. However, the screening length gets smaller as a metal is compressed, so it is on the cards that σ does so too; the variation of y may well be less rapid.

The only detailed calculations which have been reported are those of Dickey *et al.* (1967) for the alkali metals and of Animalu

TABLE 5.6. *The volume coefficient of resistivity at the melting point*

Metal	$\frac{\Omega}{\rho_L}\left(\frac{\partial\rho_L}{\partial\Omega}\right)_T$ Expt.	$\frac{\Omega}{\rho_L}\left(\frac{\partial\rho_L}{\partial\Omega}\right)_T$ Theory
Li	$\lesssim 0$	-1.1 (D)
Na	4.8	>0 (D)
K	5.1	3.7 (D)
Rb	4.2	2.5 (D)
Cs	2.1	-0.7 (D)
Hg	8.2	-2 (A), $+7$ (E)
Ga	2.9	—
Bi	3.1	—

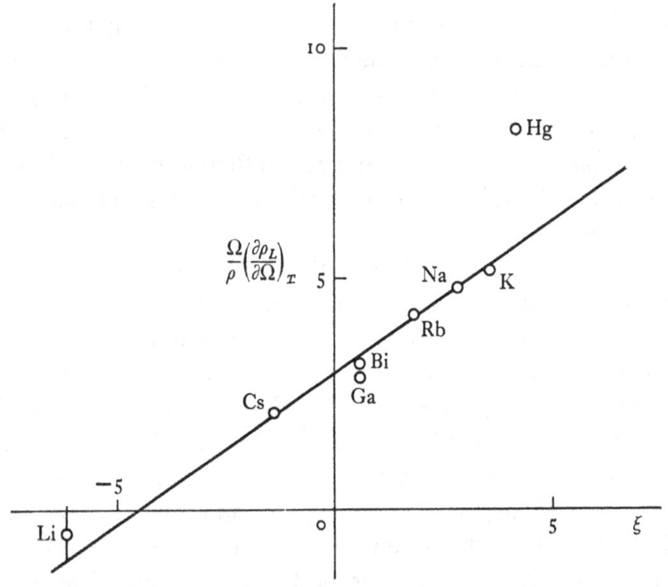

Fig. 5.6. Correlation between the volume coefficient of resistivity and the thermoelectric parameter. The line has a slope of $\frac{2}{3}$. (N.B. m^*/m has here been taken to be 1.5 for Li but 1.0 for all the other metals.)

(1967) and Evans (1970) for Hg. The first authors use a phase-shift analysis, as mentioned in §5.4 above, together with the assumption that the core potential is insensitive to volume; they determine y as a function of volume by the heuristic recipe referred to on p. 327. Animalu uses the Heine–Animalu potential, recalculated for a series of different volumes, and apparently takes y to be a constant, equal to 0.46. Evans computes $(\partial\rho_L/\partial T)_p$ in the first instance, using curves for $a(q)$ obtained experimentally at three different temperatures, and estimates $(\partial\rho_L/\partial\Omega)_T$ by combining his result with the known value of $(\partial\rho_L/\partial T)_\Omega$. The results are compared with experiment in the table. The superiority of Evans's pseudo-potential for Hg over that of Heine & Animalu is clearly evident.

If one is prepared to take ξ from experiment and to use the NFE model to relate changes of volume to changes of k_F, it is helpful to express the volume coefficient in the form

$$\frac{\Omega}{\rho_L}\left(\frac{\partial\rho_L}{\partial\Omega}\right)_T = \tfrac{2}{3}\xi - 1 + \frac{\langle\Omega(\partial|u|^2/\partial\Omega)\,a\rangle}{\langle|u|^2a\rangle} + \frac{\langle|u|^2\Omega(\partial a/\partial\Omega)_T\rangle}{\langle|u|^2a\rangle}, \quad (5.29)$$

and then to plot the left-hand side against ξ, as in fig. 5.6. A straight line with a slope of $\frac{2}{3}$ can be drawn through most of the points, as shown in the figure. Its intercept implies that the last two terms on the right-hand side of (5.29) normally add up to about 4.† The point for Li has been brought onto the line by the reasonable assumption that m^*/m is about 1.5 for this metal (see footnote to table 5.4).

It is readily shown from equation (1.67) that in the limit $q \to 0$ the quantity $\Omega(\partial|u|^2/\partial\Omega)$ should be $\frac{2}{3}|u|^2$. If the only effect of compression on the pseudo-potential arises via the change of screening, however, this quantity should go to zero for large q, where screening is of no importance. Hence we may expect

$$\langle\Omega(\partial|u|^2/\partial\Omega)\,a\rangle/\langle|u|^2a\rangle \quad (5.30)$$

to approach $\frac{2}{3}$ for a metal such as Na where the small-q region is important, but to be small for a metal such as Cs or for any of the polyvalent metals.

Near $q = 0$ the magnitude of $a(q)$ is fixed according to (2.40) by the compressibility β, and Bridgman's experiments, chiefly on solid metals, suggest that β varies with volume in such a way that $(\Omega/a)(\partial a/\partial\Omega)_T$ is about 3. Near the first peak in $a(q)$, however, the compressibility is irrelevant. Rather than appeal to the Percus–Yevick rigid-sphere model in this region, it is instructive to make

† This empirical result is the basis for some of the entries in table 5.3.

the plausible assumption, which is at least consistent with the sum rule, that the position and width of the peak scale proportional to $\Omega^{-\frac{1}{3}}$, without any change in its height or shape. This assumption implies that

$$(\Omega/a)(\partial a/\partial\Omega)_T \simeq \tfrac{1}{3}(q/a)(\partial a/\partial q)_{T,\Omega}. \tag{5.31}$$

The right-hand side of (5.31) may be determined from an experimental curve, without invoking any particular model, and for a typical liquid monovalent metal it appears to be about 4 near $2k_F$. Hence between say Na and Cs we may expect

$$\langle|u|^2\Omega(\partial a/\partial\Omega)_T\rangle/\langle|u|^2 a\rangle \tag{5.32}$$

to vary between about 3 and 4.

These considerations make it entirely reasonable that (5.30) and (5.32) together should add up to 4 for most monovalent liquids, but the argument cannot be applied with success to polyvalent ones. Though $(\partial a/\partial q)$ is large and positive to the left of the main peak it is equally large but negative to the right, and it would appear from (5.31) that (5.32) must then be negative. This seems especially likely for a divalent metal, and it is presumably on this account that Animalu's figure for Hg in table 5.6 is negative. To obtain agreement with experiment for Hg, Ga and Bi it seems necessary to suppose that (5.30) is much more substantial in the large-q region than our discussion of screening has led us to suppose. Ziman (1967) has tentatively suggested that the core pseudo-potential may change with volume because the conduction band as a whole shifts relative to the core levels; the relevance of their separation is brought out by the Phillips–Kleinmann formulation of the pseudo-potential problem discussed in §1.8. It would take us into realms of uncharted speculation to pursue this suggestion here.

The effect of a change of volume upon the thermo-electric power has been studied for liquid Hg by Bradley (1963), Schmutzler & Hensel (1968, 1971), Crisp & Cusack (1969) and Crisp, Cusack & Kendall (1970); the last of these papers include results for some liquid alkali metals as well. The results are in rather poor agreement with such theoretical predictions as exist. In the case of Hg, an extrapolation of the experimental data suggests that a compression of about 5% would be sufficient to reduce ξ to a value comparable with that observed for other divalent liquid metals. This would also bring the resistivity more into line, by reducing it from 90 to about 50 $\mu\Omega$ cm. According to Bridgman (1958), however, $(\Omega/\rho_L)(\partial\rho_L/\partial\Omega)$ would remain at its unusually high level of about 8. Whether these details of behaviour for Hg can be explained by the model which Evans (1970) has used is still to be seen.

5.9. Cs at high densities

The compressibilities of the heavier alkali metals are so large that it is possible to double their densities by the application of 50–100 kbar of pressure. Curiously enough the compression does not necessarily force them to solidify. The liquid starts off more compressible than the solid and with a volume Ω_L which is only slightly greater. As a result, Ω_L falls below Ω_S when pressure is applied, and according to the Clausius–Clapeyron equation dT_M/dp then becomes negative. Thus the curve for T_M as a function of p passes through a maximum in the case of Rb (Bundy, 1959). In the case of Cs it passes through at least two maxima, as shown in fig. 5.7. So long as the temperature of Cs is kept above about 200 °C it will remain liquid up to more than 50 kbar.

The resistivity of Cs under pressure is of particular interest because of the three phase transitions that occur under pressure in the solid (see Jarayaman *et al.*, 1967, for references and a fuller discussion). The transition from Cs I to Cs II involves a change from b.c.c. to f.c.c. packing, Cs III is also f.c.c., but the structure of Cs IV has not yet been determined. All the transitions naturally involve contraction. The resistivity of the solid behaves as shown by the lower curve in fig. 5.8,† the most notable feature being the relatively high resistivity of Cs III. It has been conjectured that in this phase the 6s valence electrons find themselves promoted into 5d states which are more compact but have higher energy,

† To be accurate the curves in fig. 5.8 show the behaviour of the measured *resistance*. Since quite large changes of volume are involved in this high pressure work it is often necessary to remember that curves of resistance and resistivity will not look identical.

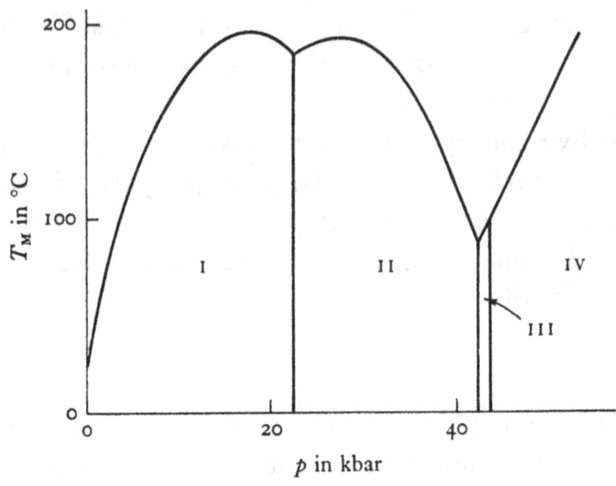

Fig. 5.7. The melting curve for Cs.

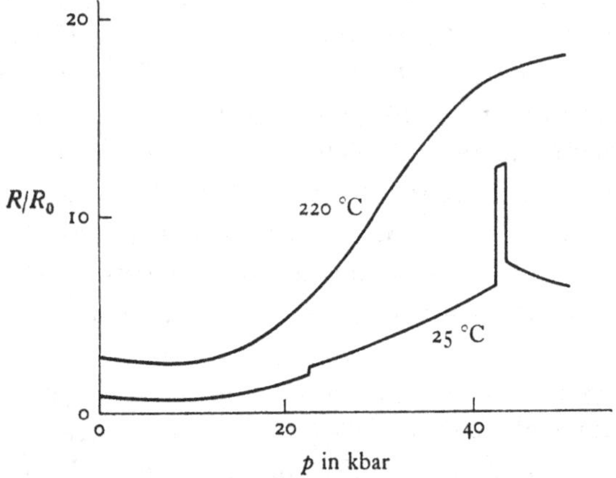

Fig. 5.8. The resistance of solid and liquid Cs relative to the resistance of the solid at NTP. Note that a pressure of 40 kbar is roughly sufficient to halve the specimen volume. (Redrawn by permission from Jayaraman, Newton & McDonough (1967), *Phys. Rev.* **159**, 527.)

but no suggestion seems to have been made to explain the Cs III–Cs IV transition.

The upper curve in fig. 5.8 shows how the resistivity of the liquid behaves over the same range of pressure. We may note first that $(\partial\rho_L/\partial\Omega)_T$ which starts off positive at low pressures as shown in table 5.5, soon becomes negative. This sort of feature is not unexpected from the earlier work of Bridgman, and a first attempt to explain the volume dependence of $(\partial\rho_L/\partial\Omega)_T$ for all the liquid alkali metals has been made by Dickey *et al.* (1967). Of more interest here, however, is the rather spectacular rise in resistivity, by a factor of 5 or more, over a range of pressures around 30 kbar. It has been suggested that the ionic and electronic structure of liquid Cs is changing continuously over this range into something more comparable with Cs III than with Cs I. Rapoport (1967) claims to fit the observations with a semi-empirical theory in which the liquid is treated as an equilibrium mixture of these two phases, but the phenomenon deserves further study.

5.10. Results at high temperatures and low densities

It is obviously of interest to examine the properties of liquid metals when their density is much less than usual, as well as when it is much greater. Low densities can be achieved by heating specimens up to the critical temperature T_C, under a pressure that is only sufficient to prevent boiling.† Since the critical temperature seems to be about 3–4 times the normal boiling point for most metals, it is the especially volatile ones like Cs and Hg which have been selected for study. For these two it has proved possible to extend the measurements right through the critical point into the vapour phase. The techniques used and results obtained have been reviewed by Ross & Greenwood (1969).

Results for the resistivity of Cs at low densities are shown in fig. 5.9, where ρ is plotted logarithmically against the distance R_A defined by (1.68). The experimental points for $4\,Å < R_A < 20\,Å$ are those obtained by Renkert, Hensel & Franck (1969) at a uniform temperature of 2000 °C, under pressures of up to 1000 bar; the critical temperature of Cs is about 1750 °C. The full curve in the range between 3 and 4 Å represents the behaviour of liquid Cs under its saturated vapour pressure as reported by Hockman & Bonilla (1964), who varied the temperature from the melting point (28.5 °C) up to 1650 °C. We have already seen that just above the melting point $(T/\rho)(\partial\rho/\partial T)_\Omega$ is $+0.7$ for liquid Cs, which means for that a fixed value of R_A such as 3 Å the resistivity is probably 3 or 4 times greater at 2000 °C than at 28.5 °C.

† At the critical point itself, according to Ross (1971), the packing fraction y is probably about 0.1 or 0.15.

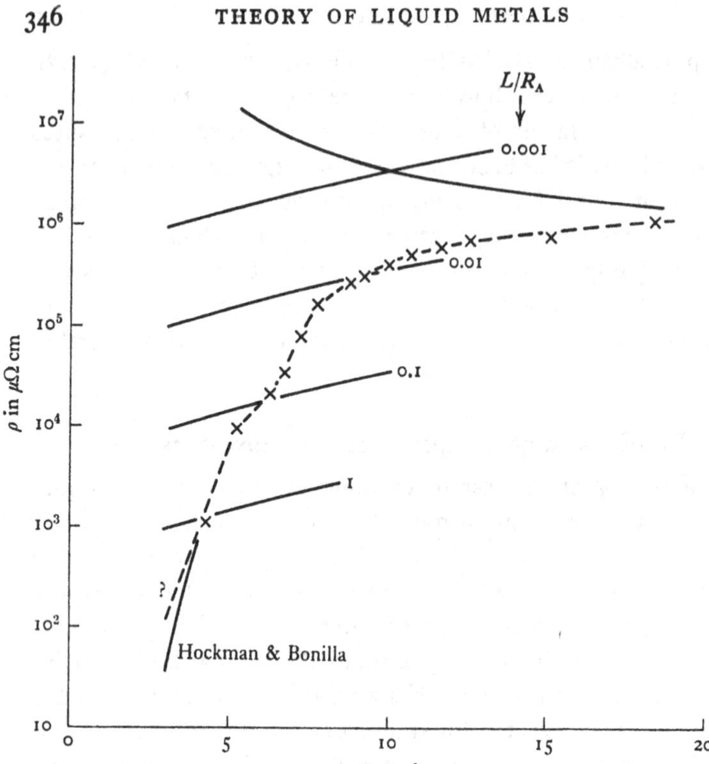

Fig. 5.9. The resistivity of liquid Cs at low densities. The broken curve through the points is an isotherm, corresponding to $T = 2000\,°\mathrm{C}$. The full curve at the top corresponds to equation (5.33).

It seems likely therefore, that the resistivity isotherm corresponding to $T = 2000\,°\mathrm{C}$ has the form indicated by the broken curve in the figure.

By the time that R_A is about 18 Å the resistivity has reached almost 1 Ω cm and we can no longer pretend to be dealing with a metal. Instead we have a fairly dense gas of Cs atoms and Cs_2 molecules,† a small proportion of them being split by thermal excitation into ions and free electrons. The theory of the resistivity of such a system is essentially the same as for an extrinsic semiconductor whose donor centres are only slightly ionised (consult e.g. Smith, 1964). If the number of atoms per unit volume is

† At the temperatures and pressures of interest here, atoms should be a good deal more common than molecules; the dissociation energy of Cs_2 is only about 0.5 eV.

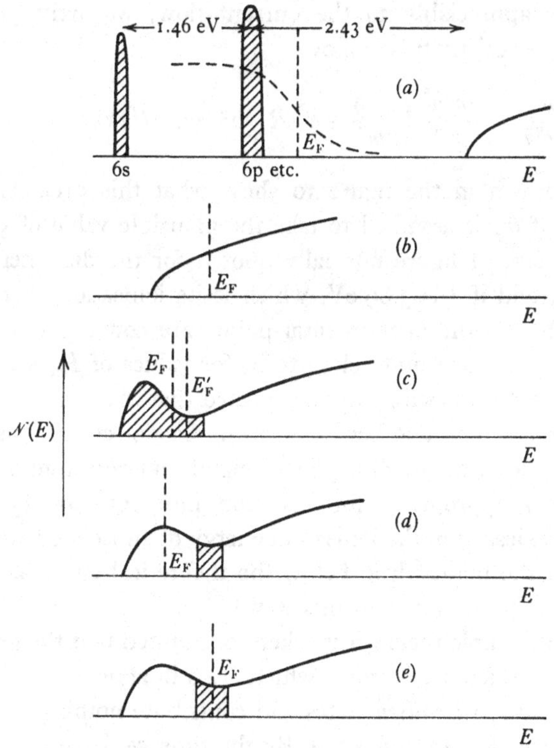

Fig. 5.10. The density of states for monovalent Cs at different densities (diagrams (*a*) to (*d*)) and for divalent Hg (diagram (*e*)).

N/Ω and I is their ionisation energy, the density of free electrons should be given by

$$n = (N/\Omega)^{\frac{1}{2}}(mk_B T/2\pi\hbar^2)^{\frac{3}{4}} \exp(-I/2k_B T);$$

this is the appropriate formula when n is small enough for the electron gas to be non-degenerate, a condition which is certainly satisfied for Cs when $R_A \sim 18$ Å. The average relaxation time for the free electrons may be written as

$$\bar{\tau} = \tfrac{4}{3}L(m/2\pi k_B T)^{\frac{1}{2}} = (16\Omega/3N\pi\sigma_A^2)(m/2\pi k_B T)^{\frac{1}{2}},$$

where $\frac{1}{4}\pi\sigma_A^2$ represents the effective scattering cross-section of an atom. Since the ions are too heavy compared with the electrons to contribute appreciably to the current flow, we arrive at a formula for the resistivity as follows:

$$\rho = \frac{m}{ne^2\bar{\tau}} = \frac{3\pi^2\hbar^{\frac{3}{2}}}{4e^2}\left(\frac{9}{\pi m k_B T}\right)^{\frac{1}{4}} R_A^{-\frac{3}{2}}\sigma_A^2 \exp(I/2k_B T). \quad (5.33)$$

A curve is drawn in the figure to show what this predicts at $T = 2000\,°C$ if σ_A is assumed to take the plausible value of 5 Å, which is the sort of figure normally quoted for the diameter of the Cs atom, and if I is 3.89 eV, which is its ionisation energy. It looks as though the experimental points are converging onto this curve, or onto one quite close to it, for values of R_A a little greater than 18 Å. It is surely to be expected that as the atoms are brought nearer to one another, and as the free-electron density builds up to a level at which it plays a significant screening role, the value of I appropriate for insertion into equation (5.33) should become less than the ionisation energy of an isolated atom. The fact that the points drop below the theoretical curve as R_A diminishes can be explained in this way.

In the above simple theory it is taken for granted that the great majority of the valence electrons, which are still attached to their parent atoms, are in *localised* states and contribute nothing to the conductivity (see §§1.12 and 4.10). By the time we have reached the metallic regime at $R_A \sim 3$ Å, however, they are all of them free or 'nearly' free. These two extreme situations are illustrated by diagrams (*a*) and (*b*) respectively in fig. 5.10, where some hypothetical curves of density of states versus energy are plotted. Hatching is used to indicate states which are localised, while broken lines suggest the position of the Fermi level and, in (*a*), the Fermi–Dirac distribution function for a temperature of 2000 °C. The reader should note that the Fermi–Dirac function applies *only* to the non-localised electrons; in (*a*) for example, it describes the small number of electrons excited into the conduction band, but the bound electrons are distributed between the 6s, 6p and higher atomic levels (to avoid confusion the higher ones are not shown on the diagram) according to a Boltzmann distribution, which means that the 6s level is almost full and the 6p level almost empty.

Diagram (*c*) in fig. 5.10 represents one plausible guess at what happens when R_A is intermediate between 18 Å and 3 Å. Due to the lowering of the 'muffin-tin zero' (see p. 61), the 6p and higher atomic levels have by this stage broadened to such an extent that they are merged into the conduction band, but there is still a recognisably distinct 6s band of localised states with a pseudo-gap between the two. Double occupancy of a localised state round a monovalent ion such as Cs^+, by electrons with opposite spins, is likely to be prevented by coulomb repulsion between electrons. It may be shown that in these circumstances the probability of a localised state being occupied by a single electron is given by the usual Fermi–Dirac function

$$(\exp{(E-E'_F)/k_BT}+1)^{-1},$$

but with $$E'_F = E_F + (\log 2)k_BT \simeq E_F + 0.14 \text{ eV}.$$

The Fermi level in fig. 5.10(*c*) has been placed with this result in mind.

What is likely to happen if, starting with the situation represented by fig. 5.10(*c*), the density is slowly increased? One possibility is that a *Mott transition* occurs (see p. 62), i.e. that the states in the 6s band become non-localised as in fig. 5.10(*d*) and that the Fermi level moves to the middle of this band because it can then accommodate two electrons per atom. It is equally possible, however, that the pseudo-gap in fig. 5.10(*c*) fills slowly up and the boundary of the hatched region moves continuously to the left without any abrupt transition occurring. Whichever of these pictures is correct it is hardly surprising, in view of the extent to which the cut-off in the Fermi–Dirac function is smeared out at these high temperatures (fig. 5.10(*a*) is to scale in this respect), that the resistivity varies between the gas-like and metallic limits in a manner that is essentially smooth. Fig. 5.9 suggests that there may be an inflexion when R_A is about 5.5 Å (at the critical density, for comparison, $R_A = 4.85$ Å) and one school of thought would attribute this to a Mott transition. But the evidence for it is a single point, which requires confirmation.

Over what portion of the experimental curve in fig. 5.9 is a theory of the Ziman type likely to prove applicable, assuming that we know how to calculate the Cs pseudo-potential, or its effective pseudo-potential if the Born approximation has broken down (see §4.11), as a function of volume? A tentative answer to this question is provided by the following chain of argument.

(*a*) So long as the theory is valid the effective mean free path L, defined in terms of the measured resistivity by the free-electron equation (5.13), should not differ greatly from the range of coherence l. The relation between these two lengths should be described, in fact, by (5.14), in which the $\overline{1-\cos\theta}$ factor cannot exceed 2, while the terms that involve m^* tend to cancel.

(*b*) However strong the scattering by the ions, an eigenfunction which is above the muffin-tin zero must surely behave like a coherent wave in the interstices between ions, i.e. within a large number of cells of order R_A in radius. The most the scattering can do is to remove all coherence between adjacent cells, in which case we may expect

$$\overline{\langle\Psi(\mathbf{r}+\mathbf{X})|\Psi(\mathbf{r})\rangle} \gtrsim \frac{\sin k_0 X}{k_0 X}\left(1-\frac{3}{4}\frac{X}{R_A}\right)$$

for small X. Taken together with equation (4.88), this suggests that l can never be less than $2R_A/3$.

(*c*) Hence we probably require

$$L/R_A \lesssim 1. \tag{5.34}$$

If this condition is not satisfied, it may mean that three-body effects are important or that $\hbar/2\tau K_F$ is no longer small compared with unity (either of these could invalidate (5.14)), or else that the Fermi level has fallen below the muffin-tin zero, or that some of the electrons are trapped in localised states; in any event the Ziman formula can no longer be applied. Some contours of constant L/R_A are plotted in fig. 5.9, from which it may be seen that when the resistivity exceeds about 1000 $\mu\Omega$ cm we have probably passed the limit.

Let us now turn from Cs to some other metals. Rather similar data have been reported, though they cover a much smaller range of density, for K (see Grosse, 1966), and Cu (Ben-Yosef & Rubin, 1969); in the latter case an ingenious technique involving the explosion of Cu wires was used. Of more interest, however, is the

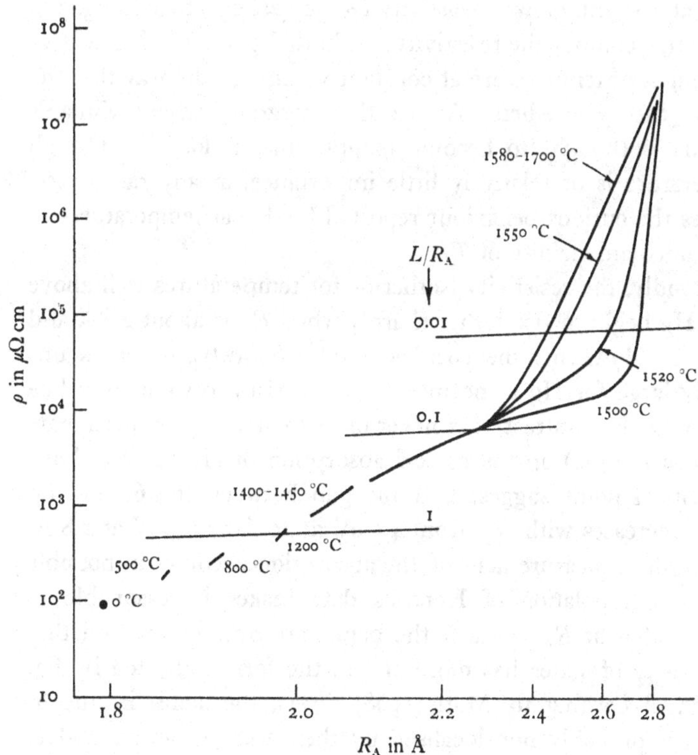

Fig. 5.11. The resistivity of liquid Hg at low densities. Experimental points have been omitted, but they lie closely on the smooth curves drawn. These correspond to various temperatures of measurement, as indicated. The abscissa scale, though labelled in terms of R_A, is linear in R_A^3. (After Hensel & Franck, 1968.)

behaviour of divalent Hg, whose density, resistivity and thermoelectric power have been studied at high temperatures and pressures by several teams of investigators (Kikoin *et al.*, 1966, 1967; Franck & Hensel, 1966; Hensel & Franck, 1968; Schmutzler & Hensel, 1968, 1971; Postill, Ross & Cusack, 1967, 1968; Crisp & Cusack, 1969). Fig. 5.11 shows some of Hensel & Franck's results for ρ plotted this time against R_A^3 rather than R_A. The range of variation of R_A is less than for Cs, but the data extend into the interesting region near the critical point, which lies at about $T = 1490$ °C, $R_A = 2.6$ Å. A number of features call for comment.

First, the contour for $L/R_A = 1$ intersects the experimental curve at a slightly lower resistivity ($\sim$ 500 $\mu\Omega$ cm) than in fig. 5.9. Below this contour the resistivity, for both Hg and Cs, is sensitive to changes of temperature at constant volume in the way that the Ziman theory predicts. Above this contour, where Ziman's formula is thought to become inapplicable, it looks as though temperature is of relatively little importance, at any rate if one ignores the curious behaviour reported for Hg at temperatures in the immediate vicinity of T_C.

Secondly, the resistivity isotherms for temperatures well above T_C in Hg begin to rise rather sharply when R_A is about 2.4 Å and ρ about 10^4 $\mu\Omega$ cm: the coefficient $(\Omega/\rho)(\partial\rho/\partial\Omega)_T$ which is unusually large for Hg at normal densities, rises to something like 16; for Cs, by contrast, it is never more than 2 or 3. Experiments by Hensel (1970) on the optical absorption of Hg vapour above the critical point suggest that the gap between the 6s and 6p bands decreases with R_A, from 4.9 eV at 20 Å to 1.3 eV at 2.8 Å. At this stage measurement of the absorption becomes impossible but an extrapolation of Hensel's data makes it reasonable to suppose that at $R_A = 2.4$ Å the bands are overlapping and that the density of states has more or less the form indicated by fig. 5.10(*e*). According to Mott (1968, 1971), the states in the 6s band are probably non-localised by then and can accommodate two electrons per atom, so that the Fermi level (Hg being divalent) must lie more or less in the middle of the pseudo-gap. Here, again according to Mott (1969), localisation is more probable and the conductivity is naturally sensitive to whether it occurs or not. Mott attributes to the onset of localisation the 'knee' in the resistivity isotherms at 2.4 Å.†

Thirdly, when R_A is increased beyond 2.4 Å the resistivity of Hg rises to 10^8 $\mu\Omega$ cm and beyond, without any sign of the

† Mott attempts to deduce a numerical value for the density of states in the middle of the pseudo-gap at the moment when localisation sets in, using a formula equivalent to

$$L = l(m^*/m)^2$$

(though he prefers to write the density of states ratio as g rather than m^*/m) together with the reasonable assumption that l is not less than R_A. His formula for L is a simplified version of (5.14), without the $(1-\overline{\cos\theta})$ factor and without the normalisation corrections. It is clear that (5.14) is no longer to be trusted when L/R_A is 0.1 or less, but the reliability of Mott's simpler alternative has not yet been firmly established.

flattening off that occurs for Cs vapour at about $10^6\ \mu\Omega$ cm. This difference is clearly due to the much higher ionisation energy of the free Hg atom – 10.4 instead of 3.89 eV.

Finally, we should note how the resistivity isotherms for temperatures close to T_C are depressed near the value of R_A that corresponds to the critical density. The effect may be a consequence of the large density fluctuations which are to be expected near the critical point, but it is not worth trying to explain it in detail until it has been confirmed by further experiments. Measurements close to the critical point are necessarily difficult.

5.11. A different approach to the calculation of conductivity

The justification for Ziman's formula outlined in §5.2 was a traditional one, based upon the Boltzmann equation, which is appropriate if the electrons can be described by a set of travelling wave states with a fairly sharp Fermi surface in **k**-space dividing the occupied ones from the unoccupied ones. Alternative methods for calculating the conductivity of a metal have been suggested by a number of authors, and the more general formula obtained by Greenwood (1958) and others will be of use to us here.

The formula applies to any set of one-electron eigenfunctions $\Psi_1, \Psi_2, \ldots, \Psi_n, \ldots$, which need not be capable individually of carrying any current. When an electric field is switched on it acts as a perturbation exciting transitions between the eigenstates and it is as a result of these transitions that a current flows. Its density may be calculated from the standard expression for a single electron,

$$\mathbf{j} = \frac{e\hbar}{2m\Omega i}\int (\Psi^*\nabla\Psi - \Psi\nabla\Psi^*)\, d\mathbf{r} - \frac{e^2}{m\Omega c}\int \mathbf{A}\Psi\Psi^*\, d\mathbf{r}, \quad (5.35)$$

where the integrals are taken over the whole volume of the specimen, which we shall assume to be in the form of a long wire of uniform cross-section lying along the x direction.

The electric field **E** may be expressed by a perturbing potential in the Hamiltonian of $-e\mathrm{E}x$, in which case the transition probabilities involve the matrix elements

$$X_{nm} = \langle\Psi_n|x|\Psi_m\rangle.$$

An alternative procedure is to suppose that the wire forms a continuous loop and that the field is generated by increasing the magnetic flux within it. In this case the perturbation is

$$\frac{ie\hbar}{mc}\mathsf{A}\frac{\partial}{\partial x},$$

where A is a uniform vector potential directed along the wire such that

$$\mathsf{E} = -\dot{\mathsf{A}}/c, \tag{5.36}$$

and the transition probabilities involve

$$X'_{nm} = \left\langle \Psi_n \left| \frac{\partial}{\partial x} \right| \Psi_m \right\rangle.$$

Certain rather trivial difficulties may arise with this second method, out of the fact that an electron moving round a continuous loop of wire is moving in a potential which is strictly speaking periodic and the first is to be preferred. For calculational purposes the matrix elements X_{nm}, which look as though they diverge for large x, may be converted to X'_{nm} using the theorem that

$$X_{nm} = -\frac{\hbar^2}{m(E_n - E_m)} X'_{nm}, \tag{5.37}$$

where E_n, E_m are the eigenvalues of energy for the two states. This theorem can readily be proved provided that the boundary conditions are *not* periodic, but are such that Ψ_n and Ψ_m vanish at the two ends of the specimen.

There is also more than one way of switching the field on. If it is switched on suddenly and kept constant thereafter, it contains high-frequency Fourier components which generate terms in the current of no physical significance. These may be avoided by switching it on adiabatically, i.e. by allowing it to increase like $\exp(\gamma t)$ and letting γ tend to zero. One may show by first-order time-dependent perturbation theory that an electron which was in the state Ψ_0 at $t = -\infty$ is represented at time t by a wave function

$$\Psi = \sum_n C_n \Psi_n, \tag{5.38}$$

with

$$C_n = -\frac{e\mathsf{E}}{i\hbar}\frac{X_{n0}}{\gamma + i\delta_n}, \tag{5.39}$$

where $$\hbar\delta_n = E_n - E_0. \tag{5.40}$$

From (5.35), the current density carried by this electron is

$$j_x = \frac{2e\hbar}{im\Omega}\sum_n \mathrm{Im}\,(C_n^* X'_{n0}),$$

which with the aid of (5.39) and (5.37) becomes

$$j_x = \frac{2e^2\hbar \mathsf{E}}{m^2\Omega}\sum_n \frac{\gamma}{\delta_n(\gamma^2+\delta_n^2)}|X'_{n0}|^2. \tag{5.41}$$

To obtain the bulk conductivity we need to sum the effects of all the electrons in the metal. Evidently, the contribution to the current caused by exciting an electron initially in the state Ψ_0 up to Ψ_n is liable to be cancelled by a contribution due to the reverse transition, since (5.41) is an odd function of δ_n. The average current due to both processes together may be obtained by including a factor

$$(f_0 - f_n) = -\hbar\delta_n(\mathrm{d}f/\mathrm{d}E),$$

in (5.41), where f is the Fermi–Dirac distribution function. After multiplication by the density of states as well, (5.41) may be summed over n and integrated over E to obtain the result

$$\sigma = \frac{2\pi e^2\hbar^3}{m^2\Omega}\mathcal{N}(E)_\mathrm{F}^2\overline{|X'(0)|_\mathrm{F}^2}, \tag{5.42}$$

where $\overline{|X'(0)|_\mathrm{F}^2}$ is the mean square matrix element between eigenstates which are very close† to one another in energy, in the immediate vicinity of the Fermi level. A factor 2 has been introduced in (5.42) to allow for the existence of two assemblies of electrons with opposite spins; $\mathcal{N}(E)$ is the density of states for one spin orientation only.

The calculation of σ now becomes a matter of evaluating $\overline{|X'(0)|_\mathrm{F}^2}$. We have seen already in §4.10 how this quantity is bound to vanish if the eigenstates are localised. The analysis that follows applies only to non-localised states.

† This means within an energy $\hbar\gamma$. In passing to the limit $\gamma = 0$ one may let the length of the specimen go to infinity at such a rate that the number of states within this energy range is always large.

5.12. Evaluation of the formula

For a liquid metal, in which there is no long-range order and the scattering potential due to the ions is on the average isotropic, it is convenient to represent the eigenfunctions by wave groups, as suggested in §4.7. Let us initially ignore the distinction between real and pseudo wave functions and write

$$\Psi_n = \Omega^{-\frac{1}{2}} \exp(-iE_n t/\hbar) \sum_{\mathbf{k}} n_{\mathbf{k}} \exp(i\mathbf{k}\cdot\mathbf{r}), \tag{5.43}$$

where the coefficients $n_{\mathbf{k}}$ are equivalent to the $c_{\mathbf{k}}$ of (4.51); a similar expression with coefficients $m_{\mathbf{k}}$ may be used to describe Ψ_m. Then

$$|X'_{nm}|^2 = \sum_{\mathbf{k}} k^2 \cos^2 \alpha_{\mathbf{k}}\, n_{\mathbf{k}} n_{\mathbf{k}}^* m_{\mathbf{k}} m_{\mathbf{k}}^* + \sum_{\mathbf{k}} \sum_{\mathbf{k}' \neq \mathbf{k}} k \cos \alpha_{\mathbf{k}} n_{\mathbf{k}} m_{\mathbf{k}}^* k' \cos \alpha_{\mathbf{k}'} n_{\mathbf{k}'}^* m_{\mathbf{k}'}, \tag{5.44}$$

where $\alpha_{\mathbf{k}}$ is the angle between $\mathbf{k}$ and the x direction.

To evaluate the ensemble average of the first term on the right-hand side of (5.44) is straightforward, if (4.51), or its three-dimensional equivalent, is accepted. The two eigenfunctions under consideration must of course be orthogonal, which requires that

$$\sum_{\mathbf{k}} n_{\mathbf{k}} m_{\mathbf{k}}^* = 0,$$

and presumably there are some rather subtle correlations of phase between $n_{\mathbf{k}}$ and $m_{\mathbf{k}}$ to ensure that this condition holds. There is no reason, however, to suspect that in so far as $n_{\mathbf{k}} n_{\mathbf{k}}^*$ and $m_{\mathbf{k}} m_{\mathbf{k}}^*$ fluctuate about the ensemble average which (4.51) describes these fluctuations are correlated, and hence it is legitimate to replace the average of their product by the product of their averages. Thus

$$\begin{aligned}\overline{\sum_{\mathbf{k}} k^2 \cos^2 \alpha_{\mathbf{k}}\, n_{\mathbf{k}} n_{\mathbf{k}}^* m_{\mathbf{k}} m_{\mathbf{k}}^*} &= \sum_{\mathbf{k}} \tfrac{1}{3} k^2 B^2 \frac{\Gamma_k^2}{((E_k - E - A_k)^2 + \Gamma_k^2)^2} \\ &= \tfrac{1}{3} B^2 \int_{-(E+A)/\Gamma}^{\infty} \mathcal{N}(E_k) \frac{k^2}{\Gamma_k^2} \left(\frac{\partial E_k}{\partial z}\right)_E \frac{dz}{(1+z^2)^2},\end{aligned} \tag{5.45}$$

where z is defined by (4.53). If Γ is small it is legitimate to regard $\mathcal{N}(E_k)(k^2/\Gamma_k^2)(\partial E_k/\partial z)_E$ as constant over the region where the integrand is significant and to replace the lower limit by $-\infty$ (cp. the treatment of (4.54)). Since we are interested in the matrix element between two states near the Fermi level we should replace k by k_F, Γ_k by Γ_F and so on. Then with the aid of (4.55) and (4.57) for the normalising constant B one may obtain the result

$$\overline{\sum_{\mathbf{k}} k^2 \cos^2 \alpha_{\mathbf{k}}\, n_{\mathbf{k}} n_{\mathbf{k}}^* m_{\mathbf{k}} m_{\mathbf{k}}^*} = \frac{k_F^2(1+\partial A/\partial E)_0}{6\pi \mathcal{N}(E)_F \Gamma_F}. \qquad (5.46)$$

If this were the only term in the mean square matrix element we could substitute it straight away into (5.42) and obtain as a first approximation to the conductivity

$$\sigma = \frac{ne^2}{m}\left(\frac{\hbar}{2\Gamma_F}\right)\left(\frac{m^*}{m}\right)\left(1+\frac{\partial A}{\partial E}\right)_0$$
$$= \frac{ne^2\tau_l}{m}\left(\frac{m^*}{m}\right)\left(1+\frac{\partial A}{\partial E}\right)_0^2, \qquad (5.47)$$

where τ_l is the lifetime of a plane-wave state at the Fermi surface as given by (3.76) and (4.81). However, the second term in (5.44) – a double sum over $\mathbf{k}$ and $\mathbf{k}'$ – is not really negligible. An elementary method for calculating it has been given by Faber (1966), based upon an argument due to Edwards (1958), but further examination has shown this method to be incorrect and a better one is outlined in an appendix to this chapter. It turns out that the result of including the second term is to multiply the answer by a factor

$$1+\overline{\cos\theta}+(\overline{\cos\theta})^2+\ldots = (1-\overline{\cos\theta})^{-1} = \tau/\tau_l, \qquad (5.48)$$

where $\overline{\cos\theta}$ as in (3.79), is the mean value of $\cos\theta$ for a scattering process which connects two plane-wave states at the Fermi surface and τ is the relaxation time of (3.80).

Since both τ_l and τ are inversely proportional to m^*, the effective mass does not affect the conductivity at this level of approximation; equation (4.60) suggests, it is true, that m^* includes the correction factor $(1+\partial A/\partial E)$ which also appears in (5.47), but this is probably close to unity in most circumstances. However, we have still to recognise the distinction between real and pseudo wave functions, which was left out of account in much of chapter 4. We should write

$$\Phi_n = \Omega^{-\frac{1}{2}} \exp(-iE_n t/\hbar) \sum_{\mathbf{k}} C_{\mathbf{k}}^{\frac{1}{2}} n_{\mathbf{k}} \exp(i\mathbf{k}\cdot\mathbf{r})$$

instead of (5.43), where C is a normalisation constant first introduced into (1.63), not to be confused with the C_n of (5.38) and (5.39). It is easy enough to carry this constant C through the argument of §4.7 and so long as it can be treated as independent of k it makes no essential difference to the conclusions reached there.† When we come to evaluate $|X'_{nm}|^2$, however, the correction factor expressed by (5.11) must be included. Our expression for the conductivity therefore becomes

$$\sigma = ne^2\tau(m^*/m_1^{*2})(1+\partial A/\partial E)_0^2. \tag{5.49}$$

Apart from the $(1+\partial A/\partial E)^2$ factor, this justifies the result already quoted as (5.12) above. In terms of the coherence length l we have, from (4.75),

$$\sigma = \frac{ne^2 l}{\hbar k_F}\left(\frac{1}{1-\overline{\cos\theta}}\right)\left(\frac{m^*}{m_1^*}\right)^2, \tag{5.50}$$

which is precisely equivalent to (5.13) and (5.14).

5.13. Corrections when the mean free path is short

There are two quite distinct ways in which the theory developed above may go wrong if the scattering becomes too strong. In the first place, the Born approximation may break down. The only correction that seems to be required in this case lies in the value to be used for Γ and has been fully discussed in §4.11; it is possible that the argument used in evaluating the cross products in (5.44) may also need some correction, but for the majority of liquid metals $\overline{\cos\theta}$ is probably quite small compared with unity, in which case errors here should be of little significance to the final answer for σ. In the second place, the assumption made at several stages in the argument that Γ is small compared with K_F may become invalid; the Fermi surface may become significantly blurred. If so, then much greater care is needed in the various integrals that determine Γ, B, $|X'_{nm}|^2$ and so on. So far as Γ is concerned we have discussed on pp. 264–6 a suggestion due to

† In principle, as is clear from (1.64), C_k could depend upon k, which might cause some distortion of the 'profile' of the pseudo-eigenfunctions in **k**-space. But since it is believed that C differs from unity, for k-values in the neighbourhood of k_F, by no more than a few parts per cent in most metals, it is unlikely that this distortion need be allowed for in practice.

Edwards, that one may obtain the right answer by suitably blurring the interference function rather than the Fermi surface, but this is only part of the story.

The worst complications disappear for the simple case of a 'gas' of δ-function potential wells, such that $U(q)$ is independent of q and also, incidentally, of the wave number of the electron on which it acts. Then the parameter β defined by (4.82) vanishes, so that one does not need to worry about the variation of Γ_k with E_k. Since A_k is also constant, the eigenfunctions have a true Lorentzian profile in **k**-space. The density of states is unaffected by the perturbation. The quantity $\overline{\cos\theta}$ vanishes, so that the corrections which might otherwise be needed in the evaluation of the cross-products in (5.44) are irrelevant. All that is required is allowance for the variation of $\mathcal{N}(E_k)$ with E_k in the integrals and appreciation that the lower limit of $-(E+A)/\Gamma$ can no longer be replaced by $-\infty$. The calculation may be carried through without any approximations and the gratifying result is that just the same answer emerges as before.† Thus in this particular case, within the limitations of the Born approximation, *Ziman's formula appears to be exact, however short the mean free path and however blurred the Fermi surface.*

It appears to be an accident that things work out in just this way, and there is little doubt that for real liquid metals, in which $U(q)$ is not independent of q, the formula for σ should include correction terms of order $(1/lk_F)^2$ or $(\Gamma_F/K_F)^2$. To see whether they are likely to be significant in practice, the analysis has been worked through on the assumption that Γ is proportional to $E_k^{\frac{1}{2}\beta}$ and at the same time to $E^{\frac{1}{2}(1+\beta)}$ (see p. 264). The variation of A has been ignored, and also the cross-terms in the matrix element, but these simplifications are not likely to vitiate the answer. It may be written in the form

$$\sigma = \sigma_0\left(1+\left(\frac{\Gamma_F}{K_F}\right)^2\left(\tfrac{19}{8}\beta^2-\beta\right)+O\left(\frac{\Gamma_F}{K_F}\right)^4\right), \qquad (5.51)$$

where σ_0 is the uncorrected conductivity given by (5.50). Now

† The correction factor quoted by Faber (1966, expression (83)), which incidentally is wrongly printed, turns out on closer examination to reduce to unity.

according to (5.25) there should be a link between β and the parameter ξ derived from the thermo-electric power; in fact

$$\beta = 2\left(1 - \frac{m^*}{m}\xi\right). \tag{5.52}$$

It follows that when $0.79 < \xi < 1.0$ the correction terms should be negative. For the majority of liquid metals ξ lies outside this range, however, and a modest enhancement of σ is to be expected.

Consider liquid Li, for which $m^*\xi/m$ is exceptionally large in magnitude, at -9.3 (see table 5.3). If Γ_F/K_F is $(0.022/1.3)$ (see table 1.1 and remarks on p. 322), then the formula suggests $\sigma/\sigma_0 \simeq 1.3$. The result for Hg is rather similar, since $m^*\xi/m$ is $+4.1$ while Γ_F/K_F lies between 0.05 and 0.1, depending on the magnitude of $\overline{\cos\theta}$. These predictions must not be taken too literally, for the thermo-electric power tells us only the first derivative of Γ at $E = E_F$, and there is no guarantee that the power law which describes this is adequate over the whole range of E_k which is relevant to the calculation of σ. But clearly one ought to bother about the corrections, in both Li and Hg. These are two exceptional cases, however. For most liquid metals σ should not differ from σ_0 by more than a few parts per cent.

The argument in §5.6 relating the thermo-electric power to $\partial\sigma/\partial E$ is valid whatever method we adopt in calculating σ, and we should still be able to trust equation (5.26), even when the mean free path is short. Thus if the blurring of the Fermi surface has a negligible effect upon σ, its effect upon P should also be negligible.

5.14. Conduction at high frequencies: the Drude equations

A primitive method for calculating the response of a metal to an oscillating electric field starts from the differential equation for current density

$$\frac{d\mathbf{j}}{dt} = -\frac{\mathbf{j}}{\tau} + \frac{\sigma(0)}{\tau}\mathbf{E}; \tag{5.53}$$

the first term on the right expresses the attenuation of the current by scattering, and the fact that it involves τ rather than τ_l may be justified by summing a set of equations as in (3.79); the second term is deliberately adjusted so that $\mathbf{j}$ equals $\sigma(0)\mathbf{E}$ in the steady-state, $\sigma(0)$ being the DC conductivity as above. If $\mathbf{E}$ is supposed to oscillate with angular frequency ω the current is liable to lag behind it in phase. We shall adopt the convention here that the in-phase component is described by a (real) conductivity $\sigma(\omega)$, while the in-quadrature component is regarded as part of the displacement current and hence associated with a (real) term in the dielectric constant of the metal $\epsilon(\omega)$. Thus we shall write

$$\mathbf{j} = \sigma(\omega)\mathbf{E} + \frac{\epsilon(\omega)-1}{4\pi}\dot{\mathbf{E}}. \tag{5.54}$$

The alternative conventions are to write

$$\mathbf{j} = \tilde{\sigma}(\omega)\mathbf{E} \quad \text{or} \quad \mathbf{j} = \frac{\tilde{\epsilon}(\omega)-1}{4\pi}\dot{\mathbf{E}},$$

where $\tilde{\sigma}$ and $\tilde{\epsilon}$ are complex, but these are liable to create confusion when quantum-mechanical calculations are attempted. The quantity ϵ, by the way, is closely related to the dielectric constants which were used to describe the phenomenon of screening in chapter 1. The relation is explored in §5.16 below.

Substitution of (5.54) into (5.53) leads immediately to

$$\left.\begin{aligned} \sigma(\omega) &= \frac{\sigma(0)}{1+\omega^2\tau^2}, \\ \epsilon(\omega)-1 &= -\frac{4\pi\sigma(0)\tau}{1+\omega^2\tau^2}. \end{aligned}\right\} \tag{5.55}$$

These are the famous *Drude equations* for a free-electron metal. Let us now see whether they can be verified using quantum mechanics, by an extension of the argument in §5.11.

We may suppose that at time $t = 0$ a field is switched on in the x direction which subsequently oscillates with a steady amplitude according to the equation

$$\mathbf{E} = \mathbf{E}_0 \cos \omega t = \tfrac{1}{2}\mathbf{E}_0 (\exp(\mathrm{i}\omega t) + \exp(-\mathrm{i}\omega t)).$$

In practice this field might be the result of exposing the metal to electromagnetic radiation, in which case $\mathbf{E}$ would oscillate in space as well as time. It turns out, however, (see §5.15) that the wavelength is normally long enough in liquid metals to justify the simplifying assumption that E is uniform. Hence instead of (5.39) we now have

$$C_n = \frac{e\mathsf{E}_0}{2\hbar} X_{n0} \times \left[\frac{\exp(\mathrm{i}\omega t) - \exp(-\mathrm{i}\delta_n t)}{\omega + \delta_n} + \frac{\exp(-\mathrm{i}\omega t) - \exp(-\mathrm{i}\delta_n t)}{-\omega + \delta_n}\right],$$

and the current density is

$$j_x = \frac{e^2\hbar}{m^2\Omega} \mathsf{E}_0 \sum_n \frac{|X'_{n0}|^2}{\delta_n} \times \left[\sin\omega t \frac{2\omega}{\omega^2 - \delta^2} + \sin\delta_n t \left(\frac{1}{\omega + \delta_n} + \frac{1}{-\omega + \delta_n}\right)\right]. \quad (5.56)$$

On the right there occurs a term $\sin\delta_n t$ which may be expanded as

$$\sin(\pm\omega + \delta_n)t\cos\omega t \mp \cos(\pm\omega + \delta_n)t\sin\omega t.$$

Only the first half of this need be retained, since the second half generates a component in j_x proportional to

$$\sum_n \cos(\pm\omega + \delta_n)t/(\pm\omega + \delta_n)$$

which becomes insignificant in the limit of large t. Hence the $\sin\delta_n t$ term in (5.56) generates a current which varies like $\cos\omega t$ and is therefore in phase with the field. It is upon this term that we need to concentrate to obtain the conductivity and it yields the result

$$\sigma(\omega) = \frac{\pi e^2\hbar}{m^2\Omega\omega}[\mathcal{N}(\omega)\overline{|X'(\omega)|^2} - \mathcal{N}(-\omega)\overline{|X'(-\omega)|^2}]. \quad (5.57)$$

Here $\mathcal{N}(\omega)$ is the density of states per unit frequency range available for excitation around $\delta_n = \omega$, which appears when the sum in (5.56) is converted into an integral; similarly, $\overline{|X'(\omega)|^2}$ is the mean square matrix element for $\delta_n = \omega$. Note how transitions into higher levels, for which δ_n is positive, tend to make σ positive, while transitions downwards tend to make it negative. Since it is σ which determines the rate at which the electron absorbs power from the field, this is entirely as expected.

The component of the current in quadrature with field comes from the $\sin\omega t$ term in (5.56). From (5.56) and (5.57) it may be seen that the effective dielectric constant is related to σ by the equation

$$\epsilon(\omega)-1 = -\frac{8\pi e^2\hbar}{m^2\Omega}\int_{-\infty}^{\infty}\frac{\overline{|X'(\delta)|^2}\,\mathcal{N}(\delta)}{\delta}\frac{d\delta}{\omega^2-\delta^2}$$

$$= 8\int_0^{\infty}\frac{\sigma(\delta)}{\delta^2-\omega^2}\,d\delta. \qquad (5.58)$$

This constitutes one of the famous *Kramers–Kronig relations*, which can also be proved by more general arguments (see (5.85) below).

The above results are for a single electron. When all the electrons in the specimen are taken into account, including both spin orientations, (5.58) remains valid but (5.57) becomes

$$\sigma(\omega) = \frac{2\pi e^2\hbar^2}{m^2\Omega\omega}\int_0^{\infty}\{f(E)-f(E+\hbar\omega)\}\mathcal{N}(E)\mathcal{N}(E+\hbar\omega)\overline{|X'(\omega)|^2}\,dE, \qquad (5.59)$$

where $f(E)$ is the Fermi–Dirac distribution function; $\overline{|X'(\omega)|^2}$ in this expression is liable to be a function of E as well as ω. It is evident that (5.59) reduces to our previous formula (5.42) for the DC conductivity in the limit when ω goes to zero.

The matrix element in (5.59) links two eigenfunctions which we envisage, in a liquid metal, as wave groups with a finite spread in **k**-space, and its magnitude depends upon the extent to which they overlap. As the frequency is increased these eigenfunctions move apart in energy and the overlap diminishes. It is on this account that $\sigma(\omega)$ falls. Since the spread of each wave group in terms of E_k is of order

$$\Gamma/(1-\partial A/\partial E_k)_0,$$

one may expect a significant reduction by the time that

$$\hbar\omega \sim \Gamma_F(\partial E/\partial E_k)_0/(1-\partial A/\partial E_k)_0 = \Gamma_F/(1+\partial A/\partial E)_0,$$

and this prediction is borne out by detailed evaluation of the formula.

So long as $\overline{\cos\theta}$ is negligible and the cross-products in the matrix element (i.e. the second term in (5.44)) may be ignored, the evaluation is straightforward. To begin with, suppose that not only Γ but also $\hbar\omega$ is small compared with K_F, which allows us, for example, to ignore the difference of Γ between the two eigenfunctions. Then we find that whereas at zero frequency the square of the matrix element involves (see (5.45)) the integral

$$\int_{-\infty}^{\infty} \frac{dz}{(1+z^2)^2} = \tfrac{1}{2}\pi,$$

it now involves

$$\int_{-\infty}^{\infty} \left\{1+\left[z-\frac{\hbar\omega(1+\partial A/\partial E)_0}{2\Gamma}\right]^2\right\}^{-1} \times \left\{1+\left[z+\frac{\hbar\omega(1+\partial A/\partial E)_0}{2\Gamma}\right]^2\right\}^{-1} dz.$$

When ω is zero this integral is just $\tfrac{1}{2}\pi$, but for finite frequencies it is reduced by a factor

$$\{1+[\hbar\omega(1+\partial A/\partial E)_0/2\Gamma]^2\}^{-1} = \{1+(\omega\tau_l)^2(1+\partial A/\partial E)_0^4\}^{-1}. \quad (5.60)$$

This factor should be attached to (5.47) to obtain an approximate formula for $\sigma(\omega)$.

When $\overline{\cos\theta}$ is not negligible the cross-products in the matrix element present a problem of some complexity which is relegated to an appendix below. It turns out that, provided the Born approximation is valid, the factor multiplying (5.47) should be

$$\frac{1-\overline{\cos\theta}}{(1-\overline{\cos\theta})^2+(\omega\tau_l)^2(1+\partial A/\partial E)_0^4} = \frac{\tau/\tau_l}{1+(\omega\tau)^2(1+\partial A/\partial E)_0^4} \quad (5.61)$$

instead of (5.60). The full answer for $\sigma(\omega)$ can be expressed most neatly in terms of yet another relaxation time, which we may call the Drude time τ_D. This is defined by

$$\tau_D = \tau(1+\partial A/\partial E)_0^2, \quad (5.62)$$

which enables us to write the answer in the Drude form,

$$\sigma(\omega) = \frac{n^* e^2\tau_D/m}{1+(\omega\tau_D)^2}, \quad (5.63)$$

with an effective electron density given by

$$n^*/n = m^* m/m_1^{*2}. \quad (5.64)$$

Complex as the above derivation may appear, it contains a number of approximations which deserve further scrutiny. To simplify the task a little, the distinction between τ_D and τ will for the moment be set aside.

5.15. The anomalous skin effect

Throughout the previous section the electric field was taken to be uniform in space. The calculation may be extended without difficulty to cover the case when **E** varies sinusoidally with wave vector **q**, as long as q is small compared with k_F. Additional factors of $\exp(\pm i\mathbf{q}\cdot\mathbf{r})$ appear in the relevant matrix elements, so that one must use

$$|X'_{nm}|^2 = \sum_{\mathbf{k}} k^2 \cos^2\alpha_{\mathbf{k}}\, n_{\mathbf{k}-\frac{1}{2}\mathbf{q}} n^*_{\mathbf{k}-\frac{1}{2}\mathbf{q}} m_{\mathbf{k}+\frac{1}{2}\mathbf{q}} m^*_{\mathbf{k}+\frac{1}{2}\mathbf{q}} + \text{cross-terms} \qquad (5.65)$$

instead of (5.44). The magnitude of the leading term now depends upon the integral

$$\int_{-\infty}^{\infty} [1+(z-\omega\tau_l - ql\cos\chi)^2]^{-1}[1+(z+\omega\tau_l+ql\cos\chi)^2]^{-1}\,dz = \tfrac{1}{2}\pi[1+(\omega\tau_l+ql\cos\chi)^2]^{-1},$$

where χ is the angle between **k** and **q**. If the disturbance is a *transverse* one such that **E** and **q** are at right angles to one another, which is of course the case for an ordinary electromagnetic wave, then (see fig. 5.12),

$$\cos\alpha_{\mathbf{k}} = \sin\chi\cos\phi.$$

Hence in evaluating the leading term in (5.65) we need to average

$$\frac{\cos^2\phi\sin^2\chi}{1+(\omega\tau_l+ql\cos\chi)^2} = \text{Re}\,\frac{\cos^2\phi(1-\cos^2\chi)}{(1-i\omega\tau_l)-iql\cos\chi}$$

instead of $\cos^2\alpha_{\mathbf{k}}$. A straightforward integration over ϕ and χ shows that

$$\frac{\sigma_\perp(q,\omega)}{\sigma(0,0)} = \text{Re}\left(\frac{1}{1-i\omega\tau_l} f_\perp\left(\frac{ql}{1-i\omega\tau_l}\right)\right)$$

where

$$f_\perp(y) = \frac{3}{2y^3}((1+y^2)\tan^{-1}y - y). \qquad (5.66)$$

The reason for expressing the answer as the real part of a function of the complex variable $(1-i\omega\tau_l)$ is that it makes it easier to establish the associated dielectric constant, by means of the Kramers–Kronig relation. The reader may satisfy himself that if the dielectric constant is incorporated in a complex conductivity $\tilde{\sigma}_\perp$, then

$$\frac{\tilde{\sigma}_\perp(q,\omega)}{\sigma(0,0)} = \frac{1}{1-i\omega\tau_l} f_\perp\left(\frac{ql}{1-i\omega\tau_l}\right). \qquad (5.67)$$

Had the disturbance been *longitudinal*, in which case χ and $\alpha_{\mathbf{k}}$ would have been the same, a similar analysis would have led to the result

$$\frac{\tilde{\sigma}_{\parallel}(q,\omega)}{\sigma(0,0)} = \frac{1}{1-i\omega\tau_l} f_{\parallel}\left(\frac{ql}{1-i\omega\tau_l}\right)$$

with

$$f_{\parallel}(y) = \frac{3}{y^3}(y-\tan^{-1}y). \tag{5.68}$$

The two suffices $\perp$ and $\parallel$ should be self-explanatory. Note that in the limit $q \to 0$ the distinction between $\tilde{\sigma}_{\perp}$ and $\tilde{\sigma}_{\parallel}$ disappears so that neither suffix need be attached to the DC conductivity $\sigma(0, 0)$.

Equation (5.66) was first obtained by Reuter & Sondheimer (1948) using semi-classical kinetic theory arguments. In their treatment the mean free path $L = l/(1-\overline{\cos\theta})$ appears in place of l (though only because they take $\overline{\cos\theta}$ to be zero). If we were to include the cross-terms in (5.65) we could probably justify this substitution, along with the substitution of τ_l by τ.

We have arrived now at a conductivity which depends upon wave vector q as well as frequency ω. To find the current due to a field which is non-uniform we need to add the effects of all the different Fourier components which are present, and the nature of the solution depends upon the range over which $|qL/(1-i\omega\tau)|$ spreads. If this quantity is always much less than unity we are in the *classical* regime, for $\tilde{\sigma}(q,\omega)$ can be replaced by $\tilde{\sigma}(0,\omega)$. If it is comparable with unity we are in the *anomalous* regime. And when it is much greater than unity we have reached the *extreme anomalous limit*. We must now demonstrate that for electromagnetic waves propagated through liquid metals the conditions are almost always classical.

Consider first the situation when $\omega\tau \ll 1$. It is well known that the classical solution of Maxwell's equations corresponding to a plane wave of frequency ω takes the form

$$\mathsf{E} = \mathsf{E}_0 \exp(-z/\delta)\cos(z/\delta-\omega t),$$

the amplitude being attenuated by absorption in any conducting medium. Here δ is the *normal skin depth*, given at low frequencies by

$$\delta = c/(2\pi\omega\sigma(0,0))^{\frac{1}{2}}. \tag{5.69}$$

The Fourier components of this attenuated plane wave lie in a range of q about δ^{-1} so that

$$\left|\frac{qL}{1-i\omega\tau}\right| \sim \frac{L}{\delta} = \frac{1}{\sqrt{2}}(\omega\tau)^{\frac{3}{2}}\frac{v_F}{c}\frac{\omega_p}{\omega}, \tag{5.70}$$

where ω_p is the free-electron plasma frequency (see §5.26) defined by the equation

$$\omega_p^2 = 4\pi ne^2/m. \tag{5.71}$$

At a wavelength in the infrared of several microns, where $\omega\tau$ is about 0.1 for a typical liquid metal, ω_p/ω is only about 10^2 while v_F/c is 10^{-2}. It follows that the right-hand side of (5.70) is small enough to be neglected when $\omega\tau$ is small; in this limit the assumption of classical behaviour is a self-consistent one.

As the frequency rises the conductivity begins to fall because of relaxation effects, in the way that the Drude equations describe, and the normal skin depth saturates when $\omega\tau \gg 1$ to

$$\delta_0 = c(\tau/4\pi\sigma(0, 0))^{\frac{1}{2}} = c/\omega_p. \tag{5.72}$$

Then $$\left|\frac{qL}{1-i\omega\tau}\right| \sim \frac{L}{\omega\tau\delta_0} = \frac{v_F}{c}\frac{\omega_p}{\omega}.$$

Since ω_p/ω is now about 10 at most, the condition for the response to be classical is still satisfied. More detailed examination shows that there is no frequency at which, for a typical liquid metal, the q-dependence of $\tilde{\sigma}_\perp(q, \omega)$ should be detectable experimentally.

5.16. Dynamic screening

In the previous section we have had occasion to distinguish between a transverse conductivity $\sigma_\perp$ and a longitudinal one $\sigma_\parallel$, identical in the limit $q \to 0$. To generate a wave in which the electric field is longitudinal, however, it is necessary to introduce into the medium some external charges, with a density ρ which varies in the appropriate sinusoidal fashion. The electrons tend to screen these charges and the response of the whole system is then complicated by space charge effects and by the periodic term which arises in U^{XC} (see §1.4). In a transverse wave the electron density remains uniform and these complications do not arise.

A plausible differential equation to describe the time variation of the polarisation $\mathbf{P}$ in the longitudinal case is

$$\dot{\mathbf{P}}+4\pi\frac{q^2\epsilon'_{\parallel}(q,0)}{q_s^2\epsilon_{\parallel}(q,0)}\mathbf{P} = \tilde{\sigma}_{\parallel}\mathbf{E}.$$

According to this equation the current density is equal to $\tilde{\sigma}_{\parallel}\mathbf{E}$ so long as the density of electrons is uniform, but $\mathbf{P}$ relaxes exponentially towards the value $(q_s^2\epsilon_{\parallel}(q,0)\mathbf{E}/4\pi q^2\epsilon'_{\parallel}(q,0))$ which we know to describe the steady state screening (see (1.30)). It suggests that if $\mathbf{E}$ is oscillatory, with angular frequency ω, then

$$\tilde{\epsilon}_{\parallel}(q,\omega) = 1+4\pi\mathbf{P}/\mathbf{E} = 1+\left(\frac{\epsilon'_{\parallel}(q,0)q^2}{\epsilon_{\parallel}(q,0)q_s^2}-\frac{i\omega}{4\pi\tilde{\sigma}_{\parallel}}\right)^{-1}. \quad (5.73)$$

The reader may satisfy himself that the corresponding expression for

$$\tilde{\epsilon}'_{\parallel}(q,\omega) = \frac{\mathbf{E}}{\mathbf{E}-\operatorname{grad} U^{XC}}\tilde{\epsilon}_{\parallel}(q,\omega)$$

is

$$\tilde{\epsilon}'_{\parallel}(q,\omega) = \frac{\left(1+\dfrac{\epsilon'_{\parallel}(q,0)}{\epsilon_{\parallel}(q,0)}\dfrac{q^2}{q_s^2}-\dfrac{i\omega}{4\pi\tilde{\sigma}_{\parallel}}\right)}{\left(\dfrac{q^2}{q_s^2}-\dfrac{i\omega}{4\pi\tilde{\sigma}_{\parallel}}\right)}. \quad (5.74)$$

He should note that neither of these dielectric constants is related to $\tilde{\sigma}_{\parallel}$ in quite the way that $\tilde{\epsilon}_{\perp}$ is related to $\tilde{\sigma}_{\perp}$ in the transverse case, because of the space charge effects. Nevertheless, in the classical regime, when $q \to 0$ and $\tilde{\sigma}_{\parallel}$ and $\tilde{\sigma}_{\perp}$ become identical, it is clear that for $\omega \neq 0$, $\tilde{\epsilon}_{\parallel}$, $\tilde{\epsilon}'_{\parallel}$ and $\tilde{\epsilon}_{\perp}$ do so too. In this regime no distinguishing suffices are needed. But when $\omega = 0$ equations (5.73) and (5.74) reduce to results already obtained in §1.4.†

We are now equipped to discuss an idea due to Hopfield (1965) which could be relevant to the calculation of the quantity in which we are primarily interested, i.e. the absorption $\sigma(\omega)$. Hopfield set out to obtain an expression for this quantity by perturbation theory, but he reversed the procedure employed in §5.14, for he treated the perturbation due to the electromagnetic wave *before* switching on the bare ionic pseudo-potential. His answer, valid only in the limit $\omega\tau \gg 1$, took the Drude form

$$\sigma(0,\omega) = ne^2/m\omega^2\tau(\omega),$$

† For a more detailed discussion of the form of $\tilde{\epsilon}_{\parallel}(q,\omega)$ and $\tilde{\epsilon}_{\perp}(q,\omega)$ consult Kliewer & Fuchs (1969).

as expected. He reached the conclusion, however, that if $1/\tau(\omega)$ is to be calculated from the Ziman formula one should use a pseudo-potential which is screened *dynamically*, i.e. which differs from the bare pseudo-potential by a factor $(\tilde{\epsilon}'_{\parallel}(q, \omega))^{-1}$ instead of $(\epsilon'_{\parallel}(q, 0))^{-1}$. The relaxation time, according to Hopfield, is therefore frequency-dependent at high frequencies.

How much difference would this make? Let us suppose that we have to deal with a polyvalent metal in which the values of q that contribute most to Ziman's integral lie close to $2k_F$. Then we might set

$$\frac{1}{\tau(\omega)} \simeq \frac{1}{\tau(0)} \frac{|\epsilon'_{\parallel}(2k_F, 0)|^2}{|\tilde{\epsilon}'_{\parallel}(2k_F, \omega)|^2}. \tag{5.75}$$

For frequencies in the near ultraviolet, say, we have

$$\left|\frac{2k_F L}{1 - i\omega\tau}\right| \simeq \frac{2k_F v_F}{\omega} \simeq \sqrt{3}\,\frac{2k_F}{v_F}\frac{\omega_p}{\omega} \sim 20,$$

i.e. we are not far from the extreme anomalous limit where

$$\tilde{\sigma}_{\parallel}(2k_F, \omega) \simeq \sigma(0, 0)\,\frac{3(1 - i\omega\tau)}{(2k_F L)^2} \simeq -\frac{i\omega}{4\pi}\left(\frac{q_s}{2k_F}\right)^2$$

according to (5.68). It follows from (5.74) and (5.75) that

$$\frac{1}{\tau(\omega)} \simeq \frac{4}{\tau(0)}\left[\frac{1 + \dfrac{\epsilon'_{\parallel}(2k_F, 0)}{\epsilon_{\parallel}(2k_F, 0)}\left(\dfrac{2k_F}{q_s}\right)^2}{1 + \left(1 + \dfrac{\epsilon'_{\parallel}(2k_F, 0)}{\epsilon_{\parallel}(2k_F, 0)}\right)\left(\dfrac{2k_F}{q_s}\right)^2}\right]^2. \tag{5.76}$$

We may replace $\epsilon'_{\parallel}(2k_F, 0)/\epsilon_{\parallel}(2k_F, 0)$ by unity, while

$$\left(\frac{2k_F}{q_s}\right)^2 = \frac{1}{r_s}\left(\frac{9\pi^4}{4}\right)^{\frac{1}{3}} \simeq 2.5.$$

Hence it seems that $\quad \dfrac{1}{\tau(\omega)} \simeq 1.4\,\dfrac{1}{\tau(0)}.$

The assumption of dynamic screening appears to lead, therefore, to a considerable enhancement of the predicted absorption.

Such an enhancement would be hard to reconcile with the sum rule to be discussed in §5.19, and it may be said at once that there is no evidence for it in the experimental results. For these reasons a *caveat* must be entered against Hopfield's analysis, which is suspect also on the following physical grounds. A free-electron assembly which is acted on by an oscillating electric field moves to and fro with an amplitude of $e\mathbf{E}/m\omega^2$. If there are fixed ions embedded within it, which are such weak scatterers that they have little effect upon the electrons' motion (this corresponds to the condition $\omega\tau \gg 1$), then to the electrons it is the ions that appear to oscillate. As shown by the argument in §2.16, therefore, each Fourier component in the static pseudo-potential appears to the electrons, before screening, to acquire two satellites at $\mathbf{q}\pm\mathbf{Q}$, where $\mathbf{Q}$ is the small wave vector of the electromagnetic field; their amplitude is

$$\frac{1}{2}\frac{e\mathbf{q}\cdot\mathbf{E}}{m\omega^2}U_{\mathrm{b}}(\mathbf{q}).$$

These satellites oscillate with the frequency ω (see §3.15) and we should certainly expect them to be screened dynamically by the factor $(\tilde{\epsilon}_{\parallel}'(q,\omega))^{-1}$. It is surely the parent Fourier components, however, which are responsible to first order in $\mathbf{E}$ for all the current that is in phase with the driving field and hence for all the absorption. Since these components do not appear to the electrons to oscillate in time at all, our original assumption that the pseudo-potential in Ziman's formula is screened by $(\epsilon_{\parallel}'(q,0))^{-1}$ appears to be correct. It will be retained in what follows, despite the wide-spread acceptance of Hopfield's results.†

5.17. Many-body effects

The theory developed in §§5.11 and 5.14 employs one-electron eigenfunctions, and to extend it to take account of correlation and exchange between electrons in a rigorous fashion would be difficult, for reasons already noted on p. 322. Landau's quasi-particle theory suggests, however, as may readily be seen by comparing equation (5.53) with (5.9), that the Drude equations should be corrected at low frequencies by replacing ω by $\omega(m/m^*_{\mathrm{XC}})$. We can probably retain our Drude-like equation (5.63), therefore, even when correlation and exchange are important, by redefining τ_{D} and n^* as follows:

$$\tau_{\mathrm{D}} = \tau(1+\partial A/\partial E)_0^2(m/m^*_{\mathrm{XC}}), \tag{5.77}$$

$$n^*/n = m^*m^*_{\mathrm{XC}}/(m^*_{\mathrm{I}})^2. \tag{5.78}$$

† It is perhaps worth adding that dynamic screening, if (5.74) is to be trusted, does not remove the $1/q^2$ singularity of the pseudo-potential near $q = 0$. Ziman's integral with dynamic screening diverges at the lower limit, in fact, unless ω is strictly zero – a physically absurd result.

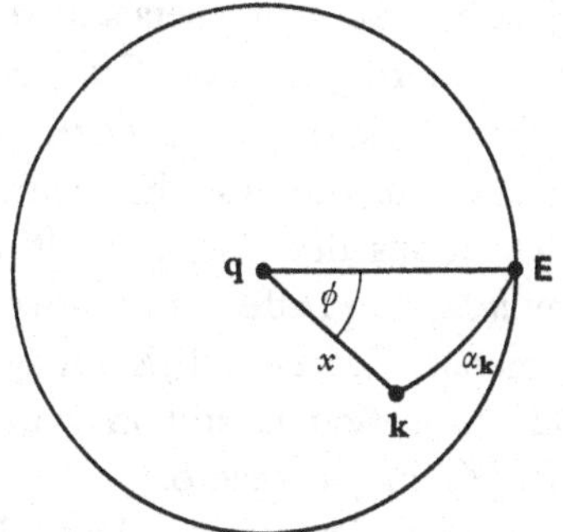

Fig. 5.12. A stereogram to show how ϕ is related to the angles α and χ.

Thus for an ideal jellium, where correlation and exchange constitute the only reasons why m^* differs from m, n^* and n are thought to be identical (Pines & Nozieres, 1966).

Further many-body corrections to our expression for $\sigma_\perp$ are undoubtedly needed at high frequencies, for when $\hbar\omega$ becomes comparable with K_F the states involved in the absorption process are liable to be too far from the Fermi level for the quasi-particle picture to be valid. An electron which has been excited above the Fermi level – and similar remarks apply to hole excitations below it – can decay by processes of the Auger type, i.e. by exciting a second electron. The further the first electron is above the Fermi level the greater the number of decay modes open to it, and hence the shorter its lifetime. When the lifetime due to electron–electron interactions of this sort becomes comparable with or less than the ordinary lifetime due to electron–ion interactions, we ought probably to add to $1/\tau$ a correction term, proportional in the first instance to $(E-E_F)^2$. A number of authors have suggested that on this account we may expect

$$\sigma(\omega) \simeq \frac{ne^2}{m\omega^2}\left(\frac{1}{\tau}+\text{constant}\times\omega^2\right) \tag{5.79}$$

at high frequencies, but few attempts to calculate the constant have yet been reported.

5.18. Residual inter-band absorption

Among the assumptions made in the derivation of equation (5.63) for $\sigma(\omega)$ is one to the effect that both Γ and $\hbar\omega$ are small compared with K_F. This is justified for solid metals at relatively low temperatures where the scattering is weak; Γ/K_F is then usually less than 0.01, while at frequencies such that $\hbar\omega/K_F$ is no longer small the *intra*-band absorption due to the conduction electrons, which the Drude equations describe, is itself very small and is normally masked by *inter*-band effects to be discussed later in this section. In liquid metals, however, Γ/K_F may be 0.1 or more, and the intra-band absorption is still in evidence in the near ultraviolet, where $\hbar\omega/K_F$ may exceed 0.3.

It is a tedious business to improve on the analysis in §5.14, but if one is content with an expansion in powers of Γ/K_F and $\hbar\omega/K_F$ up to second order only and with the sort of approximations that were used in the derivation of (5.51), then the following answers can be obtained. First, the conductivity should fall off with frequency when $\omega\tau \ll 1$ like

$$\sigma(\omega)/\sigma(0) = 1-(\omega\tau)^2\left[1-\left(\frac{\Gamma_F}{K_F}\right)^2(\tfrac{2}{3}\beta^2-\tfrac{11}{6}\beta-\tfrac{1}{4})+\ldots\right]+O(\omega\tau)^4. \tag{5.80}$$

Secondly, if $\omega\tau$ is large enough to justify expansion in powers of $(\omega\tau)^{-1}$ although $(\hbar\omega/K_F)$ is still small, then

$$\begin{aligned}\sigma(\omega)/\sigma(0) &= \left(\frac{\Gamma_F}{K_F}\right)^2[(\tfrac{13}{24}\beta^2-\tfrac{5}{4}\beta-\tfrac{1}{6})+\ldots]\\ &\quad+(\omega\tau)^{-2}\left[1-\left(\frac{\Gamma_F}{K_F}\right)^2(\tfrac{7}{2}\beta^2-\tfrac{3}{2}\beta-\tfrac{1}{2})+\ldots\right]\\ &\quad-(\omega\tau)^{-4}[(\tfrac{1}{2}\beta^2-\beta+1)+\ldots]+O(\omega\tau)^6.\end{aligned} \tag{5.81}$$

Values for β may be estimated from the thermopower using (5.52). The reservations made on p. 360 are still necessary, however, and it would be unwise to expect equations (5.80) and (5.81) to match the experimental data for $\sigma(\omega)$ in any detail. They merely indicate the sort of deviations from Drude-like behaviour for which we should be prepared. For most liquid metals, in which (Γ_F/K_F) does not exceed 0.1 while ξ is not far from 1 (i.e. $\beta \simeq 0$), the deviations should be insignificant. They might show up for metals like Li and Hg, however, for which $|\xi|$ is large.

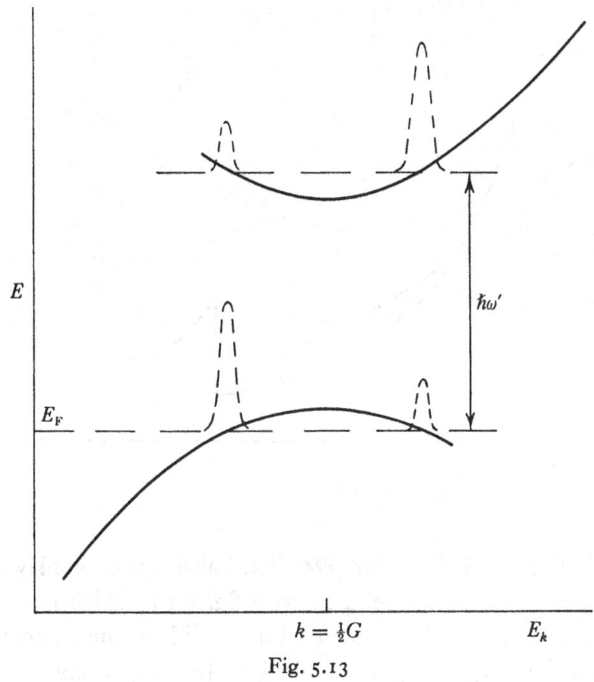

Fig. 5.13

Taken at its face value equation (5.81) suggests that, within the range of frequencies where it is applicable, $\sigma(\omega)$ contains a term which is independent of ω; for liquid Li it could amount to 5% of $\sigma(0)$. It must be remembered, however, that the steep slope of $a(q)$ in the neighbourhood of $q = 2k_F$, to which the large value of $|\xi|$ in Li is partly due, does not continue indefinitely; $a(q)$ passes through a peak and then diminishes. It may be rather misleading on this account to assume that Γ in Li is a monotonic function, proportional to $E^{\frac{1}{2}(1+\beta)}$ and $E_k^{\frac{1}{2}\beta}$. If the peak in $|u|^2a$ is a sharp one, situated at $q = q_{max}$, Γ may have a pronounced maximum for $k \simeq \frac{1}{2}q_{max}$, and the Lorentzian profile of the eigenfunctions may be distorted to such an extent that it becomes double-humped; this possibility has been illustrated for the one-dimensional case in fig. 4.11 on p. 258.

In these circumstances the excess absorption suggested by (5.81) is most unlikely to be independent of frequency, and the liquid metal may show some memory of whatever edges and

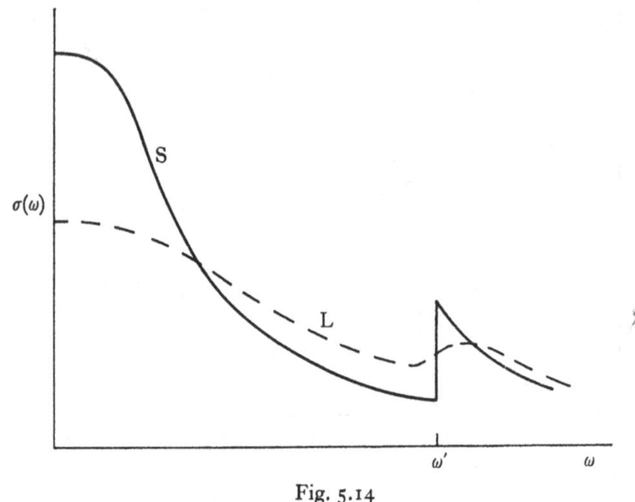

Fig. 5.14

peaks are characteristic of the *inter*-band absorption displayed by its solid phase. Compare fig. 4.11 with fig. 5.13, which illustrates the band structure to be expected for a solid in one dimension; the Fermi level is supposed to lie just below the energy gap. The profiles which are sketched for two particular energies are intended to convey, as in fig. 4.11 for the equivalent liquid, the range of k or E_k over which Fourier components are selected to make up each eigenfunction. The subsidiary maxima in each profile describe the reflected components of the Bloch wave states, and because of these there is a finite matrix element between the two states concerned. This allows electrons to be excited across the gap as soon as ω exceeds ω' and the result is an absorption curve of the type shown schematically in fig. 5.14. If the eigenfunctions retain their subsidiary maxima on melting, as suggested by fig. 4.11, the absorption curve may retain traces of the edge at ω'. A broken line in fig. 5.14 suggests the sort of behaviour that is conceivable.

If the peak in $|u|^2a$ is too weak to affect m^* appreciably in the majority of liquid metals, as the theoretical calculations discussed in §4.9 suggest, then it is probably too weak to generate an inter-band peak in the liquid phase. Some detailed computations of $\sigma(\omega)$ in liquid Na reported by Smith (1967*a*) certainly

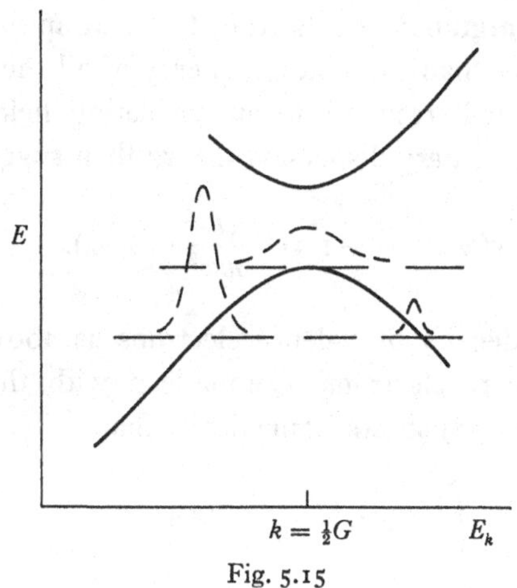

Fig. 5.15

reveal no trace of one, though it is clearly visible in his experimental curve for solid Na (Smith, 1969).

Calculations by Nettel (1966) suggest that in solid Na the inter-band peak may be presaged by a rise in the intra-band absorption above the level predicted by the Drude formulae, and there is some evidence for this in Smith's results. Nettel's formalism is different from the one adopted here, but the absorption processes which he has in mind seem to be described by fig. 5.15. This shows an E–E_k plot, similar to the one in fig. 5.13, for a metal with a small energy gap; the two profiles are for an occupied state well below the gap and an unoccupied one only just below it. The subsidiary maximum of the lower profile enhances its overlap with the upper one, and therefore enhances the absorption at the frequency where transitions between these two states can occur. Excess absorption of this type, by no means independent of frequency, is also conceivable in the liquid phase when the peak in $|u|^2a$ is a strong one.

5.19. The absorption sum rule

It is instructive to compare the predictions of the last few sections with a sum rule of great generality that governs the area under any absorption curve. The sum rule must first be derived. Three rival proofs will be presented here, each of them containing points of interest.

The simplest argument starts from the assumption that when $\hbar\omega$ is much larger than the binding energy of all the electrons in the atoms they will respond to an oscillating field as though entirely free. Elementary dispersion theory then suggests that

$$\epsilon(\omega \to \infty) - 1 = -\frac{4\pi e^2}{m\omega^2}(n + n_c),$$

where n is the density of valence electrons as above and n_c is the density of core electrons. Comparison with the Kramers–Kronig relation (5.58) shows immediately that

$$\int_0^\infty \sigma(\delta)\,d\delta = \frac{\pi e^2}{2m}(n + n_c), \tag{5.82}$$

which is the desired result.

In the second argument we start by considering the response of the system to a pulse of electric field $\mathbf{E}$, uniform in space, lasting for a very short time δt. If this is a true δ-function pulse, for which δt is allowed to tend to zero while $(\mathbf{E}\delta t)$ remains finite, the electric field is so strong that all other interactions, whether between the electrons and the ions or between the electrons themselves, are irrelevant so long as it persists. This means that the electrons accelerate as though entirely free and the current density generated is given instantaneously by

$$\mathbf{j}(0) = (n + n_c)e^2(\mathbf{E}\delta t)/m. \tag{5.83}$$

Subsequently this current density must die away to zero, and we may define a decay function $f(t)$ by the equation

$$\mathbf{j}(t) = \mathbf{j}(0)f(t). \tag{5.84}$$

Now the δ-function pulse can equally well be represented as the superposition of an infinite number of sinusoidal Fourier components of electric field, each varying like $\cos \omega t$ and with amplitudes independent of frequency. We have only to add together the sinusoidal terms in the current density which are generated by these Fourier components to see that

$$\mathbf{j}(t) = (\mathbf{E}\delta t)\frac{1}{\pi}\int_0^\infty \left[\sigma(\omega)\cos\omega t - \frac{\omega(\epsilon(\omega) - 1)}{4\pi}\sin\omega t\right] d\omega.$$

Since there is no current *before* the pulse is applied this integral must vanish when t is negative. We therefore know that for positive t

$$\int_0^\infty \sigma(\omega)\cos\omega t\,d\omega = -\int_0^\infty \frac{\omega(\epsilon(\omega)-1)}{4\pi}\sin\omega t\,d\omega, \qquad (5.85)$$

and hence that

$$\mathbf{j}(t) = \frac{(n+n_c)e^2(\mathbf{E}\delta t)}{m}f(t) = (\mathbf{E}\delta t)\frac{2}{\pi}\int_0^\infty \sigma(\omega)\cos\omega t\,d\omega. \qquad (5.86)$$

The sum rule can be obtained by letting t go to zero in equation (5.86).†

The interest of this argument lies largely in the conclusion that $f(t)$ and $\sigma(\omega)$ are Fourier transforms of one another. Equation (5.86) can of course be inverted if desired, to give

$$\sigma(\omega) = \frac{(n+n_c)e^2}{m}\int_0^\infty f(t)\cos\omega t\,dt. \qquad (5.87)$$

The classical Drude formulae expressed by (5.55) imply that

$$f(t) = \frac{n}{n+n_c}\exp(-t/\tau);$$

the current carried initially by the core electrons must be supposed to die away so quickly, oscillating as it does so, that it is irrelevant to $\sigma(\omega)$ at low frequencies, while the current carried initially by the valence electrons undergoes a straightforward exponential decay. Our modified equation (5.63) suggests a rather more subtle variation of f for very short times, while the core and valence electrons reach equilibrium among themselves, but a subsequent decay that is once more exponential, with

$$f(t) = \frac{n^*}{n+n_c}\exp(-t/\tau_D). \qquad (5.88)$$

The third argument is based upon the well-known *F-sum rule* which is proved in many standard texts. This says that

$$\sum_j F_{ij} = \frac{2\hbar^2}{m}\sum_j \frac{|X'_{ij}|^2}{E_j-E_i} = 1, \qquad (5.89)$$

† Equation (5.85), incidentally, provides a convenient starting point for the derivation of the Kramers–Kronig relations.

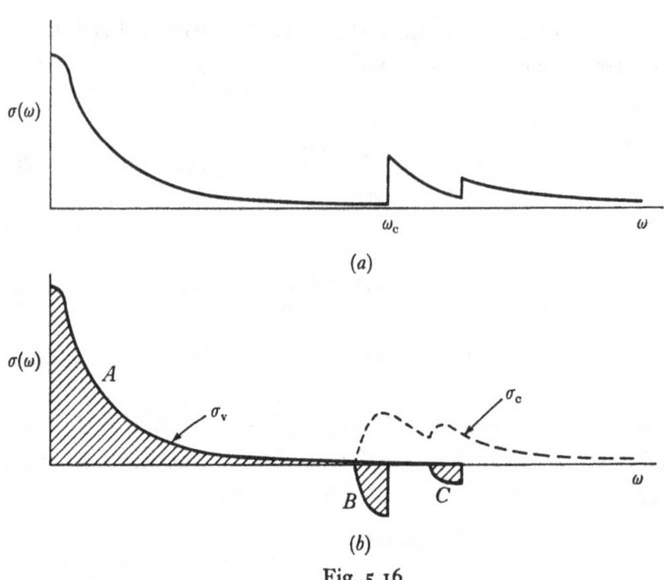

Fig. 5.16

where F_{ij} is the 'oscillator strength' for a transition from the ith state with energy E_i to the jth state with energy E_j and the sum is taken over all possible values of j for the system in question.† It is readily shown from (5.57) and (5.89) that

$$\int_0^\infty \sigma(\omega)\, d\omega = \frac{\pi e^2}{2m\Omega}, \tag{5.90}$$

for a system which contains only one electron, and (5.82) is an obvious generalisation.

The third proof is perhaps the most powerful because it shows us how to distinguish between the effects of the valence and the core electrons. In a typical metal in which the core electrons are tightly bound we may expect the absorption as a function of frequency to vary as shown schematically in fig. 5.16(*a*). The absorption at low frequencies is due to excitation of the valence electrons and should be described by (5.63); the absorption edges at ω_c and beyond are associated with excitation of core electrons.

† The F-sum rule is often stated for isolated atoms, but it holds equally well for a metal as long as its surfaces are perfectly reflecting. Some modification of the rule is required for an ideally regular solid metal with periodic boundary conditions, but this need not concern us here.

In the normal course of events the exclusion principle prevents the core electrons from being excited into the states in the conduction band which are already occupied by valence electrons, and this indeed accounts for the sharpness of the absorption edges. In calculating $\sigma(\omega)$, however, it is legitimate to ignore the exclusion principle (cp. (3.77) and (3.78)) and to view the curve in fig. 5.16(*a*) as made up of two parts, shown separately in fig. 5.16(*b*). The broken curve in the latter figure shows what the core electrons would contribute if the exclusion principle did not operate. The hatched areas A, B, C, etc., show the contribution due to the valence electrons, including now some *negative* absorption where transitions down into the core states would occur, with stimulated emission of photons. Since the F-sum rule involves a summation over all states, above and below the ith one, occupied and unoccupied, it is apparent that if we wish to write

$$\int_0^\infty \sigma_v(\omega)\, d\omega = \frac{\pi e^2}{2m} n$$

we must include these negative regions in the valence absorption σ_v. Hence the area A by itself is liable to exceed $\pi e^2 n/2m$.

With this point in mind, consider the following situations.

(*a*) *A jellium.* Here there are no ion cores and n_c is zero, while the areas B, C... all vanish; the area A must equal $\pi e^2 n/2m$ precisely. It will certainly do so, as the reader can verify by a trivial integration, if the Drude formula is obeyed over the whole frequency range with $n^* = n$, as suggested by (5.78). Presumably the excess absorption at high frequencies due to electron–electron interactions (§5.17) is balanced by a slight reduction in the absorption at low frequencies, though the latter is *not* predicted by the Landau quasi-particle theory.

(*b*) *A polyvalent liquid metal.* The peak in $|u|^2 a$ is not expected to affect the density of states, so in this case

$$m^* \simeq m_1^*$$

and

$$n^*/n \simeq m_{XC}^*/m_1^* \simeq m/m_1^* \qquad (5.91)$$

according to (5.78). Since β is likely to be close to zero the corrections discussed in §5.18 should be trivial, and the area A should therefore amount to $(\pi e^2 n/2m_1^*)$. The sum rule implies that

$m_{\rm I}^*/m$ should be slightly less than unity, which is consistent with most of the theoretical estimates for polyvalent metals which are listed in table 1.2. *It seems that in these metals the deviation of* $m_{\rm I}^*$ *from* m *can be attributed directly or indirectly to interaction with core states.*

(*c*) *A solid metal which shows inter-band absorption peaks in the visible and ultraviolet.* Orthodox theory for solid metals (e.g. Mott & Jones, 1936; see also §5.2) suggests that at low frequencies the intra-band absorption may be described by

$$\sigma(\omega) = \frac{ne^2\tau}{m_{\rm opt}(1+\omega^2\tau^2)}, \tag{5.92}$$

where
$$m_{\rm opt} = \hbar^2(m/m_{\rm XC}^*)\left(\overline{\frac{\partial^2 E_k}{\partial k_x^2}}\right)^{-1},$$

the average being taken over all occupied Bloch states. If the Fermi surface is not far from spherical, however, this optical effective mass should be given approximately (see p. 50) by

$$m_{\rm opt} \simeq (m/m_{\rm XC}^*)m^* \simeq m^*. \tag{5.93}$$

It is evident from the sum rule that *the existence of inter-band absorption at relatively high frequencies implies that* $m_{\rm opt}$ *and* m^* *must exceed* m. Experimental work on the optical properties of solid metals supports this conclusion. In solid Al, for example, $m_{\rm opt}/m$ is found experimentally to be about 1.5.

(*d*) *Solid and liquid Li.* The evidence concerning the effective mass in this particular metal has been summarised already in §4.15. It seems that $m^*/m \geqslant 1.5$, $m_{\rm I}^*/m \geqslant 1.3$, $m_{\rm XC}^*/m \simeq 1.01$, there being no significant difference between the two phases. Since $m_{\rm I}^*/m$ exceeds unity it is clearly not to be attributed to interaction with core states; some other peculiarity of the Li pseudo-potential must be to blame, e.g. the virtual state suggested by Meyer *et al.* (see p. 28). The optical effective mass for the solid phase in the infrared is found experimentally to be $1.57m$ (Hodgson, 1966) so that to satisfy the sum rule there must be quite a bit of extra absorption in the ultraviolet, some of it no doubt being inter-band absorption of a conventional kind. For the liquid phase equation (5.78) suggests

$$\frac{n^*}{n} \simeq \frac{1.5}{(1.3)^2} \simeq 0.9,$$

but we must not forget the correction factors involving $(\Gamma_F/K_F)^2$ which were discussed in §§5.13 and 5.18. These could push the absorption appreciably above the Drude value corresponding to $n^*/n = 0.9$, even at low frequencies. It follows that to satisfy the sum rule for liquid Li we need not necessarily postulate any residual interband absorption in the ultraviolet.

The case of Li is a complicated one which deserves further study.

5.20. Core polarisability

Virtual excitations of core electrons at low frequencies endow the cores with a finite electrical polarisability. They therefore contribute a small term ϵ_c to the low-frequency dielectric constant of the metal, which has been ignored above, e.g. in equations (5.55). The sum rule may be used to show that for most metals it is indeed negligible. From (5.58) we have

$$\epsilon_c = 8\int_0^\infty \frac{\sigma(\delta)-\sigma_v(\delta)}{\delta^2}\,d\delta$$

$$< \frac{8}{\omega_c^2}\int \sigma_c(\delta)\,d\delta,$$

where ω_c is the frequency at which core excitation begins. Hence the sum rule shows that

$$\epsilon_c < \frac{4\pi e^2}{m\omega_c^2}\,n_c. \tag{5.94}$$

For most metals it requires $\hbar\omega > 50$ eV to excite the 10 electrons per atom in the topmost d band, so the contribution of these electrons to ϵ_c is certainly less than 0.2. The contribution of the more tightly bound electrons is equally small or smaller.

In Cu, Ag and Au, however, the d band is only a few eV below the Fermi level and ϵ_c turns out to be significant.

5.21. Optical properties of liquid metals: experimental methods

The characteristic frequency τ_D^{-1} lies in the far infrared for solid metals at low temperatures, but for liquid metals it is usually in the near infrared or visible. Consequently, it is over the wavelength range between about 10 μ (equivalent to roughly 0.1 eV) and 0.1 μ (10 eV) that the optical properties of liquid metals are of interest experimentally. To determine both σ and ϵ it is necessary to study at least two independent properties of the specimen, but there are a wide variety to choose from. For example, one may measure its reflectivity at normal incidence (for electromagnetic radiation of known wavelength) and at the same time the phase change on reflection; or else the reflectivity at two different angles of incidence; or the modulus and phase of the quantity usually written as $\tan\psi \exp(i\Delta)$, which is the ratio between the amplitude reflection coefficients for the two principal planes of polarisation; or the reflection and transmission coefficients of a thin film; and so on. Further variety may be introduced by studying a specimen in contact with a transparent medium of known refractive index, instead of one with a free surface. One final possibility is to measure a single quantity, e.g. the reflectivity at normal incidence, over as wide a frequency range as possible, and to deduce the complementary quantity, e.g. the phase change on reflection at normal incidence, from some theoretical relation of the Kramers–Kronig type. Some applications of the many techniques available are described in a review article by Schulz (1957) and for further information the reader may consult papers referred to below. When transmission experiments are ruled out, and none have yet been performed on liquid specimens, the most sensitive method in principle is to measure ψ and Δ by an ellipsometric technique at a rather large angle of incidence θ; ideally θ should be close to the so-called 'principal' angle of incidence, where $\Delta = \frac{1}{2}\pi$, and this is usually about 80°, becoming even larger in the infrared.

It is necessary, of course, to invoke some model for the specimen's surface, to which Maxwell's equations can be applied, if readings of ψ and Δ are to be translated into values for σ and ϵ, and the customary assumption is that the surface of the specimen is a sharp one, i.e. that both σ and ϵ change discontinuously from the values characteristic of the bulk metal to those of the vapour or dielectric with which it is in contact. How realistic is this model? Standard electromagnetic theory tells us that the incident wave penetrates the metal to a *skin depth* δ (see §5.15) which is always greater than c/ω_p, and for liquid metals this lies between 130 Å and 500 Å or more. These figures give us some idea of the scale of surface contamination, roughness and so on that we may need to bother about.

One possibility is that the surface is contaminated by a layer of oxide, say, of thickness t and dielectric constant $\epsilon_{\text{ox}}(\sim 2)$. It may be shown that the effect of such contamination on ellipsometric measurements is always to *reduce* the apparent value of σ, and for $\omega\tau \gtrsim 1$ to reduce the apparent value of ϵ as well. A formula for the magnitude of the effect is quoted by Miller (1969) and it reduces to

$$\left.\begin{aligned} \sigma_{\text{app}} &\simeq \sigma\left(1-\frac{3t\omega_p}{c}\frac{\epsilon_{\text{ox}}-1}{\epsilon_{\text{ox}}}+O\left(\frac{t\omega_p}{c}\right)^2\right), \\ \epsilon_{\text{app}} &\simeq \epsilon\left(1-\frac{2t\omega_p}{c}\frac{\epsilon_{\text{ox}}-1}{\epsilon_{\text{ox}}}+O\left(\frac{t\omega_p}{c}\right)^2\right) \end{aligned}\right\} \tag{5.95}$$

when the angle of incidence is large and $\omega\tau_{\text{D}} \gg 1$; the results for $\omega\tau_{\text{D}} \simeq 1$ are not dissimilar. Roughly speaking, therefore, each atomic layer of oxide is liable to reduce the apparent optical constants of a liquid metal by about 1%. In practice, if the surface of a liquid metal such as Hg is exposed to the atmosphere the apparent optical constants do fall steadily over a period of hours in a manner that suggests the development of a relatively thick oxide layer, but if the surfaces are freshly formed, cleaned by heating or by ion-bombardment in a glow discharge (Faber & Smith, 1968) and kept in a vacuum limited only by the vapour pressure of the specimen, then the results are normally very stable and reproducible. In these circumstances it is difficult to believe that oxide contamination amounts to more than a monolayer at the worst (see especially Smith, T., 1967).

Another possibility is that minute quantities of what are called *surface-active* impurities in the specimen (see p. 477) might migrate to its surface when it is melted and form – at the worst – a monolayer there. It is harder to assess the effect that this would have, but it seems unlikely to exceed a few parts per cent, and once again the reproducibility of results obtained by different observers, using specimens derived from different sources, is encouraging.

Even if the surface is perfectly clean, however, its properties do not necessarily correspond to those of our simple model, for it is quite conceivable that the electronic properties of a liquid metal are not the same close to its surface as they are in the interior: the arrangement of the ions could well be different, and even if this is not the case the electronic states may still in principle be sensitive to the propinquity of the surface (Stern, 1967). Knowing that the range of the effective interionic potential (see §1.13) is not more than two or three times the mean separation between ions, one may reasonably suppose that the anomalous surface layer is not much more than about 10 Å deep, and this guess seems to be consistent with the surface tension data examined in §2.5. If, within this layer, the effective values of σ and ϵ are only mildly anomalous, then it is unlikely to distort the results much more than the monolayer of oxide discussed above. As we shall see, however, there is some indication that for Hg its consequences are more serious.

A lot of work has of course been done on the optical properties of solid metals, especially evaporated films. In the course of this, evidence has accumulated that the boundary conditions for a conduction electron at the surface itself may correspond to *diffuse* rather than *specular* reflection, unless the surface is unusually smooth. Diffuse reflection, or in other words surface scattering, is bound to depress the effective conductivity within a layer whose thickness is of the order of the mean free path. Its effect upon the apparent values of σ and ϵ for the specimen as a whole has been discussed by Holstein (1952), Dingle (1953) and others, and as a rough rule of thumb it may be expressed by replacing $1/\tau_D$ by

$$\left(\frac{1}{\tau_D}\right)_{\text{app}} = \frac{1}{\tau_D}\left(1+\frac{3}{8}\frac{\delta}{L}\right),$$

though it is only when $L/\delta \ll 1$, $\omega\tau_D \gg 1$ that this formula is fully reliable. Thus at high frequencies the apparent conductivity should be enhanced by

$$\left(1+\frac{3}{8}\frac{L\omega_p}{c}\right),$$

the apparent dielectric constant being unaffected. *If* diffuse reflection is the rule for liquid metals, then the enhancement should be particularly noticeable when L is large; for liquid Na and K, for example, it should amount to 18% and 15% respectively. According to Bennett, Bennett, Ashley & Motyka (1968), however, the degree of roughness at which it makes itself felt is characterised by a scale length of 45 Å. By this standard it is more than probable that liquid surfaces are very smooth indeed (see §2.5).

These are tiresome qualifications and the sad tale is not quite ended. Granted that we have a perfectly sharp boundary at which σ and ϵ change discontinuously from one pair of values to another, we have still to solve Maxwell's equations so as to obtain expressions for ψ and Δ, say, with which to compare our experimental data. The solutions which have been used for many years are based upon the assumption that the charge density inside the metal is everywhere zero. Recently this assumption has been questioned (Forstmann, 1967; Kliewer & Fuchs, 1968; Melnyk & Harrison, 1970): if an electromagnetic wave is incident upon a metal surface at a large angle of incidence with the electric field in the plane of incidence (the so-called p-polarisation), the field inside the metal may have a longitudinal component and charge density fluctuations may then be excited. The full theory is distinctly complex, for the reflection coefficient depends upon $\tilde{\epsilon}_{\parallel}$ as well as $\tilde{\epsilon}_{\perp}$. Fortunately it is only under anomalous conditions (see §5.15) that the final results are significantly different. For liquid metals the conventional theory should be good enough.

5.22. Optical properties of liquid metals: results

In view of the many reservations that have been expressed about the Drude formulae, it is remarkable how well they serve in practice to describe the high-frequency behaviour of liquid metals. The limits of error in a single measurement of $\sigma(\omega)$ or $\epsilon(\omega)$ are usually ± 5% or more. Within these limits the points lie on

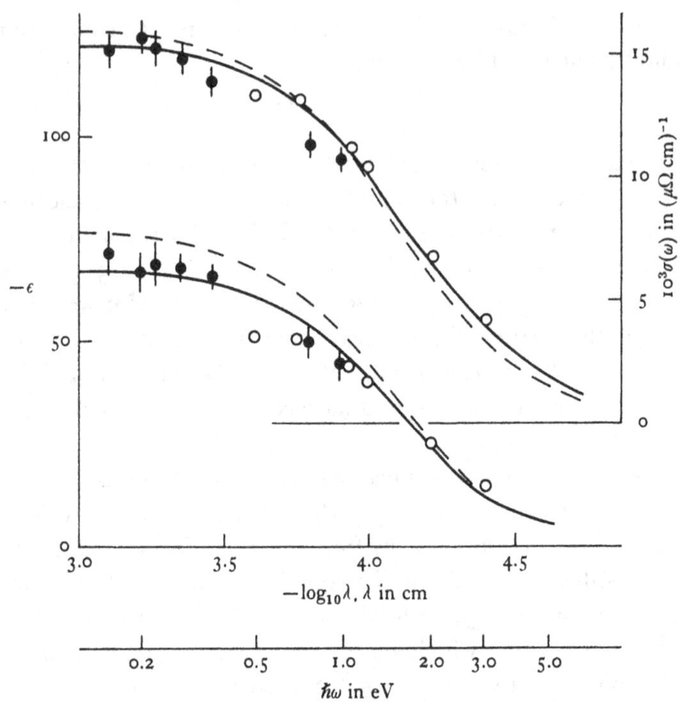

Fig. 5.17. The optical properties of liquid Sn at 800 °C: ○, Hodgson; ●, Comins. The broken curves correspond to the free-electron Drude formulae. The full curves correspond to the Drude formulae after adjustment of n^* and τ_D.

smooth curves which can be matched by the Drude formulae, over the whole frequency range between say $\omega\tau_D \simeq 0.1$ and $\omega\tau_D \simeq 3$, after some adjustment of the parameters τ_D and n^*. A typical fit is shown in fig. 5.17 which contains the results of Hodgson (1961) and Comins (1972) for liquid Sn. Results for other liquid metals may be found in the work of Schulz (1957), Hodgson (1960, 1961, 1962), Lelyuk, Shklyarevskii & Yarovaya (1964), Mayer & Hietel (1966), Smith (1967*b*), Miller (1969) and Comins (1972).

It is particularly to be noted that the curves are smooth: no liquid metal shows the sort of absorption spectrum to be expected if the conduction electrons were affected by residual band structure. Absorption edges which are associated with the excitation of core electrons up to the Fermi level are certainly preserved on melting,

TABLE 5.7. *Optical properties of liquid metals*

Metal	T (°C)	$\frac{n^*}{n}$	$\frac{\sigma(\omega \to 0)}{\sigma(0)}$	$\frac{n}{n^*}\frac{m^*_{XC}}{m}$	$\frac{m^*_I}{m}$
Na	100	0.85	—	1.21	0.99
K	65	1.0	—	1.07	0.97
Rb	40	0.8	—	(1.4)	0.89
Cs	29	0.7	—	(1.6)	—
Cu	1100	0.76	—	1.31	—
Ag	1000	1.05	—	0.95	—
Au	1100	1.01	0.98	0.99	—
Cd	450	1.06	0.79	0.94	0.95
Al	900	0.83	0.83	1.18	1.04
	1170	0.80	0.84	1.22	
Ga	600	1.03	0.96	0.95	0.96
	860	1.00	0.98	0.98	
In	813	1.06	0.95	0.93	0.93
Ge	1000	1.07	0.82	0.92	—
Sn	600	1.10	0.98	0.89	0.93
	1170	1.03	0.96	0.95	
Pb	600	1.19	0.92	0.83	0.86
	800	1.13	0.95	0.87	
Sb	654	1.22	0.985	0.80	—
	824	1.21	0.96	0.81	—
Bi	500	1.05	0.93	0.93	0.87
	695	1.10	0.98	0.89	

as we shall see, but those which are associated with band gaps *above* the Fermi level seem to disappear entirely. At one time experiments by Mayer and co-workers (Mayer & Hietel, 1966) suggested that the solid alkali metals, especially Na, displayed not only an absorption peak due to inter-band excitation but an additional, unexplained peak at lower frequencies, and both of these features appeared to survive melting; for Na, indeed, the unexplained peak was enhanced. However, the values of ψ and Δ for alkalies in the visible region are such that it is extremely hard to measure $\sigma(\omega)$ with any accuracy. The later work of Smith (1969) does not support the results of Mayer & Hietel for solid Na and therefore casts doubt upon their results for all the alkali metals, solid and liquid.

Table 5.7 contains a selection of values for n^*/n, taken from the papers quoted above, which are probably significant to within say $\pm 3\%$. For the alkali metals $\omega\tau_D$ is much greater than unity throughout the range of measurement, so that ϵ is virtually independent of τ_D. This means that n^*/n can be deduced from the results for ϵ alone and should not be affected by the errors that beset the measurement of σ. The same is almost true of the noble metals, though one does have to make use of the absorption data in the ultraviolet, in order to estimate ϵ_c by means of the Kramers–Kronig relations; because the d electrons can be excited at relatively low frequencies in the noble metals ϵ_c is no longer negligible – in Cu it is about 5 and becomes even higher as the absorption edge is approached – and one must subtract it from ϵ before attempting to fit the results to the Drude expression.

For the polyvalent metals $\omega\tau_D$ is comparable with unity, as fig. 5.17 suggests, and the results quoted for n^*/n have been deduced by fitting Drude curves to the data for both ϵ and σ simultaneously. There are grounds for mistrusting this procedure, as we shall see, but the results appear to be essentially correct nevertheless. If the structure-dependent term in the density of states is negligible, then according to (5.91) the quantities (nm^*_{XC}/n^*m) and (m^*_1/m) should be the same. They are listed in the last two columns of table 5.7 (the values used for m^*_{XC}/m were taken from the curve in fig. 4.21, while the values quoted for m^*_1/m are the more reliable of the theoretical estimates already listed in table 1.2) and the agreement is satisfactory for all the polyvalent metals except Al, which is notoriously prone to surface oxidation. Why the agreement does not extend to the monovalent alkali metals is not entirely clear; perhaps in the calculation of m^*_1 for these the effects of exchange and correlation have been inadequately recognised.

The magnitude of n^* seems to be little changed by melting in the case of Na or K (compare the figures in table 5.7 with those already quoted for the solid phase in table 4.1). This was to be expected because the inter-band absorption observed by Smith (1969) for solid Na and K is very weak; it accounts for only about 2% of the total area under the absorption curve between zero frequency and the frequency at which excitation of the core electrons begins. In Cu, Ag and Au, however, the inter-band absorption is rather stronger and its disappearance on melting results in a noticeable increase in n^*. For all these noble metals

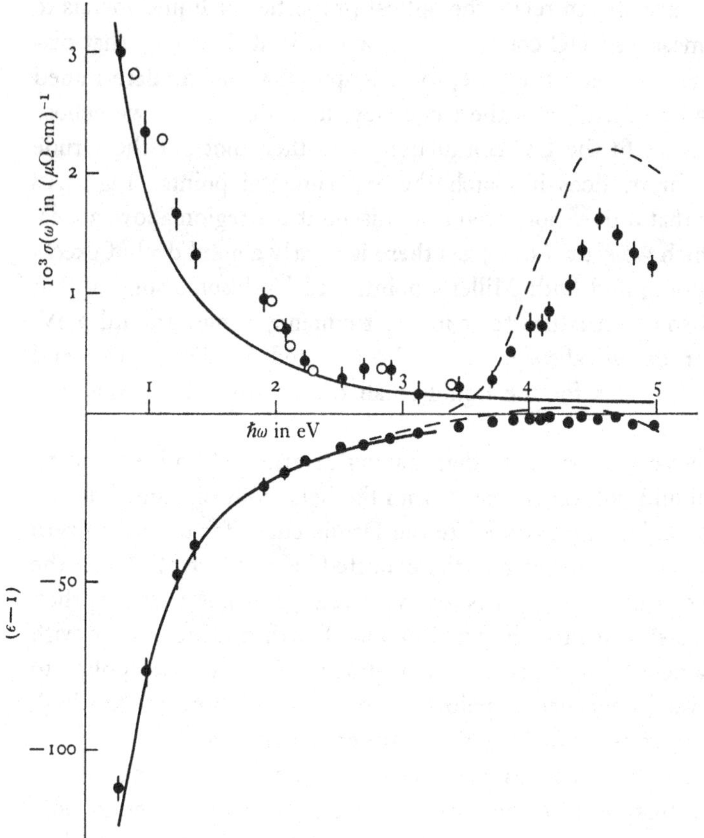

Fig. 5.18. The optical properties of liquid Ag at 1000 °C: ○, Hodgson; ●, Miller. The full curves correspond to the Drude formulae with $n^*/n = 1.05$ and τ_D adjusted to give the observed DC conductivity $(58.5 \times 10^{-3}\ (\mu\Omega\ \text{cm})^{-1})$; the full curve for $(\epsilon - 1)$ includes a correction for ϵ_c. The broken curves indicate the behaviour observed for solid Ag at lower temperatures.

the Fermi surface in the solid phase makes contact with the {111} faces of the first Brillouin zone and a small fraction p of its area is thereby neutralised, in the sense that low-frequency excitation of the electrons is impossible; de Haas–van Alphen studies show p to be about 7%, 4% and 6% in Cu, Ag and Au respectively. Since the whole Fermi surface should be free to contribute to the intra-band absorption in the liquid phase it is not surprising that these are just about the percentages by which n^* is observed to increase on melting.

We have still to relate the optical properties of liquid metals to their measured DC conductivities, and it is at this stage that discrepancies appear. Suppose, for example, that having determined n^* for liquid Ag from the frequency dependence of ϵ we choose τ_D so as to fit the DC conductivity and then plot out the Drude curve for σ, does it match the experimental points? Fig. 5.18 shows that it does not; even if we discount the region above 3.5 eV (to which we return in § 5.24) there is clearly a good deal of excess absorption, and both Miller's points and Hodgson's suggest that it has some structure to it in the frequency range around 2 eV. Similar excess absorption has been noted for liquid Cu, and probably exists for the liquid alkali metals studied by Mayer & Hietel.

Alternatively, suppose that, having determined both n^* and τ_D for a liquid polyvalent metal from the behaviour of ϵ and σ in the visible region, we extrapolate the Drude curve for σ back to zero frequency, do we arrive at the expected value for $\sigma(0)$? Again the answer is no; $\sigma(\omega \rightarrow 0)$ is always a few parts per cent less than $\sigma(0)$, as shown in table 5.7. Compared with a Drude curve with the same n^* but forced, by adjustment of τ_D, to extrapolate to $\sigma(0)$, the experimental points for σ are once more on the high side for $\omega\tau > 1$ but for $\omega\tau < 1$ are consistently low.

These discrepancies are still not understood. Some of the factors that could be involved are residual oxide contamination, surface scattering of the electrons, and the effects discussed in § 5.18. In liquid Bi at 500 °C, however, Comins has found that $\sigma(\omega)$ remains 7% below $\sigma(0)$ right down to such low frequencies that $\omega\tau_D$ is only 0.03. By this stage the skin depth is about 10^{-5} cm and errors due to the factors mentioned would not have been expected to exceed 1%. Perhaps (see the comment in paragraph (*a*) on p. 379) we should blame some many-body effect instead.

5.23. Optical properties of liquid metals: mercury

The optical properties of liquid Hg in the visible and infrared have been a bone of contention among workers in this field, and they require separate discussion. According to one school they agree exactly with the Drude free-electron formulae with $n^*/n = 1.0$; such agreement was demonstrated by Schulz (1957)

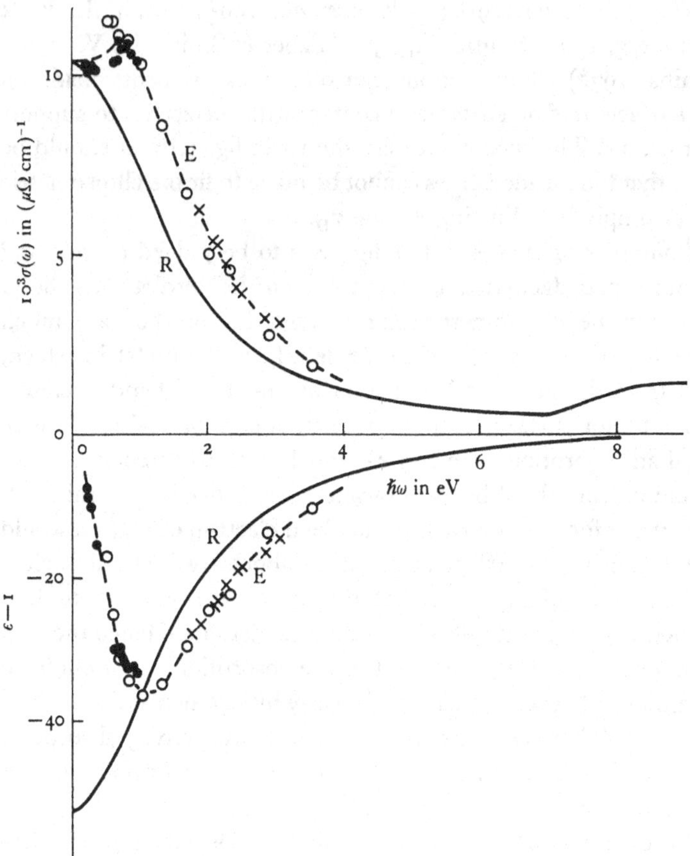

Fig. 5.19. The optical properties of Hg at room temperature. E, ellipsometric results (○, Hodgson; ×, Smith, N. V.; ●, Comins); R, reflectivity results, which are essentially coincident with the Drude free-electron curves below about 4 eV.

and has been confirmed by Wilson & Rice (1966), Boiani & Rice (1969), Bloch & Rice (1969) and Mueller (1969), who have measured the reflectivity either at normal incidence or at 45° over a very wide range of wavelengths, from 20 μ to 0.1 μ. Experimenters who have used ellipsometric methods at larger angles of incidence, however, report that $\sigma(\omega)$ is at least 20% greater than the Drude value throughout the near infrared and visible – the more care that is taken to secure a clean surface the higher it becomes – and that there are corresponding discrepancies for $\epsilon(\omega)$ (Hodgson, 1959; Lelyuk *et al.*, 1964; Smith, L. E. & Stromberg, 1966; Smith, T., 1967; Faber & Smith, N. V., 1968; Comins, 1972). Both factions can offer measurements made on free surfaces *and* on surfaces in contact with dielectrics to support their claims. The rival curves are shown in fig. 5.19. It should be noted that the Drude curves cannot be made to fit the ellipsometric points simply by adjusting n^* and τ_D.

If one pair of curves in this figure is to be judged correct and the other pair dismissed as spurious then the verdict must be in favour of the ellipsometric pair marked E, for they are much easier to reconcile with the sum rule. Hg is the metal in which, apart from the noble and transition metals, the d band is closest to the Fermi surface, and the measurements of Wilson & Rice reveal an absorption edge at 7 eV which is clearly associated with excitation from the d band to vacant states above the Fermi level. Were it not for the exclusion principle, d electron excitation would start at much lower frequencies, say at about 1 eV, and it is clear by inspection of fig. 5.19 that the area corresponding to that labelled B in fig. 5.16 must be quite considerable. Hence the area A under the low-frequency part of the absorption curve ought to be significantly greater than $\pi e^2 n/2m$, say by as much as 4×10^{-3} eV $(\mu\Omega\ \text{cm})^{-1}$. There is *no* excess area under the reflectivity absorption curve R of course, but the excess area under absorption curve E seems to be just about right.

However, the form of the curves labelled E is not easy to understand. The excess absorption is much too large to attribute to the anomalous skin effect; $3L\omega_p/8c$ is only about 1.5% in Hg. Any suggestion that we are seeing an absorption peak due to some sort of residual band structure or pseudo-gap (Mott, 1966) is untenable, because the peak comes at such a low energy – at less than 1 eV, whereas $\hbar/\tau$ or Γ_F would seem to be about 2 eV; surely all band structure effects on a finer scale than Γ_F must be smeared out? Smith (1968) has pointed out that if these curves are correct for bulk Hg they imply, according to equation (5.87) that a pulse of current does not decay monotonically but *oscillates*. He has reproduced the curves rather successfully by the *ad hoc* manoeuvre of truncating the series of equations in (3.79) so as to include only the current densities j_1 produced by direct excitation and j_2 generated by a single scattering process. This suggests a decay function of the form

$$f(t) = \frac{n^*}{n+n_c}(1+\overline{\cos\theta}t/\tau_l)\exp(-t/\tau_l) \qquad (5.96)$$

and the fit is achieved by choosing $n^*/n = 1.35$ and $\overline{\cos\theta} = -0.49$. There is no obvious justification, however, for the neglect of j_3, j_4 etc.†

In a sense, the curves labelled E are only grotesquely anomalous for frequencies below about 1 eV; above that they conform reasonably well to Drude curves, with $n^*/n = 1.5$ or so and

$$\sigma(\omega \to 0)/\sigma(0) = 1.23.$$

One possibility which is worth consideration is that there is an anomalous layer on the surface of liquid Hg of such thickness that at high frequencies, when the skin depth is small, one is effectively making measurements on the properties of this layer alone. Perhaps within this layer the DC conductivity is indeed 1.23 times the bulk conductivity, while in view of the propinquity of the d band to the Fermi level 1.5 is by no means an impossible value for n^*/n. The conductivity of Hg is unusually sensitive to compression, of course, and to raise it by 23% near the surface would require a local increase of density of only about 3%.

The trouble with this explanation as it stands is that it seems to require such a thick anomalous layer: at least as thick, presumably, as the skin depth of 'anomalous' Hg at a frequency of 1 eV, which would be about 300 Å. For reasons touched on above, one is reluctant to believe that structural anomalies persist to such a depth. It is possible, however, that we have to deal with a surface layer which is still more peculiar in its electrical properties and therefore of smaller thickness. Bloch & Rice (1969) have investigated a variety of models in the hope of finding one that will not only account for the unexpected shape of the

† The success of Smith's formulae does at least show that the pair of curves labelled E in fig. 5.19 are, like curves R, consistent with the Kramers–Kronig relations. This provides no evidence that they correctly represent the properties of bulk Hg, of course, since the Kramers–Kronig relations can be shown to be just as valid for the *apparent* optical constants of e.g. a severely contaminated surface as for the true optical constants of the underlying material.

ellipsometric curves but will also reconcile these curves with the results of their own work on reflectivity at normal incidence. One model for which they claim success (if errors of up to $\frac{1}{2}\%$ are admitted in the reflectivity data) is such that the local value of $\sigma(0)$ is about 2.5 times the bulk value over a depth of about 10 Å while the underlying bulk Hg is taken to obey the free-electron Drude equations with $n^*/n = 1.0$. The model incorporates a further assumption, however, that at the interface between the liquid and its vapour there is a region where the valence electron states are localised and where $\sigma(\omega)$ is not Drude-like at all; it is supposed in this region to display a broad peak of Lorentzian form. A good part of the excess absorption apparent in the ellipsometric curve for σ can be traced, according to Bloch & Rice, to this Lorentzian peak.

The Bloch–Rice model is an interesting one deserving further study. It is too early to pronounce upon whether or not it is essentially correct.

5.24. Absorption and emission processes involving core electrons

So far, in discussing the optical properties of liquid metals, we have concentrated upon relatively low frequencies and hence upon processes in which only the valence electrons are involved. We have already seen evidence in figs. 5.18 and 5.19, however, of absorption edges associated with excitation of d electrons, and if the experiments could be pushed further into the ultraviolet we would expect to see such edges in a wider range of metals. In the soft X-ray region we should begin to be able to excite the deeper lying core electrons. Conversely, if we excite core electrons by bombarding the metal with fast electrons from outside, we should be able to observe soft X-ray emission lines, as valence electrons jump downwards to fill the vacant states. What, if anything, can such observations contribute to our understanding of liquid metals?

From the theoretical point of view these processes involving core electrons raise problems of a substantial nature which are by no means solved (see, for example, Friedel, 1969; Hopfield, 1969). If we could trust the independent-particle model, not only for valence electrons close to the Fermi surface but also for the deeper lying electrons, then all would be straightforward, in a regular solid metal at any rate: we could describe all the electron states by non-localised wave functions of the Bloch type, extending through the whole specimen; we could determine transition probabilities by evaluating the matrix elements X'_{nm} between these states; and, as a consequence of Koopman's theorem, briefly discussed on p. 10, we could take the energy required to effect any particular transition to be the difference of two independent-particle energies $E_n - E_m$ for the states concerned.

Unfortunately this model is not valid; electron–electron interaction effects cannot simply be lumped together into a *k*-dependent pseudo-potential and thereafter ignored. When an electron is excited from one of the states towards the bottom of the conduction band, they undoubtedly have the effect of broadening the energy of the resultant hole, because Auger-type processes make its lifetime so short, and they may shift its energy too. This shows up very clearly in soft X-ray emission spectra (Skinner, 1946). Naively, one would expect each X-ray emission line to have a width corresponding to K_F, and indeed to reproduce more or less the parabolic profile of the density of states for the conduction band from which the emitting electron is drawn. In practice, as is shown by fig. 5.20, the low energy side of the emission line is broadened by 1 eV or more.

The effect of electron–electron interaction on the tightly-bound core electrons seems to be not so much to broaden their energy levels as to make the Bloch type of wave function totally inappropriate. Correlation effects force each of them to spend its whole time on a single atom, and they should be described by atomic-like wave functions which do *not* spread through the whole specimen. Their potential energy is the lower because of this, since they each sit in an attractive well which is in no way screened by similar electrons on neighbouring ions. But the energy that is required to excite them to the Fermi levels is *not* simply $E_F - E_C$: the localised hole that is left behind exerts a perturbation that is not infinitesimal on the other core electrons in the atom and one must allow for the change in energy of all of these. The hole may also perturb, locally, the wave function

Ψ_F of the excited Bloch state to which the core electron makes its transition, and one should not evaluate the matrix element

$$X'_{Fc} = \left\langle \Psi_F \left| \frac{\partial}{\partial x} \right| \Psi_c \right\rangle$$

without taking this perturbation into account. In semiconducting materials the perturbation may be so great that Ψ_F is changed altogether from a non-localised to a localised function, i.e. the excited electron may be bound to the hole to form an *exciton*. The minimum energy required for excitation is thereby reduced. It is generally supposed, however, that in metals the conduction electrons screen the potential set up by the hole so effectively that exciton formation is prevented.

It is probable that in between these two extremes, represented by the simple independent-particle model on the one hand and the model of a completely localised hole on the other, a variety of situations may occur, requiring a more subtle many-body treatment.

In view of these complexities we shall not attempt to explain the details of the few absorption and emission spectra involving core electrons that have been reported for liquid metals in the ultraviolet and X-ray regions but will concentrate solely upon the extent to which they are affected by melting. To begin with, there are the absorption edges which are well known to occur at about 2, 4 and 2 eV in Cu, Ag and Au respectively and to be due primarily to the excitation of d electrons. These have been studied in the liquid state by Otter (1961) and Miller (1969). Their results suggest remarkably little change on melting, for Cu and Ag at any rate, both in the frequency at which the edge occurs and in the level of the d electron absorption which then ensues; in the case of Au the data are incomplete and it is possible that the edge shifts to slightly lower frequencies, as suggested by the photoemission results discussed on p. 403 below. The drop in the absorption in Ag which is evident above 4 eV in fig. 5.18 is attributed to the disappearance of some inter-band absorption by the valence electrons, rather than to any change in the pattern of d electron excitation; in Cu and Au, where the valence inter-band peak does not overlie the d electron peak, the indifference of the

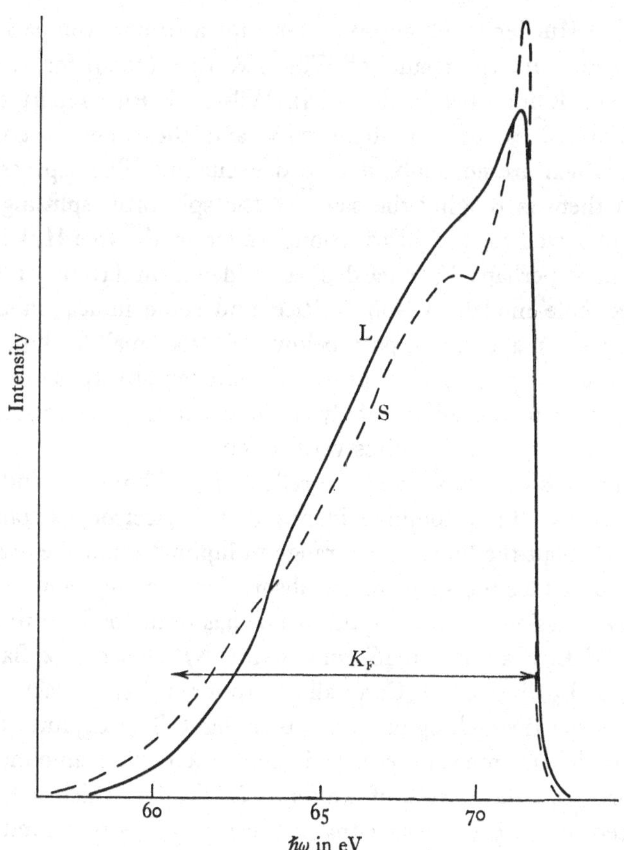

Fig. 5.20. The L_{23} emission line of Al: S, solid; L, liquid. (After Catterall & Trotter, 1963.)

latter to the melting process is very evident. According to the localised hole model some slight changes might have been expected, (*a*) because the band structure of the vacant states above the Fermi level is thought to be wiped out on melting, (*b*) because the change of volume might shift the mean energy of the d electrons relative to the Fermi level, and (*c*) because the somewhat greater density fluctuations in the liquid phase might increase the extent to which the d electron energy varies between one ion and the next. However, these effects could well be too small to show up experimentally.

There are insufficient data for the ultraviolet absorption of solid Hg (Mueller & Thompson, 1969) for a fruitful comparison to be made with the results of Wilson & Rice (1966) for liquid Hg (curve R for $\sigma(\omega)$ in fig. 5.19). Wilson & Rice report two absorption edges, one at about 7 eV and the other at about 9.4 eV, which are no doubt due to d excitation. The separation beween them is roughly the same as the spin-orbit splitting of 1.9 eV observed for the $5d^9 6s^2$ configuration of the free Hg^+ ion, which may perhaps be regarded as evidence in favour of the localised hole model. Wilson & Rice find some indication that there is a small absorption peak below 7 eV, too small to show up on the scale of fig. 5.19, which they attribute tentatively to exciton formation or to the effects of dynamic screening; the existence of this peak, however, requires verification.

Wilson & Rice also studied the reflectivity of liquid In and Bi out to 20 eV. They found evidence that d electron excitation begins at about the limit of this range in liquid In, but they were unable to deduce the shape of the absorption edge in detail.

X-ray emission work on liquid metals has been confined to the K line of Li (Skinner, 1946) and to the $K\beta$ (Fischer & Baun, 1965) and L_{23} lines of Al (Catterall & Trotter, 1963). In each case the change at the melting point is rather slight. The L_{23} line of Al is of particular interest because it shows a certain amount of structure, and the curves of Catterall & Trotter which are reproduced in fig. 5.20 suggest that a trace of this is preserved on melting. If one is to attribute it to band structure in the valence states just below the Fermi level, i.e. to a non-parabolic form for the density of states curve in Al, its partial preservation on melting, though not necessarily inconsistent with the results of §4.9, is somewhat of a surprise. It is now thought likely, however, to be an effect of perturbation by the hole on the matrix elements governing the L_{23} transition. No similar structure has been detected in the $K\beta$ line for Al, as may be seen from fig. 5.21.

Catterall & Trotter devote some attention to the breadth of the emission edge, which they observe to be 0.62 eV for liquid Al; this is the energy range over which the emitted intensity rises from 5% to 95% of its maximum value. It is accounted for completely by lifetime broadening of the hole state (0.02 eV), instrumental resolution (0.02 eV), and thermal broadening of the

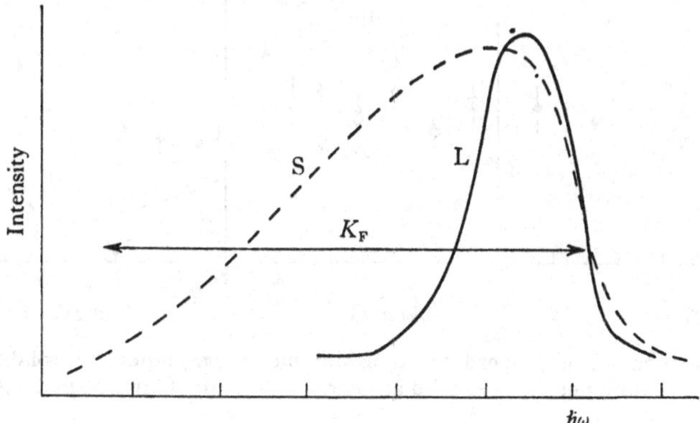

Fig. 5.21. The Kβ emission line of Al: S, solid at room temperature; L, liquid several hundred degrees above the melting point. The line occurs at a photon energy of about 1560 eV and each division on the scale for $\hbar\omega$ corresponds to 2 eV. (After Fischer & Baun, 1965.)

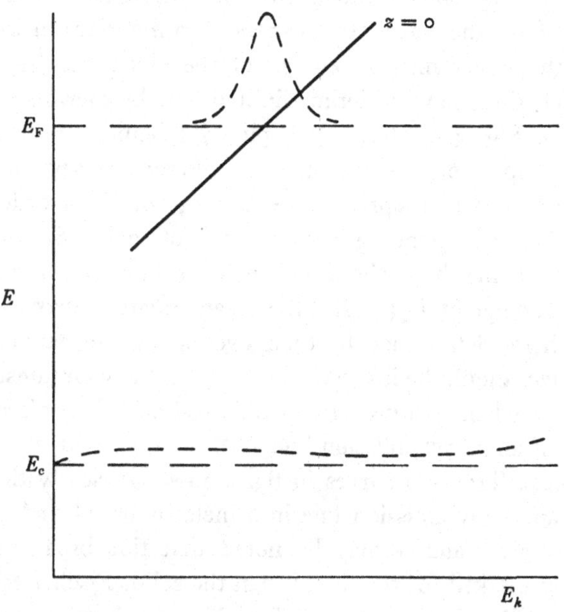

Fig. 5.22

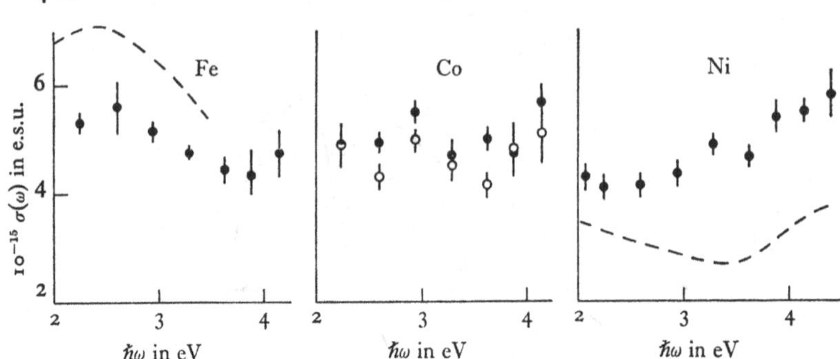

Fig. 5.23. The optical properties of transition metals: ●, liquid; ○, solid just below the melting point; – – –, solid at room temperature. (After Miller, 1969.)

cut-off in the Fermi distribution ($6k_B T = 0.58$ eV) so that there is nothing left over to be associated with lifetime broadening of the excited state from which the electron falls, which might have been expected to contribute a further $\hbar/\tau_l \sim 1$ eV. In fact this is not surprising, since although $\hbar/\tau_l$ is indeed the uncertainty in energy of a plane wave state near the Fermi surface in **k**-space it is *not* the uncertainty in energy of the eigenstates near the Fermi level; these have a lifetime limited only by thermal motion of the ions. Fig. 5.22 shows in a purely schematic fashion the profiles in **k**-space of two eigenstates of different energy, one near the Fermi level which spreads over a range of E_k of order $\hbar/\tau_l$ and the other corresponding to a core state of energy E_c. So long as the latter is highly localised and includes Fourier components over a wide range of E_k ($\gg K_F$) the mean square matrix element between them, determined by their overlap in the manner described above, should be insensitive to $1/\tau_l$. In these circumstances we are justified in treating $1/\tau_l$ as infinitesimal when discussing the shape of an X-ray emission line.

Catterall & Trotter's curves in fig. 5.20 show clearly the way in which an X-ray emission line in a metal is broadened on the *low* energy side, and it may be noted that this broadening is slightly less marked for the liquid than the solid. Fischer & Baun report that on heating 'several hundred degrees' above the melting point the Kβ line for liquid Al changes rather suddenly and becomes very much less broad; its width, as shown in fig. 5.21, shrinks from 13.4 to 5.6 eV (cp. $K_F \simeq 11$ eV). The change is reported by Powell (1968) to be reversible. This startling observation requires confirmation.

Liquid transition metals receive little attention in this book, but the present section would not be complete without a brief reference to the optical measurements of Miller (1969) on Fe, Co and Ni; her absorption data are reproduced in fig. 5.23, with some results for the solid phase for comparison.† In transition metals the d band overlaps the Fermi level, and presumably most of the absorption that we see in this figure is due to the two types of inter-band excitation that are possible, from the d band to vacant states in the sp conduction band and from the conduction band to vacant states in the d band. The level of the absorption is strikingly similar in all three liquid metals. In fact it is very similar to the level of the absorption shown by Cu (which of course lies next to Fe, Co and Ni in the periodic table of elements) above the 2 eV absorption edge, if one allows a factor 2 because in Cu only the first of the two inter-band processes described above can operate. In Fe, Co and Ni $\sigma(\omega)$ averages about 5×10^{15} e.s.u., whereas in Cu at 3.0 eV it is 2×10^{15} e.s.u.

5.25. Photo-emission and the work function

When a metal surface is illuminated with ultraviolet radiation photo-electrons are emitted, and a detailed study of their energy spectra as a function of photon frequency can supplement the information derived from optical experiments. Valuable results have been obtained for solid metals in this way, especially by Spicer and his co-workers (e.g. Berglund & Spicer (1964) and Kindig & Spicer (1965)).

The electrons are excited by ultraviolet photons in the interior of the metal. To be observed experimentally they must make their way to the surface, and they stand some chance of being scattered *en route* or reflected at the surface by the potential step which exists there. In favourable circumstances a yield of about 0.1 electrons per incident photon may be observed. Since, while they

† The curves for solid Fe and Ni were obtained at room temperature where these metals are ferromagnetic. There is evidence, however, (Shiga & Pells, 1969) that for Ni at any rate the absorption is not much changed on heating through the Curie point.

are still inside the metal, they have an energy well above the Fermi level they are scattered not only by the electron–ion interaction but by the electron–electron interaction as well, and they can therefore lose energy to other electrons in the process. The observed energy spectrum may include a spurious contribution due to inelastically scattered electrons, for which a correction may have to be applied.

According to the independent-particle model, all the electrons in a regular solid metal and all the vacant states above the Fermi level are described by unique values of **k**, and **k** is conserved during transitions; since the wave number of the photon is negligible, only excitations which are *vertical* in an E–**k** diagram are allowed. If this model is correct, it follows that the variation of the energy spectrum of the emitted electrons will normally be complex; there should be peaks in the spectrum for some frequencies, which are just right to excite electrons from a region of high density of states in one band to a region of high density of states in another, but when the frequency is changed these peaks should disappear. In practice, however, the appearance of the spectrum is often independent of frequency, though its peaks are liable to be displaced in energy as $\hbar\omega$ increases. Berglund & Spicer showed that all their results for solid Cu were consistent with the hypothesis that the number of electrons excited to energy E was simply proportional to $\mathcal{N}(E)\mathcal{N}(E-\hbar\omega)$ as would be expected if there were *no* restriction on the change of **k** during an excitation and if the matrix elements governing the transition probabilities were independent of $\hbar\omega$. They were able to unfold their results so as to obtain a curve for $\mathcal{N}(E-\hbar\omega)$ below the Fermi level; the derivation of a similar curve for $\mathcal{N}(E)$ above the vacuum level is necessarily complicated by uncertainty about the way in which the escape probability for an electron varies with E. Berglund & Spicer coined the term *non-direct* to describe transitions which are not vertical in **k**-space and yet which do not appear to require the intervention of any phonons to conserve **k**. Evidence for the importance of non-direct transitions has since been found for several solid metals besides Cu.

There is still some controversy about Berglund & Spicer's hypothesis, for there are those who claim that the results can be explained without it (e.g. Smith & Traum, 1970; Christensen, 1971). In view of what has been said in the previous section about the fallibility of the independent-particle model, however, the concept of a non-direct transition is by no means offensive. The values of E involved in these experiments are necessarily several eV above the Fermi level (though the work function of the metal is sometimes deliberately reduced by coating the surface with a thin layer of Cs) and the values of $E-\hbar\omega$ at which interesting features appear in the density of states curve may lie several eV below it. The electron is therefore liable to be perturbed by electron–electron interaction, both in its initial and excited states, and when strongly perturbed its $\mathbf{k}$ value is no longer well defined. The matters for controversy concern such details as just how far one has to go from the Fermi level before the requirement of $\mathbf{k}$ conservation becomes irrelevant, whether the matrix elements are really independent of frequency, and so on. Until these matters are settled by a sound many-body theory the density of states curves obtained by photo-emission experiments must be accepted with some reserve.

The only detailed photo-emission studies which have so far been reported for liquid metals are those of Koyama & Spicer (1971) on In, of Cotti *et al.* (private communication) on Hg and of Eastman (1971) on Au. The energy distribution curves for solid In suggest a peak in the density of states about 1.3 eV below the Fermi level, which has been interpreted by Koyama, Spicer, Ashcroft & Lawrence (1967) in terms of the crystal band structure. Rather surprisingly, this feature seems to survive the melting process. The most noticeable difference between the results for solid and liquid In is that a greater proportion of the electrons in the latter phase get scattered on the way out of the metal and emerge with reduced energies. For Hg melting seems to make very little difference of any sort. For Au the results suggest that the energy separation between the top of the d band and the Fermi level decreases by about 0.35 eV on melting. As noted on p. 396 this is not inconsistent with the optical properties of solid and liquid Au.

The photo-electric effect also provides a means, of course, for measuring the *work function* of a metal. We have discussed the theory of the work function for a jellium in §1.3, and have noted that the theoretical formula (1.15) is remarkably successful for Na and K. For a typical polyvalent metal it is not so successful; for Hg, for example, with $r_s = 2.71$ Bohr units, it predicts a work function of only 1.63 eV whereas the measured value is 4.5 eV. The difference is largely due to the surface dipole layer. The effect of this layer, for a real metal as opposed to a jellium, is liable to vary with the roughness of the surface on an atomic scale and the work function is not necessarily the same, therefore, for the different faces of a solid crystal; the calculations of Lang & Kohn (1971) suggest that variations of 0.2 eV or more are to be expected. It would be of interest to study the change of the work function on melting, starting with single crystal specimens of different orientations, for something might be learnt in this way about the structure of the liquid surface. Measurements of sufficient accuracy, however, would not be easy. In the experiments of Koyama & Spicer (1971) and Jean (1965) on In no change on melting was detected.

5.26. Plasma oscillations

The converse of a photo-emission experiment is to fire fast electrons into a metal from outside and to study their energy loss spectrum, ideally as a function of the angle through which they are scattered. This, too, can in principle yield information to supplement the results of purely optical experiments. In practice one is usually concerned with the rather limited objective of identifying peaks in the loss spectrum with the energy of the *plasmons* which a fast electron is liable to excite.

Plasma oscillations are longitudinal modes of vibration of the conduction electron gas. They can occur for values of ω and q such that $\tilde{\epsilon}_{\parallel}(q, \omega)$ (or alternatively $\tilde{\epsilon}'_{\parallel}(q, \omega)$) vanishes. Whenever this condition is satisfied an oscillatory electric field E can exist in the metal even when the induction D is zero, i.e. even though no charges have been introduced from outside (see p. 15). According to equations (5.73) and (5.74) the condition may also be written as

$$i\omega = \frac{4\pi\sigma(0, 0)}{1 - i\omega\tau} f_{\parallel}\left(\frac{qL}{1 - i\omega\tau}\right)\left(1 + \frac{\epsilon'_{\parallel}(q, 0)}{\epsilon_{\parallel}(q, 0)} \frac{q^2}{q_s^2}\right). \qquad (5.97)$$

Since the solution turns out to be such that $\omega\tau \gg 1$, approximations are possible. For small q we may write

$$\omega^2 \simeq \omega_p^2 \left(1 + \frac{q^2}{q_s^2}\left(\frac{9}{5} + \frac{\epsilon'_{\parallel}(0,0)}{\epsilon_{\parallel}(0,0)}\right) + \dots\right), \tag{5.98}$$

where ω_p is the free-electron plasma frequency defined by (5.71), usually in the neighbourhood of 10 eV. This result involves the assumption, however, that $\tilde{\sigma}(0, \omega)$ is Drude-like and that the distinctions between n^* and n, m^* and m, τ_D and τ can all be ignored. The plasma frequency for zero q may be shifted from ω_p by anything that affects $\tilde{\epsilon}(0, \omega)$, e.g. by inter-band effects in solid metals, by the core polarisability, or by electron–electron interaction. The reader will observe that since $\tilde{\epsilon}_{\parallel}(q, \omega)$ is a complex quantity (though the imaginary part is small at high frequencies) the exact solution for the plasma frequency must also be complex. The physical significance of this is that plasma oscillations are *damped*; their lifetime may be estimated experimentally from the width of the peaks in the energy loss spectrum for fast electrons.

In addition to the bulk plasma modes which (5.98) describes, there are surface modes whose dispersion curve is quite different; it is described by the equation (Ferrell, 1958)

$$\omega^2 = c^2q^2(1 + \tilde{\epsilon}_{\parallel}(q, \omega))/\tilde{\epsilon}_{\parallel}(q, \omega).$$

This means that ω is almost equal to cq for small q but saturates when $q > \omega_p/c$ (i.e. when $q/q_s > v_F/c$) to a value such that

$$\tilde{\epsilon}_{\parallel}(q, \omega) = -1. \tag{5.99}$$

In the same way that (5.97) reduces to (5.98), (5.99) reduces to

$$\omega^2 \simeq \tfrac{1}{2}\omega_p^2 \left(1 + \frac{q^2}{q_s^2}\left(\frac{9}{5} + \frac{2\epsilon'_{\parallel}(0,0)}{\epsilon_{\parallel}(0,0)}\right) + \dots\right). \tag{5.100}$$

The surface plasma frequency is therefore less than the bulk plasma frequency by a factor of about $\sqrt{2}$ over a large range of q. Both dispersion curves are shown schematically in fig. 5.24. Surface plasmons involve motion of the conduction electrons to a depth of say 100 Å, and the tendency to excite them is particularly

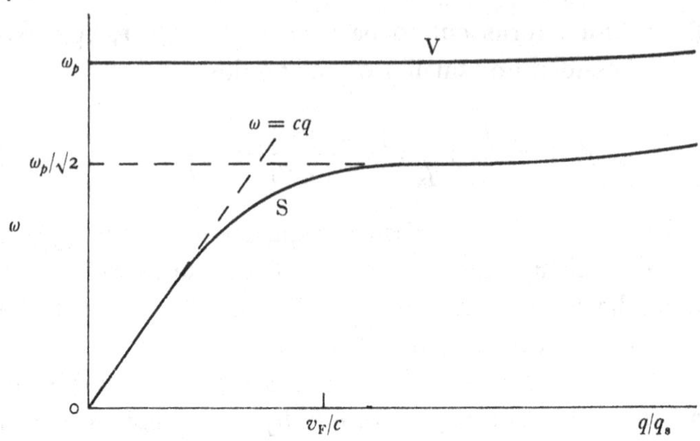

Fig. 5.24. Schematic dispersion curves for volume (V) and surface (S) plasmons.

great if the fast electrons are incident on the surface at a grazing angle.

Measurements of electron energy loss spectra for liquid metals have been reported by Powell (1968) for Au, Hg, Al, Ga, In and Bi, and he also quotes references to earlier work. In the case of Hg a detailed check of the theory is possible since the optical experiments of Wilson & Rice (1966) have provided us with a curve for $\epsilon(0, \omega)$ which extends out to high frequencies.† The curve crosses zero at about 7 eV and reaches -1 at about 6 eV, so we should expect to observe a bulk plasmon peak at one of these frequencies and a surface plasmon peak at the other. Powell's experiments reveal a single broad peak centred around 6.5 eV. It is, as yet, rather difficult to tie up his other results with what we know already from optical experiments or to learn anything fresh from them. The change in the loss spectrum on melting is most marked in the case of Ga and Bi, which is scarcely surprising. In the case of Al it is possible to detect very clearly both the bulk plasmon peak, at about 14.5 eV, and the surface one, at about 10.0 eV. Both these frequencies rise by about 0.4 eV on solidification but there is very little change in the proportions with which, for a given angle of incidence, the two types of

† It will be remembered that in the limit of small q there is no distinction between the $\epsilon_\perp$ measured in an optical experiment and the $\epsilon_\parallel$ which determines the plasma frequency (see p. 368).

plasmon are excited. These observations encourage the belief that the liquid phase is just as homogeneous as the solid over a depth of 100 Å or so below the surface.

Surface plasmon effects are not confined, incidentally, to experiments on the energy loss spectrum of fast electrons. They have been detected in photo-emission work (Smith & Spicer, 1969), in the optical emission from irradiated foils (Ritchie & Eldridge, 1962) and in reflectivity studies using solid specimens with deliberately roughened surfaces (e.g. Beaglehole & Hunderi, 1970). In a book devoted to liquid metals, however, these observations are of little relevance.

5.27. The Hall effect

One last problem concerning the transport properties of liquid metals remains to be discussed, namely the effect of a magnetic field upon them.

Let us start with the simplest available treatment, based upon the free-electron model, for which fig. 5.25 will serve as an illustration. The first picture shows, in a highly exaggerated fashion, the effect upon the electron distribution in **k**-space of an electric field **E** in the plane of the paper; it is shifted without change of form (see p. 212) through a vector

$$\Delta\mathbf{k} = \frac{m}{ne\hbar}\mathbf{j} = \frac{m\sigma}{ne\hbar}\mathbf{E},$$

which corresponds to the displacement OP in the figure. The second picture shows the effect of a magnetic field **H** perpendicular to the paper; the electrons now experience a Lorentz force $(e\hbar/mc)\mathbf{k}\wedge\mathbf{H}$ which drives them around the contours of constant energy in **k**-space, causing the whole distribution to rotate about its centre O with the cyclotron frequency

$$\omega_C = e\mathrm{H}/mc. \tag{5.101}$$

The third picture shows the effect of applying the two fields simultaneously; the distribution is shifted without change of form through a vector

$$\Delta\mathbf{k}' = \frac{m\sigma}{ne\hbar}\mathbf{E}' \tag{5.102}$$

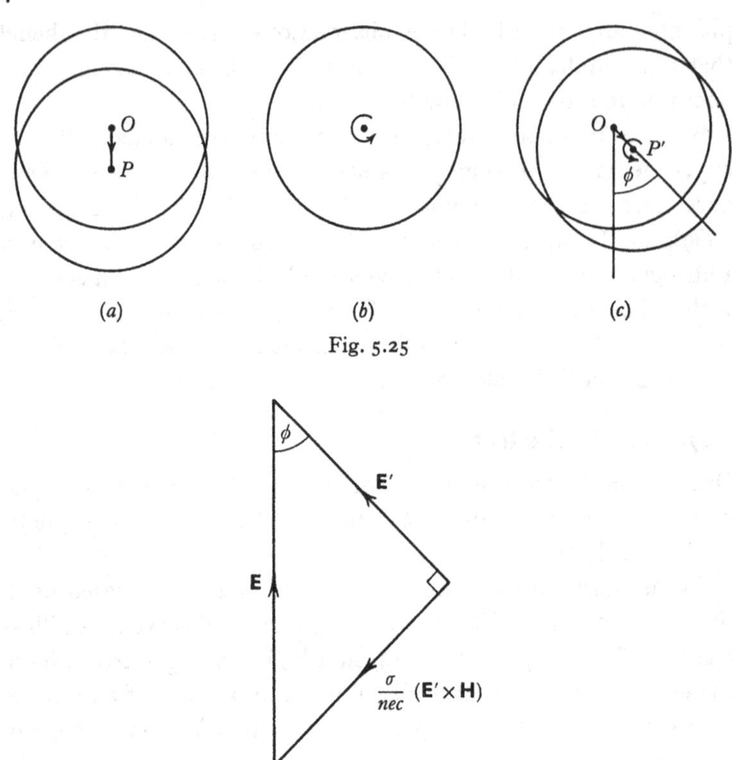

Fig. 5.25

Fig. 5.26

and is set rotating with the cyclotron frequency about its new centre at P'. The rotation does not affect the current, of course, which is given by

$$\mathbf{j} = \sigma\mathbf{E}'. \tag{5.103}$$

The magnetic field twists it away from the direction of $\mathbf{E}$ through the *Hall angle* ϕ.

To justify the final picture and to calculate the effective field $\mathbf{E}'$ is very simple. When both fields are present an electron at $\mathbf{k}$ (measured from O) experiences a net force

$$e\left(\mathbf{E}+\frac{\hbar}{mc}\mathbf{k}'\wedge\mathbf{H}+\frac{\hbar}{mc}\Delta\mathbf{k}'\wedge\mathbf{H}\right),$$

where $\mathbf{k}'$ $(=\mathbf{k}-\Delta\mathbf{k}')$ is measured from P'. The second term is the one that causes rotation about P' and this may be ignored. It is clearly a self-consistent hypothesis, therefore, to calculate the distribution on the assumption that every electron experiences the same effective field

$$\mathbf{E}' = \mathbf{E} + (\hbar/mc)\Delta\mathbf{k}' \wedge \mathbf{H}.$$

From (5.102) it follows that

$$\mathbf{E}' = \mathbf{E} + (\sigma/nec)\mathbf{E}' \wedge \mathbf{H},$$

a relationship which is displayed geometrically in fig. 5.26. Evidently

$$\tan\phi = \sigma \mathrm{H}/nec = \omega_C \tau. \tag{5.104}$$

Since ω_C is still only about 10^{12} sec^{-1} in a field of 60 kgauss, while τ for a typical liquid metal is of order 10^{-15} sec, the Hall angle is very much smaller in practice than the diagram suggests.

The usual experimental arrangement is such that the direction and magnitude of $\mathbf{j}$, and therefore of $\mathbf{E}'$, are fixed. Fig. 5.26 shows that the actual electric field $\mathbf{E}$ must then acquire a transverse component given by

$$\mathrm{E}_\perp = j\mathrm{H}/nec.$$

The coefficient $(1/nec)$ is of course the *Hall coefficient*, usually denoted by R_H. Its magnitude depends only upon the density and charge of the carriers, not upon their relaxation time and mass. If the electrons have an effective mass m^* which differs from m, the answer, as given by the conventional transport theory of §5.3, is still the same: the Lorentz force, being determined by the group velocity, is altered by a factor m/m^* and so are the cyclotron frequency and the Hall angle, but m^* vanishes from the expression for R_H.

For many solid metals the equation

$$R_H = R_H(\mathrm{FE}) = \frac{1}{nec} \tag{5.105}$$

proves to be quite unsatisfactory, but that is largely because the Fermi surface is non-spherical. The Hall coefficient is a sensitive quantity which is easily upset by Fermi surface distortion, so sensitive in fact that it may vary appreciably with temperature on heating the metal up to its melting point. At the melting point, however, there is usually a big change in R_H – the precise factor varies widely from metal to metal and need not concern us –

TABLE 5.8. *The Hall coefficient in liquid metals*

Metal	z	$necR_H$
Na	1	0.99, 0.98
Cu	1	1.00
Ag	1	1.02, 1.97
Au	1	1.00
Zn	2	1.01, 1.01, 1.00, 1.00
Cd	2	0.99, 0.98, 0.96, 1.04
Hg	2	0.99, 0.98, 1.00, 1.00, 1.00, 0.98, 0.96, 1.20, 1.22
Al	3	1.00
Ga	3	0.97, 0.99, 1.00, 1.04
In	3	0.93, 1.00, 0.98, 1.04, 0.95, 0.80
Tl	3	0.96, 0.76
Ge	4	1.00
Sn	4	1.00, 1.00, 1.00, 1.00, 0.98, 1.07
Pb	4	0.88, 0.88, 0.88, 0.73, 0.38
Sb	5	0.92, 1.14
Bi	5	0.95, 0.95, 0.69, 0.60
Te	6	3.3 (decreases on heating)

and in the liquid phase (5.106) is obeyed with remarkable precision. This was one of the first bits of information about liquid metals to attract the attention of theorists and to persuade them that the NFE model should be applicable.

Measurement of the Hall coefficient for a liquid metal is not easy, because it is small, because the specimen is hard to contain in exactly the required shape, and because the force exerted by $\mathbf{H}$ on $\mathbf{j}$ is liable to set the liquid in motion, with disturbing results. In recent years, however, much patience and technical ingenuity has been devoted to the problem and reliable results are now available for most liquid metals that melt at accessible temperatures. Table 5.8 shows results collected by Busch & Güntherodt (1967) from papers published by six or seven different schools, supplemented by the more recent results of Enderby & Walsh (1966), Aldridge (1968), Hasan (1969), Crozier (private communication), and Shackle (1970). They are presented in the form of ratios to the free-electron value predicted by (5.106). Some effort has been made to place first those results which seem to be most reliable, though it is of course difficult to compare measurements that have often been made by quite different methods. Since few authors claim better than 5% accuracy, the spread that exists in the figures is hardly surprising. By and large they are astonishingly close to unity, with Pb and Te as the only marked exceptions.† In one or two cases it has been shown that $|R_H|$ increases slowly on heating above the melting point, at about the rate to be expected from the known coefficient of thermal expansion.

The agreement is remarkable because the theory undoubtedly has its limitations. In view of the reservations expressed in §5.3 about the group velocity concept, are we really justified in eliminating m^* from the answer? Are there no normalisation corrections to be included? Are there no corrections of order $(\Gamma_F/K_F)^2$ or $(1/Lk_F)^2$ when the mean free path is short?‡ And finally, what are the effects, especially for heavy elements such as Pb and Bi, of spin-orbit interaction inside the ion cores? A more sophisticated quantum mechanical theory is needed to answer these questions. Attempts at a better theory have been described by Evans (1966), Banyai & Aldea (1966), Aldea (1967), Springer (1967), Fukuyama, Ebisada & Wada (1969) and Ramakrishnan (1971), but none of them is conclusive. Fukuyama *et al.* arrive at a result which, for a system of nearly-free electrons subjected to weak perturbations, may be expressed in the simple form

$$R_H/R_H(\text{FE}) = (m/m^*)^2. \tag{5.106}$$

This may be compared with an earlier conjecture due to Ziman (1967),

$$R_H/R_H(\text{FE}) = m/m^*. \tag{5.107}$$

Neither formula appears to be consistent with the experimental results.

In principle it should be straightforward enough to extend the theory developed in §§5.11–5.14 so as to include the effects of a magnetic field. Suppose, for example, that one imagines an

† It should be said, however, that since it is more difficult to make an absolute than a relative measurement of R_H, some experimenters have chosen to calibrate their apparatus with Hg or Ga. There seems to be good evidence that $necR_H$ is very close to 1.00 for both these liquid metals, but if this evidence is ever overturned it will upset many of the other readings listed in table 5.8.

‡ The suggestion made by Greenfield (1964) that a fair correlation exists between $necR_H$ and $1/Lk_F$ has not been borne out by subsequent experiments.

oscillating electric field to be applied to the metal in the x direction, with

$$\mathsf{E}_x = \mathsf{E}_0 \cos \omega t$$

and at the same time a magnetic field in the z direction described by a vector potential (in a convenient linear gauge) of the form

$$\mathsf{A}_y = \mathsf{H}_z x, \quad \mathsf{A}_x = 0. \tag{5.108}$$

One may pick out immediately one contribution to the current density in the y direction, which arises from the term

$$-\frac{e^2}{m\Omega c}\int \mathbf{A}\Psi^*\Psi \,\mathbf{dr} \tag{5.109}$$

in (5.35); the transitions excited by the electric field 'polarise' the electronic wave functions Ψ along the x direction in such a way that this term does not vanish. It may be shown to amount to

$$\begin{aligned} i_y &= \frac{2e^2\hbar\mathsf{H}_z}{m^2c\Omega}\sum_n \frac{1}{\delta_n}\operatorname{Re}(C_n X'^*_{n0}) \\ &= -\frac{e^3\hbar}{m^3c\Omega}\mathsf{H}_z\,\mathsf{E}_0\sum_n \frac{|X'_{n0}|^2}{\delta_n^2}(\cos\omega t - \cos\delta_n t)\left(\frac{1}{\omega+\delta_n}+\frac{1}{-\omega+\delta_n}\right) \end{aligned}$$

for a single electron. If we concentrate upon the part of this expression which yields a current in phase with E_x and sum over all the conduction electrons in the specimen, we obtain

$$\begin{aligned} j_y = \frac{4e^3\hbar^2}{m^3c\Omega}\mathsf{H}_z\mathsf{E}_x\int_0^\infty \frac{d\delta}{\delta(\omega^2-\delta^2)} \\ \times\int_0^\infty \{f(E)-f(E+\hbar\delta)\}\mathcal{N}(E)\mathcal{N}(E+\hbar\delta)\,\overline{|X'(\delta)|^2}\,dE, \end{aligned}$$

which reduces with the aid of (5.59) to

$$j_y = \frac{2e}{\pi mc}\mathsf{H}_z\mathsf{E}_x\int_0^\infty \frac{\sigma(\delta)\,d\delta}{\omega^2-\delta^2}.$$

If this were to be the *only* term in j_y proportional to H_z and E_x and in phase with the latter, then the Hall angle would be (see (5.58))

$$\frac{j_y}{j_x} = \frac{2e\mathsf{H}_z}{\pi mc}\int_0^\infty \frac{\sigma(\delta)\,d\delta}{\sigma(\omega)(\omega^2-\delta^2)} = \frac{e\mathsf{H}_z}{4\pi mc}\,\frac{1-\epsilon(\omega)}{\sigma(\omega)}. \tag{5.110}$$

This reduces to

$$\frac{j_y}{j_x} = \frac{e\mathsf{H}_z\tau_D}{mc}$$

if σ and ϵ obey the Drude formulae, which immediately suggests

$$|R_H/R_H(\text{FE})| = n/n^* = (m_1^*)^2/m^*m_{XC}^*. \qquad (5.111)$$

This result looks so very plausible that one is tempted to accept it as the full answer to the problem. It would justify us in ignoring the corrections of order $(\Gamma_F/K_F)^2$ so long as they are negligible in $\sigma(\omega)$. And it could be claimed that the values for $necR_H$ in table 5.8 do tend to be rather less than unity, which would tie up with the observation that n^* generally exceeds n; in this respect (5.111) is noticeably more successful than (5.106) or (5.107). However, a detailed comparison with the figures in table 5.7 shows that there is no real correlation between $necR_H$ and the results of optical experiments on the same metal. And honesty compels the author to admit to a more serious failing of the theory: the sign of the Hall effect predicted by (5.110) appears to be incorrect.

The fact is that when we switch on the magnetic field a perturbing term

$$\frac{ie\hbar}{mc} \mathsf{H}_z \left(x \frac{\partial}{\partial y}\right)$$

appears in the Hamiltonian. This can excite transition between states in the same way as the electric field, and a complete calculation of j_y requires consideration of a whole range of second-order processes in which the electrons are perturbed by both E and H in succession. Had we chosen a different gauge for the vector potential, namely

$$\mathsf{A}_x = -\mathsf{H}_z y, \quad \mathsf{A}_y = 0,$$

then we would have obtained no transverse current from (5.109) and we would have been forced to discuss these second-order processes. There seems no justification for ignoring them when (5.108) is used instead. Unfortunately, they complicate the calculation so severely that a solution is not yet in sight.

It has recently been reported by Even & Jortner (1972*a,b*) that the Hall coefficient of Hg, which conforms so closely to the free-electron value at normal densities, begins to rise when the density is diminished by heating towards the critical point; it is said to have risen by a factor 3 by the time that R_A has reached about

2.1 Å, i.e. well before the 'knee' in the resistivity curve that is plotted in fig. 5.11. To understand this behaviour, if it is confirmed, we may have to turn to theories based upon the tight-binding approximation, such as Matsubara & Kaneyoshi (1968) and Friedman (1971) have developed. These theories appear to confirm the sort of dependence on density of states that is indicated by (5.107) and they therefore suggest that the rise in R_H observed by Even & Jortner is associated with the development of a pseudo-gap, the existence of which in Hg at low densities has already been postulated on p. 352.

All the theories mentioned so far imply that the sign of the Hall coefficient in liquid metals should always be negative. Recent experimental work at Zürich, however, has demonstrated that many transition and rare earth metals are exceptions to this rule; the Hall coefficient is clearly positive, i.e. it has the sign characteristic of conduction by holes rather than electrons, in liquid Fe, Co, La, Pr, Nd, Ce and U (Busch, Güntherodt, Künzi & Schapbach, 1970; Busch, Güntherodt & Künzi, 1970, 1971; Güntherodt, private communication). The change in R_H at the melting point in these metals, like the change in resistivity (see p. 330), is rather slight. These results still await explanation.

5.28. Magneto-resistance

The resistivity of a metal with a spherical Fermi surface ought not to be affected by the application of a transverse magnetic field for, as may be seen from fig. 5.26, the component of electric field along the direction of the applied current $\mathbf{j}$ should remain equal to $\mathbf{j}/\sigma$; it is only this component which is measured, of course, in a typical experiment to determine ρ. The transverse magneto-resistance which has frequently been reported for liquid metals, and for liquid Hg in particular, can safely be dismissed as a spurious effect, due to motion of the specimen (Williams, 1925). A longitudinal magneto-resistance effect has been reported for liquid K by Kikoin & Fadikov (1935) (see also Armstrong (1935) for similar results on liquid Na–K). This too is presumably spurious, though it is by no means clear what was wrong with the observations.

5.29. Appendix

We take up here the problem of how to evaluate the cross-product terms in the square of the matrix element that determines $\sigma(\omega)$, i.e. the double sum

$$\sum_{\mathbf{k}}\sum_{\mathbf{k}'\neq\mathbf{k}} k\cos\alpha_{\mathbf{k}}n_{\mathbf{k}}m_{\mathbf{k}}^{*}k'\cos\alpha_{\mathbf{k}'}n_{\mathbf{k}'}^{*}m_{\mathbf{k}'} \tag{5.112}$$

which occurs on the right-hand side of (5.44). To make the argument easier to follow we shall adopt a number of simplifying assumptions: (*a*) that m^*/m is unity and that $A_{\mathbf{k}}$ is small enough to be ignored; (*b*) that $\Gamma_{\mathbf{k}}$ and $\hbar\omega$ are both small compared with K_{F}; (*c*) that the Born approximation is valid; (*d*) that the pseudo-potential is Hermitian and independent of E. Of these, (*a*) and (*b*) are not essential but (*c*) undoubtedly is; it is possible that a breakdown of the Born approximation could seriously affect the answer obtained.

The fact that we are interested only in the ensemble average of (5.112) allows us to achieve some slight simplification at the start by invoking the theorem that

$$\cos\alpha_{\mathbf{k}'} = \cos\alpha_{\mathbf{k}}\cos\theta+\sin\alpha_{\mathbf{k}}\sin\theta\cos\phi, \tag{5.113}$$

where the angles involved are defined by the stereogram in fig. 5.27. Evidently, when the sum over $\mathbf{k}'$ is completed any term that involves $\cos\phi$ is bound to vanish on averaging. Since assumption (*b*) allows us to replace both k and k' by k_{F}, we may concentrate attention on the average of

$$\tfrac{1}{3}k_{\mathrm{F}}^{2}\sum_{\mathbf{k}}\sum_{\mathbf{k}'\neq\mathbf{k}} n_{\mathbf{k}}m_{\mathbf{k}}^{*}n_{\mathbf{k}'}^{*}m_{\mathbf{k}'}\cos\theta. \tag{5.114}$$

To see that this does not necessarily vanish one need only observe that $n_{\mathbf{k}}$ contains an infinitesimal part which has arisen by scattering direct from $n_{\mathbf{k}'}$ and is therefore proportional to $U_{\mathbf{k}'\mathbf{k}}n_{\mathbf{k}'}$, where $U_{\mathbf{k}'\mathbf{k}}$ is the matrix element of the pseudo-potential, as used in §4.7. Similarly, $m_{\mathbf{k}}^{*}$ contains an infinitesimal part proportional to $U_{\mathbf{k}'\mathbf{k}}^{*}m_{\mathbf{k}'}$. The product of these two infinitesimal terms survives averaging because it involves $(n_{\mathbf{k}'}m_{\mathbf{k}'}^{*}n_{\mathbf{k}'}^{*}m_{\mathbf{k}'})$, which is essentially positive, and the *closed ring* of matrix elements $U_{\mathbf{k}'\mathbf{k}}U_{\mathbf{k}\mathbf{k}'}$.

The secret of evaluating the average of (5.114) is to sum all

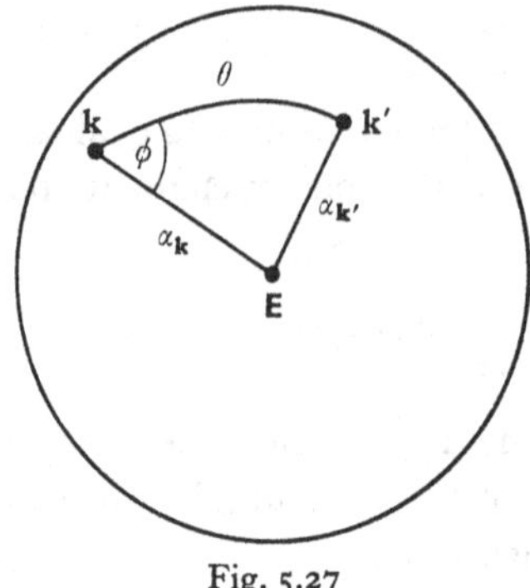

Fig. 5.27

the infinitesimal terms in which a closed ring of matrix elements is involved. If the Born approximation is valid, however, we may assume that the contribution associated with the ring

$$U_{\mathbf{k}'\mathbf{k}''}U_{\mathbf{k}''\mathbf{k}}U_{\mathbf{k}\mathbf{k}'},$$

even after summation over $\mathbf{k}''$, is negligible compared with the contribution associated with $U_{\mathbf{k}'\mathbf{k}}U_{\mathbf{k}\mathbf{k}'}$, and similarly we may neglect $U_{\mathbf{k}'\mathbf{k}''}U_{\mathbf{k}''\mathbf{k}}U_{\mathbf{k}\mathbf{k}'''}U_{\mathbf{k}'''\mathbf{k}'}$ so long as $\mathbf{k}''$ and $\mathbf{k}'''$ are different. In fact all rings like those shown schematically in figs. 5.28(*b*), (*c*) and (*d*) may be forgotten; it is only the ones which are completely *collapsed*, like (*a*), (*e*) and (*f*) which matter. In our previous calculation of Γ we have allowed fully for the higher-order collapsed rings such as (*e*) and (*f*) by associating a complex energy $E_{\mathbf{k}}-\mathrm{i}\Gamma_{\mathbf{k}}$ with each $\mathbf{k}$. This device is still useful. In particular, it enables us to treat as a single case both (*e*) and those rings such as (*f*) which differ from (*e*) only by the addition of side arms.

Let us now focus on the infinitesimal contribution to

$$n_{\mathbf{k}}m_{\mathbf{k}}^{*}n_{\mathbf{k}'}^{*}m_{\mathbf{k}'} \tag{5.115}$$

which involves a collapsed ring of matrix elements with $2l$ links, not counting side arms. It simplifies the notation to label the states which are connected by this ring by numbers, 0 (representing $\mathbf{k}$), 1, 2, etc., up to l (representing $\mathbf{k}'$). Let

$$\nu_0 = \frac{1}{(E_0-E_n-\mathrm{i}\Gamma)} = \frac{1}{\Gamma}\frac{1}{(z_0-\omega\tau_l-\mathrm{i})} \tag{5.116}$$

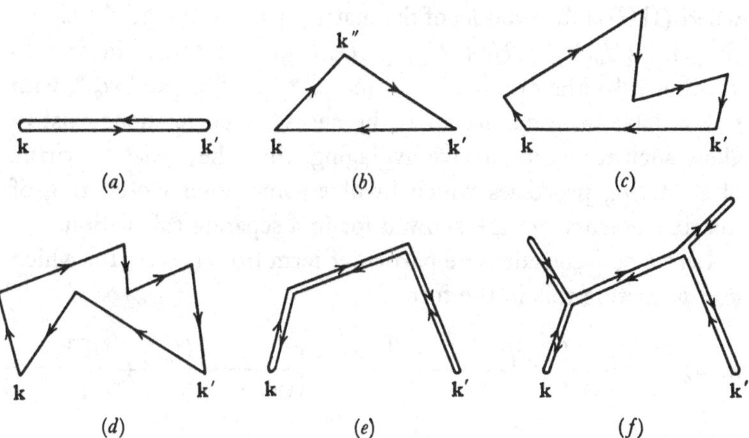

Fig. 5.28. Closed rings in **k**-space linking **k** and **k**′.

and
$$\mu_0 = \frac{1}{(E_0 - E_m - i\Gamma)} = \frac{1}{\Gamma}\frac{1}{(z_0 + \omega\tau_l - i)}, \tag{5.117}$$

and define ν_1, μ_1, ν_2, μ_2, etc. similarly for the other states on the ring. The energies of the two eigenfunctions, E_n and E_m, differ by $\hbar\omega$ and z has the same meaning as on p. 251. Then we may rewrite (5.115), in the light of (4.37), in the form

$$\begin{aligned}
&[n_{\bar{0}} + n_{\bar{1}}(\nu_0 U_{10}) + n_{\bar{2}}(\nu_1\nu_0 U_{21}U_{10}) + \dots n_{\bar{l}}(\nu_{l-1}\dots\nu_0 U_{l,l-1}\dots U_{10})] \\
&\quad\times [m_{\bar{0}} + m_{\bar{1}}(\mu_0 U_{10}) + \dots m_{\bar{l}}(\mu_{l-1}\dots\mu_0 U_{l,l-1}\dots U_{10})]^* \\
&\quad\times [n_{\bar{0}}(\nu_1\dots\nu_l U_{01}\dots U_{l-1,l}) + n_{\bar{1}}(\ ) + \dots n_{\bar{l}}]^* \\
&\quad\times [m_{\bar{0}}(\mu_1\dots\mu_l U_{01}\dots U_{l-1,l}) + m_{\bar{1}}(\ ) + \dots m_{\bar{l}}].
\end{aligned} \tag{5.118}$$

The various infinitesimal contributions to n_0 that arise by scattering along the chain of states under consideration are written out explicitly in the first of these square brackets, $n_{\bar{0}}$ being the residue of n_0, and the other three brackets describe $m_{\bar{0}}^*$, n_l^* and m_l in the same detail. The relation between n_1 and $n_{\bar{1}}$, n_2 and $n_{\bar{2}}$ etc., is similar to that between n_0 and $n_{\bar{0}}$. In each case the difference is an infinitesimal one, so that $n_{\bar{0}}n_{\bar{0}}^*$ can safely be replaced by $n_0 n_0^*$. We can therefore rewrite (5.118) as

$$\begin{aligned}
&(\Pi U)\sum_{p=0}^{l}\sum_{q=0}^{l} n_p n_p^* m_q m_q^* \\
&\quad\times(\nu_1\nu_2\dots\nu_{p-1}\nu_{p+1}^*\dots\nu_{l+1}^*)(\mu_1^*\mu_2^*\dots\mu_{q-1}^*\mu_{q+1}\dots\mu_{l+1}),
\end{aligned} \tag{5.119}$$

where (ΠU) is the product of the matrix elements round the closed ring, i.e. $(U_{01}U_{12}\ldots U_{l-1,l}U_{l,l-1}\ldots U_{21}U_{10})$. All terms in (5.118) which involve the product of n_p^- and n_q^{-*}, or of m_p^- and m_q^{-*}, with $p \neq q$, have been discarded. If the necessary correlations exist to allow such terms to survive averaging, then they exist by virtue of scattering processes which involve some other closed ring of matrix elements and are allowed for in a separate calculation.

Let us now consider one particular term from (5.119) for which $q = p$, writing this in the form

$$n_p n_p^* m_p m_p^* \left[\frac{|U_{01}|^2/\Gamma^2}{(z_0-\omega\tau_l-\mathrm{i})(z_0+\omega\tau_l+\mathrm{i})}\right] \times \ldots \times \left[\frac{|U_{p-1,p}|^2/\Gamma^2}{(z_{p-1}-\omega\tau_l-\mathrm{i})(z_{p-1}+\omega\tau_l+\mathrm{i})}\right]$$
$$\times \left[\frac{|U_{p,p+1}|^2/\Gamma^2}{(z_{p+1}-\omega\tau_l+\mathrm{i})(z_{p+1}+\omega\tau_l-\mathrm{i})}\right] \times \ldots \times \left[\frac{|U_{l-1,l}|^2/\Gamma^2}{(z_l-\omega\tau_l+\mathrm{i})(z_l+\omega\tau_l+\mathrm{i})}\right]. \quad (5.120)$$

In the course of evaluating (5.114) we are obliged to sum (5.120) over all possible values for $\mathbf{k}$ and $\mathbf{k}'$ and we may at the same time sum over all possible intermediate states along the $2l$-fold ring. In principle these states can lie anywhere in $\mathbf{k}$-space, but in practice, when Γ is small, (5.120) is only significant when they all lie close to the Fermi sphere. It then follows from (4.44) that when $\omega\tau_l$ is zero each square bracket in (5.120) yields a factor of unity when the summation is carried through. When $\omega\tau_l$ is not zero a simple contour integration of expressions such as

$$\int_{-\infty}^{\infty} \frac{\mathrm{d}z}{(z-\omega\tau_l-\mathrm{i})(z+\omega\tau_l+\mathrm{i})}$$

shows the answer to be

$$\sum_p n_p n_p^* m_p m_p^* \left(\frac{1}{1-\mathrm{i}\omega\tau_l}\right)^p \left(\frac{1}{1+\mathrm{i}\omega\tau_l}\right)^{l-p}. \quad (5.121)$$

We should not lose sight, however, of the vital $\cos\theta$ factor which occurs in (5.114). This can be broken down, according to (5.113), into

$$\cos\theta = \cos\theta_{0l} = \cos\theta_{01}\cos\theta_{1l} + \sin\theta_{01}\sin\theta_{1l}\cos\phi', \quad (5.122)$$

where θ_{01} is the angle of scattering in going from state 0 to 1 and similarly for θ_{1l}; ϕ' is analogous to ϕ in fig. 5.27. The term in

$\cos\phi'$ cannot survive the summation over states and a subsequent ensemble average, so that we may write

$$\cos\theta = \cos\theta_{01}\cos\theta_{1l}$$

and, by an extension of the argument,

$$\cos\theta = \cos\theta_{01}\cos\theta_{12}\ldots\cos\theta_{l-1,l}.$$

A factor $(\overline{\cos\theta})^l$ should therefore be added to (5.121).

A similar analysis is now required for each term in (5.119) for which $q \neq p$. Terms for which $|q-p| \geqslant 2$ fortunately vanish on summation, because they all involve at least one integral of the form

$$\int_{-\infty}^{\infty}\frac{dz}{(z-\omega\tau_l-i)(z+\omega\tau_l-i)} \quad \text{or} \quad \int_{-\infty}^{\infty}\frac{dz}{(z-\omega\tau_l+i)(z+\omega\tau_l+i)}$$

which is clearly zero. A term for which $(q-p) = 1$ yields

$$|U_{pq}|^2[n_p n_p^* \mu_p][m_q m_q^* \nu_q]\left[\frac{|U_{01}|^2/\Gamma^2}{(z_0-\omega\tau_l-i)(z_0+\omega\tau_l+i)}\right]$$

$$\times\ldots\times\left[\frac{|U|_{l-1,l}|^2/\Gamma^2}{(z_l-\omega\tau_l+i)(z_l+\omega\tau_l-i)}\right] \quad (5.123)$$

in place of (5.120). Since the coefficients n and m are proportional to $(z\pm\omega\tau_l-i)^{-1}$, it is legitimate to rewrite the first part of (5.123) in the form

$$[n_p n_p^* m_p m_p^*(z_p+\omega\tau_l-i)]$$

$$\times\left[\frac{|U_{pq}|^2/\Gamma^2}{(z_q-\omega\tau_l+i)(z_q+\omega\tau_l-i)(z_q+\omega\tau_l+i)}\right]\times\ldots$$

It is then identical with (5.120), apart from the factors $(z_p+\omega\tau_l-i)$ and $(z_q+\omega\tau_l+i)^{-1}$. On integration over z_p and z_q these turn out to introduce factors of $-i(1+i\omega\tau_l)$ and $1/2i$ respectively, so that we arrive at

$$-\tfrac{1}{2}\sum_p n_p n_p^* m_p m_p^*\left(\frac{1}{1-i\omega\tau_l}\right)^p\left(\frac{1}{1+i\omega\tau_l}\right)^{l-1-p} \quad (5.124)$$

in place of (5.121). For every term like this there is of course an identical one for which the roles of p and q are reversed and these mirror terms may be allowed for by deleting the factor $\frac{1}{2}$ from (5.124).

We now have to sum over all values of p, which runs from 0 to l if $q = p$, but only from 0 to $(l-1)$ if $|q-p| = 1$. Finally we must sum over all sizes of ring from $l = 1$ to ∞. The basic expression (5.114) then reduces to

$$\tfrac{1}{3}k_F^2 \sum_{\mathbf{k}} n_{\mathbf{k}} n_{\mathbf{k}}^* m_{\mathbf{k}} m_{\mathbf{k}}^* \left\{ \sum_{l=1}^{\infty} (\overline{\cos\theta})^l (A_l - A_{l-1}) \right\}, \tag{5.125}$$

with
$$A_l = \sum_{p=0}^{l} \left(\frac{1}{1-i\omega\tau_l}\right)^p \left(\frac{1}{1+i\omega\tau_l}\right)^{l-p}. \tag{5.126}$$

The effect of including the cross-terms in the square of the matrix element in (5.44) is therefore to multiply the answer by a factor

$$1 + \sum_{l=1}^{\infty} (\overline{\cos\theta})^l (A_l - A_{l-1}) = (1-\overline{\cos\theta}) \sum_{l=0}^{\infty} (\overline{\cos\theta})^l A_l = (1-\overline{\cos\theta})\, S. \tag{5.127}$$

The series S defined in (5.127) may be summed by making use of the recurrence relation

$$A_l = \left(\frac{1}{1-i\omega\tau_l}\right) A_{l-1} + \left(\frac{1}{1+i\omega\tau_l}\right) A_{l-1} - \left(\frac{1}{1-i\omega\tau_l}\right)\left(\frac{1}{1+i\omega\tau_l}\right) A_{l-2},$$

or
$$A_l - \frac{2A_{l-1} - A_{l-2}}{1+\omega^2\tau_l^2} = 0 \quad (l \geqslant 2), \tag{5.128}$$

which may be seen to hold between its coefficients. From this it follows that

$$\left(1 - \frac{2\,\overline{\cos\theta} - (\overline{\cos\theta})^2}{1+\omega^2\tau_l^2}\right) S = A_0 + \overline{\cos\theta}\left(A_1 - \frac{2A_0}{1+\omega^2\tau_l^2}\right) = 1.$$

Hence
$$(1-\overline{\cos\theta})\, S = \frac{(1-\overline{\cos\theta})(1+\omega^2\tau_l^2)}{(1+\omega^2\tau_l^2 - 2\,\overline{\cos\theta} + (\overline{\cos\theta})^2)} \tag{5.129}$$

is the correction factor required. This serves to justify the results quoted above as (5.48) and (5.61), at any rate within the limits set by our initial approximations.

When $\sigma(\omega)$ is calculated by the primitive method outlined on p. 361 it is obvious almost immediately that if $\overline{\cos\theta} \neq 0$ it is necessary to use τ rather than τ_l in both the enumerator *and* the denominator of the Drude formula. It is curious how elaborate the proof of this result becomes, once the traditional kinetic theory approach to conduction is abandoned.

CHAPTER 6

LIQUID ALLOYS

6.1. Prologue

Most metals are free to dissolve in one another once they are liquid, and the number of different alloy systems whose physical properties have now been studied as a function of concentration in the liquid state is very considerable. The papers listed in the valuable bibliographies prepared by Wilson (1965), B. W. Mott *et al.* (1966) and B. W. Mott (1968), to which the reader is frequently referred below, cover roughly 200 systems between them, though they are restricted almost exclusively to binary ones. Since the properties that have been studied range from straightforward mechanical and thermodynamic ones – structure, viscosity, heats of mixing, and so on – through electrical transport properties to such extras as the Soret effect or the Knight shift, it will be clear that the quantity of data to be surveyed in this final chapter is formidable. A frame of reference must be established at the start.

The lynch-pins of the theory developed for pure liquid metals in previous chapters are as follows:

(*a*) The NFE model for the conduction electrons.

(*b*) The idea that the scattering of electrons at the Fermi level by a single ion may be represented by some self-consistent screened pseudo-potential, $u(q)$, which we hope is weak enough to justify the use of perturbation theory and to make it unnecessary to go beyond the second order except in special cases.

(*c*) The use of an interference function $a(q)$ to allow for coherence effects due to the short range order of the ions.

(*d*) The use of an effective pair interaction between ions, $w(R)$, to express that part of the energy which depends upon structure rather than volume; the most essential feature of $w(R)$ has proved to be its repulsive core, and many properties of pure liquid metals that depend upon the interaction of ions with one another can often be described by means of simple rigid-sphere models.

To extend these concepts to liquid alloys is trivial from a formal point of view. It becomes necessary, of course, to distinguish between the pseudo-potentials of the different species of ion which are present, and for a given system these are liable to vary with concentration (§6.2). A set of *partial* inference functions $a_{\alpha\beta}(q)$ must be defined, as in §2.12, and a set of pair potentials $w_{\alpha\beta}(R)$. Fortunately, there are reasons to suppose (§6.4) that in many binary systems the range at which the pair potential w_{01} between a solvent and a solute ion becomes strongly repulsive is the arithmetic mean of the range of the repulsive cores in w_{00} and w_{11}. This observation entitles us to continue to make use of rigid-sphere models where appropriate. They may be used, for example, to predict the form of $a_{\alpha\beta}(q)$ (§6.5) and we may then go on to extend Ziman's theory and to reach quite powerful conclusions, of a semi-quantitative nature, about the resistivity and thermo-electric power of liquid alloys (§6.10).

By the end of §6.10 a comprehensive picture of how a liquid alloy *ought* to behave has begun to emerge, though some of the thermodynamic details are still obscure. This is because a knowledge of the pair potentials is insufficient to determine the total internal energy of a metal or its equilibrium volume, and little work has yet been done on how these should vary with concentration in alloys (§6.6). A crude elastic-continuum model has been applied to solid alloys by Friedel and others with some success and the theory of this is outlined in §6.7, though it is out of key with the philosophy of the rest of the book.† But Friedel's theory involves an assumption that to determine the equilibrium volume, say, it is sufficient to minimise internal energy rather than free energy. It is pointed out in §6.8 that the entropy of mixing, which can be estimated from the rigid-sphere model, is liable to deviate too far from its 'ideal' value to justify this approximation.

† Chemists and metallurgists instinctively think of alloys as assemblages of *atoms* and place much emphasis on the difference of atomic size between solvent and solute; this difference plays an essential role in Friedel's theory too. Once we start to think of assemblages of *ions* immersed in a sea of electrons, however, the emphasis changes completely. Each ion has an effective rigid sphere diameter, it is true, but this is determined largely by screening considerations (see §1.14) and may alter drastically when the ion is transplanted into a foreign solvent.

Unfortunately, our idealised picture of alloy behaviour rests on some rather shaky foundations. To begin with it is far from clear, as shown in §6.3, just how the Fermi radius in an alloy ought to be defined. More serious are the doubts expressed in §6.2 as to whether the pseudo-potentials in alloys are weak enough, especially when there is a large difference in valence or perhaps in electro-negativity† between solvent and solute, to justify the use of second-order perturbation theory. There may be alloy systems in which it fails so completely that the NFE model must be abandoned in favour of a tight-binding model of some sort. We must be prepared for the distribution functions g_2, g_3 etc. which describe the structure in such alloys to be so different from those of a rigid-sphere fluid that a description in terms of *chemical bonds* between neighbouring atoms may be required. It is even conceivable that the atoms may associate to form distinct *molecules* in the melt. It is still beyond our powers to predict when this situation will arise, and no general theory of the thermodynamical or electrical properties of non-free-electron liquid alloys is yet in existence. From an empirical point of view, however, our idealised picture is especially suspect for systems which form compounds in the solid phase (see §6.9).

The second part of the chapter is devoted to a systematic exposition of the experimental data that are available: §6.11 concerns the results of diffraction experiments; §§6.12 and 6.13, thermodynamic properties; §§6.14 and 6.15, surface properties; §§6.16 and 6.17, diffusion and viscosity; §6.18, electromigration; §6.19 magnetic susceptibility; §6.20, nuclear magnetic resonance; §6.21, the Hall effect; §6.22, optical properties; §§6.23–6.26, resistivity; §6.27, thermo-electric properties. Under each heading we shall be concerned first with the extent to which many liquid alloys conform to the NFE rigid-sphere model – the theory contained in the earlier sections is extended to cover particular problems as they arise – and secondly with the anomalies that may be attributed to 'compound formation'. In one or two cases the

† *Electro-negativity* is another concept which is frequently used by chemists and metallurgists but which seems elusive to those who approach metals from the point of view of the NFE model. An attempt to define it within the framework of this model is made in §6.2.

anomalies are so severe that the alloys must be classified as *semi-conductors* rather than metals. Some speculative models to account for the transport properties of these are described in §6.28, though here we are poaching on ground already covered by Mott & Davis (1971).

The argument is restricted to binary alloys, though many of the theoretical results could be extended to systems with more than two components if necessary. The convention is adopted that where one component is written in bold type, as in **Au**–Ag, this is to be regarded as the *solvent*; i.e., it is only near the **Au**-rich end of the concentration range that the properties of the system are under discussion. Where something like **Au**–M is written, M stands for a variety of metallic solutes. Concentrations are quoted in *atomic* proportions throughout. They are denoted by c_0 and c_1 ($c_0+c_1 = 1$) as in §2.12, and where the argument concerns a dilute alloy it is generally c_0 that refers to the solvent and c_1 to the solute. A super-script zero is sometimes used to indicate the limiting value of a concentration-dependent quantity when $c_1 \to 0$; thus u_1^0 represents the pseudo-potential of a solute ion at infinite dilution, and similarly u_1^1 represents the pseudo-potential of a solute ion in the pure solute. These superscripts are so clumsy, however, that they are employed only when the equations would be seriously ambiguous without them.

6.2. Pseudo-potentials in alloys

The ionic pseudo-potential $u(q)$ has played a central role in much of the theory we have developed for pure liquid metals. If we are to use it for binary alloys as well, we must evidently distinguish between u_0 and u_1, the pseudo-potentials of the two constituent species. What may not be so evident is that both u_0 and u_1 are liable to vary with concentration.

The diagrams in fig. 6.1 are designed to illustrate this point. The one on the left represents schematically, as in fig. 1.12, the potential experienced by an electron in the neighbourhood of an ion of species 0, when the concentration of species 1 is zero. At $R = R_A$ the true electrostatic potential (as measured by an infinitesimal test charge) is zero, because the atom as a whole is neutral, but for an electron the potential at P and P' is depressed

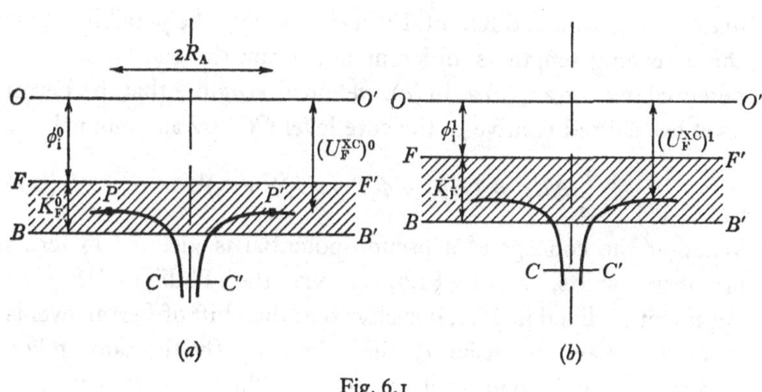

Fig. 6.1

below the vacuum level OO' by the effects of exchange and correlation (see § 1.3); the relationship between the mean pseudopotential of an electron at the Fermi level, its kinetic energy K_F, the exchange+correlation energy U_F^{XC}, and the inner† work function ϕ_i is as shown. Now suppose that elsewhere in the specimen some of the 0 ions are replaced by ions of species 1. As long as c_1 is very small, we may suppose the situation around our 0 ion to remain unchanged; even if the impurity ions have a quite different valency from the solvent ions we know that they are screened (see § 1.4) and that the electronic charge density and Fermi energy at a good distance from them are unaffected by their presence.

But the Friedel oscillations of charge density around the impurity ions are thought to extend over a range of four or five times R_A. This means that by the time c_1 is a few parts per cent our 0 ion is liable to have at least one impurity neighbour which is close enough to affect it. As c_1 is further increased the mean charge density around the 0 ion is bound to change continuously. Moreover, the Fermi level is bound to shift, towards the position that it occupies in the pure metal of species 1. Fig. 6.1(*b*) illustrates the situation of the 0 ion when it is completely surrounded by 1 ions, i.e. in the limit $c_1 \to 1$. The separation between the levels PP', BB', FF' and VV', being determined entirely by the surrounding

† No allowance is made in fig. 6.1 for the effect on the potential inside the specimen of a dipole layer at its surface.

matrix, is now quite different. Even if we ignore the possibility that the screening length is different and draw the same curve for potential in diagram (*b*) as in (*a*), we must recognise that the Fermi level has shifted relative to the core level CC' by an amount†

$$\Delta E_{\mathrm{F}} = (U_{\mathrm{F}}^{\mathrm{XC}}+\phi_1)^0-(U_{\mathrm{F}}^{\mathrm{XC}}+\phi_1)^1. \tag{6.1}$$

Whether the concept of a pseudo-potential is justified in terms of phase shifts, as in §1.7, or via the Phillips–Kleinman argument outlined in §1.8, it is clear that this shift of Fermi level is more than likely to make u_0^1 differ from u_0^0. By the same token u_1 may vary with concentration, between the two extremes of u_1^1 and u_1^0.

Experimental results are available for the total work function ϕ and theoretical estimates are available for $U_{\mathrm{F}}^{\mathrm{XC}}$. They suggest that ΔE_{F}, as given by (6.1), does not exceed a few tenths of an eV when the two constituents are homovalent and have similar atomic volumes. It is when z_1 differs greatly from z_0, or Ω_1 from Ω_0, that ΔE_{F} is likely to be large and the concentration-dependence of the pseudo-potentials most apparent. The same principle should hold even in the limit of small q, where the magnitude of ΔE_{F} is irrelevant; an extension of (1.67) suggests that

$$u_1^0(0) \simeq z_1\Omega_0/2\mathcal{N}(E)_{\mathrm{F},0}, \tag{6.2}$$

and hence that $$u_1^0(0)/u_1^1(0) \simeq (\Omega_0 z_1/\Omega_1 z_0)^{\frac{1}{3}}.$$

A warning has already been issued in §6.1 that when the two species differ greatly in valency the pseudo-potentials may be too strong to justify a linear screening theory. In a pure metal the atom as a whole is neutral even before screening, and the shift of charge density which is necessary to establish the self-consistent screened state is relatively small on this account. A solute atom at infinite dilution in a foreign solvent, however, if it occupies a volume Ω_1^0/N, carries a net charge of $(z_1-z_0\Omega_1^0/\Omega_0)\,e$ before screening. Since screening should render it more or less neutral,‡ the factor $\gamma(R)$

† The quantity $(U_{\mathrm{F}}^{\mathrm{XC}}+\phi_1)$ which occurs in (6.1) is probably as good a measure as any, for a metallic element, of what is sometimes called its 'electro-negativity'.

‡ But see some remarks concerning 'charging' in alloys on p. 444.

which describes the extent to which an electronic wave function is modulated by the pseudo-potential must be such that

$$\frac{1}{3R_A^3}\int_0^{R_A} \overline{|1+\gamma(R)|^2}\, R^2\, dR \simeq (z_1\Omega_0/z_0\Omega_1^0), \tag{6.3}$$

the average being taken over all values of k up to k_F. When $z_1 = 4$ and $z_0 = 1$, say, it is clearly impossible to satisfy this condition without $\gamma(R)$ being substantially greater than unity for some values of R and k. The dielectric constant method, however, rests on the assumption that γ is always small.

Presumably it is the low energy electrons, with $k \ll k_F$, which do most of the screening and whose wave functions are most severely modulated. Properties such as the resistivity are primarily determined by the electrons at the Fermi level, however, and for these γ may still be small enough, with luck, to justify the use of the pseudo-potential concept and even the Born approximation. In the language of phase shifts, there is still some hope that the reduced phase shifts at $k = k_F$ remain small compared with $\frac{1}{2}\pi$, even for an ion dissolved in a metal of very different z.

So long as the reduced phase shifts are small at the Fermi level and all the valence electrons remain at large in the conduction band, we may justify (1.67) and (6.2) by invoking the Friedel sum rule, as in the discussion of equation (1.56) on p. 32. It is not essential to use the dielectric constant method to prove these results. When z_1 significantly exceeds z_0, however, the attractive potential of the solute ion may in principle be strong enough to force one or more of the conduction electrons to drop out of the bottom of the conduction band, into *localised* states, more or less tightly bound about the solute ion and analogous to the states occupied by its core electrons. This possibility has been discussed by Friedel (1954) and others. If localised states do occur, then the effective valency of the solute becomes $(z_1 - 1)$ or $(z_1 - 2)$ etc. and (6.2) must be modified accordingly.

The above remarks are necessarily qualitative because reliable calculations of pseudo-potentials for alloy systems are still few and far between. Some phase shift calculations reported by Dickey, Meyer & Young (1966), however, for Li, K and Rb dissolved in **Na**, show up very clearly the large distinction that may exist between u_1^0 and u_1^1. According to these authors the three principal reduced phase shifts at the Fermi level for an Rb ion, for example, change as follows when it is transplanted from pure Rb into otherwise pure Na: η_0 decreases from 0.57 to 0.22; η_1 decreases from 0.19 to 0.09; and η_2 increases from 0.08 to 0.21. These are two metals which differ in atomic volume by a factor of more than 2, of course, though their valencies are identical. It is reassuring that despite this difference, which makes the right-hand side of (6.3) something like $\frac{1}{2}$ and implies that $\gamma(R) \sim -1$ for a substantial proportion of the conduction electrons, the reduced phase shifts at the Fermi level do remain reasonably small compared with $\frac{1}{2}\pi$.

6.3. The NFE model for alloys

In the previous section we have touched briefly on the problem of what happens to the extra valence electrons which are introduced into a metal whenever we replace one of its atoms by an atom of a different species, with $z_1 > z_0$. Do they remain in localised states around the impurity or do they join the conduction band of non-localised states? The question is not an easy one to answer. If one assumes from the start that none of the valence electrons is bound and treats the impurity ion as a point charge, one is led to expect a potential around it of the form $-(z_1 e^2/R) \exp(-q_s' R)$, and this is weak enough to make binding seem unlikely, largely because the screening length $(q_s')^{-1}$ in metals is normally rather less than the Bohr radius a_H. But if an electron were to adopt a localised state the attractive potential it would experience would be a good deal stronger than that seen by the rest of the electrons – strong enough, conceivably, to prevent it from escaping.

In the absence of an authoritative theoretical treatment of this question we must leave it to be settled by experimental evidence. In one or two extreme cases localisation is almost certainly occurring: when the solute ion is a halide, for example, it is far more likely to capture one extra electron from the conduction band so as to complete a shell of eight localised electrons than to release the other seven, and the effect of small amounts of halide impurity on the resistivity of liquid metals seems to be consistent with that picture (see p. 518); the diamagnetic susceptibility of liquid **Na** alloys suggests, according to Rigert & Flynn (1971), that dissolved Te has a similar tendency to capture electrons, and perhaps Bi as well. But, such cases apart, the bulk of the evidence seems to point against localisation.

So we are normally to think of a conduction band containing a total of $N(c_0 z_0 + c_1 z_1)$ electrons, and it is natural to describe these by the NFE model which has proved so helpful for pure liquid metals. The question now arises, what is the radius of the Fermi sphere?

The simplest answer is obtained by writing

$$8\pi k_F^3 \Omega / 3(2\pi)^3 = N(c_0 z_0 + c_1 z_1), \tag{6.4}$$

a straightforward extension of (1.8). This implies that for very dilute alloys

$$\left(\frac{1}{k_F}\frac{dk_F}{dc_1}\right)^0 = \frac{1}{3}\left(\frac{z_1}{z_0} - 1 - \left(\frac{1}{\Omega}\frac{d\Omega}{dc_1}\right)^0\right). \tag{6.5}$$

Yet we know that in a very dilute alloy the total Fermi energy is unaffected by the presence of the solute ions; this principle, first enunciated by Friedel (1954), was used as the basis for our derivation of the Friedel sum rule in §1.5 and has been referred to again in passing in §6.2. The total energy is of course made up of kinetic and potential parts, but the screening is so effective in a metal that it is only within a range of order R_A that a solute ion has any effect upon the latter. The constancy of E_F implies, therefore, that so long as an electron at the Fermi level does not approach too close to one of the solute ions it propagates with the same kinetic energy and the same velocity as in the pure solvent. How can this be consistent with the change of k_F which (6.5) implies?

The apparent paradox here may be resolved by an analogy with the case of a light wave propagating through a gas. The gas molecules do not change the frequency of the light and in the interstices between them it travels with the free-space velocity, so that if we fix our attention on a small portion of the disturbance it may appear to have the free-space wavelength. Nevertheless the *mean* wavelength, averaged over a region that contains many molecules, must be reduced by the refractive index of the gas. The reason for the reduction is that wavelets radiated by the molecules combine with the main wavefront in such a way as continually to retard its phase. It is a standard exercise in dispersion theory to relate the change of mean wavelength or mean wave vector, and hence the refractive index, to the forward scattering amplitudes of the molecules.

The necessary theory has in fact been given in §1.11. Let us apply equation (1.81) to the case of a very dilute alloy, treating the solvent matrix as a jellium for the sake of simplicity. It immediately suggests that

$$\left(\frac{1}{k_F}\frac{dk_F}{dc_1}\right)^0 = \frac{2}{3\pi z_0}\sum_l (2l+1)\sin\eta_l \exp(i\eta_l), \tag{6.6}$$

where the η_l are reduced phase shifts that describe the scattering cross-section of a solute ion, set in the solvent jellium, for an electron at the Fermi level. The imaginary term in k_F which (6.6) implies is of no direct interest to us here; it describes the attenuation which scattering by the impurity ions inevitably brings about. For the real part alone we have

$$\mathrm{Re}\left(\frac{1}{k_F}\frac{dk_F}{dc_1}\right)^0 = \frac{1}{3\pi z_0}\sum_l (2l+1)\sin 2\eta_l. \tag{6.7}$$

If the phase shifts are small enough for $\sin 2\eta_l$ to be replaced by $2\eta_l$, we can use the sum rule (1.45) to obtain the result

$$\mathrm{Re}\left(\frac{1}{k_F}\frac{dk_F}{dc_1}\right)^0 = \frac{1}{3z_0}\left(\frac{Q}{|e|}\right),$$

where Q is the *excess* charge centred upon the solute ion in its bare or unscreened condition.

In the case of a *substitutional* alloy, such that a solute atom can replace a solvent atom without causing the neighbours to move or the volume of the specimen as a whole to change, it is evident that Q is just $(z_1-z_0)|e|$ and (6.7) is then entirely equivalent to (6.5). In general, however, the solute atom is likely to cause some dilatation or contraction. To evaluate Q in these circumstances it is helpful to imagine a sphere to be drawn about the solvent atom that is due for replacement, of such a radius that the solvent matrix outside it may safely be treated as jellium of uniform density. When the replacement occurs the amount of positive charge which gets pushed out through the surface of this sphere is just $(Nz_0|e|/\Omega)\,d\Omega$, so that we can set

$$\frac{Q}{|e|} = \left(z_1 - z_0 - z_0\left(\frac{1}{\Omega}\frac{d\Omega}{dc_1}\right)^0\right),$$

however the solvent matrix is distorted in the immediate neighbourhood of the impurity. Once again (6.7) is equivalent to (6.5).

Dispersion theory therefore substantiates the elementary approach so long as the impurity phase shifts are small, but it clearly generates a different answer when one or more of them is comparable with $\frac{1}{2}\pi$. Is the answer necessarily a more reliable one? This is a difficult point which has not been properly resolved; it all depends, perhaps, on just how k_F is to be defined. In an idealised system in one-dimension with reflecting ends the eigenfunctions would be standing waves and it would be possible to label them by counting their nodes; then if n_F were the number of nodes in a large length L for an eigenfunction corresponding to the Fermi energy, we could define k_F unambiguously as $\pi L/n_F$. In that case there is no doubt that the one-dimensional equivalent of (6.5) would be rigorously correct. But what we are interested in for real liquid metals in three dimensions is the phase difference between the pseudo wave function on the ith and jth ions, separated by a distance $\mathbf{R}_{ij}$; we want to be able to write this, for an electron at the Fermi level, as $(\mathbf{k}_F \cdot \mathbf{R}_{ij})$. For application in such a context, equation (6.7) seems to be more soundly based.

It is of interest to consider what would happen to k_F in a very dilute alloy if we could somehow distort the potential around each solute ion until it was strong enough to give rise to a bound state – an s state, let us say, able to accommodate two electrons with opposite spins. As soon as these bound states appeared the reduced phase shift at the Fermi level, η_0, would change suddenly by π. According to (6.5) there would also be a sudden jump in k_F, of $-(2c_1k_F/3z_0)$, since the charge on each solute ion would change from $z_1|e|$ to $(z_1-2)|e|$. There would be no sudden jump in k_F according to (6.7), however, since $\sin 2\eta_0$ would remain unaffected.

Our hope is, of course, that in many alloy systems the phase shifts are small enough for (6.5) and (6.7) to be identical, and for the size of the Fermi sphere at large concentrations to be calculated with confidence from (6.4). The reader may be reminded that experiments on solid alloys provide some evidence for the swelling of the Fermi surface which (6.4) describes. The most direct evidence is provided by de Haas–van Alphen measurements (e.g. on **Cu**–M by Chollet & Templeton (1968) and on **Al**–M by Shepherd & Gordon (1968)) and positron annihilation experiments (e.g. on Li–Mg by Stewart (1964)). For many solid alloy systems the way in which the electronic specific heat or magnetic susceptibility varies with concentration can be explained successfully if (6.4) is coupled with the *rigid-band hypothesis*, that $\mathcal{N}(E)$ remains the same function of kinetic energy K whatever the solute and whatever (within limits) its concentration; where this line of explanation fails it is the rigid-band hypothesis which is customarily blamed. Finally, we may note the success of the *Hume-Rothery rules*, which define certain critical values for the mean number of valence electrons per atom in an alloy, i.e. for $(c_0 z_0 + c_1 z_1)$, at which the addition of further solute is liable to bring about a change of crystal structure; attempts to account for these rules hinge on the assumption that the Fermi surface has to expand to accommodate more electrons when $(c_0 z_0 + c_1 z_1)$ increases.

In liquid alloys it is impossible to observe the de Haas–van Alphen effect or to measure the electronic specific heat, and work on positron annihilation is in its infancy.† We shall take the NFE model, coupled with equation (6.4), for granted in what follows, while looking out all the time for indirect evidence that may support or contradict it. Since the effective mass ratio for most pure liquid metals seems to be so nearly equal to unity, we shall make the simplification for liquid alloys of ignoring the distinction between m^* and m. The calculations of Itami & Shimoji (1972) provide some support for this.

† Results on liquid **Hg**–Na and **Hg**–In have been reported by Tsuji, Fukushima, Oshima & Endo (1969), but they are not very revealing.

6.4. The rigid-sphere model

Our next task is to consider how best to describe the ionic structure of a liquid alloy and how this is likely to differ from the structure of a pure liquid. For the ideal *substitutional* alloy, of course, there is no difference; solute ions are free to replace solvent ions at random, without affecting the disposition of neighbouring ions in any way. In the great majority of systems, however, the two constituents have different atomic volumes and the addition of one to the other is liable to cause some dilatation or contraction. It is clear that the alloy cannot be strictly substitutional in such a case.

So far as pure description is concerned we can choose between *models* of one sort or another and the use of *distribution functions*, and the latter approach seems to be much more powerful. If the pseudo-potentials are weak enough to justify a second-order theory throughout, then all the structural information that we really need is comprised in the pair distribution functions. For a binary system, as pointed out in §2.12, there must be three of these, $g_{00}(R)$, $g_{01}(R)$ and $g_{11}(R)$, and correspondingly we may define three *partial* interference functions, $a_{00}(q)$, $a_{01}(q)$ and $a_{11}(q)$. If it is necessary to go to the third order then, naturally, the three-body distribution functions may become important; there will be four of these, $g_{000}(\mathbf{R}, \mathbf{R}')$, $g_{001}(\mathbf{R}, \mathbf{R}')$, $g_{011}(\mathbf{R}, \mathbf{R}')$ and $g_{111}(\mathbf{R}, \mathbf{R}')$.† One could go on to define four-body distribution functions and so on.

In §1.13 we saw that, provided a second-order theory is sufficient, the energy required to rearrange the ions while keeping the total volume constant may be expressed in terms of a pair potential. This principle may readily be extended to alloys. It is apparent from (2.60) that the term in the energy of the conduction electrons which depends upon structure must be given to second order by

$$-\frac{N}{8\pi e^2\Omega}\sum_{\mathbf{q}\neq 0} q_s^2\epsilon'(q)\sum_{\alpha}\sum_{\beta} c_\alpha c_\beta u_\alpha(\mathbf{q})u_\beta(-\mathbf{q})a_{\alpha\beta}(q) \qquad (6.8)$$

† The three suffices in, for example, g_{001} are used to denote the species of the three ions at the origin, at $\mathbf{R}$ and at $\mathbf{R}'$ respectively. Just as $g_{01}(R)$ is the same as $g_{10}(R)$, so $g_{100}(\mathbf{R}, \mathbf{R}')$, say, is clearly the same as $g_{001}(\mathbf{R}'-\mathbf{R}, -\mathbf{R})$. Thus there is no need to introduce g_{100} or g_{010} as functions distinct from g_{001}, and similarly g_{011}, g_{101} and g_{110} are all equivalent.

instead of by (1.92). We can represent this by an *indirect* pair potential provided we distinguish, in the case of a binary system, between w'_{00}, w'_{01} and w'_{11}. Evidently we must equate (6.8) to the structure-dependent part of (cp. (1.93) and (2.58))

$$\tfrac{1}{2}\sum_{i\neq j} w'_{\alpha\beta}(\mathbf{r}_j-\mathbf{r}_i) = \frac{1}{2\Omega}\sum_{\mathbf{q}\neq 0}\sum_{\alpha,\beta} w'_{\alpha\beta}(\mathbf{q})\sum_{\substack{i(\alpha)\\ i\neq j}}\sum_{j(\beta)}\exp(i\mathbf{q}\cdot\mathbf{R}_{ji})$$

$$= \frac{N}{2\Omega}\sum_{\mathbf{q}\neq 0}\sum_{\alpha,\beta} w'_{\alpha\beta}(\mathbf{q})\,c_\alpha c_\beta(a_{\alpha\beta}(q)-1).$$

It follows that

$$w'_{\alpha\beta}(\mathbf{q}) = -\frac{q_s^2\epsilon'}{8\pi e^2}(u_\alpha(q)u_\beta(-q)+u_\alpha(-q)u_\beta(q)). \tag{6.9}$$

To this indirect interaction must be added the direct term

$$w''_{\alpha\beta}(\mathbf{q}) = 4\pi z_1^* z_2^*\, e^2/q^2. \tag{6.10}$$

The appropriate generalisation of (1.98), based upon the Ashcroft empty-core model and some rather drastic simplifications concerning ϵ', then turns out to be

$$w_{\alpha\beta}(R) \simeq \frac{z_\alpha z_\beta e^2}{R}\cosh(q_s R_{M,\alpha})\cosh(q_s R_{M,\beta})\exp(-q_s R). \tag{6.11}$$

A slightly better approximation may perhaps be achieved by replacing q_s by q'_s.

We have seen in earlier chapters that there is still some uncertainty about the behaviour of the effective pair potential in pure liquid metals at large values of R, but that the gross features of the structure are normally determined by its repulsive core; for many purposes it is a good approximation to treat the ions as rigid spheres. We may hope to use the rigid-sphere picture for liquid alloys too, *provided* that one essential condition is satisfied: if $w_{00}(R)$ and $w_{11}(R)$ become strongly repulsive when R is less than σ_0 and σ_1 respectively, then clearly the range at which $w_{01}(R)$ becomes strongly repulsive should be just $\frac{1}{2}(\sigma_0+\sigma_1)$. Fortunately, equation (6.11) seems to be reasonably consistent with this requirement, as may be checked by following through the arguments of §1.14. The effect of the finite core radii is to add $2R_{M,0}$ to σ_0, to add $2R_{M,1}$ to σ_1, and to add $(R_{M,0}+R_{M,1})$ to the closest distance of approach for two different ions.

The additivity of rigid-sphere diameters has been examined more thoroughly for the particular case of liquid Na–K by Ashcroft & Langreth (1967*b*). These authors used the empty-core model and adjusted $R_{M,Na}$ and $R_{M,K}$, which they took to be independent of concentration in the alloy and also independent of temperature, so as to fit the resistivities of pure Na and K respectively. Curves for w_{NaNa}, w_{KK} and w_{NaK} were computed using the formulae quoted above, and the values of R at which these three functions had reached an energy $\frac{3}{2}k_B T$ above their respective minima were identified as σ_{NaNa}, σ_{KK} and σ_{NaK}. The computations were repeated for several different concentrations, using the dielectric constant believed to be appropriate for the mean electron density in the alloy under consideration. Neither σ_{NaNa} nor σ_{KK} were entirely independent of concentration at constant temperature, but the ratio between them varied very little and, which is of more importance here, their arithmetic mean was always close to σ_{NaK}.

6.5. Partial interference functions for rigid spheres

Supposing that a rigid-sphere description is appropriate for a liquid alloy, with $\sigma_0 \neq \sigma_1$, what sort of behaviour should we expect for the three partial interference functions? A plausible answer is provided by the Percus–Yevick theory, which has been extended to binary mixtures of rigid spheres by Lebowitz (1964). The theory involves three parameters, concentration c_1, packing fraction y $(= N\pi(c_0\sigma_0 + c_1\sigma_1)^3/6\Omega)$ and misfit ratio α $(= \sigma_1/\sigma_0)$, and solutions for various values of these have been plotted by Ashcroft & Langreth (1967*b*) and by Enderby & North (1968).† Some typical curves, appropriate for the alloy NaK, are shown in fig. 6.2. The points to note are the following:

(*a*) For large values of q the curves are similar to those obtained for $a(q)$ in a pure liquid at the same value of y.

(*b*) As was surely to be expected, the main peak for $a_{01}(q)$ lies midway between the peaks for $a_{00}(q)$ and $a_{11}(q)$.

(*c*) In the neighbourhood of the main peak the curves for constant y and constant α are rather insensitive to concentration.

† Beware! The quantities labelled S_{ij} and a_{ij} by these two sets of authors are closely related to $a_{\alpha\beta}$ as defined above but they are *not* identical.

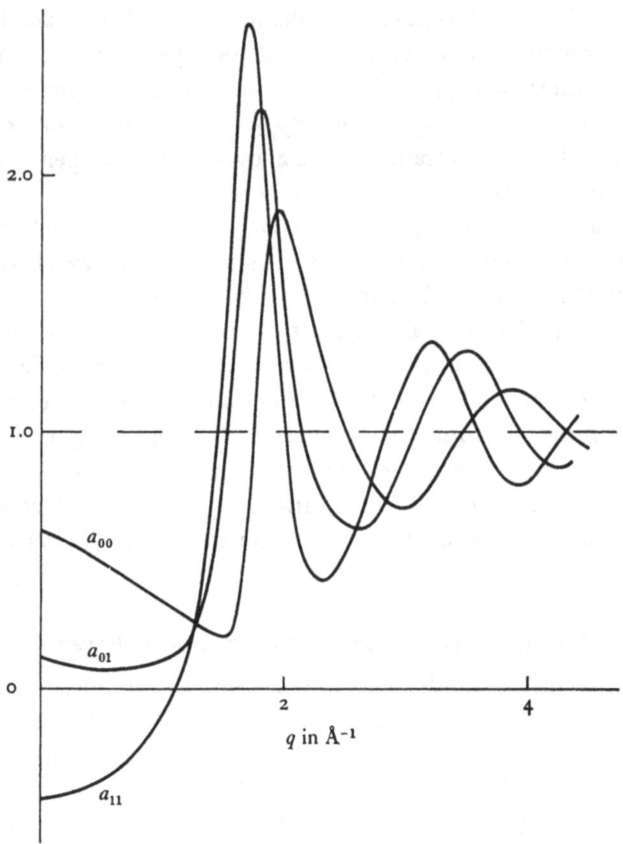

Fig. 6.2. Partial interference functions according to the Percus–Yevick rigid-sphere model for $c_1 = \frac{1}{2}$, $y = 0.45$, $\alpha = 1.24$. (After Enderby & North, 1968.)

(*d*) Near $q = 0$ it is possible for all three partial interference functions to be relatively large, i.e. a substantial fraction of unity, at any rate if α is not close to unity itself. They may also be negative in sign. a_{01} is still midway between a_{00} and a_{11} in this region, but all three curves shift if the concentration is changed.

The results listed under (*d*) may surprise the reader at first sight, so it is worth demonstrating that they are consistent with the line of argument used to prove that $a(0)$ for a pure liquid is always positive and small. Let us set out to calculate the number of ions of either species, m_0 or m_1, which are instantaneously to be found in some volume ω which is large compared with atomic dimensions but smaller than the size of the specimen Ω. By the method used on p. 108 we have

$$\overline{m_0^2} = \overline{m_0}\left(1+\frac{Nc_0}{\Omega}\int^{\omega} g_{00}(|\mathbf{r}'-\mathbf{r}|)\,d\mathbf{r}'\right)$$

$$= \overline{m_0}(1+\overline{m_0}+c_0(a_{00}(0)-1)),$$

whence, if m is the sum of m_0 and m_1,

$$\overline{\Delta m_0^2}/\overline{m} = c_0c_1+c_0^2a_{00}(0). \tag{6.12}$$

Similarly

$$\overline{\Delta m_1^2}/\overline{m} = c_0c_1+c_1^2a_{11}(0), \tag{6.13}$$

while it follows by a straightforward extension of the argument that

$$\overline{\Delta m_0\Delta m_1}/\overline{m} = c_0c_1(a_{01}(0)-1). \tag{6.14}$$

Let us now suppose that the compressibility of the liquid is negligible. Since each atom of species 1 occupies a volume α^3 times greater than one of the species 0 we must surely expect that

$$\alpha^3\Delta m_1 \simeq -\Delta m_0, \tag{6.15}$$

in which case, assuming α to be independent of concentration,

$$\alpha^{-3}(c_0c_1+c_0^2a_{00}(0)) \simeq \alpha^3(c_0c_1+c_1^2a_{11}(0)) \simeq c_0c_1(1-a_{01}(0)). \tag{6.16}$$

For a very dilute alloy in which c_1 is vanishingly small, it is readily shown from (6.16) that

$$a_{01}(0) \simeq 1-\alpha^3, \quad a_{00}(0) \simeq c_1(\alpha^3-\alpha^{-3}), \tag{6.17}$$

while at the other end of the concentration range, where c_0 is vanishingly small,

$$a_{01}(0) \simeq 1-\alpha^{-3}, \quad a_{11}(0) \simeq -c_0(\alpha^3-\alpha^{-3}). \tag{6.18}$$

The sort of concentration dependence that would be consistent with these results is shown in fig. 6.3. Evidently it is also consistent with the curves drawn in fig. 6.2. The argument may be refined, if required, to yield an expression for the compressibility. According

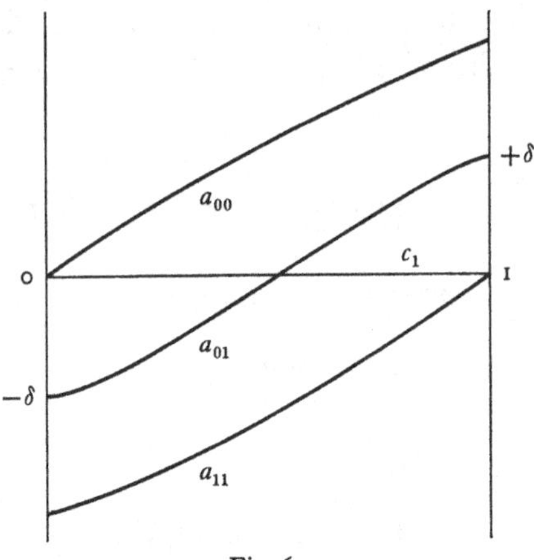

Fig. 6.3

to Kirkwood & Buff (1951) it is true for any binary mixture, whether it is composed of rigid spheres or not, that

$$\frac{Nk_B T\beta}{\Omega} = \frac{(c_0 a_{00}(0)+c_1)(c_1 a_{11}(0)+c_0)-c_0 c_1(a_{01}(0)-1)^2}{1+c_0 c_1(a_{00}(0)+a_{11}(0)-2a_{01}(0))}. \quad (6.19)$$

If the equations in (6.16) were entirely correct the compressibility would evidently vanish. It is not large enough in practice for the schematic curves in fig. 6.3 to be seriously misleading.

For liquid alloys that are very dilute an extremely simple model, first suggested by Faber & Ziman (1965), is capable of yielding much the same results for $a_{01}(q)$ as the Percus–Yevick theory. The model is based on an elementary result of classical elasticity theory, to which we shall return in the following section, that when a small spherical hole is cut in a homogenous isotropic medium and dilated by internal pressure the density of the medium is unaffected by the dilatation, even in the immediate neighbourhood of the hole. If this result can be applied on an atomic scale, it suggests that the replacement of a single ion in the pure solvent by an impurity ion of larger size will cause each neighbouring solvent ion, initially at a distance R, to be shifted outwards through a distance

$$\delta\Omega/4\pi R^2 N. \quad (6.20)$$

Here δ is the dilatation associated with the impurity, defined in such a way that $c_1\delta$ is the increase in volume of a specimen per

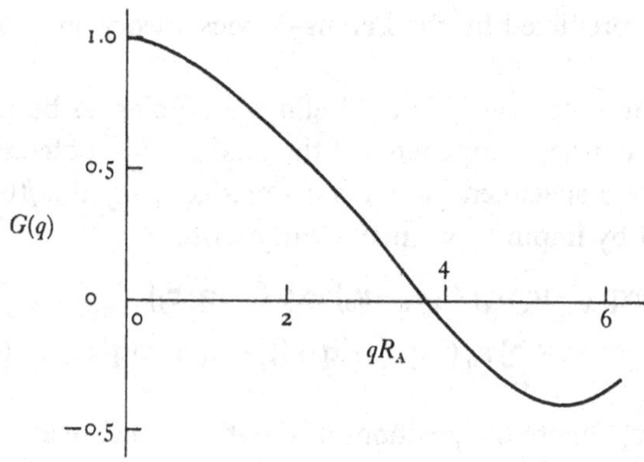

Fig. 6.4

unit volume, when c_1 is small; it can be determined experimentally by density measurements on the dilute alloy of interest, and values in the range between +0.4 and −0.4 are commonly observed. The shift described by (6.20) implies that

$$g_{01}(R) \simeq g_{00}(R) - (\delta\Omega/4\pi R^2 N)\frac{dg_{00}}{dR} \tag{6.21}$$

if δ is small, where because c_1 is infinitesimal g_{00} is essentially the pair distribution function for the pure solvent. Hence

$$a_{01}(q) \simeq a_{00}(q) - \delta G(q), \tag{6.22}$$

with
$$G(q) \simeq \int_0^\infty \frac{dg_{00}}{dR}\frac{\sin qR}{qR}\,dR. \tag{6.23}$$

The function $G(q)$ can be computed from experimental data for $g_{00}(R)$, and a typical curve for it is shown in fig. 6.4. The curve extrapolates to unity in the small q limit, and it is easy to show that it must always do so, whether (6.21) is exactly correct or not.

To make contact with the rigid-sphere model we may identify δ with $(\sigma_1^3-\sigma_0^3)/\sigma_0^3$, i.e. with (α^3-1). With this identification it may be seen that (6.22) is essentially the same result near $q = 0$ as (6.17); the small term $a_{00}(q)$ is missing from the latter formula, but only because the compressibility of the solvent was neglected. For large q, moreover, the oscillatory nature of the function G should generate just the type of curve for $a_{01}(q)$, according to

(6.22), that is predicted by the Percus–Yevick theory and plotted in fig. 6.2.

Many of the properties of liquid alloys are going to be determined by the Fourier components of the total pseudo-potential U, and if we have a specimen in which a single ion, say the jth, has been replaced by impurity we may clearly write

$$U(q) = \sum_i u_0 \exp(-i\mathbf{q}\cdot\mathbf{r}_i) + (u_1 - u_0)\exp(-i\mathbf{q}\cdot\mathbf{r}_j) + \sum_{i\neq j} u_0 (\exp(-i\mathbf{q}\cdot\mathbf{r}'_i) - \exp(-i\mathbf{q}\cdot\mathbf{r}_i)), \quad (6.24)$$

where $\mathbf{r}_i$ and $\mathbf{r}'_i$ denote the positions of the ith ion before and after the replacement. Evidently the ensemble average of (6.24) is

$$\sum_i u_0 \exp(-i\mathbf{q}\cdot\mathbf{r}_i) + (\overline{u'_1} - u_0)\exp(-i\mathbf{q}\cdot\mathbf{r}_j),$$

where

$$\begin{aligned} \overline{u'_1} &= u_1 + u_0 \overline{\sum_{i\neq j} (\exp(i\mathbf{q}\cdot\mathbf{R}'_{ji}) - \exp(i\mathbf{q}\cdot\mathbf{R}_{ji}))} \\ &= u_1 + u_0(a_{01}(q) - a_{00}(q)) \\ &\simeq u_1 - u_0 \delta G(q). \end{aligned} \quad (6.25)$$

Faber & Ziman suggested, on the basis of this result, that for very dilute alloys the dilatation effects associated with the solute, which are all included in the third term of (6.24), might be described by using in place of u_1 the effective pseudo-potential defined by (6.25). In the limit of small q, where u_1 and u_0 are determined (see (6.2)) by the charge on the solute and solvent ions and where $G(q)$ is unity, this corresponds to replacing z_1 by an effective valency

$$z'_1 \simeq z_1 - z_0 \delta. \quad (6.26)$$

Equation (6.26) may be justified more directly by the simple physical argument that what reduces the apparent charge on a solute ion is the number of ions, δ, which it displaces outwards. It had been suggested previously by Harrison (1958, 1960) and applied to solid alloys by Blatt (1957).

The Faber–Ziman dilatation model has simplicity and convenience to recommend it, but it can never be exact in a context where it is $\overline{U^2}$ that matters, rather than $\overline{U}$. The trouble is that the difference between $(u'_1)_j$ and u_1 depends upon the environment of the jth ion and is liable to fluctuate from one value of j to the next. The situation is reminiscent of our attempt in §4.11 to include the higher-order terms for a pure liquid metal in the effective pseudo-potential of (4.102). To estimate the importance of fluctuations to $\overline{U^2}$ it is necessary in both cases (see Faber & Ziman, 1965) to consider three-body effects.

The exact expression for $\overline{U^2}$ in a binary alloy is of course† (see (2.62))

$$\overline{|U(q)|^2} = N\{c_0u_0^2a_{00}+c_1u_1^2a_{11}+c_0c_1[u_0^2(1-a_{00})+u_1^2(1-a_{11}) -2u_0u_1(1-a_{01})]\}. \quad (6.27)$$

It may be shown by manipulation of this result that to use the Faber–Ziman effective pseudo-potential of (6.25) is equivalent, when terms of order c_1^2 can be ignored, to invoking (6.22) for a_{01} and in addition to setting

$$\frac{da_{00}(q)}{dc_1} = \delta G(q)(2-2a_{00}(q)+\delta G(q)). \quad (6.28)$$

This is nicely consistent near $q = 0$ with (6.17), and for large q it is consistent with the Percus–Yevick solution, at least to the extent that it suggests a rather slow concentration dependence of a_{00} because both G and $(1-a_{00})$ are small. It is not likely, however, to be trustworthy in detail.

6.6. The volume and energy of mixing

In the previous section we have discussed the probable ionic structure of liquid alloys and have established the basic features of a rigid-sphere model, to provide a norm with which the structure of real systems can later be compared. Can we extend this model in any way, so as to describe such thermodynamic properties as volume, internal energy and entropy? The discussion that follows is somewhat inconclusive, but it will indicate the difficulties involved.

Accurate density measurements reveal that the mean atomic volume for most liquid alloy systems is not exactly a linear function of concentration at constant temperature. There is a volume change on mixing,

$$\Delta\Omega = \Omega-(c_0\Omega_0+c_1\Omega_1), \quad (6.29)$$

† For the sake of simplicity it has been assumed in writing (6.27) that u_0 and u_1, are real. The generalisation to complex pseudo-potentials is straightforward.

such that $\Delta\Omega/\Omega$ is usually a few parts per cent in the middle of the concentration range, sometimes positive and sometimes negative. The rigid-sphere model tells us without further elaboration that

$$\Omega = N\pi(c_0\sigma_0+c_1\sigma_1)^3/6y, \tag{6.30}$$

but we cannot use this formula to compute Ω, and to explain the observations for $\Delta\Omega/\Omega$, without an accurate knowledge of the concentration dependence of the parameters involved. Both σ_0 and σ_1 may vary if the screening properties of the conduction electron gas are concentration-dependent. As for y, we may take this to be 0.46 just above the melting point for most pure liquids, but it decreases on heating, as the volume of the specimen expands and as the effective rigid-sphere diameter decreases (see § 1.14). Consequently, if the two constituents have different melting points, then the value of y is different at the two ends of the concentration range for an isothermal set of measurements, and how do we know what happens in between? Ashcroft & Langreth (1967*b*) have suggested that y may be constant along the liquidus curve, though there is reason to suppose that when the misfit ratio α ($= \sigma_1/\sigma_0$) is different from unity a rigid-sphere fluid can tolerate values of y greater than 0.46 before it is forced by geometrical considerations to crystallise. But it would be neater to use this principle, if indeed it is valid, to estimate the liquidus temperature from the volume rather than the reverse.

Thermodynamics shows that a linear variation of Ω is to be expected if the free energy obeys the so-called *ideal law of mixing*, which may be written as

$$\mathscr{F}(\Omega) = c_0\mathscr{F}_0(\Omega_0')+c_1\mathscr{F}_1(\Omega_1')+Nk_BT(c_0\log c_0+c_1\log c_1), \tag{6.31}$$

with

$$\Omega = c_0\Omega_0'+c_1\Omega_0'; \tag{6.32}$$

the final term in (6.31) derives, of course, from an entropy of mixing that is inevitable even if the constituent ions are randomly disposed, as in perfect gas. The condition for equilibrium at a constant temperature T and a constant applied pressure p_{ext} is

$$\partial\mathscr{F}/\partial\Omega = p_{ext}.$$

It therefore follows from (6.31) and (6.32) that

$$(\partial\mathscr{F}_0/\partial\Omega_0') = (\partial\mathscr{F}_1/\partial\Omega_1') = p_{ext}$$

in equilibrium, and this is sufficient to prove that Ω_0' and Ω_1' must be identical for all concentrations with Ω_0 and Ω_1 respectively, the equilibrium volumes of the pure solvent and pure solute.

Equation (6.31) implies, however, that there is no internal energy or heat of mixing,† i.e. that

$$\mathscr{U}(\Omega) = c_0\mathscr{U}_0(\Omega_0)+c_1\mathscr{U}_1(\Omega_1), \tag{6.33}$$

and it is easy to see on theoretical grounds that this condition is most unlikely to be obeyed. Consider first the structure-dependent part of the internal energy, expressed by

$$\frac{N}{2\Omega}\sum_{\mathbf{q}\neq 0}\sum_{\alpha}\sum_{\beta} c_\alpha c_\beta w_{\alpha\beta}(\mathbf{q})(a_{\alpha\beta}(q)-1). \tag{6.34}$$

For a binary alloy this can be rearranged as

$$\frac{N}{2\Omega}\sum_{\mathbf{q}\neq 0}[c_0 w_{00}(a_{00}-1)+c_1 w_{11}(a_{11}-1)$$
$$+c_0c_1\{2w_{01}(a_{01}-1)-w_{00}(a_{00}-1)-w_{11}(a_{11}-1)\}], \tag{6.35}$$

which has little chance of being a linear function of concentration unless the term in c_0c_1 vanishes, and it does not appear to do so, even for the rigid-sphere model. In any case, it is only the energy of interaction between neighbouring atoms which is included in (6.35) and the energy of each atom by itself ought to be considered in addition. An inspection of the two schematic diagrams in fig. 6.1 will show that unless ΔE_F, the Fermi level shift given by (6.1), happens to be small, the energy of the solvent atom is most unlikely to remain the same when it is transplanted into the solute; the charge density within it may be little altered, but the electrons – in the situation that is illustrated – have surely acquired greater kinetic energy. Add to these theoretical arguments the observation that heats of mixing are rarely negligible in practice, and it becomes clear that ideal solutions must be the exception rather than the rule.

An attempt to estimate the internal energy of liquid Na–K and of various other combinations of alkali metals, and subsequently to calculate $\Delta\Omega$, has been made by Christman (1967*a*, 1967*b*; see also Christman & Huntington, 1965). He supposes that

† The heat of mixing which can be measured experimentally is strictly speaking the change of *enthalpy* rather than internal energy. For liquid alloys at atmospheric pressure, however, the distinction can be ignored: $p_{\text{ext}}\,\Delta\Omega$ is minute compared with $\Delta\mathscr{U}$.

the alloy may be treated as an assembly of spherical 'cells' of two sizes, corresponding to the two different kinds of atom it contains, and uses the quantum defect method to describe the behaviour of the electronic wave functions within each cell. The condition that the wave functions in one cell must join smoothly on to those in the next is found to dictate that the total conduction electron charge on the smaller atom exceeds e by a few parts per cent, and vice versa.† In his final expression for the total energy Christman includes a correction for the Coulomb interaction between the charged 'atoms', but there is no attempt to analyse the structure-dependent terms in detail. In this respect a calculation for Na–K alone by Tamaki & Shiota (1966) is more ambitious. The results of both calculations, which are in moderate agreement with experiment, are discussed in §6.12.

6.7. The elastic continuum model

Since there have been few other attempts to discuss the heat of mixing problem from first principles it is worth describing an indirect approach which Friedel (1954) has suggested; a related theory has been worked out by Varley (1954). Friedel sets out to calculate the energy released or absorbed when a small number of solute atoms are dissolved in the pure solvent, by treating separately the following stages:

(*a*) the atoms are sublimed from the pure solute;
(*b*) they are each ionised z_0 times;
(*c*) the ionised atoms are swapped for solvent ions;
(*d*) the excess electrons are added to these solvent ions;
(*e*) the neutral solvent atoms are returned to the solvent metal, which is now slightly contaminated by impurity.

Every stage except (*c*) is easily dealt with, since the sublimation energies and ionisation energies of metallic elements are known from experiment. To evaluate the energy change involved in (*c*), Friedel uses a primitive elastic continuum model.

Suppose that we cut a small spherical hole of volume ω_0 in an

† This so-called 'charging' effect is not necessarily inconsistent with the NFE model adopted above, though it would need a careful discussion of screening, and in particular of the Friedel oscillations about each ion, to demonstrate its existence.

isotropic homogeneous elastic medium of compressibility β_0 and Poisson's ratio π_0, and force into it a sphere of volume ω_1 ($> \omega_0$), cut from a different elastic medium described by β_1 and π_1. How much does the specimen as a whole dilate and how much energy is stored in the distortion both of the surrounding matrix and of the included sphere? These questions can be answered by standard elasticity theory (Eshelby, 1954). It is found that the strain in the matrix must be a combination of a shear strain which falls off like R^{-3} and a uniform compression independent of R. The relative magnitude of the two terms is fixed by the boundary condition that at the surface of the whole specimen the radial component of stress vanishes. In a very large specimen of volume Ω_0 containing only one inclusion the uniform compression is infinitesimal and the strain is a pure shear which leaves the density of the matrix unaltered. If there is a small but finite number of inclusions randomly distributed, say $c_1\Omega_0/\omega_0$ of them altogether, it turns out that

$$\frac{1}{\Omega_0}\left(\frac{d\Omega}{dc_1}\right)^0 = \left\{1 - \frac{2(1-2\pi_0)(\beta_1-\beta_0)}{2\beta_1(1-2\pi_0)+\beta_0(1+\pi_0)}\right\}\frac{(\omega_1-\omega_0)}{\omega_0}. \quad (6.36)$$

The stored elastic energy in this situation is given by

$$\frac{(1-2\pi_0)}{2\beta_1(1-2\pi_0)+\beta_0(1+\pi_0)}\,\frac{(\omega_1-\omega_0)^2}{\omega_0} \quad (6.37)$$

per inclusion, provided that $(\omega_1-\omega_0) \ll \omega_0$.

Friedel's claim is that this model may be applied on the atomic scale and that, in fact, (6.37) provides a valid measure of the energy change during stage (*c*), if ω_1 and ω_0 are replaced by Ω_1/N and Ω_0/N respectively. Although no rigorous justification is apparent (but see Mott, 1962), the theory gives an adequate explanation of the heats of mixing reported for a number of very dilute alloys in the solid state, and equation (6.36) is not unsuccessful in describing the volume of mixing. Since it is generally the case that $\beta_1 > \beta_0$ when $\Omega_1 > \Omega_0$ and vice versa, while $2\pi_0$ is necessarily less than unity for a solid, the theory suggests that $\Delta\Omega/\Omega$ should normally be *negative* throughout the concentration range.

An ideal liquid cannot support any shear stress and has an effective Poisson's ratio which is $\frac{1}{2}$ exactly. If Friedel's theory is extended to liquid alloys by setting $2\pi_0 = 1$, however, it predicts that $\Delta\Omega/\Omega$ should always vanish and so should the energy change associated with stage (*c*). Perhaps it is more plausible to suppose that a real liquid *can* support shear stress locally, say within the first few coordination shells about an impurity ion. In that case the only conclusion we can base upon the elastic model is that the magnitude of both $\Delta\Omega/\Omega$ and the heat of mixing should tend to fall when an alloy melts.

6.8. The entropy of mixing

A serious weakness of both Christman's theory and that of Friedel, at any rate as applied to liquid alloys, is that they ignore the influence of entropy on the equilibrium configuration; energy is minimised rather than free energy. Let us have a closer look at the entropy of mixing of liquid alloys, to see what errors the neglect of it may introduce. It is, of course, a quantity which can be measured directly, like the volume and heat of mixing, so that it is of interest in its own right. A cluster expansion for the entropy of a pure liquid has been quoted as equation (2.36), and to extend this to alloys is not difficult. If we denote by S'_α the entropy that the system would have if all the N ions in the volume Ω were of the same α species and were disposed randomly, then we may write

$$S = -Nk_B \sum_\alpha c_\alpha \log c_\alpha + \sum_\alpha c_\alpha S'_\alpha + \sum_\alpha\sum_\beta c_\alpha c_\beta S'_{\alpha\beta} + \sum_\alpha\sum_\beta\sum_\gamma c_\alpha c_\beta c_\gamma S'_{\alpha\beta\gamma} + \dots \quad (6.38)$$

The first term is the ideal entropy of mixing referred to above, and the first two terms together constitute the 'perfect gas' entropy of the mixture; the subsequent terms constitute the 'excess' entropy of chapter 2, a consequence of the order that exists in the ionic structure. By analogy with (2.36) it may be seen that

$$-S'_{\alpha\beta}/Nk_B = (N/2\Omega)\int g_{\alpha\beta} \log (g_{\alpha\beta})\, d\mathbf{R}, \quad (6.39)$$

$$-S'_{\alpha\beta\gamma}/Nk_B = (N^2/6\Omega^2) \times \iint g_{\alpha\beta\gamma}(\mathbf{R}, \mathbf{R}') \log\left(\frac{g_{\alpha\beta\gamma}(\mathbf{R}, \mathbf{R}')}{g_{\alpha\beta}(R)g_{\alpha\gamma}(R')g_{\beta\gamma}(|\mathbf{R}-\mathbf{R}'|)}\right) d\mathbf{R}\, d\mathbf{R}', \quad (6.40)$$

and so on. For a binary system it is suggestive to rearrange the terms in (6.38) in the same way that we rearranged (6.34), i.e. to write

$$\begin{aligned} S = & -Nk_B(c_0 \log c_0 + c_1 \log c_1) \\ & + c_0(S_0' + S_{00}' + S_{000}' + \ldots) + c_1(S_1' + S_{11}' + S_{111}' + \ldots) \\ & + c_0 c_1(2S_{01}' - S_{00}' - S_{11}') + c_0^2 c_1(3S_{001}' - 2S_{000}' - S_{111}') \\ & + c_0 c_1^2(3S_{011}' - S_{000}' - 2S_{111}') + \ldots \end{aligned} \quad (6.41)$$

We shall base our discussion on this result.

Let us first recall that, for a rigid-sphere alloy at any rate, $a_{01}(q)$ is half way between $a_{00}(q)$ and $a_{11}(q)$, and presumably $g_{01}(R)$ is more or less half way between $g_{00}(R)$ and $g_{11}(R)$. This surely implies, in view of (6.39), that $(2S_{01}' - S_{00}' - S_{11}')$ is fairly small, and for many systems it may be legitimate to ignore this term completely. By the same token let us ignore the subsequent terms in (6.41). Now S_{00}', S_{000}' and so on are determined by the effective rigid-sphere diameter σ_0 in the alloy. While this is not necessarily quite the same as the value that σ_0 would take if all the ions were of species 0 the difference should not be a large one, and it is therefore reasonable to replace $(S_0' + S_{00}' + S_0''' + \ldots)$ by $S_0(\Omega)$, and similarly to replace $(S_1' + S_{11}' + S_{111}' + \ldots)$ by $S_1(\Omega)$. Granted these approximations, we may reduce (6.41) to

$$S = -Nk_B(c_0 \log c_0 + c_1 \log c_1) + c_0 S_0(\Omega) + c_1 S_1(\Omega). \quad (6.42)$$

If the variation of Ω with concentration is not far from linear, then

$$\begin{aligned} S \simeq & -Nk_B(c_0 \log c_0 + c_1 \log c_1) \\ & + c_0 S_0(\Omega_0) + c_1 S_1(\Omega_1) - c_0 c_1(\Omega_1 - \Omega_0)\left(\frac{\alpha_1}{\beta_1} - \frac{\alpha_0}{\beta_0}\right), \end{aligned} \quad (6.43)$$

where α_1, α_0 are the volume expansion coefficients of the two pure constituents at constant pressure. The final term in (6.43) represents a non-ideal entropy of mixing which could have either sign; for typical values of the parameters involved it may amount to something like $c_0 c_1 Nk_B$, so that it may be comparable with the ideal entropy of mixing and must not be neglected.

The essential point is that $S_0(\Omega)$ and $S_1(\Omega)$ appear in (6.42), rather than $S_0(\Omega_0')$ and $S_1(\Omega_1')$ as the ideal law of mixing (see (6.31)) would suggest. Suppose we modify the ideal law to take account of this by writing

$$\mathscr{F}(\Omega) = c_0\mathscr{F}_0(\Omega_0')+c_1\mathscr{F}_1(\Omega_1')+c_0c_1T(\Omega_1'-\Omega_0')\left(\frac{\alpha_1}{\beta_1}-\frac{\alpha_0}{\beta_0}\right) + Nk_\mathrm{B}T(c_0\log c_0+c_1\log c_1). \quad (6.44)$$

The conditions for equilibrium then become

$$\left(\frac{\partial^2\mathscr{F}_0}{\partial\Omega_0'^2}\right)(\Omega_0'-\Omega)-c_1T\left(\frac{\alpha_1}{\beta_1}-\frac{\alpha_0}{\beta_0}\right) = \left(\frac{\partial^2\mathscr{F}_1}{\partial\Omega_1'^2}\right)(\Omega_1'-\Omega)+c_0T\left(\frac{\alpha_1}{\beta_1}-\frac{\alpha_0}{\beta_0}\right) = 0$$

(the volume dependence of (α/β) has been ignored in the interests of simplicity), from which it follows that

$$\frac{\Delta\Omega}{\Omega} \simeq \frac{c_0c_1T(\beta_0\Omega_0-\beta_1\Omega_1)\left(\frac{\alpha_1}{\beta_1}-\frac{\alpha_0}{\beta_0}\right)}{(c_0\Omega_0+c_1\Omega_1)}. \quad (6.45)$$

Equation (6.45) is not a realistic result, of course, because in modifying the ideal law we have corrected for the entropy without bothering about the internal energy of mixing at all. Nevertheless, it is of interest that if typical values for α_0, β_0, Ω_0, etc. are inserted, it leads to values of the order of 1% for $\Delta\Omega/\Omega$, quite comparable in magnitude with the values observed experimentally. This is enough to show that the neglect of the entropy of mixing in a calculation of $\Delta\Omega/\Omega$ is dangerous for liquid alloys. The right-hand side of (6.45) is usually positive, so presumably the entropy of mixing tends to enhance $\Delta\Omega$.

6.9. Compound formation

The time has now come to recognise the possible limitations of the NFE model for liquid alloys, upon which our discussion of their structure and thermodynamic properties has largely been based. In solid alloys it is known that *intermetallic compounds* may occur, and we should be prepared for them to exist in the liquid phase also.

Evidence for compound formation in solid alloys is provided by *phase diagrams*, by *heats of formation*, by their *structure*, and by their *electrical properties*. These points will be reviewed in turn.

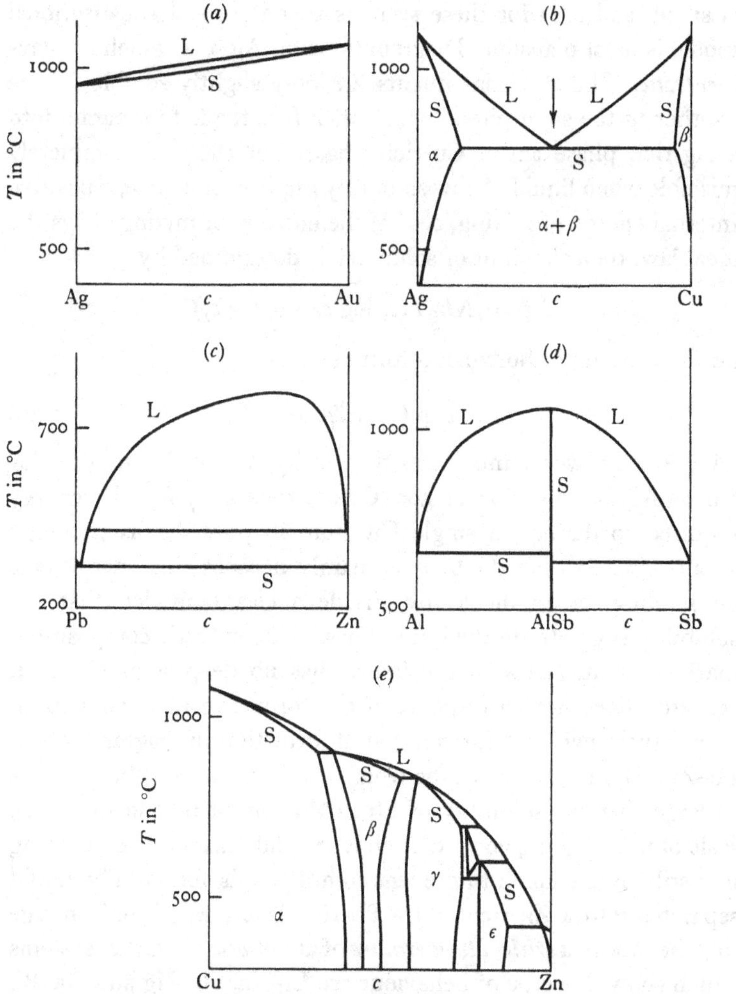

Fig. 6.5. Typical phase diagrams for binary alloys (L, liquidus; S, solidus).

To remind the reader of the various types of behaviour that the phase diagram for a binary alloy may exhibit, some relatively simple but nevertheless characteristic examples are illustrated in fig. 6.5. Diagram (*a*) is for Ag–Au, a system which shows a complete range of solutions in both the solid and the liquid phase, with no hint of compound formation. Such behaviour is confined to systems for which the disparity of size between the two constituents is small, and it is for these systems that the ideal substitutional model is most plausible. Diagram (*b*) is for Ag–Cu, which displays a *eutectic*. The two constituents are only slightly soluble in one another in the solid phase, which therefore tends to separate into a Ag-rich phase and a Cu-rich phase, but they are completely miscible when liquid. A low solubility implies, of course, a positive internal energy of mixing, $\Delta\mathscr{U}$. If the entropy of mixing obeys the ideal law, then the limit of solubility is determined by

$$\Delta\mathscr{U} = Nk_{\mathrm{B}}T(c_0 \log c_0 + c_1 \log c_1),$$

i.e. by the simple Boltzmann formula

$$c_1 \simeq \exp(-\Delta\mathscr{U}/c_1 Nk_{\mathrm{B}}T) \tag{6.46}$$

when this answer is much less than unity. When the limit, as for Cu in Ag, is only 0.02 at 400 °C say, then $\Delta\mathscr{U}/Nc_1$, the energy required to dissolve a single Cu atom in pure Ag, is probably about +0.2 eV. No doubt it is mainly because this energy falls on melting, as predicted by Friedel's elastic model, that the solubility is greater in the liquid phase. The eutectic composition, marked by an arrow in the figure, has no deep significance; it certainly does not correspond to the formation of a compound. Nor is there evidence for compound formation in diagram (*c*), for Pb–Zn. Here is a system for which $\Delta\mathscr{U}/Nc_1$ in the solid phase is so large that the solubility of Pb in Zn cannot be shown on the scale of the diagram, and although it certainly decreases on melting it is still large enough to prevent complete miscibility. The liquid separates into a Zn-rich and a Pb-rich layer, which only merge together above a *critical temperature* of about 800 °C. Other systems which show this type of behaviour are Li–Na, Ga–Hg and Ga–Bi.

Now compare the above examples with diagram (*d*), for Al–Sb. The solid can exist as Al, Sb, or as AlSb, and the solubility of all three forms in one another is too limited to be shown. The high melting point of AlSb is a preliminary indication that this form is particularly stable, and since its composition corresponds to exactly one atom of Sb to every one of Al it looks very like a compound in the chemical sense. It must be clearly distinguished from the so-called *electron compounds* which are formed by a number of alloy systems. These are solid phases with characteristic crystal structures which become stable over a limited range of c; examples are to be seen in diagram (*e*) for Cu–Zn, which forms three electron compounds with the characteristic β-brass (b.c.c.), γ-brass (complex cubic), and ϵ-brass (h.c.p.) structures. These compounds do not necessarily correspond to any neat stoichiometric composition; their range of stability seems rather to be determined by $(c_0 z_0 + c_1 z_1)$, the electron/atom ratio, in a manner first described by Hume-Rothery. Their melting point is never especially high and sometimes they are destroyed by heating before the melting point is ever reached.

The idea that true compounds such as AlSb are especially stable is confirmed by measurements of the heat of formation, which corresponds closely (at atmospheric pressure) to $-\Delta\mathscr{U}$. For many alloy systems, like Ag–Cu, mixing is an *endothermic* process; $\Delta\mathscr{U}/N$ is positive and would be of order 0.05 eV in the middle of the concentration range. For many others $\Delta\mathscr{U}/N$ is negative but still reasonably small, i.e. rarely less than -0.1 eV in the middle of the concentration range, even when an electron compound is formed. But the formation of AlSb is quite strongly exothermic; $\Delta\mathscr{U}/N$ is -0.5 eV.†

As for their structure, most pure metals crystallise with lattices that are either close-packed or are recognisable as distortions of close-packed forms, and the same is true for electron compounds in alloy systems. But true compounds crystallise in such a way that the coordination number is relatively low. A number of them (including AlSb, GaSb, InSb, MgTe, CdTe, ZnTe) adopt the zinc blende or wurtzite structures in which the coordination number is only 4, while others (including SnTe, PbTe, Mg_2Si, Mg_2Ge, Mg_2Sn, Mg_2Pb) adopt the sodium chloride or calcium fluoride structures with coordination numbers of 6 or of 4 and 8 respectively; those in the first group are sometimes classified as covalent compounds and those in the second as ionic ones. It should be added that the distribution of the two sorts of atom among the lattice sites is always ordered for a true compound, whereas electron compounds are often disordered except at low temperatures.

† Note that N still denotes the number of *atoms* in the specimen, even when compounds are under discussion; it is *not* the number of formula molecules.

Finally, whereas most alloys, electron compounds included, are every bit as metallic in their electrical properties as the pure metals of which they are composed, compounds such as InSb, CdTe and so on are well known to be semiconductors. In particular, their resistivity ρ *decreases* on heating. As might be expected, there seems to be a correlation between the size of the energy gap (obtained from the slope of a plot of log ρ versus $1/T$) and the heat of formation. For the compound Mg_2Sn in which $\Delta\mathscr{U}/N$ is only about -0.27 eV the gap is only 0.26 eV (Robertson & Uhlig, 1949). For Mg_2Pb, which shows up as a true compound on the phase diagram in much the same way as Mg_2Sn and has the same crystal structure, but for which $\Delta\mathscr{U}/N$ is only -0.18 eV, the gap seems to have closed completely, for the resistivity of this alloy is only 220 $\mu\Omega$ cm at room temperature and it *increases* linearly with T (Robertson & Uhlig, 1949). There is more of a gradation, in fact, between compounds and non-compounds than is suggested by the stark summary above. The phase diagram for Au–Sn, for example, clearly suggests the formation of a stable compound with the formula AuSn; but $\Delta\mathscr{U}/N$ is only -0.16 eV for AuSn, its nickel arsenide crystal structure (with axial ratio 1.28) is almost close-packed, and its resistivity at room temperature is only about 10 $\mu\Omega$ cm.

Numerical data concerning a number of intermetallic compounds whose properties are of interest in the liquid phase are collected in table 6.3 on p. 475 below, and this table includes such information as is available concerning their latent heats and entropies of fusion (Hultgren & Orr, 1967). A quick glance at it will show that whereas the entropy of fusion for most pure metals is roughly Nk_B (see table 2.1) it is $2Nk_B$ or more for compounds; it looks as though most of the structural ordering, which must lower the entropy of a 50/50 compound by $Nk_B \log 2$ in the solid phase, is wiped out on melting. Add to this that the energy absorbed on melting is often comparable with the binding energy in the solid phase, and it becomes distinctly improbable for most of the compounds listed that they survive in the liquid in any real sense. But one or two, such as AlSb, Mg_3Sb_2 or KHg_2 may be more tenacious than the rest. What observable effects might they produce?

To answer this question we need a clear idea of what the structure of a liquid intermetallic compound might look like. If we say that liquid AlSb is a compound, are we to picture the atoms as joined together to form molecules, as in liquid O_2 for example, with all the valence electrons engaged in localised bonding orbitals of a covalent nature? Are we to compare liquid Mg_2Si, since this compound crystallises with an ionic structure, with liquid CaF_2, i.e. to suppose that all its valence electrons are in localised orbitals about ions? If so, we should clearly expect these substances to remain semiconductors on melting, though they might show the sort of electrical behaviour that is associated with a high density of defects in solid semiconductors. No doubt an association of atoms to form covalent molecules would also be detectable by X-ray or neutron diffraction. The heat of formation would surely remain anomalously large at the stoichiometric composition. Anomalies would probably show up in the volume of mixing and in properties such as the viscosity as well.

Now that is an extreme point of view, too extreme to be plausible for AlSb and Mg_2Si, though there are compounds outside our terms of reference, such as BeS (for which $\Delta\mathscr{U}/N$ is -1.2 eV in the solid phase), to which it may well apply. It is possible to envisage a situation, however, in which some *fraction* of the valence electrons would be localised at any instant of time. Wherever the configuration of atoms was favourable a localised bonding state would occur, and the configuration would be stabilised in consequence; the bonding energy might be too small compared with $k_B T$ for it to be stabilised indefinitely, but it would be reasonable to think of the structure, as containing *some* molecules, though of a transient character. There could be enough free electrons to make the alloy metallic, but the Hall coefficient would be anomalous, and no doubt some of the other properties mentioned above.

Intermetallic compounds, if they occur at all in the liquid phase, are evidently more willing than solid compounds to depart from the stoichiometric composition by dissolving an excess of one or other constituent; thus liquid compounds are not observed to separate out from an alloy of arbitrary composition in the way that solid alloys do. This is hardly surprising if the idea of transient

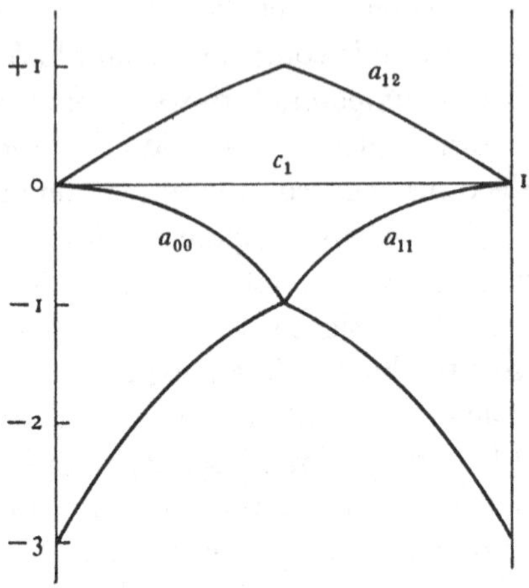

Fig. 6.6.

molecule formation is correct. Presumably we could expect molecules to go on being formed from time to time whatever the composition. The observable effects of compound formation in liquids may be the less conspicuous on this account; they need not be confined to a narrow composition range. Enderby & Collings (1970) have made the interesting point that in some cases the observable effects may be most apparent at compositions which do not correspond to solid compounds at all. Solid compounds have to crystallise, and the composition that allows a stable structure for the crystal as a whole need not necessarily correspond to the molecules which are energetically favoured in the relative freedom of the liquid phase. It turns out that in the Tl–Te system two solid compounds are formed, at concentrations corresponding to TlTe and Tl_3Te_2, yet it is at the concentration corresponding to Tl_2Te that the electrical properties of the liquid are markedly anomalous. Isolated molecules of Tl_2Te, of course, are easier to reconcile with the standard rules of valency than isolated molecules of TlTe or Tl_3Te_2.

Even if there are no electrons in localised states and the concept of transient molecule formation is no longer strictly tenable, it is quite conceivable that some of the *non*-localised states in a liquid inter-metallic compound, particularly the low energy ones near the bottom of the conduction band, are too strongly perturbed for the NFE model to be appropriate. The electrons may then impose constraints on the two-body and three-body distribution functions of the ions and hence produce a structure with more local order than is typical of a pure liquid metal. But to talk of compound formation in a situation of this sort would surely be misleading.

Molecule formation, by the way, could affect the three partial interference functions for a binary liquid in a particularly conspicuous fashion near $q = 0$. To take an extreme case, imagine that the concentration is just right to allow *all* the atoms to be absorbed into molecules, and that these are too tightly bound ever to dissociate. Then, if the compressibility is negligible, one has

$$\Delta m_0 = \Delta m_1 = 0$$

in place of (6.15), and it follows from (6.12), (6.13) and (6.14) that

$$a_{00}(0 = a_{11}(0) = -1, \quad a_{01}(0) = +1, \qquad (6.47)$$

whatever the molecular formula. It appears from the results of Bhatia & Thornton (1970) that curves like those sketched in fig. 6.6 (cp. fig. 6.3) should be expected in such circumstances.

6.10. Electrical transport properties: the Faber–Ziman theory

There is one more step to be taken in this somewhat artificial discussion of how liquid alloys *ought* to behave; a model must be constructed for their electrical transport properties. The following treatment is based upon the NFE model for the valence electrons, and we shall appeal to the rigid-sphere model for the ions where necessary, but some speculative remarks concerning the effects of compound formation are included at the end of the section. The argument is based throughout upon the work of Faber & Ziman (1965); the alternative approach suggested by Bhatia & Thornton (1970, 1971) appears to be completely equivalent.

Granted the validity not only of the NFE model but of the Born approximation, it is a simple matter to extend Ziman's theory of the resistivity of pure liquids to cover liquid alloys. The necessary formula for $\overline{|U(q)|^2}$, which determines the probability of scattering and hence the relaxation time τ, has been quoted already as (6.27). From this formula it follows immediately that

$$\rho = \frac{3\pi m^2}{\hbar^3 e^2}\frac{N}{\Omega k_F^2}\langle c_0 u_0^2 a_{00} + c_1 u_1^2 a_{11} + c_0 c_1[u_0^2(1-a_{00}) + u_1^2(1-a_{11}) - 2u_0 u_1(1-a_{01})]\rangle, \quad (6.48)$$

where the triangular brackets $\langle\ \rangle$ have the same significance as in (5.3).

For the purposes of a qualitative discussion it will be helpful to re-express this in terms of a *mean interference function* $\overline{a(q)}$, defined by

$$\bar{a} = \frac{c_0^2 u_0^2 a_{00} + 2c_0 c_1 u_0 u_1 a_{01} + c_1^2 u_1^2 a_{11}}{c_0^2 u_0^2 + 2c_0 c_1 u_0 u_1 + c_1^2 u_1^2}. \quad (6.49)$$

Encouraged by the result of the rigid-sphere model that a_{01} is mid-way between a_{00} and a_{11}, we shall assume that $\overline{a(q)}$ has the same general appearance as the interference function for a pure liquid metal, i.e. that it rises from a low value at small to a prominent peak at q_{max} and then levels off to unity for large q. The magnitude of $2k_F$ relative to q_{max} should be determined in the usual way by the number of conduction electrons which have to be accommodated within the Fermi sphere, i.e. by $(c_0 z_0 + c_1 z_1)$. Near the values of q at which u_0 and u_1 have nodes, $\bar{a}$ is evidently tied to a_{11} and a_{00} respectively, and for this reason its peak may be broader and have more structure in it than is normal for a pure liquid, but we shall leave such subtleties on one side for the moment. Following Faber & Ziman, it is also helpful to split the resistivity into two parts. Let us write

$$\rho = \rho' + \rho'',$$

with

$$\rho' = \frac{3\pi m^2}{\hbar^3 e^2}\frac{N}{\Omega k_F^2}\langle (c_0 u_0^2 + c_1 u_1^2)\bar{a}\rangle \quad (6.50)$$

and

$$\rho'' = \frac{3\pi m^2}{\hbar^3 e^2}\frac{N}{\Omega k_F^2}\langle c_0 c_1 (u_1 - u_0)^2 (1-\bar{a})\rangle. \quad (6.51)$$

As in writing (6.27), we are supposing u_1 and u_0 to be real; the generalisation to complex pseudo-potentials is straight-forward.

Of the parameters in (6.50) and (6.51), Ω, k_F, u_1, u_0 and $\bar{a}$ are all of them liable to vary with concentration, and the variation of k_F

may be particularly important because it affects the average expressed by $\langle\ \rangle$. If the two constituents do not differ much in valency or atomic volume, however, the concentration dependence of ρ' and ρ'' should be determined mainly by the c_0 and c_1 which appear explicitly in the equations. Hence ρ' should vary in a more or less *linear* fashion between ρ_0 and ρ_1, the resistivities of the pure solvent and pure solute, while ρ'' should give rise to a *parabolic* deviation.

In a regular solid $a(\mathbf{q})$ is small, except near each reciprocal lattice vector where it rises to a sharp spike. Since the spikes do not contribute directly to the resistivity (see p. 319), we may expect ρ'' in a solid alloy to be always positive and to dominate ρ' in the middle of the concentration range; isothermal curves for resistivity versus concentration should show pronounced humps and indeed, unless the alloy forms an ordered phase, they do so. The situation is not dissimilar for a liquid alloy composed of two monovalent metals; $\bar{a}(q)$ is not necessarily small but it should be less than unity over the whole range of q between 0 and $2k_F$, so although ρ'' may not dominate ρ' it is certainly likely to make the isotherms convex. If the two metals are divalent, however, the main peak in $\bar{a}$, where it exceeds unity, occurs in a range of q that is heavily weighted in the average expressed by $\langle\ \rangle$. Hence it is possible in this case for ρ'' to be negative and for the isotherms to be concave. If the two metals are such that $z_0, z_1 > 2$, then presumably $\bar{a}$ is close to unity over most of the range of q that matters, and in this case ρ'' should be rather small. These qualitative predictions, which are illustrated in fig. 6.7, are in agreement with experimental data, as we shall see below.

The argument may be extended to the thermo-electric power. In §5.5 we found it possible to interpret the thermo-electric properties of most pure liquid metals on the simple assumption that $\partial u^2/\partial k$ is negligible. Granted this assumption, we may deduce from (5.27) that a quantity

$$r = (3-\xi)\rho \tag{6.52}$$

should be proportional to $u(2k_F)^2 a(2k_F)$ for a pure liquid. For a liquid alloy it should be proportional to

$$c_0 u_0(2k_F)^2 \bar{a}(2k_F) + c_1 u_1(2k_F)^2 \bar{a}(2k_F) + c_0 c_1 [u_1(2k_F) - u_0(2k_F)]^2 [1 - \bar{a}(2k_F)]. \tag{6.53}$$

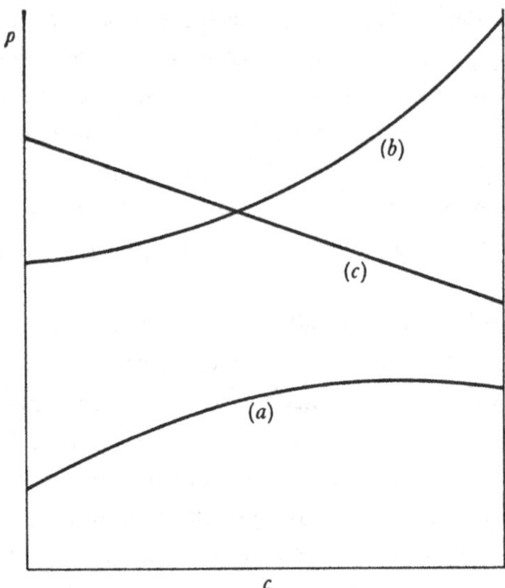

Fig. 6.7. Resistivity isotherms to be expected for mixtures of (a) two monovalent metals, (b) two divalent metals, (c) two polyvalent metals in the liquid state.

Hence in cases where the concentration dependence of k_F, etc., may be ignored, we may expect

$$r = c_0 r_0 + c_1 r_1 + c_0 c_1 (r_0^{\frac{1}{2}} - r_1^{\frac{1}{2}})^2 (1 - \bar{a}(2k_F))/\bar{a}(2k_F), \quad (6.54)$$

where r_0 and r_1 are the values of r for the pure solvent and pure solute. Measurement of the thermo-electric power for an alloy yields a value for ξ from (5.26) if one is prepared to use (6.4) to describe k_F. Then if the resistivity is also known, r may be deduced and plotted as a function of concentration. Since the quantity $(1 - \bar{a}(2k_F))/\bar{a}(2k_F)$ is expected to be rather small, whether the constituents are monovalent or polyvalent, a more or less straight line is to be expected.

The results are inevitably going to be more complicated when the difference between z_0 and z_1 is too large for the variation of k_F to be ignored. Suppose that z_0 is only 1, for example, while z_1 is 4 or 5. The magnitude of $2k_F$ is determined, as stated previously, by $(c_0 z_0 + c_1 z_1)$, and as long as this is large we may ignore ρ''. Hence the resistivity isotherm should be more or less linear for large c_1. It should *not* extrapolate to ρ_0 as $c_1 \to 0$, however, since the resistivity of the solvent is reduced by the coherence effects which

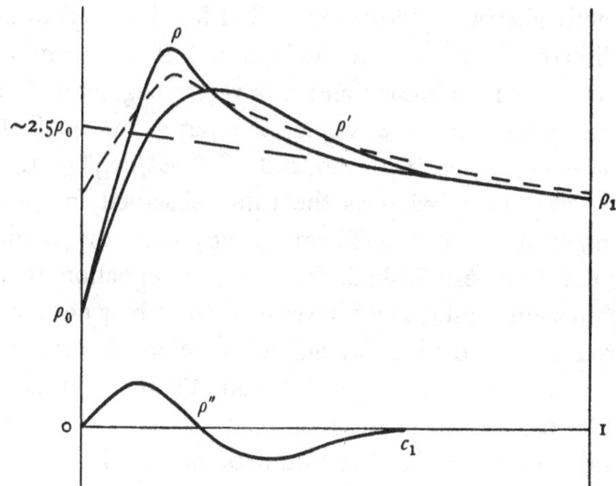

Fig. 6.8. Resistivity isotherms for an alloy with one monovalent and one polyvalent constituent. The total resistivity ρ is the sum of ρ' and ρ''. The broken curve indicates the behaviour to be expected for ρ at a higher temperature.

reduce $\langle a \rangle$, for a monovalent liquid metal, from unity to about 0.4; an extrapolated value of something like $2.5\rho_0$ is therefore to be expected. But as $(c_0 z_0 + c_1 z_1)$ falls below 2 the value of $2k_F$ should pass through the region where $\bar{a}(q)$ exceeds unity, and coherence effects may then enhance ρ' sufficiently for it to rise to a peak, as suggested by fig. 6.8, before falling sharply to ρ_0. The likely behaviour of ρ'' and of the total resistance ρ are also illustrated in fig. 6.8.

It must be emphasised that all three curves in fig. 6.8 are highly schematic; detailed predictions cannot be made without a full knowledge of how u_0, u_1, and $\bar{a}$ depend upon both q and c_1. Faber (1967) has suggested that use be made of the thermo-electric parameter for the pure solvent, ξ_0, to obtain a numerical estimate for the steep slope of the resistivity isotherm near $c_1 = 0$, but the argument is of little help in practice. From (6.48) it follows that

$$\left(\frac{d\rho}{dc_1}\right)^0 = \left(\frac{\partial \rho_0}{\partial k_F}\right)_\Omega \left(\frac{dk_F}{dc_1}\right)^0 - \rho_0 \left(\frac{1}{\Omega}\frac{d\Omega}{dc_1}\right) - 2\rho_0 + \frac{3\pi m^2}{\hbar^3 e^2}\frac{N}{(\Omega k_F^2)^0}\langle (u_1^0)^2 + (u_0^0)^2 - 2u_1^0 u_0^0(1-a_{01})\rangle, \quad (6.55)$$

and the suggestion is to set

$$\left(\frac{\partial \rho_0}{\partial k_F}\right)_\Omega \left(\frac{dk_F}{dc_1}\right)^0 = -\tfrac{2}{3}\xi_0\rho_0\left(\frac{z_1}{z_0} - 1 - \left(\frac{1}{\Omega}\frac{d\Omega}{dc_1}\right)^0\right). \quad (6.56)$$

This certainly allows the terms in the first line of (6.55) to be calculated directly from experimental data, but the term in the second line remains unknown and it is rarely negligible. To take the **Cu**–Sn system as an example, the terms in the first line (with $\xi_0 = -3.6$, $\rho_0 = 21\ \mu\Omega$ cm and $(\Omega^{-1}\, d\Omega/dc_1)^0 = 1.2$) add up to only 23 $\mu\Omega$ cm, whereas the value observed for $(d\rho/dc_1)^0$ experimentally is about 200 $\mu\Omega$ cm. In any case, the validity of (6.56) is not fully established. Not only is equation (6.5) for $(dk_F/dc_1)^0$ in some doubt, as we have seen, but care is needed when ξ^0 is introduced: a distinction ought to be made between the $(\partial/\partial k_F)_\Omega$ that occurs on the left of (6.56) and the $(\partial/\partial k)_F$ that occurs in (5.25). By the second of these derivatives we mean the rate of change with respect to k_F that would be observed if the Fermi sphere could be expanded without altering the screening and without altering the level BB' in fig. 6.1. If k_F is changed by adding impurity, however, BB' *does* shift; it is FF' which remains unaltered. Because of this distinction (6.56) is not to be trusted unless the solvent pseudo-potential happens to be insensitive to energy.

When Ω and k_F are liable to change with concentration it is clearly better, when considering the thermo-electric power to plot a quantity

$$r' = (3-\xi)(\Omega k_F^2/\Omega_0 k_{F,0}^2)\rho \tag{6.57}$$

versus concentration instead of r. The shape of the resultant curve should still be described by (6.53), but it is liable to be affected by the concentration dependence of $u_0(2k_F)$, $u_1(2k_F)$ and particularly $\bar{a}(2k_F)$. Equation (6.54), even with r' in place of r throughout, is not necessarily valid.

What should happen to the resistivity, in the situation illustrated by fig. 6.8, when the temperature is increased? At the polyvalent end of the concentration range $(\partial\rho/\partial T)_p$ is small and at the monovalent end it may be relatively large. In between, where $(c_0 z_0 + c_1 z_1)$ is in the range between say 1.5 and 2.0, there is the interesting possibility suggested by the properties of pure divalent liquid metals (see §5.4) that $(\partial\rho/\partial T)_p$ may become negative. Hence at a higher temperature the isotherm for ρ may look like the broken curve in the figure, and in fact this does describe qualitatively the temperature variation observed for a number of systems (see §6.23). The curve as drawn, however, implies that $(\partial\rho/\partial c_1)_T^0$

should decrease steadily on heating. This is just the reverse of what is normally reported for very dilute alloys involving monovalent solvents, though many of the experimental results are of dubious accuracy. Any temperature dependence of $(\partial\rho/\partial c_1)^0_T$ represents a departure from *Matthiessen's rule*, of course, but there is no good reason to expect this rule to be accurately obeyed for any liquid alloy; the thermal expansion of liquid metals is quite considerable and many of the parameters in the formula obtained above are sensitive to changes of volume.

The Faber–Ziman theory presented in this section refers only to alloys for which the NFE model is a good approximation. Naturally it is unreliable if the alloy forms a compound in the liquid state, for then the Born approximation is suspect, to say the least. Suppose, however, that only some of the valence electrons are localised in molecular orbitals, and that the rest remain at large in the conduction band. Perhaps it may be possible to use a modified version of the theory to calculate a relaxation time for these remaining electrons, provided that we use a suitable effective pseudo-potential to describe the scattering by each 'molecule' as a whole. Unless the composition is exactly the stoichiometric one there will also be scattering by the excess ions of one or other species to be considered, so that we have in effect a binary alloy whose constituents are AB (or A_2B, etc.) molecules – and A (or B) ions. This simple model has been pursued by Schaich & Ashcroft (1970). They point out that the molecules should carry no net charge and suggest that they may be treated, in a first approximation, as electric dipoles. Assuming the orientation of these dipoles to be random they find, after a simple averaging process, that $1/\tau$ should consist of two essentially independent terms, an incoherent scattering term due to the molecules alone and a coherent term due to the ions alone. The original paper should be consulted for further details.

6.11. Results of diffraction experiments

We turn from the construction of theoretical models for liquid alloys to examine the experimental evidence, starting with the results obtained by X-ray or neutron methods.

The diffracted intensity in an X-ray experiment consists of

three terms: (*a*) the incoherent background due to Compton scattering which occurs even for pure liquids, (*b*) the additional incoherent term proportional to $(f-\bar{f})^2$, or to $c_0c_1(f_0-f_1)^2$ for a binary alloy, which appears in (2.61), and (*c*) the coherent part proportional to $\bar{f}^2\bar{a}_{XR}$, where

$$\bar{a}_{XR} = \frac{c_0^2f_0^2a_{00}+2c_0c_1f_0f_1a_{01}+c_1^2f_1^2a_{11}}{c_0^2f_0^2+2c_0c_1f_0f_1+c_1^2f_1^2}. \tag{6.58}$$

Since X-ray form factors are determined by atomic number there are some systems (e.g. Hg–Tl) for which term (*b*) is negligibly small, but in general it has to be eliminated in the same way as term (*a*) (see §2.13) before a curve for $\bar{a}_{XR}$ can be obtained. The task of analysis may be complicated by the fact that $\bar{a}_{XR}(q\rightarrow 0)$ is not known and is not necessarily small compared with unity (see §6.5), but this is not an insuperable difficulty. Fortunately, the absolute magnitude of $\bar{a}_{XR}$ is of less interest than the structure – or lack of structure – which appears in its main peak, and this is scarcely affected by whatever allowance has to be made for terms (*a*) and (*b*).

Because f_0 and f_1 are smooth functions of q which, unlike u_0 and u_1, do not have nodes, the mean interference function described by (6.58) has less reason to show structure in its main peak than the $\bar{a}$ for electron scattering described by (6.49). If the rigid-sphere model is correct and a_{01} is intermediate between a_{00} and a_{11}, it is likely that $\bar{a}_{XR}$ will display a single rather broad peak, shifting smoothly as the concentration is altered between the positions characteristic of the pure solvent and pure solute. Resolution into three sub-peaks is scarcely to be expected. Qualitatively at any rate this prediction is satisfied by a number of the systems which have been investigated by means of X-rays, including:

(i) Na–K (Orton, Williams & Shaw, 1960);

(ii) Hg–Tl (Halder, Metzger & Wagner 1966; Smallman & Frost, 1956) and Hg–In (Kim, Standley, Kruh & Clayton, 1961; Halder & Wagner, 1967*a*; Orton & Street, 1970);

(iii) Mg–Al (Steeb & Woerner, 1965);

(iv) Cd–Sn (Alekse'ev & Evse'ev, 1959);

(v) In–Bi (Isherwood & Orton, 1969).

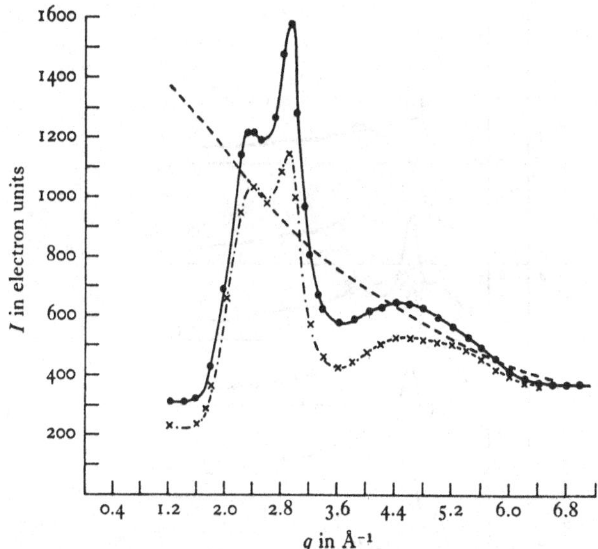

Fig. 6.9. The twin-headed diffraction peak for liquid Cu_6Sn_5. The full curve was calculated from the partial interference functions shown in fig. 6.11. ●, calculated; ×, observed. (Redrawn by permission from Enderby, North & Egelstaff, 1966.)

It is probable that Cd–Bi and Sn–Bi could be added to the list, but the evidence for these systems is still incomplete while for many others which are more than likely to obey the rigid-sphere model it is non-existent. In the case of NaK ($c_0 = c_1 = \frac{1}{2}$) the Percus–Yevick rigid-sphere model has been shown by Enderby & North (1968) to fit the curve for $\bar{a}_{XR}$ with reasonable quantitative accuracy and not just qualitatively.

There are a number of systems for which the rigid-sphere model is clearly inadequate, however, because the main peak in $\bar{a}_{XR}$ is not structureless; at intermediate concentrations it is observed to split, usually into two sub-peaks rather than three. Examples include:

(i) Cu–Sn (Orton & Williams, 1960). The main peak is twin-headed for $c_{Sn} = 0.45$ and 0.60, as shown in fig. 6.9, though not at lower concentrations. Cu–Ge (Isherwood & Orton, 1970) gives similar patterns.

(ii) Ag–Sn (Orton & Williams, 1960; Joshi & Wagner, 1965). The measurements extend only up to $c_{Sn} = 0.5$. By this stage

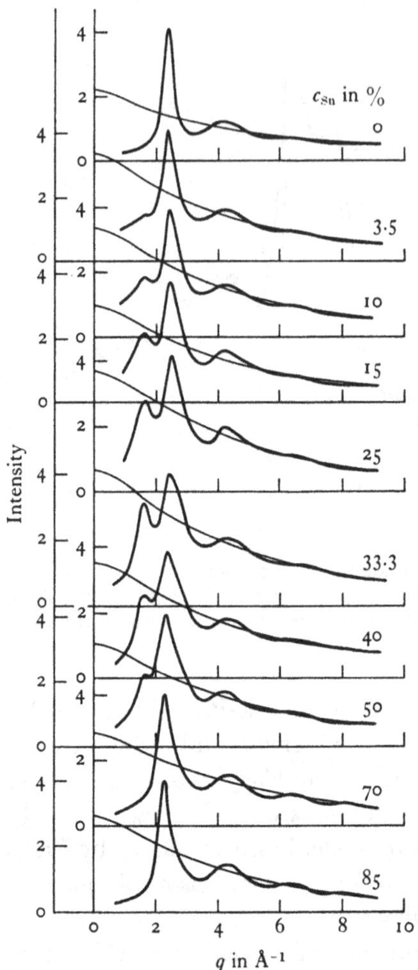

Fig. 6.10. X-ray diffraction curves for liquid Mg–Sn. (Redrawn by permission from Steeb & Entress, 1966.)

the peak is broad and at higher concentrations splitting would no doubt become apparent.

(iii) Au–Sn (Kaplow *et al.*, 1965*b*; Waghorne, Rivlin & Williams, 1967). The two sets of data are not in complete agreement, but for $c_{Sn} > 0.4$ the peak in $\bar{a}_{XR}$ certainly broadens in an anomalous fashion, even if it does not split. It is of interest that the highest point of $\bar{a}_{XR}$ in the middle of the concentration range appears to lie at $q \sim 2.8$ (Å)$^{-1}$, whereas q_{max} for pure Au and pure Sn is about 2.6 and 2.25 (Å)$^{-1}$ respectively.

(iv) Au–Al, Au–Ga, Au–Tl (Waghorne *et al.*, 1967). The evidence for these systems is incomplete.

(v) Hg–Na, Hg–K (Schuhmann, 1962). Some structure in the main peak is visible for these, though not a clear-cut split.

(vi) Mg–Sn and Mg–Pb (Steeb & Entress, 1966; Steeb, Dilger & Höhler, 1969). In both cases a subsidiary peak is reported on the low-q side of the main one as shown in fig. 6.10 and it is most marked at the concentrations corresponding to the solid compounds Mg_2Sn and Mg_2Pb.

(vii) Ag–Mg (Steeb & Hezel, 1966). Here again there is said to be a subsidiary peak at low q, most prominent when $c_{Mg} \sim 0.75$.

It is hardly surprising that several of the systems in this second group are ones which form distinct compounds in the solid phase, whereas the only clear example of compound formation in the first group is provided by In–Bi – and the heats of formation of both InBi and In_2Bi are less than 0.01 eV per atom.† Attempts have been made to conclude from these X-ray data that the structures associated with solid Mg_2Sn, Mg_2Pb and AuSn are indeed still present in the liquid phase, despite loss of long-range order and despite the presence of many 'dissociated' ions, particularly at concentrations which do not correspond to the stoichiometric formula. Until it can be shown, however, that the structural models invoked are capable of explaining not only $\bar{a}_{XR}$ but also the three partial interference functions, these arguments remain unconvincing. Knowledge of the partial interference functions is also essential if we are to interpret the anomalies observed for Cu–Sn, Ag–Sn, and Ag–Mg, whose only compounds in the solid phase are probably electron ones.‡

The first indication of how it might be possible to unfold the partial interference functions in practice was given by Keating (1963). He observed that if one could vary the form factors f_0 and f_1 without altering the structure, and hence obtain at least three

† The solid phase which is stable for the Hg–Tl system over a range of concentrations near Hg_5Tl_2 would seem to be best classified as an electron compound.

‡ The ϵ phase of Ag–Mg is known to be stable over a range of concentrations around $AgMg_3$, but its structure and therefore its classification are still uncertain.

different curves for $\bar{a}$ corresponding to three different values for the ratio f_1/f_0, one could extract values for a_{00}, a_{11} and a_{01} by solving three simultaneous equations for each value of q. He also pointed out that $\bar{f}_1/\bar{f}_0$ is not the same for neutron scattering as for X-rays. If one is prepared to eliminate one of the three unknowns, e.g. by the assumption that a_{00} and a_{11} are equal, then it should be possible to find the other two by combining the results of X-ray and neutron experiments. Just this method of analysis has been applied to Na–K by Henninger, Buschert & Heaton (1966).

The assumption that a_{00} and a_{11} are equal has little to recommend it in general, and to avoid it a third experiment is required; the neutron experiment must be repeated using a specimen in which either $\bar{f}_1$, $\bar{f}_0$, or both have been altered by isotopic replacement. Enderby, North & Egelstaff (1966) have applied this idea to liquid Cu_6Sn_5. They used three specimens in fact, one containing natural Cu and the others Cu enriched to 99% with ^{63}Cu or ^{65}Cu, and were therefore able to determine the three unknowns from neutron data alone. The variation achieved in the form factor for coherent neutron scattering by the Cu ions – it was in the ratio 1:0.73:1.96 for the three specimens – was not large enough to ensure much accuracy, so in choosing the final curves, which are reproduced in fig. 6.11, the authors were guided by a desire to achieve reasonable consistency with the X-ray data and by the necessity of satisfying the inequalities in (2.63). They were not able to match the curve for $\bar{a}_{XR}$ exactly (see fig. 6.9), but the latter may be incorrectly normalised. The most interesting point to emerge is that while the peaks in a_{CuCu} and a_{SnSn} are in almost the same place for the alloy as they are for pure Cu and pure Sn respectively, the peak in a_{CuSn} does *not* lie half way between them; it is virtually coincident with the peak in a_{CuCu}, which is quite inconsistent with the rigid-sphere model. Near $q = 0$, if the curves are to be trusted in this region, a_{CuSn} is about 0.35, while a_{CuCu} and a_{SnSn} are both small. This is also inconsistent with the rigid-sphere model, as illustrated by fig. 6.3, but it agrees no better with the speculative curves for a tightly-bound liquid compound which are shown in fig. 6.6.

It should be of interest to extend the techniques of neutron diffraction and isotopic replacement to other liquid alloys besides

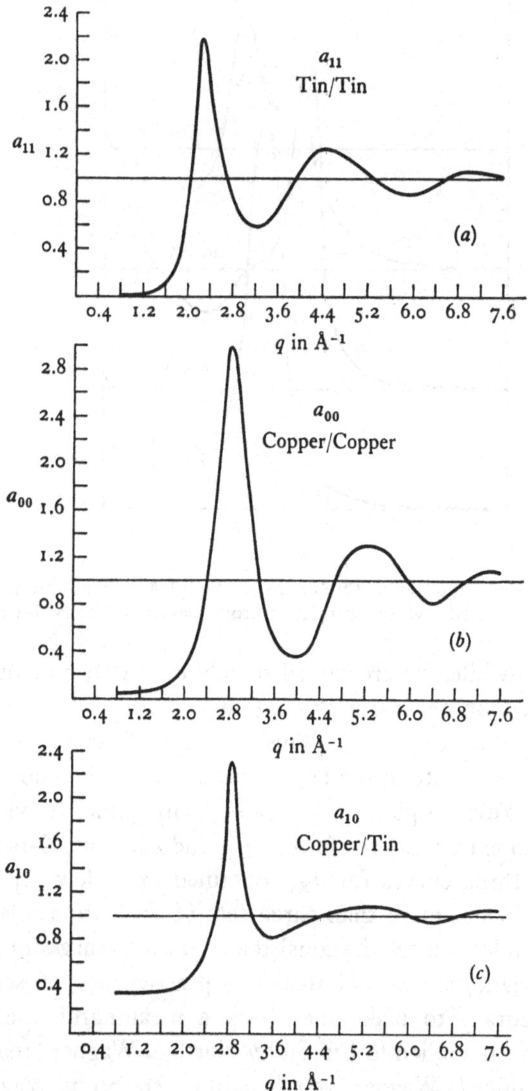

Fig. 6.11. Partial interference functions for liquid Cu_6Sn_5. (Redrawn by permission from Enderby, North & Egelstaff, 1966.)

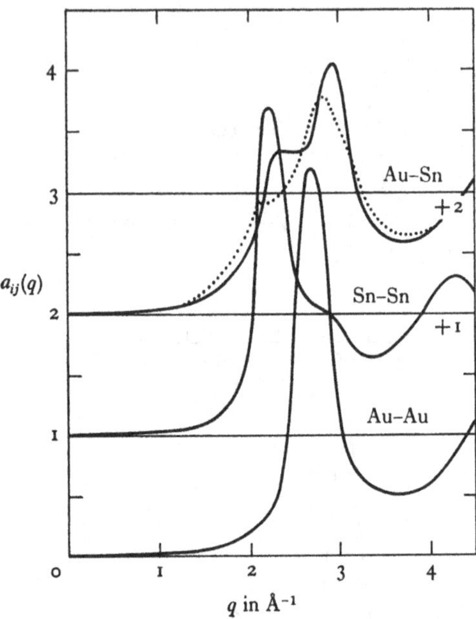

Fig. 6.12. Partial interference functions for liquid Au–Sn according to Wagner *et al.* (Redrawn by permission from Davies & Leach, 1969.)

Cu_6Sn_5. Meanwhile, we are forced to rely on another method of analysis, based on the assumption that a_{00}, a_{11} and a_{01} are all independent of concentration. This is clearly false for small q, but it may be adequate in the large-q region, near the main peak, if the Percus–Yevick rigid-sphere model is any guide. It evidently enables one to extract curves for a_{00}, a_{11} and a_{01} from X-ray data alone, given three curves for $\bar{a}_{XR}$ obtained at widely separated values of c_1. When more than three sets of data are available it becomes possible to check the consistency of the basic assumption, though the data that are available in practice are of scarcely sufficient accuracy to make the check a meaningful one. The method has been applied to Cu–Sn by North & Wagner (1970), to Ag–Sn by Halder & Wagner (1967*b*), and to Au–Sn by Wagner *et al.* (1967), with plausible results. These authors conclude that Ag–Sn is similar to Cu–Sn, in that the peak in a_{AgSn} is pulled towards the peak in a_{AgAg}. They obtain a curve of unusual shape for a_{AuSn}, however, whose maximum has been pulled beyond the peak in a_{AuAu}, as shown in fig. 6.12. The dotted curve in this figure is derived from a crude model based upon the nickel arsenide structure of solid AuSn, and the match is distinctly suggestive.

The theoretical argument that was advanced to support the rigid-sphere model in §6.4 has not been verified by detailed calculations except in the case of Na–K, and the discovery that this model sometimes breaks down in practice does not necessarily mean that the NFE approach must be abandoned, or even seriously modified. It is not impossible that accurate calculations of w_{00}, w_{11} and w_{01} for Cu–Sn, say, using the theoretical expressions quoted above, would make it seem quite natural that the separation between adjacent Cu and Sn ions is less than the arithmetic mean of the Cu–Cu and Sn–Sn separations. But whether we can retain the NFE approach for Au–Sn, let alone for Mg–Sn and Mg–Pb is more debatable.

6.12. Thermodynamic properties: normal systems

The variety of experimental methods that may be used to derive information about the thermodynamic properties of liquid alloys is very great, and a vast amount of data has been accumulated. A selection of the more interesting and reliable results is presented here, but the reader who wishes to pursue the subject further should consult compilations and reviews by Kubaschewski & Catterall (1956), Kleppa (1958), Hultgren *et al.* (1963), Wilson (1965), Pratt (1967) and Kubaschewski & Slough (1969).

It will illustrate how difficult the task of interpretation may prove to be if we start by considering in some detail the system Na–K. Here is a relatively simple system, in which the chance of compound formation in the liquid phase is remote, and it has been investigated theoretically with more care than most. Figures are given in table 6.1 to show how much the internal energy $\mathcal{U}$,† the volume Ω, the entropy and the mean compressibility β change when equal quantities of liquid Na and liquid K, both at the same temperature, are mixed to form NaK; ΔS_E is the amount by which the entropy of the mixture exceeds the 'ideal' value described by (6.42). A comparison is made with the results of various theories outlined in §6.6.

† It is really the change of heat content or enthalpy which is measured, but the distinction between this and $\Delta\mathcal{U}$ is negligible at atmospheric pressure.

TABLE 6.1. *Thermodynamic properties of liquid NaK at about* 100 °C

Property	Experiment	Theory (*a*)	(*b*)	(*c*)	(*d*)
$\Delta\mathcal{U}/N$ (eV)	0.008	0.01	0.0045	—	—
$\Delta\Omega/\Omega$ (%)	−1.2	−1.9	—	+2.4	—
$\Delta S_{\mathrm{E}}/Nk_{\mathrm{B}}$	+0.05	—	—	+0.49	+0.4
$\Delta\beta/\beta$ (%)	−6.4	−12.1	—	—	—

Theoretical estimates obtained as follows: (*a*) Christman & Huntington (1965) or Christman (1967*a*); (*b*) Tamaki & Shiota (1966); (*c*) Equations (6.45) and (6.43); (*d*) equation (6.43) corrected for volume change on mixing.

The success of Christman's calculation as far as $\Delta\mathcal{U}$ is concerned may be to some extent fortuitous, since he was not able to match the data so closely at other values of c, nor for other systems composed of alkali metals. His method enables one to predict the correct sign for $\Delta\mathcal{U}$ but the numerical value may be wrong by a factor of 2 or more, and the same seems to be true for $\Delta\Omega$ and $\Delta\beta$. His neglect of the entropy of mixing, already criticised in §6.6, may be partly responsible for the discrepancies; in particular it may explain why his value for $\Delta\Omega$ is too low. But if the entropy of mixing blows up the volume by only $(1.9-1.2) = 0.7\%$, this is a good deal less than our crude equation (6.45) would suggest. It seems that the assumptions upon which (6.45) is based are quite unreliable for NaK, since (6.43), which is based upon the same assumptions, clearly exaggerates ΔS_{E}. Even if (6.43) is corrected in a straight-forward manner to take account of the fact that Ω is not exactly a linear function of concentration it still exaggerates ΔS_{E}, as the figure in column (*d*) of table 6.1 will show.

The energy of mixing is known as a function of concentration for many liquid systems besides Na–K. It is roughly proportional to $c_0 c_1$ in a number of cases, though there are many others where the experimental curves are distinctly asymmetric. Some data are presented in table 6.2 for a number of very dilute alloys for which a comparison is possible between the solid phase (usually near room temperature) and the liquid (usually not far above the solvent melting point). The quantity tabulated, $\Delta\mathcal{U}/Nc_1$, is the energy

TABLE 6.2. *Energy of mixing and volume of mixing for dilute alloys*

Solvent	Solute	$\Delta\mathscr{U}/Nc_1$ (eV) Solid	Liquid	$\Delta\Omega/\Omega c_1$ (%) Solid	Liquid
Ag	Au	0.15	0.15	−2.5	0
Ag	Zn	0	−0.2	—	—
Cu	Zn	0.35	0.35	−10	−6.5
Mg	Cd	0.16	0.23	−14	—
Cd	Mg	0.17	—	6	—
Zn	Cd	0.30	0.09	—	+9
Cd	Zn	0.18	0.08	—	0
Zn	Al	0.25	0.08	—	−1.5
Al	Zn	0.15	0.1	—	0
Tl	Pb	−0.26	−0.05	−20	—
Pb	Tl	−0.02	−0.03	−0.5	—
Bi	Tl	−0.06	−0.06	—	—

required to dissolve a single solute atom when c_1 is small. According to Friedel's elastic model this energy should decrease on melting, and in most cases – though not all – it is observed to do so. There is an interesting contrast between systems like **Ag**–Au or **Cu**–Zn on the one hand and **Zn**–Cd and **Cd**–Zn on the other. For the first pair the misfit-energy described by (6.37) is negligible, and Friedel and Varley would agree that the observed values of $\Delta\mathscr{U}/Nc_1$ in the solid phase are to be attributed almost entirely to other terms of a purely electronic nature. The misfit between Zn and Cd is relatively large, however, while the other terms are thought to be particularly small; (6.37) would be sufficient to account for the whole of the observed $\Delta\mathscr{U}/Nc_1$ in solid **Zn**–Cd and **Cd**–Zn if Poisson's ratio were about 0.15, a value that is on the low side but not impossibly low. It is therefore hardly surprising that the fall in $\Delta\mathscr{U}/Nc_1$ on melting is too small to be detectable in the case of **Ag**–Au and **Cu**–Zn, but relatively large in **Zn**–Cd and **Cd**–Zn.

Table 6.2 includes some results for the volume change on mixing in these dilute alloys, but the data are very incomplete.

Friedel's formula, equation (6.36), would suggest values for $\Delta\Omega/c_1\Omega$ of order -10% for most of the systems listed, but it is unlikely that the elastic model will suffice to explain the results in detail. For the majority of dilute liquid alloy systems which do not form compounds $\Delta\Omega/c_1\Omega$ is positive, being as much as $+10\%$ or $+20\%$ in some cases. The elastic model implies as has been noted previously, that it should be either negative, as for a solid, or zero if the shear strain around each impurity ion is free to relax completely. Positive values could perhaps be attributed to the influence of the entropy of mixing, but there seems to be no correlation between the observed results and equation (6.45).

A number of straightforward liquid alloy systems are exceptions to the rule just stated and contract on mixing, Na–K, **Cu**–Zn and **Zn**–Al among them. The most notable group of exceptions are the Hg amalgams investigated by Kleppa, Kaplan & Thalmeyer (1961). These authors measured $\Delta\Omega$ for liquid Hg mixed with liquid Zn, Cd, In, Tl, Sn, Pb or Bi and found it to be negative in every case; $\Delta\Omega/c_1\Omega$ is typically -10% or so for dilute solutions, and in the middle of the concentration range $\Delta\Omega/\Omega$ reaches a low of about -2%. The compressibility of amalgams is also anomalously low. Ultrasonic measurements by Khodov (1960), Gordon (1961) and Hill & Ruoff (1965) have shown that the adiabatic compressibility is an almost linear function of concentration for Cd–Sn, Cd–Pb, Cd–Bi, Sn–Pb, Sn–Bi and Pb–Bi. According to Abowitz & Gordon (1963*a*), however, the adiabatic compressibility of pure Hg is sharply reduced by the addition of any of the seven solutes already quoted; $\Delta\beta/c_1\beta$ in the limit $c_1 \to 0$ is of order -200%, so that the fall is much more dramatic than in the Na–K system (see table 6.1). These anomalies are no doubt the fault of the unusual pseudo-potential of Hg (see p. 40) but they are not yet understood.

Abowitz & Gordon (1963*b*) have extended their ultrasonic work so as to study the attenuation in the case of Hg–Tl, and they find it to be anomalously large. The attenuation in Na–K has been measured by Jarzynski & Litowitz (1964) and in K–Rb and Na–Cs (see p. 524) by Kim & Letcher (1971).

The entropy of mixing for a large number of non-compound forming systems is displayed graphically in fig. 6.13. The quantity

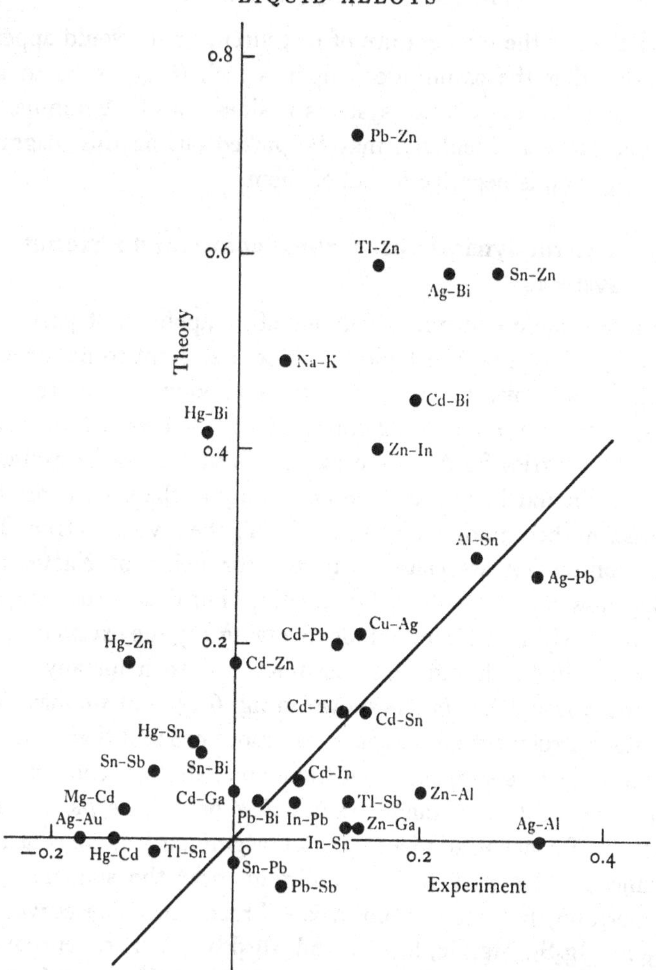

Fig. 6.13. Theoretical and experimental values for $(\Delta S_E/Nk_B)$.

plotted as the abscissa here is the maximum value observed for $\Delta S_E/Nk_B$ in the liquid phase, usually in the middle of the concentration range at $c_1 \simeq 0.5$. Plotted as the ordinate is the maximum value of $\Delta S_E/Nk_B$ according to equation (6.43). The correlation is poor and it is not improved by making allowance for the non-linear dependence of Ω on concentration. Some of the experimental results may be wrong and there is room for error in the theoretical estimates as well, because the values used for Ω, α and β are those that apply at the melting point of the metal concerned rather than at the temperature of measurement. It would appear, however, that the assumptions upon which (6.43) is based are inadequate for many other systems besides Na–K. A number of Hg amalgams, incidentally, may be picked out on this diagram, and $\Delta S_E/Nk_B$ is negative for all of them.

6.13. Thermodynamic properties: compound-forming systems

Thermodynamic data for a number of compounds of particular interest are presented in table 6.3. The first point to be noted is that $\Delta\mathscr{U}$ rises towards zero when the compounds melt. In most cases, however, it remains a significant fraction of what it was in the solid. The entries for $\Delta\Omega$ are erratic; it is surprising for example, that Mg_2Sn and Mg_2Pb expand on mixing in the solid phase but contract in the liquid, whereas for Mg_3Bi_2 the reverse is true. The large contraction associated with the formation of $NaHg_2$ and KHg_2, however, is unaffected by melting. For these two amalgams and for nearly all the compounds involving Mg the excess entropy of mixing in the liquid phase is much less than for any of the systems whose behaviour is shown in fig. 6.13, and we may infer that their structures are in some way more ordered than is usual.

It must not be supposed that these anomalies are confined to a narrow concentration range. If $\Delta\Omega$, $\Delta\mathscr{U}$ or ΔS_E is plotted as a function of concentration a somewhat asymmetric curve is usually obtained, which may have a minimum near the stoichiometric composition, but the minimum is a broad one. The curves for ΔS_E in Mg_2Si, Mg_2Ge, Mg_2Sn and Mg_2Pb which are reproduced in fig. 6.14 – they are due to Eldridge, Miller & Komarek (1967) – will serve to illustrate this point. Some investigators prefer to accentuate the anomalies and make them look more localised by plotting partial rather than integral quantities,† and indeed it is

† The partial energies of mixing, say, are defined for an alloy as the energy required to dissolve N atoms of either constituent in a specimen containing very many more than N atoms. They are written as $\overline{\Delta\mathscr{U}}_0$ and $\overline{\Delta\mathscr{U}}_1$ and are functions of the concentration of the specimen. Partial and integral quantities are related by equations such as

$$\overline{\Delta\mathscr{U}}_1 = \Delta\mathscr{U}+(1-c_1)(\mathrm{d}(\Delta\mathscr{U})/\mathrm{d}c_1).$$

The relation between the partial and integral excess entropies for Mg–Si is shown graphically for an arbitrary concentration in fig. 6.14.

TABLE 6.3. *Properties of some intermetallic compounds*

Formula	Crystal structure	Melting point (0 °C)	Latent heat of melting (eV/atom)	Entropy of melting (k_B/atom)	Volume change melting (%)	Energy of mixing Solid (eV/atom)	Energy of mixing Liquid (eV/atom)	Volume change on mixing Solid (%)	Volume change on mixing Liquid (%)	Entropy of mixing Liquid (k_B/atom)
AuSn	nickel arsenide	418	0.13	2.2	—	−0.16	−0.11	+1.7	—	+0.26
$NaHg_2$	aluminium boride	353	—	—	—	−0.3	−0.3	−19	−18	−2.7
KHg_2	aluminium boride	270	0.07	1.6	—	−0.27	−0.23	−25	−24	−(2.0)
ZnSb	deformed diamond	546	—	—	—	−0.08	−0.03	+7	+3.4	+0.09
CdSb	orthorhombic	456	0.17	2.9	—	−0.07	−0.02	+7.5	+5	+0.11
AlSb	zinc blende	1050	—	—	—	−0.5	—	—	—	—
GaSb	zinc blende	706	—	—	—	—	—	—	—	—
InSb	zinc blende	530	0.27	3.8	—	−0.19	−0.04	—	—	+0.07
ZnTe	zinc blende	1240	—	—	—	−0.62	—	—	—	—
MgTe	wurtzite	—	—	—	—	—	—	—	—	—
SnTe	sodium chloride	790	—	—	—	—	—	—	—	—
PbTe	sodium chloride	917	—	—	—	—	—	—	—	—
Mg_2Si	calcium fluoride	1102	—	—	−23	−0.28	−0.17	—	—	−0.75
Mg_2Ge	calcium fluoride	1115	—	—	−(3.5)	—	−0.28	—	—	−0.9
Mg_2Sn	calcium fluoride	778	0.22	2.4	−2	−0.27	−0.14	+6.7	−4.1	−0.3
Mg_2Pb	calcium fluoride	550	0.14	2.3	−(3.5)	−0.18	−0.09	+3.4	−1.6	−0.15
Ag_2Te	f.c.c. (at high temp.)	959	—	—	—	−0.13	—	−1.6	—	—
$AuTe_2$	two forms: monoclinic & orthorhombic	464	—	—	—	—	—	—	—	—
Mg_3Sb_2	lanthanum oxide (at low temp.)	1228	—	—	—	−0.6	−0.36	−1.8	—	−(1.7)
Mg_3Bi_2	lanthanum oxide (at low temp.)	823	0.19	2.0	—	−0.32	−0.23	−2.1	+1.9	−(1.1)
Sb_2Te_3	rhombohedral	622	—	—	—	—	—	—	—	—
Bi_2Te_3	rhombohedral	585	—	—	—	—	—	—	—	—

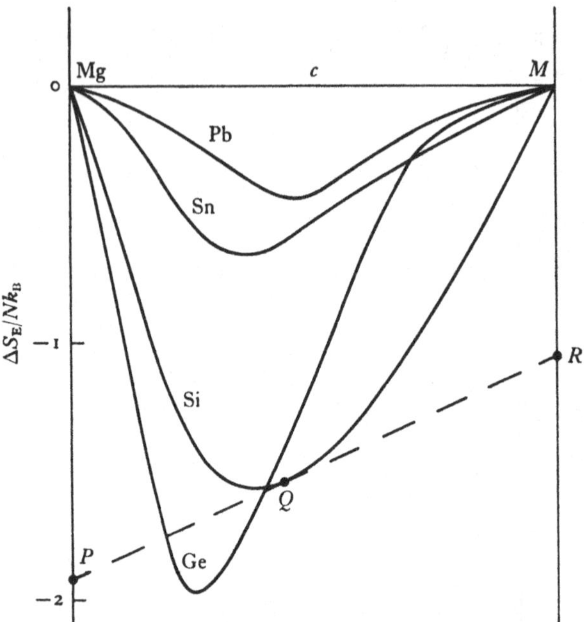

Fig. 6.14. The excess entropy of mixing for four compound-forming systems based on Mg. The intercepts at P and R correspond to the partial excess entropies of mixing, $\overline{\Delta S_E}/Nk_B$, for Mg and Si respectively in liquid Mg–Si with the concentration corresponding to the point Q.

often the partial quantities that are measured experimentally. Whether such plots are helpful for liquid alloys is, however, a debated point. They encourage the construction of fanciful models for which the data provide little evidence of real substance.

Reliable results concerning the specific heat of liquid alloys are confined to a rather small number of systems, and for most of them Kopp's law seems to be obeyed, i.e. C_p is an essentially linear function of concentration for a fixed number of atoms. The In–Sn and In–Bi systems are perhaps the best substantiated examples (Orr, Giraud & Hultgren, 1962; Hultgren & Orr, 1967). The specific heat of liquid compounds might be expected to be anomalously large, since the ordering that tends to lower their entropy at temperatures just above the melting point must presumably disappear on heating. Small (8%) positive deviations from Kopp's law have in fact been reported for liquid Au–Sn by Kubaschewski (1943).

6.14. Surface tension

The literature concerning the surface tension of liquid alloys is extensive, but since many of the published results are of doubtful accuracy, while the theory is still in a primitive state, the subject will be treated only briefly here. Semenchenko (1961) has given a detailed account of Russian work up to 1959 or so; papers published outside Russia or after this date may be traced through Wilson's (1965) bibliography.

It is hard enough to explain the magnitude of the surface tension observed in pure liquid metals, and to provide an absolute theory for a liquid alloy is not yet feasible. But one might hope to devise an interpolation formula, to explain how at a given temperature ℓ varies with concentration between ℓ_0 and ℓ_1, the surface tensions of pure solvent and pure solute respectively. What makes the task so difficult is that the concentration in the surface layers may be quite different from what it is in the bulk. If $\ell_1 < \ell_0$ there is a tendency for the impurity ions to congregate at the surface, thereby lowering the free energy of the specimen; the impurity is said to be *adsorbed* on the surface, and to be *surface-active* with respect to the solvent. If $\ell_1 > \ell_0$, the impurity avoids the surface and is said to be *surface inactive*. In extreme cases, when $\ell_1 \ll \ell_0$, the free energy required to detach an impurity ion from the surface may be considerably more than $k_B T$, and the impurity may virtually cover the surface with a monolayer, even when the bulk concentration is very small. In **Hg**–K, for example, $\ell_0 \sim 500$ and $\ell_1 \sim 100$ dyne cm^{-1}. Since the area of surface occupied by a single K ion may be about 1.5×10^{-15} cm^2, one may expect to find

$$\frac{\text{surface concentration of K}}{\text{bulk concentration of K}} \sim \exp\left(\frac{(500-100)\times 1.5\times 10^{-15}}{1.4\times 10^{-16}\times 300}\right) \sim 10^6,$$

i.e. to find that a bulk concentration of only 1 in 10^6 is sufficient to allow a monolayer to form. This prediction is in fact consistent with the observations for **Hg**–K.

Let us explore the consequences of the following set of assumptions, crude though they evidently are.

(*a*) The alloy is substitutional.

(*b*) Of the N sites which are occupied by ions a small number N_s may be distinguished as surface sites; there are n_s of these per unit area of surface.

(*c*) Ions of both species are distributed at random among the surface sites with concentrations c_0' and c_1' $(= 1-c_0')$ and also at random among the remaining 'bulk' sites with concentrations c_0'' and c_1'' $(= 1-c_0'')$. If c_0, c_1 are the concentrations in the specimen as a whole, then

$$c_0 N = c_0' N_s + c_0''(N-N_s), \text{ etc.}$$

(*d*) The system obeys (6.31), the ideal law of mixing, in the bulk and in the surface. The entropy of mixing is therefore just

$$k_B \log\left[\frac{N_s!}{(c_0'N_s)!(c_1'N_s)!}\times\frac{(N-N_s)!}{(c_0N-c_0'N_s)!(c_1N-c_1'N_s)!}\right]$$
$$= -Nk_B\,(c_0\log c_0+c_1\log c_1)-N_s\left(c_0'\log\left(\frac{c_0'}{c_0}\right)+c_1'\log\left(\frac{c_1'}{c_1}\right)\right) + O(N_s^2/N), \quad (6.59)$$

while there is no internal energy of mixing to be considered.

It follows from these premises that the term in the free energy of the specimen which is proportional to N_s and hence to surface area is

$$N_s\left(c_0'(\ell_0/n_s)+c_1'(\ell_1/n_s)+k_BT\left(c_0'\log\left(\frac{c_0'}{c_0}\right)+c_1'\log\left(\frac{c_1'}{c_1}\right)\right)\right).$$

It is a minimum when

$$\ell_0+n_sk_BT\log\left(\frac{c_0'}{c_0}\right)=\ell_1+n_sk_BT\log\left(\frac{c_1'}{c_1}\right),$$

i.e. when
$$\frac{c_0}{c_0'} = c_0+c_1\exp\left((\ell_0-\ell_1)/n_sk_BT\right). \quad (6.60)$$

The equilibrium value of the surface free energy per unit area, or surface tension, is therefore given by

$$\ell = \ell_0+n_sk_BT\log\left(\frac{c_0'}{c_0}\right) = \ell_1+n_sk_BT\log\left(\frac{c_1'}{c_1}\right)$$
$$= \ell_0-n_sk_BT\log\left(1+c_1\left(\exp\left((\ell_0-\ell_1)/n_sk_BT\right)-1\right)\right)$$
$$\simeq c_0\ell_0+c_1\ell_1-c_1(\ell_0-\ell_1)^2/2n_sk_BT+\ldots \quad (c_1 \ll 1). \quad (6.61)$$

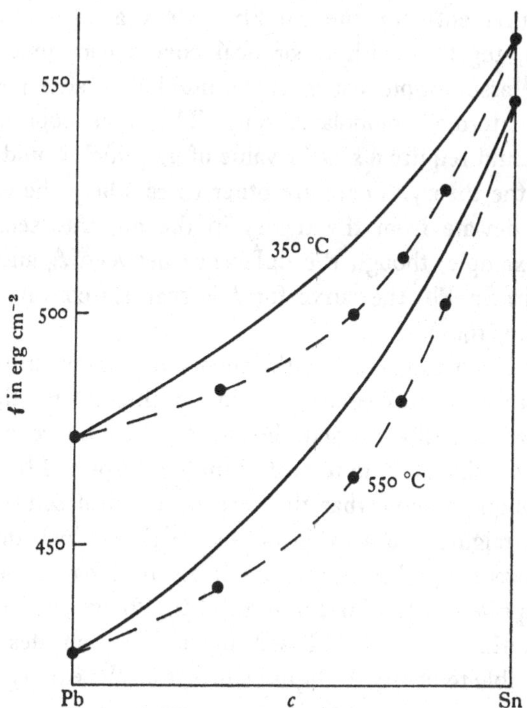

Fig. 6.15. The surface tension of liquid Sn–Pb. The full curves correspond to equation (6.61).

These formulae will be found in Semenchenko's book, and he quotes examples of systems composed of inert organic liquids for which they give an adequate fit to the data. They have also been derived by Butler (1932) and others.

This simple theory predicts that isothermal curves for ℓ as a function of concentration should be concave downwards. In extreme cases, when $(\ell_0 - \ell_1) \gg k_B T$, the surface tension should fall to ℓ_1 for very small concentrations of the solute and should remain at that value right up to $c_1 = 1$. In no circumstances should ℓ fall below ℓ_1, or below ℓ_0 if that is the smaller.

There are a number of liquid alloys which conform to these predictions in a qualitative fashion, including Cu–Sb, Zn–Sb, Cd–Sb, In–Pb, Sn–Pb, Pb–Sb, Pb–Bi, and several amalgams. At a quantitative level, however, the theory is rarely successful. Fig. 6.15 shows the results obtained by Hoar & Melford (1957) in a careful set of measurements for the Sn–Pb system at two different temperatures, together with theoretical curves computed from (6.61) on the assumption that n_s is the number of sites per unit area in the outermost monolayer only. The fit is poor and to improve it would require a *smaller* value of n_s, which would make nonsense of the theory. There are other cases where the experimental data deviate from the theory in the opposite sense; in Sn–Bi, or example, though the difference between ℓ_0 and ℓ_1 is larger than in Sn–Pb, the curve for ℓ is scarcely concave at all (Semenchenko, 1961).

It is not difficult to generalise the theory to non-substitutional alloys, such that the surface area per ion is different for the two species. What is really needed, however, is to improve upon assumption (*d*), since the finite heat of mixing displayed by liquid alloys is enough to show that they are never ideal solutions in practice. A straightforward extension of (6.61) in which the heat of mixing plays a role has been worked out by Hoar & Melford (1957) and applied to the Sn–Pb and the In–Pb systems. It does fit the data rather better. It is based, however, on an idea which seems impossible to justify for liquid alloys, that the energy can be expressed entirely in terms of 'bond energies' between nearest neighbours.

Are there features in the data which support the idea of compound formation in some liquid systems? An anomalous maximum in the curve of ℓ versus concentration for the Au–Sn system was reported by Kaufman & Whalen (1965); it was situated near $c_{\mathrm{Sn}} = 0.7$ and became steadily more prominent as the temperature was lowered from 1150 °C to 500 °C. Since there is no sign of it, however, in the curves which Pokrovsky, Pugachevich & Ibramigov (1967) have published for Au–Sn at temperatures between 500 °C and 250 °C, it may have been spurious. Lasarev (1964) and Dashevskii, Kukuladze, Lasarev & Mirgalovskii (1967) have suggested that the isothermal curves for ℓ in liquid In–Sb and Ga–Sb have cusps at the 50/50 composition, but the evidence is not impressive. Korol'kov & Igumnova (1961) have obtained the results for liquid Mg–Sn and Mg–Pb which are reproduced in fig. 6.16. They suggest that the minima displayed by these curves indicate the formation of compounds which are surface-active

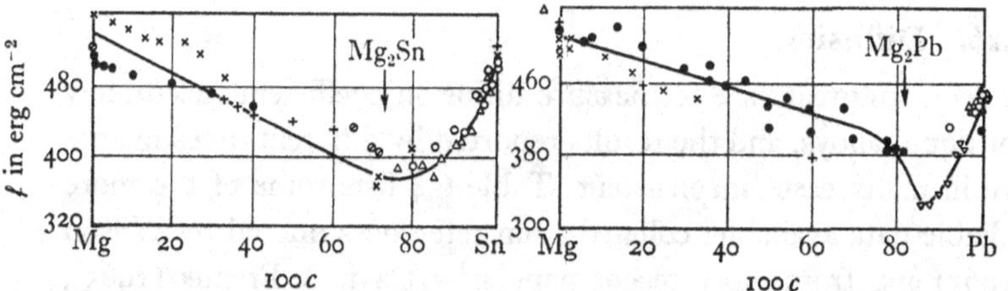

Fig. 6.16. The surface tension of liquid Mg–Sn and Mg–Pb. (Redrawn by permission from Korol'kov & Igumnova, 1961.)

with respect to the two constituents. No minima are apparent in the curves drawn by the same authors for liquid Sb–Te and Bi–Te.

6.15. Other surface properties

The electronic *work function* has been measured photo-electrically for very dilute amalgams of Hg with various alkali metals by Lasarev & Malov (1965*a*,*b*). It falls very rapidly with concentration to begin with, as the surface active character of these particular solutes would lead one to expect; according to Malov & Lasarev (1968) there is a linear relation between work function and surface tension, for **Hg**–Na and **Hg**–Cs at any rate. Turner (1970) has confirmed the results by measuring the effect of small amounts of Na and K on the *contact potential* of liquid Hg; the change of contact potential and change of work function should be equal. Turner extended his observations to **Hg**–Cd and **Hg**–In, in which the solutes are believed to be surface inactive. Small amounts of surface inactive impurities would be expected to have rather little effect, because screening should prevent them from shifting the Fermi level and therefore from altering the inner work function ϕ_1 until c_1 exceeds 5% or so, while they can hardly affect the surface dipole contribution if they are excluded from the surface layer. In fact Turner observed a linear change of contact potential with concentration for both **Hg**–Cd and **Hg**–In. Either Cd and In are not sufficiently surface inactive with respect to Hg to be excluded from the surface layer at all completely, or else they promote a change of ϕ_1 in an indirect fashion, by causing the solvent Hg to contract. Either hypothesis can be used to explain Turner's data in a semi-quantitative fashion.

6.16. Diffusion

It requires great care to measure diffusion coefficients accurately for liquid alloys, and the results reported by different investigators are in many cases inconsistent. Table 6.4 lists some of the more reliable data available, collated from references quoted by Wilson (1965) and from more recent papers by Davis & Fryzuk (1965), Döge (1965) and Gupta (1966, 1967). Figures are quoted for the diffusion coefficient of various solute ions through five different solvents at fixed temperatures. They all refer to very dilute solutions, so that it is meaningful to include values for the self-diffusion coefficient of the solvents for comparison. The variations with temperature of the self-diffusion and solute-diffusion coefficients are usually very similar.

There is a fair amount of consistency in these data: the diffusion coefficient varies rather little within each column of the table, but the minor fluctuations from row to row do show some regularity. Thus the diffusion coefficients for Pb, Bi, Sn, and Sb ions respectively tend to lie roughly in the ratio 1.0:1.05:1.2:1.35. According to the rigid-sphere model this progression should reflect the variation of σ_1; the theory embodied in (3.10) for example suggests

$$D^{-1} \propto (\sigma_0+\sigma_1)^2 g_{01}\left(\frac{\sigma_0+\sigma_1}{2}\right). \tag{6.62}$$

It should be possible to estimate $g_{01}((\sigma_0+\sigma_1)/2)$ with the aid of the Percus–Yevick equations, but the data are scarcely of sufficient accuracy as yet to warrant a detailed discussion. Errors of 50% or more are not out of the question for many of the entries in the table.

There are a few systems for which an inter-diffusion constant†

$$D_{01} \simeq c_1 D_0 + c_0 D_1 \tag{6.63}$$

has been determined as a function of concentration. Such measurements might be of interest for compound-forming systems, since

† Equation (6.63) does not describe the measured quantity exactly unless the system is substitutional, i.e. unless ions of both species occupy the same volume.

TABLE 6.4. *Diffusion coefficients in dilute liquid alloys* *($cm^2\ sec^{-1} \times 10^{-5}$)*

Solvent ...		**Ag**	**Hg**	**Sn**	**Pb**	**Bi**
Temperature (°C) ...		1200	20–25	500	500	500
Solute						
Na	—	—	0.8	—	—	—
K	—	—	1.5	—	—	—
Rb	—	—	0.5	—	—	—
Cs	—	—	0.6	—	—	—
Ag	—	*4.2*	1.1	5.6	—	9.7
Au	—	4.6	0.8	5.4	3.7	5.2
Zn	—	—	2.0	—	—	—
Cd	—	—	1.5	—	5.0	—
Hg	—	—	*1.8*	—	—	—
In	—	5.7	0.7	—	—	—
Tl	—	—	1.0	—	—	—
Sn	—	5.6	1.1	*5.5*	3.7	6.5
Pb	—	—	1.4	3.7	*3.2*	5.5
Sb	—	6.3	—	—	4.1	—
Bi	—	—	1.5	4.6	3.6	*5.0*

Note. Self diffusion coefficients are in italics.

the existence of 'molecules', even of a transient nature, would surely impede diffusion. It has been claimed (Foley & Reid, 1963) that in Hg–Tl the inter-diffusion constant passes through a minimum at a concentration corresponding to Hg_5Tl_2, which appears to be an electron compound of sorts, but the claim was not substantiated by later work (Foley & Liu, 1964). For the more interesting compounds whose thermodynamic properties are listed in table 6.3 no data are available. It is worth noting, however, that whether or not 'molecules' exist in liquid Hg_2Na or Hg_2K, they can scarcely do so in dilute **Hg**–Na or **Hg**–K; to judge by the entries in table 6.4, the size of the diffusing units which determine D_1 is no bigger in these dilute amalgams than it is in say **Hg**–Au or **Hg**–In. In view of the relatively large volume occupied by the ions in pure Na and K this is a surprising observation, even if the hypothesis of molecule formation is to be discarded. It is consistent, however, with the large negative volumes of mixing for **Hg**–Na and **Hg**–K which have previously been noted (see table 6.3).

6.17. Viscosity

Much experimental work has been done on the viscosity of liquid alloys (see Wilson, 1965, for a review), but here again there are many sources of error and the results are frequently inconsistent. For simple systems such as Na–K, Ag–Au, Sn–Pb, Pb–Bi, Sb–Bi, and others the viscosity isotherms, plotted against concentration, are usually either linear or slightly concave. This is probably consistent with the rigid-sphere model. The appropriate extension for a mixture of rigid spheres of the result obtained for a pure liquid by Longuet-Higgins & Pople (see §3.4) is

$$\eta = \tfrac{4}{15}(\pi k_B T)^{\frac{1}{2}} \left(\frac{N}{\Omega}\right)^2 \sum_{\alpha}\sum_{\beta} c_\alpha c_\beta (m_A^{\frac{1}{2}})_\beta \left(\frac{\sigma_\alpha+\sigma_\beta}{2}\right)^4 g_{\alpha\beta}\left(\frac{\sigma_\alpha+\sigma_\beta}{2}\right),$$

which suggests that for a binary alloy

$$\eta \propto \Omega^{-2}\left[c_0\sigma_0^4 m_{A,0}^{\frac{1}{2}} g_{00}(\sigma_0) + c_1\sigma_1^4 m_{A,1}^{\frac{1}{2}} g_{11}(\sigma_1) + c_0 c_1\left\{(m_{A,0}^{\frac{1}{2}} + m_{A,1}^{\frac{1}{2}}) \times \left(\frac{\sigma_0+\sigma_1}{2}\right)^4 g_{01}\left(\frac{\sigma_0+\sigma_1}{2}\right) - m_{A,0}^{\frac{1}{2}}\sigma_0^4 g_{00}(\sigma_0) - m_{A,1}^{\frac{1}{2}}\sigma_1^4 g_{11}(\sigma_1)\right\}\right]. \tag{6.64}$$

The term in (6.64) proportional to $c_0 c_1$ seems more likely to be negative than positive; it is certainly negative if $m_{A,0} \simeq m_{A,1}$ and $g_{01} \simeq g_{00} \simeq g_{11}$. It is sometimes stated that negative deviations in the viscosity isotherms are most noticeable for eutectic systems, but there are exceptions to this rule, e.g. the results of Kanda & Colburn (1968) for Sn–Pb. The evidence that at the eutectic composition itself there tends to be a minimum in η is unconvincing.

If compound formation should impede diffusion it should equally impede viscous flow. For the Mg–Pb and Mg–Sn systems the viscosity isotherms do appear to pass through maxima in the concentration range corresponding roughly to the compounds Mg_2Pb and Mg_2Sn (Gebhardt, Becker *et al.*, 1955*a*, 1955*b*; Glazov, Glagoleva & Romantseva, 1966), as shown in fig. 6.17. Maxima have also been reported for the Hg–Na and Hg–K systems (Degenkolbe & Sauerwald, 1952, see fig. 6.18). For the

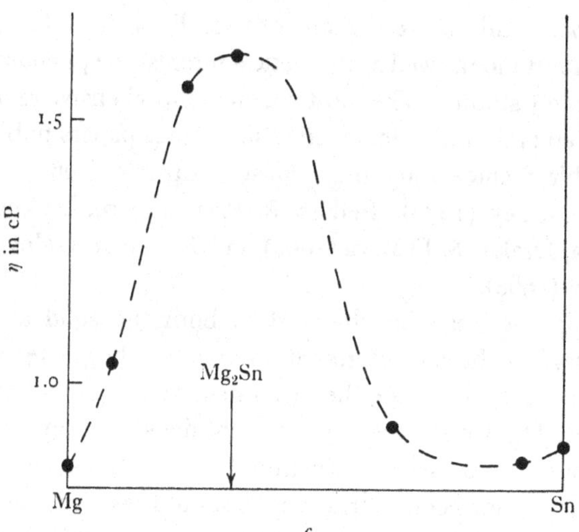

Fig. 6.17. The viscosity of liquid Mg–Sn in centipoise.

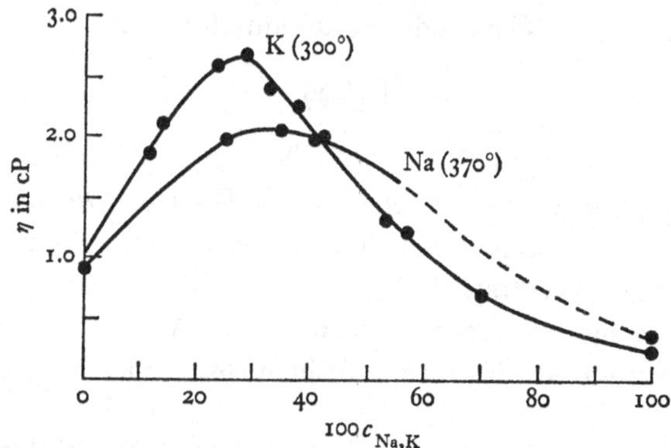

Fig. 6.18. The viscosity of liquid Hg–Na and Hg–K. (Redrawn by permission from Degenkolbe & Sauerwald, 1952.)

III–V systems such as In–Sb the experimental situation is still rather confused. No conspicuous maxima appear in the isotherms for Au–Sn (Gebhardt, Becker & Kostlin, 1956).

6.18. Electromigration, thermal diffusion and related effects

Electromigration is one of several terms that have been used to describe the process whereby the constituents of an alloy tend to separate when a DC current is passed through it; *electrotransport*, *electrodiffusion*, and *electrolysis* are others. Because it has some practical importance as well as theoretical interest the phenomenon has been widely studied. The most recent comprehensive review is by Verhoeven (1963). Readers who wish to trace papers published on the subject since 1963 may consult Epstein (1967, 1968), Epstein & Dickey (1970), Sellors & Pratt (1970), Mikhaylov, Polovinkina, Drakin & Frolova (1966) and Rudenko, Golovinskiy & Khar'kov (1968).

Electromigration can be observed in both the solid and the liquid phase, but the rate of migration is naturally greater in the latter. To take a typical case, the mean drift velocity v_D of the Cd ions in dilute **Hg**–Cd induced by a current density of say 10^3 amp cm^{-2} is about 10^{-4} cm sec^{-1} in the liquid phase.† This means that in a specimen a few centimetres long it should take only an hour or two for an equilibrium concentration gradient to be set up, such that the current-induced drift is balanced by diffusion in the reverse direction. The condition for equilibrium is

$$-D_1(\mathrm{d}c_1/\mathrm{d}x) = v_D c_1$$

or

$$c_1 \propto \exp(-v_D x/D_1). \tag{6.65}$$

Since D_1 is about 1.5×10^{-5} cm^2 sec^{-1} in **Hg**–Cd at room temperature, it may be seen that a current density of 10^3 amp cm^{-2} is sufficient to concentrate the solute ions rather effectively, into the part of the specimen that lies within a millimetre or so of one or other electrode. In this case, as it happens, the solute goes to the cathode.

The theory of the effect would appear to be straightforward in principle. Each solute ion experiences a force $-z_1 e\mathbf{E}$ due to the applied electric field and an opposing force of mean value $\mathbf{F}_1$ due to the electrons that collide with it, i.e. due to the current rather than the field. The direction in which the ion migrates depends upon which of these terms is the bigger. To evaluate $\mathbf{F}_1$, from the first principles we would need to integrate

$$e^2\psi\psi^*\nabla V \tag{6.66}$$

over the volume of the ion, where V is the potential that it presents

† The mean drift velocity of the conduction electrons, for comparison, is then about 10^{-1} cm sec^{-1}.

to the conduction electrons, and to sum over all the occupied wave functions ψ. In the case of a pure metal, however, we can obtain the answer much more simply as follows. Since the conduction electron assembly does not accelerate, once the steady state has been reached, the overall reaction of the ions upon the electrons must balance the effect of the field upon them. This means that

$$-N\mathbf{F}/\Omega + ne\mathbf{E} = 0$$

or, if we invoke Ziman's formula for the resistivity, that

$$\mathbf{F} = ze\mathbf{j}\frac{3\pi m^2}{\hbar^3 e^2}\frac{N}{\Omega k_F^2}\langle u^2 a\rangle. \tag{6.67}$$

It is not surprising that $\mathbf{F}$ depends upon the second power of u in this way. An electron in a pure plane wave state exerts no force upon an ion, for the integral of (6.66) is zero if ψ varies like $\exp(i\mathbf{k}\cdot\mathbf{r})$; the force arises only because each ψ is modulated in its passage through the specimen by a factor $(1+\gamma)$ (see p. 31) and $\gamma(\mathbf{R})$ is not symmetrical with respect to the centre of the ion. The two factors of u in (6.67) are presumably associated with (*a*) with the ion that is responsible for modulating ψ, and (*b*) with the ion which is the centre of the potential V and upon which the force $\mathbf{F}$ is exerted. These two ions may or may not be one and the same.

The argument used in describing (6.67) may readily be extended to binary alloys to show that (see (6.48))

$$\begin{aligned}\mathbf{F} &= c_0\mathbf{F}_0 + c_1\mathbf{F}_1\\ &= \bar{z}e\mathbf{j}\frac{3\pi m^2}{\hbar^3 e^2}\frac{N}{\Omega k_F^2}\langle c_0 u_0^2 + c_0^2 u_0^2(a_{00}-1) + c_1 u_1^2 + c_1^2 u_1^2(a_{11}-1)\\ &\qquad + 2c_0 c_1 u_0 u_1(a_{01}-1)\rangle,\end{aligned} \tag{6.68}$$

and there is no difficulty in picking out the physical meaning of the various terms inside the bracket on the right of this equation. The term in $c_0 u_0^2$ expresses the force exerted on each solvent ion due to the modulation of the electronic wave functions by the ion itself; the term in $c_0^2 u_0^2(a_{00}-1)$ expresses the effect of modulation due to neighbouring solvent ions; and so on. Hence we can surely write

$$\mathbf{F}_1 = \bar{z}e\mathbf{j}\frac{3\pi m^2}{\hbar^3 e^2}\frac{N}{\Omega k_F^2}\langle u_1^2 + c_1 u_1^2(a_{11}-1) + c_0 u_0 u_1(a_{01}-1)\rangle.$$

The net force on a solvent ion should therefore be given by

$$-z_1 e\mathbf{E}+\mathbf{F}_1 = -(z_1)_{\text{eff}}\, e\mathbf{E}$$

where

$$(z_1)_{\text{eff}} = \left\{ z_1 - \bar{z}\frac{\langle u_1^2 + c_1 u_1^2(a_{11}-1) + c_0 u_0 u_1 (a_{01}-1)\rangle}{\begin{array}{l}\langle c_0 u_0^2 + c_0^2 u_0^2 (a_{00}-1)\\ \quad + c_1 u_1^2 + c_1^2 u_1^2(a_{11}-1) + 2c_0 c_1 u_0 u_1 (a_{01}-1)\rangle\end{array}} \right\}. \tag{6.69}$$

If the alloy is very dilute then

$$(z_1)_{\text{eff}} \simeq z_1 - z_0 \frac{\langle u_1^2 + u_0 u_1 (a_{01}-1)\rangle}{\langle u_0^2 a_{00}\rangle}, \tag{6.70}$$

i.e. *the solute should migrate with the electrons to the anode if*

$$z_0^{-1}\langle u_0^2 a_{00}\rangle < z_1^{-1}\langle u_1^2 + u_0 u_1 (a_{01}-1)\rangle \tag{6.71}$$

and vice versa. The speed of migration may be estimated from the formula

$$\mathbf{v}_{\text{D}} \simeq -(z_1)_{\text{eff}}\, e\mathbf{E}/\zeta_1 = -(z_1)_{\text{eff}}\, e\mathbf{E} D_1/k_{\text{B}}T, \tag{6.72}$$

where ζ_1 is the friction coefficient of equation (3.8). Hence in equilibrium, from (6.65), we may expect a Boltzmann-type distribution of solute, with

$$c_1 \propto \exp\left((z_1)_{\text{eff}}\, e\mathbf{E}\cdot\mathbf{r}/k_{\text{B}}T\right) \tag{6.73}$$

independent of D_1. This equation enables $(z_1)_{\text{eff}}$ to be determined directly from experiment, without prior knowledge of D_1.

It has often been noted (Skaupy, 1914) that the experimental results suggest a connection between the direction in which the solute migrates and the sign of $d\rho/dc_1$. In dilute Hg amalgams, for example, for which $d\rho/dc_1$ is almost always negative, the solute almost always goes to the cathode; in **Hg**–Na and **Hg**–K, for which $(d\rho/dc_1)^0$ is positive, the Na and K ions go to the anode; but when the concentration of Na in **Hg**–Na exceeds about 0.5% the resistivity isotherms bend over and $d\rho/dc_1$ becomes negative, and the direction of migration is then observed to reverse (Angus & Hücke, 1961). In the Na–K system the Na migrates to the anode when c_{Na} is small but to the cathode when c_{Na} is large, and the change of sign occurs at just about the concentration where the resistivity is a maximum (Drakin & Maltsev, 1957). In **Cd**–Zn alloys the sign of $d\rho/dc_1$ and the direction of migration are both reported to change with temperature between 400 and 600 °C (Rudenko *et al.*, 1968). And so on.

Theories have been put forward to explain this correlation (e.g. Khar'kov, 1966; Epstein & Paskin, 1967; Sinha, 1972; Olson, Blough & Rigney, 1972) but the argument outlined above is more explicit. Let us restrict ourselves to very dilute alloys and suppose that the pseudo-potentials are sufficiently insensitive to energy to justify equation (6.56). We may then use (6.55) to eliminate the unknown function a_{01} from (6.70) with the following result:

$$(z_1)^0_{\text{eff}} = -\frac{z_0}{2\rho_0}\left(\frac{d\rho}{dc_1}\right)^0 - \frac{z_0}{2}\frac{\langle u_1^2-u_0^2\rangle}{\langle u_0^2 a_{00}\rangle} + (z_1-z_0)(1-\tfrac{1}{3}\xi_0) - \frac{z_0}{2}\left(\frac{1}{\Omega}\frac{d\Omega}{dc_1}\right)^0(1-\tfrac{2}{3}\xi_0). \quad (6.74)$$

For a monovalent solvent such as **Na** or **K**, $\rho_0^{-1}(d\rho/dc_1)^0$ is usually so large that the first term dominates the others and we may expect

$$(z_1)^0_{\text{eff}} \sim -\frac{1}{2\rho_0}\left(\frac{d\rho}{dc_1}\right)^0. \quad (6.75)$$

If both solvent and solute are polyvalent, the analysis in §6.10 suggests

$$\frac{1}{\rho_0}\left(\frac{d\rho}{dc_1}\right)^0 \sim \frac{\rho_1-\rho_0}{\rho_0} \sim \frac{\langle u_1^2-u_0^2\rangle}{\langle u_0^2 a_{00}\rangle},$$

so that if $|\Omega_1-\Omega_0| \ll \Omega_0$ and $\xi_0 < 1$ the formula

$$(z_1)^0_{\text{eff}} \sim z_1 - z_0(\rho_1/\rho_0) \quad (6.76)$$

seems a better guess. These crude predictions are compared with experiment in table 6.5 for the very few systems for which such a comparison seems to be possible at the present time. The data for $(z_1)^0_{\text{eff}}$ are from sources quoted by Verhoeven (1963).

The theory describes the trend of the results successfully, but it overestimates the magnitude of $(z_1)^0_{\text{eff}}$ in every case and it is unlikely that more careful calculations, based directly upon (6.70) will remove the discrepancies. Experimental error may be partly to blame, but there is little doubt that the simple theory fails to tell the whole story. Its limitations are further exposed by the observation of two effects for which it provides no explanation, the *Haeffner* effect and one which Klemm (1958) has christened *electro-osmosis*.

TABLE 6.5. *Electromigration in dilute liquid alloys*

Solvent	Solute	Temperature (°C)	$(z_1)^0_{\text{eff}}$ Expt.	Eq. (6.75)	Eq. (6.76)
Na	K	100	−0.8	−5	—
	Cd	110	−13	−28	—
	Hg	110	(−18)	−30	—
	Tl	—	(−18)	—	—
	Pb	110	−21	−60	—
K	Na	100	−0.5	−4	—
	Hg	100	−10	−18	—
	Tl	110	−20	—	—
	Pb	100	−22	—	—
Sn	Ga	300	+0.6	—	+0.8
	Bi	350	−0.8	—	−5.7

Haeffner (1953) discovered that the passage of a large DC current density ($\sim 10^4$ amp cm^{-2}) for a long period of time (say 2000 hours) would establish a concentration gradient for each *isotope* in a specimen of pure liquid Hg. The effect has since been observed in liquid Li, K, Rb, Zn, Cd, Ga, In, and Sn (Verhoeven, 1963; Lodding, 1963, 1967) and in every case it is the lighter isotope which tends to collect at the anode. The effect is surprising because the force on an ion should depend primarily upon its charge, its pseudo-potential and its environment, none of which should be influenced by its isotopic mass. It is true that if one is to allow for the inelastic nature of the scattering, the $a(q)$ that occurs in Ziman's formula for resistivity should be replaced by

$$a(q) - \frac{1}{6}\frac{\hbar^2 q^2}{m_A k_B T}$$

(see (3.85)) and (6.67) should be modified in the same way. But this suggests that for a given current density the drag force **F** should increase with isotopic mass, which is not what is needed to explain the observations.

The customary explanation is based upon experience with solid metals, in which a phenomenon known as *self transport* occurs: if the surface of a solid rod is marked in some way at regular intervals, the separation between the marks is found to increase at one end and to decrease at the other when a large current is passed through. This effect implies that a concentration gradient of vacancies in the lattice is established, and hence that the vacancies tend to drift when the current is first switched on. If the vacancies drift from left to right, say, this must be because the ions on either side of the vacancy experience a mean force from right to left, though the mean force on all the ions in the specimen must of course be zero.

It is argued by analogy that the drag force exerted by the current on each ion in a pure liquid metal is liable to vary about the mean value **F** which (6.67) describes and that it tends to be above the mean when the ion's environment permits it to diffuse with relative ease. In that case every ion will tend to diffuse through the rest of the specimen towards the anode as soon as the current is switched on. No observable consequences are to be expected so long as the specimen is homogeneous, for the mass transport due to diffusion must be exactly compensated by bulk flow in the reverse direction. But when there is a mixture of isotopes the lighter one, on account of its greater mobility, should diffuse the faster and tend to collect at the anode in the way that is observed. The explanation is plausible enough, though a convincing quantitative theory has yet to be constructed. Evidently a satisfactory theory of electromigration in alloys ought also to include some discussion of fluctuations in the drag force and the way that these are liable to correlate with fluctuations in mobility.

If the net force exerted on the ions by both field and current combined is on the average zero, then there is no reason to expect any *pressure gradient* to be associated with the current in a liquid specimen. According to Klemm, however, a gradient is detectable in narrow tubes containing Hg, the pressure being higher at the cathode than the anode. He attributes this electro-osmosis effect to scattering of the conduction electrons by the containing walls; if the electrons, drifting from cathode to anode, are continually imparting drift momentum to the sides of the tube, then a pressure difference between its ends is essential if momentum is to be conserved. Klemm's theory predicts the magnitude of the effect with fair success as well as its sign. It implies that in a thin layer near the walls the current density should be less than elsewhere, and within this layer the mean force on each ion should not vanish. A pattern of convective flow is therefore to be expected, with the liquid moving slowly towards the cathode round the perimeter of the tube and back down its middle. Such convection may occur in liquid alloys as well as pure metals, and in some electromigration experiments it may influence the results obtained.

The constituents of an alloy may be partially separated by passing a heat current through it instead of an electric current, though the concentration gradients that may be achieved in practice are a lot smaller. This is the *Soret effect*, which has been studied in a number of liquid alloys based on Sn by Ballay (1928), Kawakami (1951, 1955), and Winter & Drickamer (1955). No detailed theory is available, though Fiks (1964) has suggested that the drag force exerted on the ions by the conduction electrons, whose momentum distribution becomes unsymmetrical when a heat current is switched on, may again be a major factor. Ott & Lundén (1964) reported the thermal equivalent of the Haeffner effect in liquid Li, i.e. a separation of the two isotopes by a heat current, and it has since been observed in liquid K, Rb and Ga (Lodding & Ott, 1966; Löwenberg, Nordén-Ott & Lodding, 1968). In every case the lighter isotope is driven towards the hot end of the specimen. The results have been discussed by Lodding (1966).

One last method whereby a concentration gradient may be induced in a liquid alloy is to centrifuge it, as has been shown by Vertman, Samarin & Yakobson (1960), Kumar (1965), Singh & Kumar (1966), and Kumar, Singh & Sivaramakrishnan (1967). An important factor that distinguishes this phenomenon from the ones discussed above is that the concentration gradient set up in a centrifuge represents a state of thermodynamic equilibrium, and it follows that nothing is to be learnt from the experiments that could not also be learnt by measuring the free energy of the alloy as a function of concentration and pressure. The interpretation which Kumar and his co-workers lay upon their results is not unique.

6.19. Magnetic susceptibility

We now turn from the properties of liquid alloys which are primarily determined by the interaction between the ions to those in which the central role is played by the conduction electrons.

Magnetic susceptibility is one of these. The total susceptibility χ does of course include a contribution due to the ion cores (see §4.15) but this should vary in an almost linear fashion with concentration – precisely linear, if the quantity plotted, as in fig. 6.19, is the susceptibility for a fixed number of atoms rather than per unit volume or per unit mass (see p. 295). Here we are concerned mainly with the significant departures from linearity that have been reported for a few systems, and these must be attributed to the electronic terms, χ_p and χ_d.

It is unfortunate that there is no direct method for separating these two, since electron spin resonance cannot normally be detected; some measurements have been made on the effect of impurity on the resonance line width in liquid Na alloys (Alekseyeva, Nikitin, Khabibullin & Kharakhashyan, 1968; Cornell & Slichter, 1969) but except in the case of the Na–K system (Helman & Devine, 1971) such work is restricted to very dilute solutions. If an isothermal curve for χ versus concentration displays an obvious minimum, therefore, perhaps at some concentration that corresponds to an intermetallic compound, we shall have to choose between two different explanations. Either the density of states at the Fermi level is anomalously low at this concentration so that χ_p is reduced, or else the magnitude of χ_d is for some reason enhanced. It seems quite probable, though the theory of this has yet to be worked out, that the sort of distortion of the low-energy states in the conduction band which was discussed in tentative terms on p. 427 could enhance χ_d, even in circumstances where $\mathcal{N}(E)_F$ remained true to the free-electron model. If, in an extreme case, some of the low-lying electrons were to become localised, they would presumably add to the diamagnetic polarisability of the ions about which they were situated, and the effective value of χ_d would be enhanced for this reason. Rigert & Flynn (1971) have interpreted their susceptibility data for liquid **Na**–Te and **Na**–Bi along these lines (see p. 429).

Localised states able to accommodate only one electron each would be expected to have a marked effect on the temperature dependence of the susceptibility. The localised electrons would be able to respond to an applied field by lining up their spins with nothing but thermal agitation to prevent them, and they should

TABLE 6.6. *The magnetic behaviour of liquid alloys*

(*a*)	(*b*)	(*c*)	(*d*)	(*e*)
Na–K	K–Cs	Na–Cs	**Cu**–Cr	**Cu**–Ti
Mg–Al	Cu–Zn	Cu–Sn	**Sb**–Cr	**Al**–V
Ga–In	Zn–Sn	Ag–Sn	**Cu**–Mn	**Al**–Cr
Sn–Sb	Hg–Au	Cd–Sb (CdSb)	**Zn**–Mn	**Al**–Mn
Pb–Tl	Hg–In	Zn–Sb (ZnSb)	**Ga**–Mn	**Al**–Fe
Pb–Sn		Sn–Te (SnTe)	**Sn**–Mn	**Sb**–Fe
Pb–Sb		Bi–Te (Bi_2Te_3)	**Sb**–Mn	**Al**–Co
Pb–Te		Sb–Te (Sb_2Te_3)	**Bi**–Mn	**Sb**–Co
Bi–Ag		Tl–Te (?Tl_2Te)	**Cu**–Fe	**Cu**–Ni
Bi–Cd			**Cu**–Co	**Al**–Ni
Bi–In			**Au**–Co	**Sb**–Ni
Bi–Sn				**Bi**–Ni
Bi–Pb				
Bi–Sb				

(*a*) Smooth concentration-dependence of χ consistent with NFE model.
(*b*) Minor anomalies.
(*c*) Pronounced minimum for χ in the middle of the concentration range; the corresponding compound formula is noted where appropriate.
(*d*) Evidence for localised moments.
(*e*) No evidence for localised moments.

therefore contribute a paramagnetic term obeying Curie's law, i.e. proportional to T^{-1}; in normal circumstances χ_p for a metal is independent of T. Anderson (1961) has shown that a *virtual* bound state in the neighbourhood of the Fermi level, split by spin-orbit interaction, may in some circumstances have the same effect; it may endow the metal with a *localised magnetic moment* which can be oriented by an applied field, even though the electron states are all non-localised. The theory of localised moments is now well-established, but it would take us too far afield to discuss it. Experimental evidence for their existence in liquid alloys is confined, as we shall see, to dilute systems where the solute is a transition or rare earth metal.

The results of a large number of susceptibility measurements listed in the bibliographies of B. W. Mott *et al.* (1966, 1968) are cursorily summarised in table 6.6. For all the systems in the first column the susceptibility isotherms are smooth curves with no minima and there is little reason to doubt that the NFE model is applicable; the results do not always fit the 'theoretical' curves

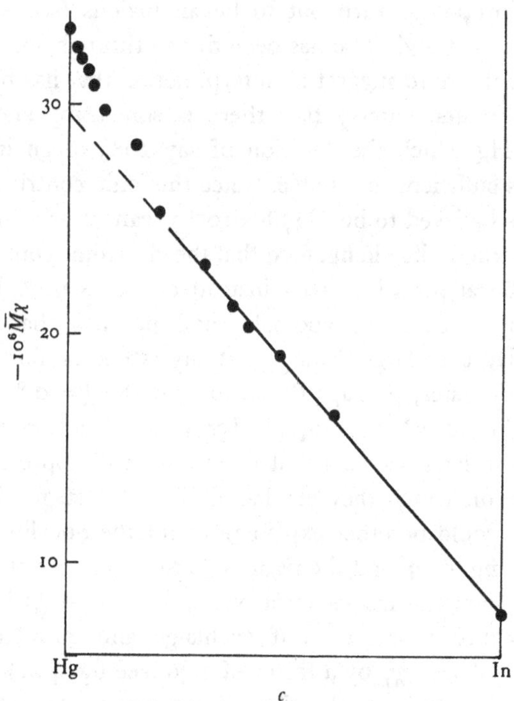

Fig. 6.19. The susceptibility of liquid Hg–In at 20 °C. The quantity plotted is the product of the mass susceptibility and the mean atomic weight. The point for pure In has been estimated from measurements at higher temperatures on the assumption that liquid In could be supercooled to 20 °C.

with which they have been compared by experimenters, but exchange effects in χ_p, usually ignored, may be partly responsible for the discrepancies.

The systems in column (*b*) display minor anomalies which may or may not be significant: shallow minima in the case of K–Cs, Cu–Zn and Hg–Au; irregular kinks (for which experimental error may be to blame) in the case of Zn–Sn; and the behaviour shown in fig. 6.19 for Hg–In. This latter case is illustrated in detail, though the anomaly is not especially dramatic, because the Hg–In system has been studied by three sets of investigators with reasonably consistent results† and because its resistivity, Hall coefficient

† The points plotted in fig. 6.19 are due to Collings (1967). The data provided by Güntherodt *et al.* (1966) and Menth & Wullschleger (1967) are for temperatures of 160 and 1070 °C respectively.

and optical properties turn out to be anomalous (see pp. 522, 511 and 514). A straight line has been drawn through some of the points in the figure to suggest an interpretation that has been put upon these results, namely that there is something anomalous about pure Hg which the addition of say 20% of an impurity such as In is sufficient to remove. Since the ionic contribution to $M\chi$ in Hg is believed to be -47.8×10^{-6} it can be argued on the basis of the straight line in fig. 6.19 that the electronic contribution 'ought' to be about 18.2×10^{-6} instead of 15.6×10^{-6}. Perhaps heating also removes the anomaly, and perhaps that is why $(d\chi/dT)$ in Hg is so large? Since χ_p at any rate is proportional to the density of states, perhaps the ratio $15.6/18.2$ ($= 0.86$) represents the value of m^*/m in liquid Hg? We shall return to these speculations in later sections, but the reader will appreciate that the evidence on which they are based is at this stage distinctly weak. There could be other explanations for the non-linearity of the points in fig. 6.19, and the figure of 15.6×10^{-6} is not in itself suspiciously low; the free-electron value for $M(\chi - \chi_i)$ in Hg at 20 °C is 9.5×10^{-6}, and even if exchange and correlation are supposed to enhance χ_p by a factor of 1.36 (see fig. 4.21) without affecting χ_d the theoretical value corresponding to $m^*/m = 1$ becomes only 14.5×10^{-6}.

The susceptibility data available for other amalgams besides Hg–Au and Hg–In (e.g. Bates & Somekh, 1944) are confined to very dilute solutions and seem too erratic to be trusted.

For all the systems in column (*c*) of table 6.6 the susceptibility isotherms are reported to have distinct minima, deviating by up to 0.1×10^{-6} gm^{-1} from simple theoretical curves. In this list the reader will recognise some compound-forming systems, and for each of these the position of the minimum in χ does seem to correspond to the most prominent compound.† As for Na–Cs, Cu–Sn, and Ag–Sn, these are all systems in which the NFE model in its simplest form may well be suspect because of the large difference

† The principal reference here is to Honda & Endo (1927). Their measurements cannot be trusted in detail because later work on the pure constituents has in some cases yielded rather different results. In particular, it is very unlikely that the minima are as sharp as Honda & Endo supposed. The Tl–Te system has been studied by Brown, Moore & Seymour (1971).

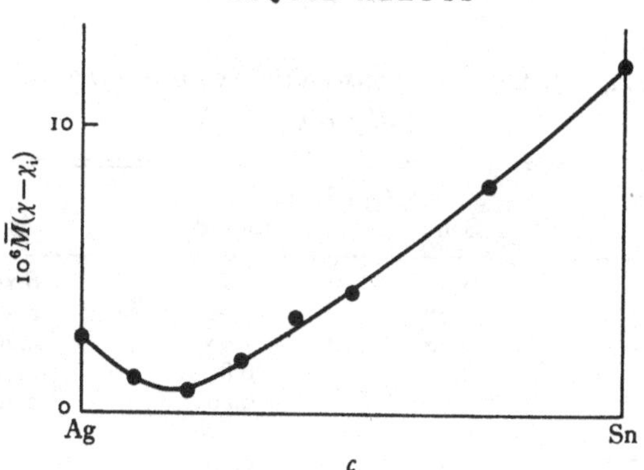

Fig. 6.20. The susceptibility of liquid Ag–Sn at 1000 °C. In this case the ionic contribution has been removed. (After Menth & Haeberlin, 1967.)

of size or valence between the constituents, and Cu–Sn and Ag–Sn have already featured as anomalous in §§6.11 and 6.13. The data for Ag–Sn, which are more recent and reliable (Menth & Haeberlin, 1967) than the Na–Cs (Bohm & Klemm, 1939) and Cu–Sn data, are shown in fig. 6.20.

There are many other compound-forming systems for which χ almost certainly displays a minimum, but these have not been systematically investigated as a function of concentration. Glazov, Krestovnikov & Glagoleva (1965) and others, however, have measured the susceptibility of the compounds themselves in the solid and liquid phases, with results shown in table 6.7. It is of interest that in both phases the compounds with 'ionic' structures are less diamagnetic than the others. The relatively small increase in χ that is observed can be attributed largely to an increase in χ_p – most of the compounds are semiconducting in the solid state and for these χ_p should be almost zero – rather than to changes in χ_d. It is probable that the susceptibility for Ga–Sb and In–Sb continues to increase on heating above the melting point, though the available results are conflicting. Heating certainly increases the susceptibility of liquid Tl–Te (Brown *et al.*, 1971).

Finally, columns (*d*) and (*e*) of table 6.6 summarise the conclusions of Gardner & Flynn (1967), Flynn, Rigney & Gardner (1967), Tamaki (1968*a*) and others, who have looked for the existence of localised moments in dilute liquid alloy systems involving transition metal solutes. Odle, Becker & Sotier (1969)

TABLE 6.7. *Change of susceptibility on melting for intermetallic compounds*

	$-\chi$ (gm^{-1} × 10^{-6}) Solid	Liquid	$\Delta\chi$
AlSb	1.31	1.16	0.15
GaSb	1.27	1.00	0.27
InSb	1.32	1.06	0.26
GaAs	1.24	0.85	0.39
InAs	1.20	0.90	0.30
Mg_2Si*	0.50	0.40	0.10
Mg_2Ge*	0.53	0.43	0.10
Mg_2Sn*	0.55	0.52	0.03
Mg_2Pb*	0.32	0.28	0.04
PbTe*	0.17	0.13	0.04
ZnTe	1.37	1.25	0.12
CdTe	1.45	1.30	0.15
Ga_2Te_3	1.34	1.21	0.13
In_2Te_3	1.36	1.22	0.14

Note. Asterisks denote compounds which crystallise with 'ionic' structures.

have shown that the localised moment associated with Mn ions in liquid **Cu**–Mn is still present when 60% of the Cu is replaced by Al, but vanishes if the proportion of Al is raised to 80%. Stupian & Flynn (1968), Rigney *et al.* (1969) and Blodgett & Flynn (1969) have reported susceptibility measurements on liquid Cu, Al and Cu–Al containing rare earth impurities, but the reader who is interested in their results must consult the original papers.

At the other end of the concentration range, needless to say, any alloy involving the metals Fe, Co or Ni – which are all ferromagnetic at room temperature – obeys the Curie–Weiss law whether it is liquid or solid. The Curie temperature seems always to be below the melting point so that no example of a ferromagnetic liquid alloy has yet been discovered. The claim by Busch & Güntherodt (1968) to have observed ferromagnetism in liquid Au–Co near its eutectic composition has been disputed by Nakagawa (1969) and by Wachtel & Kopp (1969).

6.20. Nuclear magnetic resonance

The *Knight shift* (see §4.16) has now been studied for a large number of non-compound-forming systems in the liquid state.† Attention has usually been concentrated upon the effect of the solvent shift of relatively small amounts of impurity, i.e. upon the quantity $(\mathscr{K}_0^{-1}\,\mathrm{d}\mathscr{K}_0/\mathrm{d}c_1)^0$, and some of the results obtained for this are listed in table 6.8. The table is not exhaustive but it contains enough information to illustrate most of the following generalisations.

(*a*) $(\mathscr{K}_0^{-1}\,\mathrm{d}\mathscr{K}_0/\mathrm{d}c_1)^0$ is small compared with unity, except for a few systems in which there is a large difference in valence or atomic size between solute and solvent. (We shall not consider here the effect of transition metal or rare earth impurities. If the impurity ion is the centre of a localised moment, its effect on the solvent Knight shift can be spectacular; values of order -15 may be observed for $(\mathscr{K}_0^{-1}\,\mathrm{d}\mathscr{K}_0/\mathrm{d}c_1)^0$, and it varies with temperature roughly like T^{-1}).

(*b*) Among the alkali metals a systematic dependence of $(\mathscr{K}_0^{-1}\mathrm{d}\mathscr{K}_0/\mathrm{d}c_1)^0$ on the size ratio (Ω_1/Ω_0) is immediately apparent, but there seems to be no universal correlation with (z_1/z_0). Thus polyvalent solutes decrease $\mathscr{K}_0$ in **Na** but increase it in **Cu**.

(*c*) However, if $(\mathscr{K}_0^{-1}\,\mathrm{d}\mathscr{K}_0/\mathrm{d}c_1)^0$ is positive then $(\mathscr{K}_1^{-1}\,\mathrm{d}\mathscr{K}_1/\mathrm{d}c_0)^1$ is almost always negative. This is true not only for the alkali metals but for polyvalent systems too, as is shown by the curves for four homovalent systems which are reproduced (from Moulson & Seymour, 1967) in fig. 6.21. In fact if $(\mathscr{K}_0-\mathscr{K}_0^0)/\mathscr{K}_0^0$ and

† The references are not adequately listed in bibliographies elsewhere. For work on **Na** solutions and other systems involving alkali metals consult: Oriani & Webb (1959), Rimai & Bloembergen (1960), Hanabusa & Bloembergen (1966), Kellington & Titman (1967), Titman & Kellington (1967), Van der Molen, Van der Lugt, Draisma & Smit (1968), Thornton, Young, Van der Molen & Van der Lugt (1968), Kaeck (1968). For **Cu** solutions consult: Odle & Flynn (1966), Gardner & Flynn (1967), Rigney *et al.* (1969), Blodgett & Flynn (1969), Odle *et al.* (1969). For solutions in **Al**, **In** and **Sb** consult: Rigney & Flynn (1967), Flynn *et al.* (1967), Stupian & Flynn (1968). For miscellaneous polyvalent systems consult: Seymour & Styles (1966), Moulson & Seymour (1967), Moulson & Styles (1967), Van der Lugt & Van der Molen (1967), Havill (1967), Heighway & Seymour (1971*b*), Radhadkrishna Setty & Mungurwadi (1971). Some of the systems covered by the papers in the last group include In or Sb and the results have been taken into account in drawing up table 6.7.

TABLE 6.8. *Solvent Knight shift in liquid alloys: experimental data for* $(\mathscr{K}_0^{-1}\, d\mathscr{K}_0/dC_1)^0$

Solvent Solute	Na	K	Rb	Cs	Cu	Al	In	Sb
Na	*	−0.27	−0.27	−0.5	—	—	—	—
K	0.31	*	−0.15	−0.31	—	—	—	—
Rb	0.52	0.20	*	−0.23	—	—	—	—
Cs	1.0	0.48	0.24	*	—	—	—	—
Cu	—	—	—	—	*	−0.05	—	—
Ag	—	—	—	—	0.09	0.20	0.25	−0.12
Au	−0.5	—	—	—	0.07	0.17	0.17	—
Mg	—	—	—	—	—	0.01	—	—
Zn	—	—	—	—	0.19	0.08	—	—
Cd	−0.9	—	—	—	—	—	0.07	−0.24
Hg	−0.6	—	—	—	—	—	0.08	—
Ga	—	—	—	—	0.60	0.14	0.05	—
In	—	—	—	—	—	—	*	−0.17
Tl	−1.5	—	—	—	—	—	−0.15	−0.10
Si	—	—	—	—	—	0.07	—	—
Ge	—	—	—	—	1.07	0.21	—	—
Sn	—	—	—	—	—	—	0.08	−0.09
Pb	−3.0	—	—	—	—	—	−0.09	—
As	—	—	—	—	0.99	—	—	—
Sb	—	—	—	—	—	—	0.27	*
Bi	—	—	—	—	—	—	(0.07)	0.15
Se	—	—	—	—	0.35	—	—	—

$(\mathscr{K}_1 - \mathscr{K}_1^1)/\mathscr{K}_1^1$ are plotted on the same diagram as a function of concentration, as in this figure, the two curves are often roughly parallel.

(*d*) The curves are often surprisingly linear too. Hence the figures quoted in the table for $(\mathscr{K}_0^{-1}\, \mathrm{d}\mathscr{K}_0/\mathrm{d}c_1)^0$ represent also, for the majority of systems and especially for the alkali metals, the quantity $(\mathscr{K}_0^1 - \mathscr{K}_0^0)/\mathscr{K}_0^0$, i.e. the fractional change in the Knight shift for a solvent nucleus when it is transplanted into the pure solute.

To explain these results is still a challenge. According to the theory outlined in §4.16 (see (4.118) in particular) there are two major ingredients to $\mathscr{K}$. The first is the spin susceptibility per unit volume which is proportional to the density of states per unit volume, i.e. to k_F and to the exchange enhancement correction plotted in fig. 4.21. The second ingredient is the extent to which

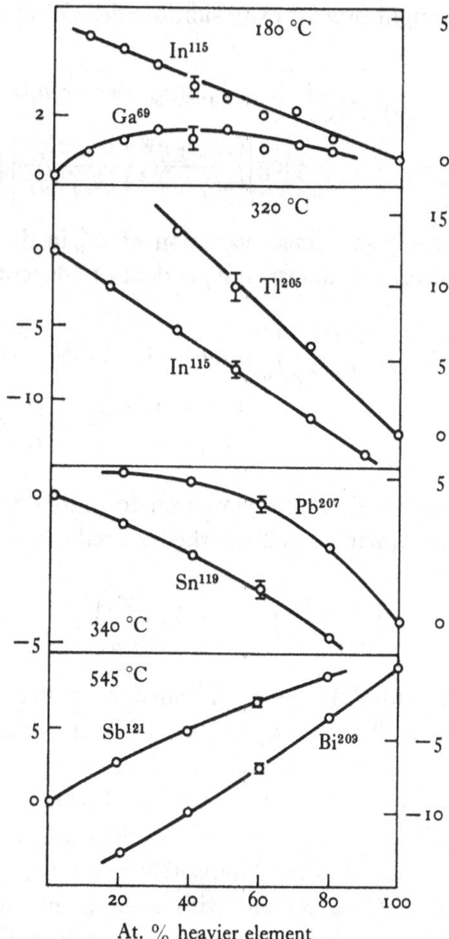

Fig. 6.21. The Knight shift in four homovalent liquid alloys. The quantities plotted are $(\mathcal{K}_0-\mathcal{K}_0^0)/\mathcal{K}_0^0$ and $(\mathcal{K}_1-\mathcal{K}_1^1)/\mathcal{K}_1^1$ in %. (Redrawn by permission from Moulson & Seymour, 1967.)

the wave functions of electrons at the Fermi level are peaked at the nucleus. Let us confine our attention initially to systems for which $z_1 = z_0$ and $\Omega_1 \simeq \Omega_0$ so that the change of k_F on alloying can be ignored. It is then only the peaking factor P_F which we need to bother about, and we should be able to express the concentration dependence of $\mathcal{K}$ by writing (Faber, 1967; Ashcroft & Schaich, 1970)

$$\mathcal{K}_0 \propto 1+2\,\mathrm{Re}\,\overline{\gamma(0)_F},$$

where, by a straightforward extension to alloys of (4.106),†

$$\operatorname{Re}\overline{\gamma(0)_F} = -\frac{m^*}{(2\pi)^2\hbar^2 k_F}\int_0^\infty (u_0 + c_0 u_0(a_{00}-1) + c_1 u_1(a_{01}-1)) \times \tfrac{1}{2}\log\left|\frac{(2k_F+q)^2+(k_F/2lq)^2}{(2k_F-q)^2+(k_F/2lq)^2}\right| q\,dq. \quad (6.77)$$

Hence we may expect a linear variation of $\mathscr{K}_0$ in this instance so long as a_{00} and a_{01} are almost independent of concentration, with

$$\left(\frac{1}{\mathscr{K}_0}\frac{d\mathscr{K}_0}{dc_1}\right)^0 \simeq -\frac{2m^*}{(2\pi)^2\hbar^2 k_F}\int_0^\infty (u_1(a_{01}-1) - u_0(a_{00}-1)) \times \tfrac{1}{2}\log\left|\frac{(2k_F+q)^2+(k_F/2lq)^2}{(2k_F-q)^2+(k_F/2lq)^2}\right| q\,dq. \quad (6.78)$$

An equivalent expression may be written down for $(\mathscr{K}_1^{-1}\,d\mathscr{K}_1/dc_0)^1$ and it is at once apparent that the theory predicts

$$\left(\frac{1}{\mathscr{K}_0}\frac{d\mathscr{K}_0}{dc_1}\right)^0 \simeq -\left(\frac{1}{\mathscr{K}_1}\frac{d\mathscr{K}_1}{dc_0}\right)^1,$$

in accordance with (*c*) above, provided only that the alloy is substitutional, i.e. that a_{00}, a_{11} and a_{01} are more or less the same. If $\Omega_0 \simeq \Omega_1$ this condition is likely to be satisfied.

Equation (6.78) has been shown to yield answers of the right order of magnitude for the four homovalent systems whose behaviour is shown in fig. 6.21 (Moulson & Seymour, 1967; Halder, 1969), but complete agreement with experiment has still to be achieved. Unfortunately, the integral is highly sensitive to the details of the pseudo-potentials employed. It must also be remembered that (6.77), like (4.106), is valid only if the pseudo-potentials are spherically symmetric and local, which may not be the case in practice.

For heterovalent systems, or for systems composed of alkali metals where there is a large difference between Ω_0 and Ω_1, we must think again; the change of k_F upon alloying introduces more

† It is open for discussion whether the initial u_0 in the integrand of (6.77) is really needed, at any rate for large q; it expresses the perturbing effect on the pseudo wave functions of the pseudo-potential due to the central ion. But to exclude it would not affect (6.78).

complications than some authors are willing to admit. First, we should add to the right-hand side of (6.78) a term

$$\frac{1}{3}\left(\frac{z_1}{z_0}-1-\left(\frac{1}{\Omega}\frac{d\Omega}{dc_1}\right)^0\right)$$

to allow (see (6.5)) for the change in the susceptibility per unit volume according to the free-electron model. Secondly, we should include a correction for the exchange enhancement factor, since this depends upon electron density. Thirdly, we need to differentiate (6.77) with respect to k_F and not just with respect to c_0 and c_1, and there is the hidden complication that both u_0 and u_1 may depend upon k_F. Finally, the ratio between the pseudo wave function at the nucleus, $\phi(0)$, and the true wave function, $\psi(0)$, which we have tacitly assumed to be a constant, may be affected by the change of k_F; according to Watson, Bennett & Freeman (1968), who have discussed the effect of impurity on the Knight shift of solid Cu and Ag, this last effect is far from negligible.

Attempts have of course been made to explain the data, for the alkali metals in particular, and attention should be drawn to the work of Daniel (1960), of Thornton *et al.* (1968), of Flynn and his co-workers (see footnote on p. 499) and of Halder (1970). Except for the last named, these authors all employ the phase shift analysis, originally proposed by Blandin, Daniel & Friedel (1959), which has become standard in discussions of the Knight shift in dilute solid alloys; in some cases the impurity phase shifts are estimated from the resistivity and in others computed more or less from first principles. The basic formula used, however, would appear to be entirely equivalent to a simplified version of (6.78), and little if any attention is paid to the complications listed in the previous paragraph. It remains to be demonstrated that such agreement with experiment as has been achieved is not fortuitous.

There is little that can profitably be added at this stage about the few systems which are clearly anomalous. In–Bi is one such, for the curves in fig. 6.22, based on measurements by Styles (1967), are far from linear and do not run parallel to one another. It is remarkable how the Knight shift of In dissolved in Bi remains constant until its concentration reaches almost 50%, while the same is true for Bi dissolved in In. Styles has pointed out that this sort of

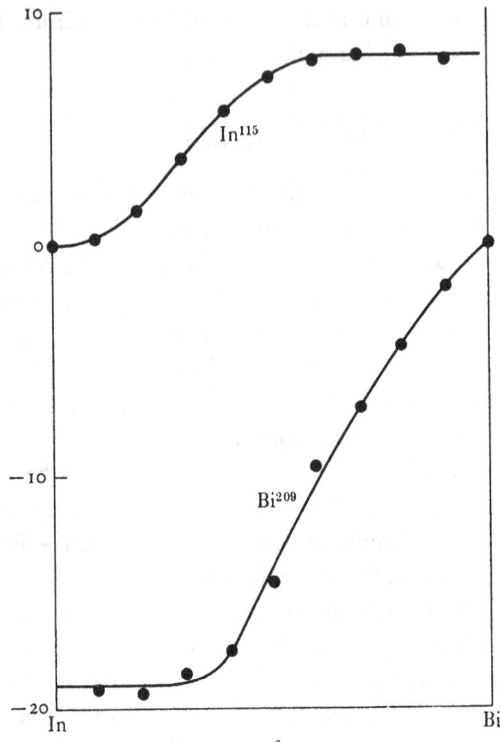

Fig. 6.22. The Knight shift of In and Bi in liquid In–Bi at 306 °C. The same quantity is plotted (% change in $\mathscr{K}$) as in fig. 6.21.

behaviour would be expected if the alloy consisted of two separate phases, one composed of pure InBi and the other of either pure In or pure Bi, depending upon the concentration; two distinct resonance lines, corresponding to the two phases, are not to be expected for the majority species if the nuclei are free to move from one phase to the other with a lifetime of much less than T_2 (say $\ll 10^{-3}$ sec) in each. But InBi is a weakly bound compound, and the diffraction, specific heat, and susceptibility data (see pp. 462, 476 and table 6.6 respectively) make it hard to believe that it survives in the liquid phase in quite the way supposed.

Another anomalous system is In–Hg, for which the curves of Moulson & Styles (1967) are reproduced in fig. 6.23; their work has since been extended to pressures of up to 6.5 kbar by Oshima *et al.* (private communication). The rather rapid rise of $\mathscr{K}_{\text{In}}$ which

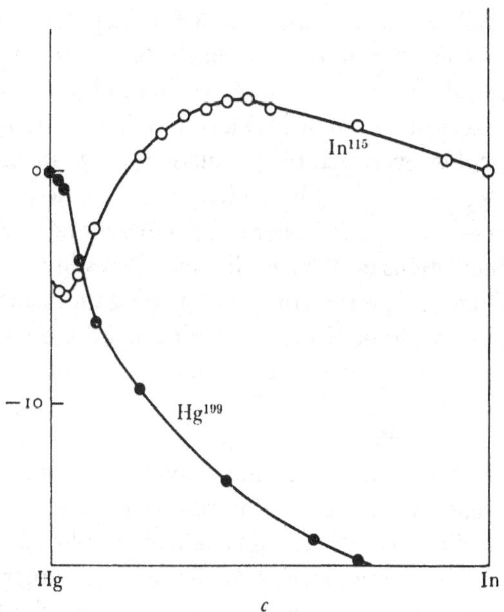

Fig. 6.23. The Knight shift of Hg and In in liquid Hg–In at 20 °C. The same quantity is plotted (% change in $\mathscr{K}$) as in figs 6.21 and 6.22. (After Moulson & Styles, 1967.)

occurs when the In concentration is increased from zero to 20 % or so has sometimes been attributed to an increase of χ_p as suggested by the susceptibility data in fig. 6.19. This does not help to explain, however, why $\mathscr{K}_{Hg}$ is *falling* over the same concentration range. Havill (1967) has continued the measurements to very small In concentrations and finds that the slope of both curves changes sign below $c_{In} \sim 0.01$. According to Havill, $\mathscr{K}_{Hg}$ is also a non-linear function of concentration in **Hg**–Cd, **Hg**–Pb and **Hg**–Bi. In the last of these systems, incidentally, $d\mathscr{K}_{Hg}/dc_{Bi}$ and $d\mathscr{K}_{Bi}/dc_{Bi}$ are both positive, so that the anomalous feature of fig. 6.23, that the two curves have opposite slopes, is not universal for all dilute amalgams.

Little attention has yet been paid to the *nuclear spin relaxation times*, T_1 and T_2, in liquid alloys. Rimai & Bloembergen (1960) point out that if the Knight shift for a given nucleus is liable to vary significantly over the time T_2 which is required for measurement, then the relaxation rate described by the modified Korringa relation (equation (3.86)) is proportional to $\overline{\mathscr{K}^2}$ rather than $\overline{\mathscr{K}}^2$. The difference between $\overline{\mathscr{K}^2}$ and $\overline{\mathscr{K}}^2$ is likely to be proportional to c_0c_1 in a binary alloy, so one might expect an isothermal curve for $1/T_1$ versus concentration to display a convex hump. Rough calculations suggest, however, that the fluctuations of $\mathscr{K}$ are unlikely to be sufficient for the hump to be noticeable in practice, and in the few measurements that have been reported there is no obvious sign of it. If the fluctuations of $\mathscr{K}$ have a lifetime τ, an additional contribution to $1/T_2$ and $1/T_1$ arises from a de-phasing mechanism which has nothing to do with the Korringa relation and this too should be proportional to $(\overline{\mathscr{K}^2}-\overline{\mathscr{K}}^2)$ and hence to c_0c_1; it is of order

$$\overline{(\Delta\omega)^2}\,\tau = \gamma_n \mathrm{H}^2\tau(\overline{\mathscr{K}^2}-\overline{\mathscr{K}}^2), \tag{6.79}$$

where $\overline{(\Delta\omega)^2}$ is the mean square fluctuation in the instantaneous value of the angular frequency for resonance, and it therefore depends upon the strength of magnetic field employed. It would require implausibly large values of τ, however, of order 10^{-8} sec or more, to make (6.79) significant compared with the Korringa term.

There remains the possibility of quadrupole relaxation (see §3.17). This has been detected unambiguously for Sb in liquid InSb by Warren & Clark (1969), and for Ga in $AuGa_2$ and In in $AuIn_2$ by Warren & Wernick (1971). The measurements of Styles (1967) and of Heighway & Seymour (1971*b*) make it almost certain that it exists for Bi in liquid In–Bi and Pb–Bi as well. In all these systems the quadrupole relaxation rate seems to decrease quite rapidly on heating above the melting point.

Solid InSb is of course a semiconducting compound in which the Knight shift is virtually absent, but when it is melted a shift abruptly appears (Allen & Seymour, 1965). Its magnitude (0.93% and 0.64% for the In and Sb nuclei respectively) is comparable with that of the Knight shift in pure liquid In and Sb, which confirms what we knew from other evidence, that liquid InSb is an ordinary metal. The behaviour of the semiconducting compounds In_2Te_3 and Ga_2Te_3 is very different (Warren, 1970*a,b*). The Knight shift for the In nucleus in one case, and for the Ga nucleus in the other, changes very little if at all in the melting process, but it rises smoothly to a typically metallic value on further heating, as

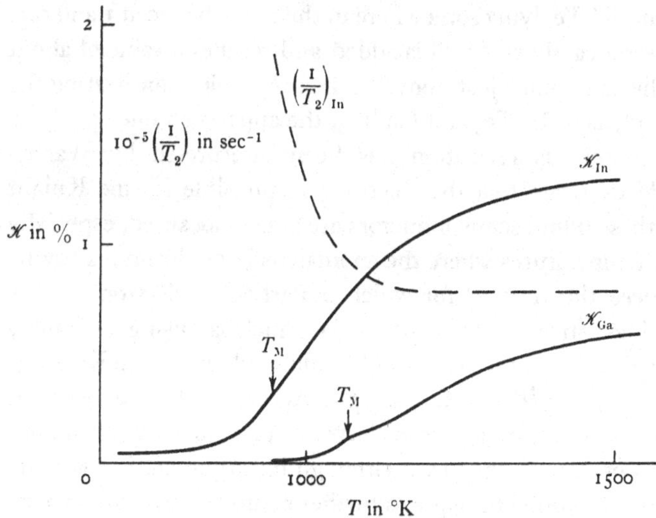

Fig. 6.24. NMR in liquid In_2Te_3 and Ga_2Te_3.

shown in fig. 6.24. The curves strongly suggest that the effect of heating is to release more free electrons, i.e. to increase the density of states at the Fermi level, though since a gradual break-up of the compound is more than likely to affect P_F it would be rash to assume that the variation of χ_p is mirrored exactly by the variation of $\mathscr{K}$. The Knight shift of liquid Tl_2Te also increases on heating above the melting point (Brown *et al.*, 1971) and in this instance we have evidence from the susceptibility measurements reported on p. 497 that an increase in χ_p is largely responsible.

Fig. 6.24 includes a curve for the relaxation rate of In^{115} in In_2Te_3, as measured by Warren. To reconcile these observations with the modified Korringa relation (3.86) would require a value of 2 or more for α, which seems to be out of the question on theoretical grounds, whatever value may be thought appropriate for r_s (see fig. 3.11). Anomalous relaxation is also observed for Ga^{69} and Ga^{71} in Ga_2Te_3. From the difference between the relaxation rates for these two isotopes Warren infers that a quadrupole mechanism is playing some part, but one cannot dispose of the entire anomaly with this hypothesis. In any case, it is shown also by Te^{125} in Tl_2Te, an isotope which has no quadrupole moment. According to Brown *et al.* the apparent value of α is normal enough in pure liquid Te, lying somewhere in the range between 1 and 0.5, but it rises steadily when Tl is added and reaches a value of about 10 for the compound just above its melting point; on heating the compound, as in In_2Te_3 and Ga_2Te_3, the apparent value of α falls.

The anomalous relaxation has been interpreted by Warren (1970*a*,*b*) as a sign that the electrons responsible for the Knight shift in these liquid semiconductors are almost localised, especially at lower temperatures where the anomaly is large. From his results he deduces the time τ_e for which a particular electron can be regarded as attached to a particular nucleus, using a frankly intuitive expression for $1/T_1$ or $1/T_2$ which reduces to the Korringa relation if $\tau_e \sim \hbar/K_F \sim \hbar\mathcal{N}(E)_F/N$. According to Warren τ_e is many times larger, because the electrons have difficulty in tunnelling from one ion to the next. Although his argument is distinctly suggestive the problem requires further scrutiny and more experiments, e.g. a determination of relaxation rates at a function of field, are also desirable.

6.21. The Hall effect

The elementary theory of the Hall effect suggests a connection between R_H and the Fermi radius of the form

$$R_H = 3\pi^2/k_F^3 ec, \tag{6.80}$$

and we have seen in §5.27 that this is well substantiated by experiments on pure liquid metals. A more sophisticated theory might reveal the existence of correction terms, e.g. proportional to $(Lk_F)^{-2}$, but it seems that these are almost negligible in practice, even for liquid Sb and Bi in which $(Lk_F)^{-1}$ approaches 0.2. Now according to the theory in §6.10 the quantity $(Lk_F)^{-1}$ should not rise much when impurity is added to a liquid metal unless the solvent is one of those monovalent metals for which it is particularly small in the first place; it should rarely exceed 0.2, whatever the concentration. Hence we may start with the presumption that equation (6.80) applies not only to metals but to liquid alloys too.

There is, however, the problem discussed in §6.3 of just how k_F is to be defined. The simple equation (6.4) suggests

$$R_H = R_H(\mathrm{FE}) = \Omega/(c_0 z_0 + c_1 z_1)Nec \tag{6.81}$$

and this is often assumed to be the free-electron value of the Hall coefficient for a binary alloy; deviations from this equation are only to be expected, it is said, if some of the valence electrons in the alloy are trapped in bound states, in which case R_H should be enhanced. But equation (6.7) implies that (6.81) may fail whether or not some states are bound, if the solute scattering is strong enough for the phase shifts η_l to be comparable with unity. We do not understand the Hall effect well enough to be certain whether (6.4) or (6.7) will give the better answer.

It is possible that both of them are misleading in the present context and that we will not achieve the right result for R_H unless we recognise that alloys are *inhomogeneous* systems. Consider, for example, a very dilute alloy in which the mean distance separating impurity centres is large compared with L_0, the solvent mean free path. It is surely false to suppose that any one impurity centre can affect the local value of R_H except within a spherical region round about itself with radius of order L_0. Perhaps, therefore, we should represent the alloy by a crude model of spherical inclusions set in a matrix of the pure solvent, the resistivity and Hall coefficient inside each sphere being $\rho_0+\Delta\rho$ and $R_{H,0}+\Delta R_H$ respectively. It may be shown by straightforward macroscopic arguments (related problems are discussed by Herring (1960) and in earlier papers referred to by him) that for such an inhomogeneous system

$$\left.\begin{aligned} \rho &= \rho_0+\beta f\Delta\rho, \\ R_H &= R_{H,0}(1-\tfrac{1}{3}f(1-f)\beta^2(\Delta\rho/\rho_0)^2)+\beta^2 f\Delta R_H, \end{aligned}\right\} \quad (6.82)$$

where $$\beta^{-1} = 1+2(1-f)\Delta\rho/3\rho_0.$$

Here f is the fraction of the total volume which is taken up by spheres; if the volume of each one is ω we may set

$$f = Nc_1\omega/\Omega \simeq 4\pi Nc_1L_0^3/3\Omega.$$

Now let us use (6.81) within each sphere to determine ΔR_H. Manipulation of the above equations then yields the result

$$\left(\frac{dR_H}{dc_1}\right)^0 = \left(\frac{dR_H}{dc_1}\right)^0_{FE}(1-\tfrac{2}{3}\gamma)^2 - R_H\frac{N\omega}{3\Omega}\gamma^2, \quad (6.83)$$

with $$\gamma = \frac{\Omega}{N\omega}\left(\frac{1}{\rho}\frac{d\rho}{dc_1}\right)^0 \simeq \left(\frac{R_A}{L_0}\right)^3\left(\frac{1}{\rho}\frac{d\rho}{dc_1}\right)^0. \quad (6.84)$$

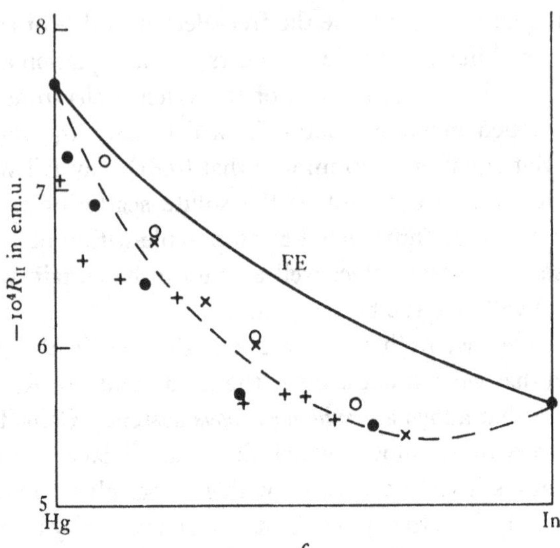

Fig. 6.25. The Hall coefficient in liquid Hg–In, compared with free-electron theory. (●, Cusack & Kendall; ○, Andre'ev & Regel; ×, Güntherodt *et al.*; +, Benkirane & Robert.)

The model implies corrections to the free-electron formula, arising entirely out of the inhomogenity of the alloy, which could be significant in particular cases where L_0 is small and $(\rho^{-1}\, d\rho/dc_1)$ nevertheless large.

The experimental situation is also somewhat confused. According to Busch & Güntherodt (1967) and Güntherodt & Künzi (1969) the results are in good agreement with (6.81) over the whole of the concentration range for Cu–Sn, Ag–Sn, Au–Sn, Ag–In, In–Sn and Al–Ga, though an exact check is possible only for the first two of these systems; for the other three the density has not been studied in detail, so that the variation of Ω with concentration is a matter for speculation. The results reported by Güntherodt *et al.* (1966) for Ga–In also agree with (6.81), apart from a systematic deviation which may be attributable to a small calibration error. Measurements on the Cu–Sn system by Enderby, Hasan & Simmons (1967), however, at a lower temperature than those of Busch & Güntherodt (ca. 600 instead of 1100 °C) suggest that dR_H/dc_1 at the **Sn**-rich end is something like four times as big as (6.81) predicts; by the time that c_{Sn} has reached 0.4 the Hall coefficient, according to Enderby *et al.*, is 20% greater than the free-electron value. Values of $R_H/R_H(\text{FE})$ *greater* than unity have also been reported for Hg–Sn, by Andre'ev & Regel (1967); in this case too, the ratio mounts to about 1.2 near the middle of the concentration range. Values significantly *less* than unity have been reported for Hg–In (see fig. 6.25) by Cusack & Kendall (1963), Andre'ev & Regel (1966), Güntherodt *et al.* (1966) and Benkirane & Robert (1967); for Hg–Tl by Andre'ev & Regel (1966); for Ga–Sn by Dutchak, Stets'kiv & Klyus (1966); and for **Hg**–Na and **Hg**–K by Aldridge (1968) and Davies, Leach & Draper (1971).

Enderby *et al.* have explained their results for Cu–Sn in terms of the bound state hypothesis; the effective valence of Sn is supposed to vary in a linear fashion with concentration from 4 at the Sn-rich end to 2 at the Cu-rich end, which would correspond to there being two bound states round every Sn ion at infinite dilution in Cu. But bound states will not help to explain the negative deviations observed for Hg–In and others, and neither will equation (6.7) it seems. Equation (6.83) looks promising at first sight for it is well-known (see §6.24 below) that $(\rho^{-1}\, d\rho/dc_1)$ is unusually large in amalgams, though negative in sign. In dilute **Hg**–In the quantity γ defined by (6.84) could be about -0.1, and the rather scattered experimental points on the left side of fig. 6.25 may be fitted by (6.83) with fair success if this value for γ is assumed. But the equation does not seem to work on the right side of the diagram, for **In**–Hg, where γ should be negligible, nor does it explain the difference between the behaviour of **Hg**–In and **Hg**–Sn.

These discrepancies must be set aside until a more sophisticated theory of the Hall effect is developed and we may turn to the much larger discrepancies which have been reported for some liquid compounds. AuSn is of course a compound but not a tightly-bound one and it was not to be suspected from the data listed in table 6.3 that it would survive melting in any real sense; its structure may not conform to the rigid-sphere model (p. 464) but its surface tension (p. 480) and viscosity (p. 485) appear to be normal, so it is scarcely surprising that no anomalies show up in its Hall coefficient either. But the compounds whose Hall coefficients are shown in the upper half of table 6.9 are another matter. With the exception of Tl_2Te, which is the subject of special

TABLE 6.9. *Anomalous Hall coefficients for liquid alloys*

	T (°C)	$-R_H$ (e.m.u. $\times 10^{-4}$)	$-R_H$(FE) (e.m.u. $\times 10^{-4}$)	$\frac{d\|R_H\|}{dT}$	Ref.
Compounds					
*ZnSb	590	5.3	4.5	0	EW
*CdSb	470	8.3	5.5	+ve	EW
$AuTe_2$	500	8	~4	0	EHS, ES
In_2Te_3	700	11	~5	−ve	B
Tl_2Te	535	$\gtrsim 250$	~4	−ve	ES
*SnTe	820	12	~4	?	EW
*Sb_2Te_3	625	14	~3.6	0	BT, EW
*Bi_2Te_3	585	8.7	~3.9	+ve	EW
Non-compounds					
$Cu_{0.5}Te_{0.5}$	650	16	~4	−ve	EHS
$Ag_{0.5}Te_{0.5}$	600	100	~4	−ve	ES

References: BT = Busch & Tièche (1963); EW = Enderby & Walsh (1966); EHS = Enderby, Hasan & Simmons (1967); ES = Enderby & Simmons (1969); B = Blakeway (1969).

Note. Asterisks denote compounds which feature also in table 6.6.

comment on p. 454, they are all quite tightly-bound in the solid phase, and clear evidence that they remain anomalous in the liquid has been provided already for those marked with an asterisk, by susceptibility measurements (see table 6.6). In every case R_H is substantially greater than the free-electron value given by (6.81).

Most of these compounds involve Te as one component and for pure liquid Te the ratio R_H/R_H(FE) is already about 3.3. Where it is less than 3.3 for the compound it is probable that it varies monotonically with c and possesses no obvious peak at the stoichiometric composition; this speculation has been confirmed for the Bi–Te system by the work of Enderby and Simmons (1969) and Blakeway (1969). For the Tl–Te system, however, a peak must clearly exist, and measurements at a variety of concentrations by Donally & Cutler (1968) and Enderby & Simmons (1969) reveal that it has the shape shown in fig. 6.25. To judge by the results quoted for $Ag_{0.5}Te_{0.5}$ in the lower half of the table 6.9, there may be an equally sharp peak in R_H/R_H(FE) for the compound Ag_2Te; Hall measurements on liquid Ag_2Te have so far been prevented by

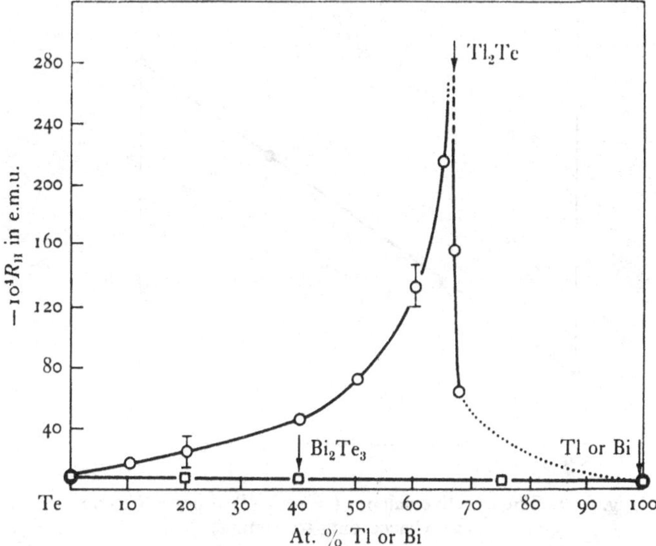

Fig. 6.26. The Hall coefficient in liquid Te–Tl at 535 °C and liquid Te–Bi at 585 °C. (Redrawn by permission from Enderby & Simmons, 1969.)

its high melting point. It looks as though liquid Tl_2Te and Ag_2Te are to be classified as *semiconductors* rather than metals. We shall discuss them further in §6.28 below.

Finally some interesting results have been reported by Busch *et al.* (1971*b*) for liquid Cu–Ce and Fe–Ge. Both of these systems involve a metal for which, in the pure liquid state, the Hall coefficient is *positive* (see p. 414). In both cases it varies monotonically as a function of concentration, changing sign half way across the diagram.

6.22. Optical properties

Before we embark on a detailed discussion of the conductivity of liquid alloys under DC conditions, it is worth considering whether the behaviour of $\sigma(\omega)$ at high frequencies adds anything to the evidence we have already accumulated in support of the NFE model. Can the Drude equations be fitted to the results as well as they can for pure liquid metals (see § 5.22) and does the effective carrier density n^* vary smoothly with concentration, more or less in line with $\bar{n}$ $(= (N/\Omega)(c_0 z_0 + c_1 z_1))$?

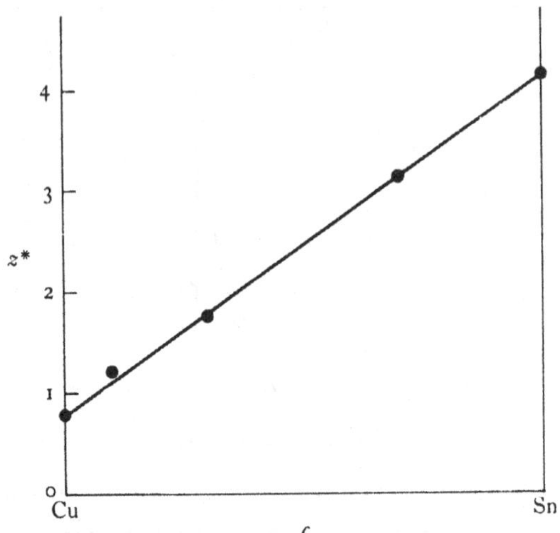

Fig. 6.27. The effective valency ($= n^*\Omega/N$) of liquid Cu–Sn at 1200 °C according to Comins.

Unfortunately, reliable data on the optical properties of liquid alloys are still scanty and they are mostly confined to Hg amalgams, where the situation is obviously complicated by the anomalous behaviour which pure Hg displays. The early work of Kent (1919) indicated good agreement with the NFE model over a wide concentration range for Cd–Pb, Sn–Pb and Bi–Pb, and so do the reflectivity measurements of Schulz (1957) on Ga–In. Comins (1972) has recently reported some accurate ellipsometric measurements on liquid Cu–Sn at 1200 °C and here again the agreement with the NFE model seems excellent (see fig. 6.27); there is no evidence for the bound states hypothesised by Enderby, Hasan & Simmons (1967) to explain their Hall effect data at a lower temperature. For Hg–In and Hg–Tl, however, Schulz found himself unable to reconcile the frequency variation of the reflectivity with the Drude formulae, except, strangely enough, for the limiting case of pure Hg. The Hg–In system has since been studied using ellipsometric techniques by Hodgson (1967) and by Busch & Guggenheim (1968). For small In concentrations the anomalous features in the curves obtained ellipsometrically for $\sigma(\omega)$ and $\epsilon(\omega)$ in pure Hg are still visible, but they tend to disappear as more In is added. The curves may be fitted by the semi-empirical equations of Smith, based upon (5.96), if the two adjustable parameters $n^*/\bar{n}$ and $\overline{\cos\theta}$ are supposed to change smoothly from 1.35 to 1.06, say, and from -0.49 to 0 respectively, as the In concentration is increased from zero to about 15%; solutions containing more than 15% In seem to obey the Drude equations as accurately as pure In does. Experiments on liquid Hg containing Zn, Cd, Sn and Bi (Guggenheim, 1970) suggest that these impurities have a similar effect.

The optical reflectivity of the liquid compounds CdSb and Bi_2Te_3 is said to be metallic (Tauc, unpublished) but detailed results are not available.

6.23. Resistivity: normal systems

Resistivity measurements are useful in the determination of phase diagrams for alloy systems, and this may explain why so many measurements have been made. The results selected for discussion in what follows are a small selection of those available in the literature. Fortunately, the field is adequately covered by the bibliographies which Faber & Ziman (1965), Wilson (1965) and B. W. Mott *et al.* (1966, 1968) have prepared.

Setting aside some compound-forming systems to which we shall revert in §6.26, it is a fair generalisation that the resistivity of binary liquid alloys is a *smooth* function of concentration at constant temperature. Isothermal curves have from time to time been reported with curious peaks and troughs in them, and attempts have been made to associate such anomalies with certain critical values of the electron/atom ratio. With the accumulation of more accurate data, however, it has become clear that they are spurious. One or two of the curves reproduced in fig. 6.28, e.g. the one for Cu–Sb (taken from Steeb, Maier & Godel, 1969), show a limited amount of structure, but no more than the Faber–Ziman theory should be capable of explaining on the basis of the NFE model.

If a comparison is made with the schematic theoretical curves in figs. 6.7 and 6.8, it will be seen that the theory gives a good qualitative account of what is observed. The isothermals are indeed convex for monovalent systems like Na–K and almost linear for polyvalent ones like Sn–Pb and Sb–Bi. That the resistivity of a

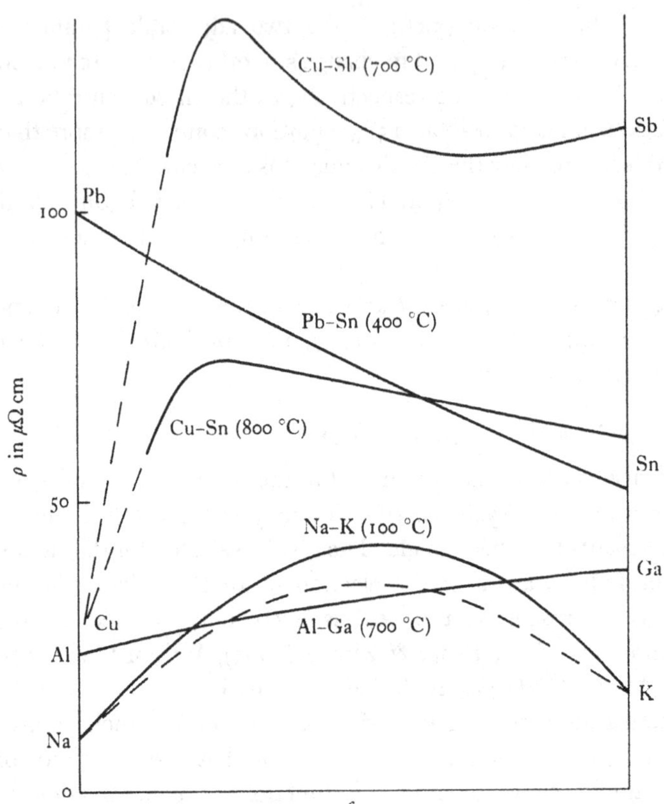

Fig. 6.28. Experimental resistivity isotherms for some typical liquid alloys. The broken curve for Na–K is a theoretical one due to Ashcroft & Langreth (1967*b*).

metal such as liquid Pb should be *lowered* by the addition of an impurity such as Sn occasioned some surprise when it was first discovered. It was also surprising that in a number of systems like Cu–Cd, Cu–Zn, Cu–Al, Cu–Ga, Cu–Sn, Cu–Sb, Ag–In, Ag–Sn, Au–Ga (Bornemann *et al.*, 1912, 1913; Matuyama, 1927; Busch & Güntherodt, 1967; Güntherodt & Tièche, 1968), the temperature coefficient of the resistivity at constant pressure is *negative* over part of the concentration range, usually where $(c_0 z_0 + c_1 z_1)$ is between 1.5 and 2.0. The discussion in §6.10 provides a simple explanation for both these features.

Attempts to apply the theory in a quantitative fashion may be divided into two classes; some are concerned with very dilute alloys, where the problem is to explain the magnitude of $(d\rho/dc_1)^0$, and others with the overall shape of the resistivity curves. An example of the first class is the work of Faber (1967) on **Li**–Mg. This system is expected to be substitutional, for Mg is known to dissolve freely in Li in the solid phase and to cause exceptionally little distortion of the lattice when it does so, so that we do not need to bother about the distinction between a_{00} and a_{01}. Faber used Heine–Animalu pseudo-potentials, correcting the one for Mg, in perhaps too simple a fashion, to the volume and electron density of pure Li. He also made use of (6.56), which represents an important term in $(d\rho/dc_1)^0$ because the thermo-electric parameter ξ_0 is unusually large for Li. It turns out that if ξ_{Li} is indeed -9.3, as listed in table 5.4, there is a serious discrepancy between theory and experiment, which no plausible adjustment of the pseudo-potentials will serve to eliminate. Equation (6.56) may be to blame, if the Li pseudo-potential is energy-dependent. Another possibility is that the apparent value of ξ_{Li} is enhanced by an unusually large effective mass ratio, m^*/m. If the true value is only about -6, as suggested in a footnote to table 5.4, the discrepancy largely disappears.

Dickey *et al.* (1966) have calculated values for $(d\rho/dc_1)^0$ in **Na**–Li, **Na**–K and **Na**–Rb. In these systems the dilatation is relatively large ($\delta \simeq -0.45$, $+0.8$ and $+1.2$ respectively) but it turns out, if Faber & Ziman's equation (6.25) is to be trusted, to contribute remarkably little to the answer, not much more than 10% of the whole. Dickey *et al.*, as mentioned on p. 427, found it essential to recalculate their solute phase shifts for the Fermi energy of liquid Na, but having done this they obtained satisfactory agreement with the results of Freedman & Robertson (1961). It would be of interest to see the calculations extended to the other dilute Na alloys for which Freedman & Robertson have provided accurate figures, namely **Na**–Ag, Au, Cd, Sn, Pb. *Linde's rule* in its simplest form, which states that for a given solvent $(d\rho/dc_1)^0$ is proportional to $(z_1 - z_0)^2$, gives an inadequate description of this set of data. Its failure need not surprise us. In a monovalent solvent such as Na, where ρ'' is expected to dominate ρ', $(d\rho/dc_1)^0$ may indeed be determined by $(u_1 - u_0)^2$, but it is only for small q that $u(q)$ is directly proportional to the valency of the ion. For large

q, where $u(q)$ reflects the configuration of the ion core, the *row* of the Periodic Table to which the solute belongs is quite as important as the *column*.

Except, it would seem, when the solute forms negatively charged *anions*. Ichikawa & Shimoji (1969, 1970*a*,*b*) have measured the resistivity of liquid Hg and Bi containing small concentrations of a halide, introduced by dissolving in the pure metal the appropriate salt, e.g. $HgCl_2$ in Hg. They find that $(d\rho/dc_1)^0$, which is positive and large in every case, is much the same whether the halide is Cl, Br or I. Presumably these elements exist as anions in solution (see p. 428), and Ichikawa & Shimoji have demonstrated that values of $u(q)$ obtained from a simple point charge model lead to reasonable agreement with experiment when inserted into the Faber–Ziman formulae. More realistic calculations are needed to show why the solute core configuration is relatively unimportant in these systems.

An interesting application of the theory to dilute alloys is that of Tamaki (1968*b*), who has studied the effect of small amounts of Fe, Co, Ni and Mn on the resistivity and thermo-electric power of liquid Sn. The values observed for $(d\rho/dc_1)^0$ vary from 70 $\mu\Omega$ cm in the case of Ni to 195 $\mu\Omega$ cm in the case of Fe and are therefore large compared with the values observed for non-transition metal impurities in Sn. Tamaki's explanation is that Fe, Co, or Ni ions dissolved in Sn have a virtual bound d state a little below the Fermi level, which is responsible for strong but very energy-dependent scattering (see §1.7). If one makes the assumptions that u_1 is much larger than u_0 on this account, and that the phase shift η_2 for a solute ion is much larger than η_0, η_1 etc., it is easy to show from (6.50), (6.51) and e.g. (1.56) that

$$\left(\frac{d\rho}{dc_1}\right)^0 \simeq \frac{3\pi m^2}{\hbar^3 e^2}\frac{N}{\Omega k_F^2}\left(\frac{2\pi\hbar^2}{m^* k_F}\right)^2 25 \sin^2 \eta_2. \tag{6.85}$$

Tamaki fits this expression to his results by choosing $\eta_2 = 0.86\pi$, 0.88π and 0.92π for Fe, Co and Ni respectively. But if all the d and s electrons of these solute ions are supposed to join the conduction band of Sn, their effective valencies should be 8, 9 and 10, and it follows from the Friedel sum rule (1.45) that if η_2 is the only significant phase shift it should equal 0.8π, 0.9π and 1.0π

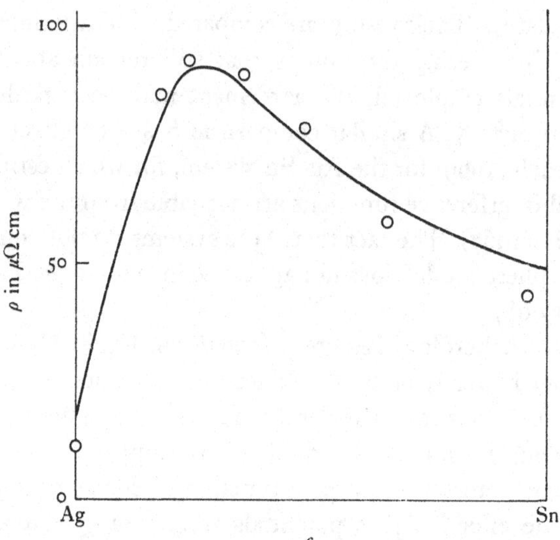

Fig. 6.29. Theoretical points compared with an experimental curve for the resistivity of liquid Ag–Sn at 825 °C. (After Halder & Wagner, 1967*b*).

respectively. In view of the approximations made, the agreement is not unsatisfactory. The case of **Sn**–Mn is complicated by the occurrence of a localised moment (see table 6.6) which implies that there are two virtual bound states, split by spin-orbit interaction, one above the Fermi level and the other below.

We now turn to papers that fall into the second class, being concerned with the whole concentration range.

One of the first attempts to use the Faber–Ziman theory to explain the shape of a resistivity isothermal in detail was that of Halder *et al.* (1966) for Hg–Tl. Two approximations of doubtful validity were involved, (*a*) that the $\bar{a}$ which occurs in (6.50) and (6.51) could be equated with the $\bar{a}_{XR}$ of (6.58), (*b*) that the quantities u_0 and u_1, represented by pseudo-potentials of the Heine–Animula type, could be regarded as independent of concentration. The agreement with experiment was poor for Hg–Tl. For Ag–Sn (Wagner & Halder, 1967) it was more impressive, though partly fortuitous perhaps.

In a subsequent analysis of their X-ray diffraction results, Halder & Wagner (1967*b*) extracted curves for the *partial* interference functions in Ag–Sn by the method discussed on p. 468, and used these to calculate ρ. Their results are compared with experiment in figure 6.29. Considering the doubts that still remain about the pseudo-potentials employed, the agreement may be regarded as distinctly satisfactory. A similar comparison has been effected by Davies & Leach (1969) for the Au–Sn system, for which estimates of the partial interference functions are available from the work of Wagner *et al.* (1967). The fact that these systems do not conform to the rigid-sphere model does not appear to invalidate the Faber–Ziman theory of ρ.

The work of Ashcroft & Langreth (1967*b*) on liquid Na–K has been mentioned already on p. 435. These authors obtained partial interference functions from the Percus–Yevick rigid-sphere model, having previously justified the rigid-sphere approach, and estimated the parameters y and α as a function of concentration, by calculating the effective pair potentials w_{NaNa}, w_{KK} and w_{NaK}. Their theoretical isotherm is included in fig. 6.28 and may be held to agree adequately with experiment, though it should be remembered that the two adjustable parameters $R_{\mathrm{M,Na}}$ and $R_{\mathrm{M,K}}$ have been chosen to secure exact agreement for pure Na and pure K.

Ashcroft & Langreth extended their ambitious programme to K–Rb, Ag–Au, Pb–Sn and to a number of amalgams. To reduce the amount of computation involved they assumed, on the basis of their experience with Na–K, that σ_{01} is always equal to $\frac{1}{2}(\sigma_0+\sigma_1)$, that y is a linear function of concentration at constant T, and that α is independent of concentration, without justifying these points afresh for each system. For K–Rb, Ag–Au and Pb–Sn the agreement with experiment may be regarded as adequate, considering the drastic simplifications involved, but to secure any sort of agreement for the amalgams the theory had to be stretched rather beyond the limits of plausibility. Ashcroft & Langreth proposed a value for $R_{\mathrm{M,Hg}}$ which implies that at room temperature in pure Hg y is 0.36 and that it falls, roughly like $T^{-0.35}$ at constant pressure, to only 0.15 at 350 °C; extrapolated to lower temperatures their calculations suggest that at the melting point y is only 0.39. Such values of y are inconsistent with the observed diffraction behaviour (Halder & Wagner, 1968).

6.24. Resistivity: amalgams

The resistivity of Hg amalgams presents a puzzle of long standing. It has been known since the work of Fenninger (1914) that small amounts of almost any metallic impurity – Na and K are among the few exceptions – lower the resistivity of Hg to a remarkable degree. A selection of values for $(d\rho/dc_1)^0$, mostly taken from the thorough work of E. J. Evans and his collaborators (see Davies & Evans (1930) and earlier papers referred to there), is listed in table 6.10, and some typical curves for ρ as a function of concentration are plotted in fig. 6.30. We are used by now to the idea that $(d\rho/dc_1)^0$ may be negative in liquid polyvalent alloys, and we have also seen that for a divalent alloy the curvature of ρ as a function of c could easily be concave (see p. 457). It is no good pretending, however, that the results displayed in table 6.10 and fig. 6.30 can be fitted by the Faber–Ziman formulae, unless there is something rather unusual about one or more of the parameters involved. The early idea (Skaupy, 1920; Clay, 1940) that the fall in ρ could be attributed to electrolytic action does not stand up to quantitative examination; impurity ions do migrate when an electric field is switched on (see §6.18) but their contribution to the current is too small to be significant in the present context.†

We have previously discussed the volume of mixing for Hg amalgams (p. 472), their compressibility (p. 472), susceptibility (p. 496), Knight shift (p. 505), Hall coefficient (p. 511) and optical properties (p. 514), and in every case have had occasion to note some anomalous features. In most cases the data suggest that there is something unusual about pure Hg which the addition of 20% or so of impurity is sufficient to remove. This impression is supported by the resistivity curves in fig. 6.30, which would look quite normal if only they behaved as shown by broken lines at low concentrations of impurity. The gap between the actual resistivity of pure Hg and the value that it 'ought' to have evidently increases on heating, so it is not altogether surprising to find from table 6.10 that $(d\rho/dc_1)^0$ does so too. Of course the resistivity of pure Hg is unusually sensitive to volume (see §5.8) and it is thermal expansion that makes it rise on heating at the rate that it does. Bradley (1966)

† See footnote on p. 486.

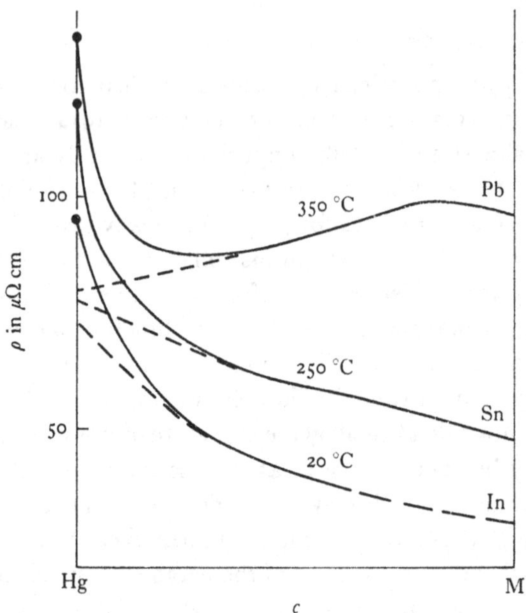

Fig. 6.30. Resistivity isotherms for three amalgams.

has shown that the volume coefficient of ρ, and incidentally of ξ as well, decreases steadily as In is added; here are two more anomalous properties of pure Hg which become normal when the impurity concentration exceeds about 20%.

According to Wallace (1955) the thing that is peculiar about pure Hg is that some of its valence electrons are locked up in bonding states, and he envisaged the impurity as breaking bonds and releasing more carriers. According to Mott (1966) there is a pseudo-gap near the Fermi level in Hg (see p. 328) which the impurity tends to fill in; the alternative hypothesis, that the impurity moves the Fermi level away from the pseudo-gap by altering the number of electrons in the conduction band, is scarcely tenable because it implies a correlation between $(d\rho/dc_1)^0$ and z_1 which is not observed experimentally (Adams, 1968). According to Ashcroft & Langreth (1967*b*) the low value of the packing fraction y in pure Hg is to blame. None of these hypotheses seems to be consistent with all the data that have now been assembled.

TABLE 6.10. *Effect of impurity on the resistivity of Hg*

Solute	T (°C)	$\left(\frac{d\rho}{dc_1}\right)^0$	$\frac{1}{\rho_0}\left(\frac{d\rho}{dc_1}\right)^0$
Na	0	+75	+0.8
	255	+280	+2.3
Ag	15	−380	−4.0
Cu, Au	100	−580	−5.6
Cu, Ag, Au	300	−1150	−9.0
Cd	14	−415	−4.4
	300	−670	−5.2
Zn	15	−400	−4.2
Ga	300	−705	−5.5
In	14	−390	−4.1
	184	−550	−4.9
Ge	300	−590	−4.6
Pb	250	−400	−3.3
Sb	300	−850	−6.6

The recent work of Evans (1970), who has devised an energy-dependent pseudo-potential for Hg which accounts for its resistivity successfully (see p. 329), is more promising. A feature of this pseudo-potential, apparent from fig. 1.7, is that instead of passing through a node and becoming positive just below $q = 2k_F$ it turns downwards and becomes even more negative. Consequently, the difference $(u_1 - u_0)$, when Hg is the solvent, is unusually large in just that region of q where $\bar{a}$ exceeds unity and where the q^3 weighting factor in Ziman's integral is also large. The means that the term ρ'' described by (6.51), which is proportional to $c_0 c_1$, is not only negative but much bigger than would otherwise have been expected. Clearly, this feature of the amalgam situation will help to explain the observed results. Evans's computed values for $-\rho_0^{-1} (d\rho/dc_1)^0$ in **Hg**–Zn and **Hg**–Pb, however, at a temperature of 27 °C, are only 2.0 and 1.2 respectively; the observed slopes are about twice as steep. For **Hg**–In and **Hg**–Tl his calculations, based upon the substitutional model for lack of information about the partial structure factors, explain the shape of the resistivity isotherms quite adequately in the middle of the concentration range, but here again he cannot match their steep slope near $c_1 = 0$.

Takeuchi (1971) has calculated the resistivity for a wider range of amalgams than Evans, but the pseudo-potential he uses for Hg is an empirical one; both its q-dependence and energy-dependence have been adjusted in an arbitrary fashion in the effort to secure agreement with experiment. He too finds difficulty in matching the experimental curves near $c_1 = 0$ for polyvalent solutes such as Zn, In, Tl, Sn, Pb and Bi. He is not unsuccessful in matching the data for **Hg**–Na and **Hg**–K, however, in which $\rho_0^{-1}(\mathrm{d}\rho/\mathrm{d}c_1)^0$ is large and positive.

6.25. Resistivity: systems with miscibility gaps

It is well known that just above the critical point for a liquid–gas transition the compressibility of the vapour is very high. This means that there are large density fluctuations and that $a(q)$ is large for small values of q; $a(0)$ tends to infinity, in fact, like some power of $(T-T_C)$. Similarly, in any binary liquid mixture whose phase diagram displays a miscibility gap (see p. 450) there are large concentration fluctuations just above the critical temperature, T_C', at which the gap closes; from the discussion on p. 437 we may infer that at the critical concentration both $a_{00}(0)$ and $a_{11}(0)$ tend to plus infinity, presumably like some power of $(T-T_C')$, while $a_{01}(0)$, which reflects the cross-correlation between Δm_0 and Δm_1, tends at the same rate to minus infinity.

Such concentration fluctuations have a number of observable effects; they enhance the scattering of light, for example, and also the attenuation of ultrasonic waves and the specific heat. For metallic mixtures the results available are still rather limited. Critical opalescence has been observed in liquid Li–Na, however, using X-ray diffraction techniques (Brumberger, Alexandropoulos & Claffey, 1967) and in liquid Zn–Bi, Ga–Bi and Ga–Pb using neutrons (Wignall & Egelstaff, 1968). Anomalous ultrasonic attenuation has been reported by Kim & Letcher (1971) in liquid Na–Cs; there is no true miscibility gap for this system and the liquidus curve has no maximum, but it does have an inflexion and its slope appears to vanish at just the concentration (corresponding to about 25% Cs) where the attenuation is greatest. An enhancement of C_p has been observed above the critical point for liquid Ga–Hg by Schürmann & Parks (1972). Finally, there are some interesting results available concerning the resistivity, for which experiments on non-metallic mixtures provide no parallel: as the temperature is lowered towards T'_C, the resistivity of a specimen with the critical concentrations is liable to drop below the straight line that describes its variation at higher temperatures (Schürmann & Parks, 1971*b*,*c*).

According to Schürmann & Parks the anomaly in ρ is too small to be detected in the case of Ga–Hg; some results reported previously for this system by Adams (1970) appear to be erroneous. In Li–Na and Ga–Bi, however, it is clearly visible over a temperature range of at least 20° C above T'_C and over a considerable range of concentration as well, even though – for Li–Na at any rate – it never amounts to more than 1 %. Incidentally, Schürmann & Parks (1971*a*) have used resistivity measurements at temperatures just *below* T'_C to map out in detail the shape of the liquidus curve. In non-metallic mixtures the width of the miscibility gap, i.e. the difference in concentration between the two immiscible layers, is known to vary like $(T'_C - T)^\beta$, where the critical exponent β is very close to $\frac{1}{3}$. In Ga–Hg, apparently, β is 0.335 ± 0.005.

To understand the behaviour of ρ, consider the effect of a sinusoidal concentration fluctuation of small wave vector $\mathbf{Q}$ on the three partial interference functions. From the argument in §2.11 it is evident that at $\mathbf{q} = \pm\mathbf{Q}$ this must generate spikes in both $a_{00}(\mathbf{q})$ and $a_{11}(\mathbf{q})$, while to satisfy the sum rule it must reduce these functions in the neighbourhood of their main peaks; a modified form of (2.55) could be used to describe the reduction. A simple extension of the argument shows that $a_{01}(\mathbf{q})$ must acquire negative spikes at $\mathbf{q} = \pm\mathbf{Q}$ and be correspondingly enhanced near its main peak. For an idealised model in which the critical concentration is exactly $\frac{1}{2}$ and in which the two species of ion occupy virtually the same volume the spikes are equal in magnitude, i.e.

$$\Delta a_{00}(\mathbf{Q}) = \Delta a_{11}(\mathbf{Q}) = -\Delta a_{01}(\mathbf{Q});$$

moreover, the peaks in a_{00}, a_{11} and a_{01} occur at the same value of q, i.e. $q_{\max}$. It follows from (6.48) and from the sum rule that if similar concentration fluctuations are excited for all possible values of $\mathbf{Q}$ within a spherical shell in $\mathbf{Q}$-space of thickness dQ, then provided $2k_F > q_{max}$ the change in the resistivity should be given approximately by

$$\Delta\rho \simeq \frac{3\pi m^2}{\hbar^3 e^2}\frac{N}{\Omega k_F^2}\left(\frac{Q}{2k_F}\right)^2 dQ\,\Delta a_{00}(Q)\left\{\left(\frac{Q}{2k_F}\right)(u_0(Q)-u_1(Q))^2 - \left(\frac{q_{max}}{2k_F}\right)(u_0(q_{max})-u_1(q_{max}))^2\right\}. \quad (6.86)$$

The model is too crude for (6.86) to be applied as it stands to the Li–Na system, in which the two species of ion do not occupy the the same volume and for which $2k_F < q_{max}$. It does suggest, however, why $\Delta\rho$ is almost bound to be negative. Because concentration fluctuations are not excited except where $Q \ll q_{max}$, any increase in the scattering for small values of q is trivial compared with the decrease for large values.

6.26. Resistivity: compound-forming systems

The behaviour of the Hg–Na system over the whole concentration range is illustrated by fig. 6.31, and it provides our first example of an anomaly that could be associated with the formation of a compound. At constant temperature there seems to be a trough in the resistivity at the concentration corresponding to the compound $NaHg_2$. Similar behaviour is displayed by the Hg–K system, though the trough is less marked. The troughs almost disappear, however, if the resistivity is plotted not for constant temperature but for the liquidus temperature. The Faber–Ziman formulae may prove capable of describing them without modification, once we know how to allow for the effect of heating above the liquidus on the structure in a_{00}, a_{11} and a_{01}. For an alloy for which $(c_0 z_0 + c_1 z_1)$ lies between 1 and 2, this structure is particularly important.

In other, polyvalent, systems the resistivity tends to pass through a *maximum* at the compound concentration, rather than a minimum. In liquid Cd–Sb, for example, ρ reaches a maximum of about 200 $\mu\Omega$ cm at the 50/50 composition corresponding to CdSb, as shown in fig. 6.32 (Matuyama, 1927; Oleari & Fiorani, 1959; Miller, Paces & Komarek, 1964). It no longer appears that there is a sharp cusp, as Matuyama believed, but the data still provide some support for his idea that one should treat the system as an alloy of CdSb+Cd on one side of the maximum and as an alloy of

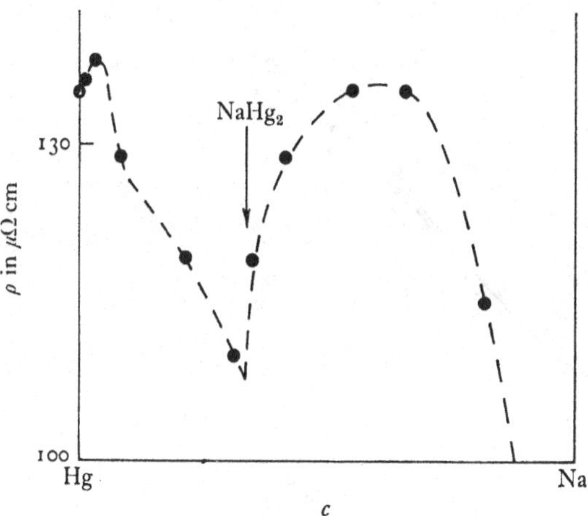

Fig. 6.31. Resistivity isotherm for Hg–Na at 350 °C according to Müller (1910).

CdSb + Sb on the other side. Similar behaviour is shown by the Zn–Sb system.

At the compound composition, and for a considerable range on either side of it, the temperature coefficient $\rho^{-1}(\partial\rho/\partial T)_p$ is negative ($\simeq -10^{-3}$ °C^{-1}) in both CdSb and ZnSb (Enderby & Walsh, 1966). This has suggested to some authors that liquid CdSb and ZnSb should be treated as semiconductors, but the Hall data presented in table 6.9 are not consistent with such a picture; $|R_H|$ is indeed larger than the free-election value, but it does not diminish on heating as it should if extra carriers are then released. Yet if the semiconductor interpretation is abandoned and the number of carriers regarded as constant, one must attribute the sign of $(\partial\rho/\partial T)_p$ to some effect that heating has upon the structure of the liquid and hence upon the relaxation time τ. It is not obvious how the Faber–Ziman formulae are to explain it, since when $(c_0 z_0 + c_1 z_1)$ is as much as 3.5 the temperature dependence of a_{00}, a_{11} and a_{01} should be relatively unimportant. Of course when ρ is as much as 200 $\mu\Omega$ cm the apparent mean free path L is very short – only about 2 Å, in fact, in liquid CdSb. Improvements on the Born approximation are almost certainly required for an accurate theory.

Discordant results have been reported for the In–Bi and In–Sb systems. Some authors have reported resistivity maxima, supposed to indicate the formation in the liquid phase of $InBi_2$ and $InSb_2$,

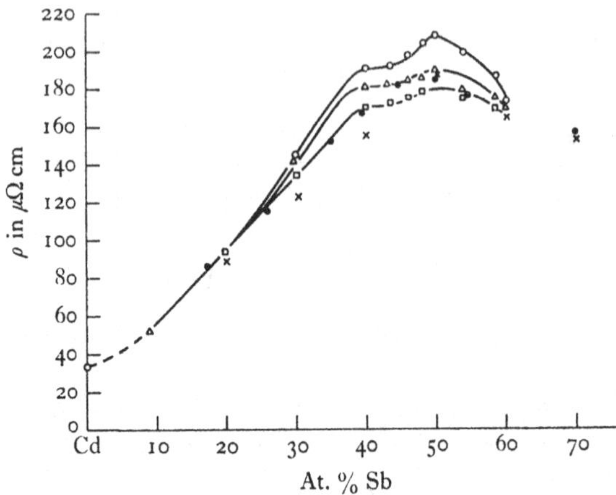

Fig. 6.32. The resistivity of liquid Cd–Sb. □, ρ at 500 °C (cooling); ▲, ρ at 25 °C above stable liquidus; ○, ρ at 25 °C above metastable liquidus; ●, Oleari & Fiorani; ×, Matuyama, (both at 500 °C). (Redrawn by permission from Miller, Paces & Komarek (1964), *Trans. Met. Soc. AIME*, **230**, 1557.)

but the work of e.g. Takeuchi & Endo (1962), Lambert (unpublished work on In–Bi) and Blakeway (1969) suggests that these results are spurious. A maximum may exist for the Mg–Sn and Mg–Pb systems at about the composition corresponding to Mg_2Sn or Mg_2Pb (Wilson, 1966; Steeb and Entress, 1966) but it is quite broad, ρ does not rise above 100 $\mu\Omega$ cm, and the temperature coefficient $(\partial\rho/\partial T)_p$, although small, remains positive.

That leaves us with a number of systems containing Te, especially those whose Hall effect has been discussed in §6.21, and the very interesting case of Mg–Bi. The tellurides have been investigated by Stoneburner (1965, Tl_3Te_2), Dancy (1965, Cu–Te, Ag–Te, Sn–Te), Cutler & Mallon (1966, Tl–Te), Cutler & Field (1968, Tl_2Te doped with a third element), Cutler & Peterson (1970, Tl–Te), Enderby & Walsh (1966, Bi_2Te_3, Sb_2Te_3), Enderby, Hasan & Simmons (1967, CuTe, AgTe, $AuTe_2$), Enderby & Simmons (1969, Ag–Te, Tl–Te, Bi–Te), Nakumura & Shimoji (1969, Tl–Te) and Blakeway (1969, In–Te, Sb–Te, Bi–Te). Since liquid Te is only semi-metallic itself, we can hardly expect these systems to behave like normal liquid alloys. They all of them have high resistivities, greater than 500 $\mu\Omega$ cm, unless the Te concentration is less than about 30%, and the temperature coefficient of

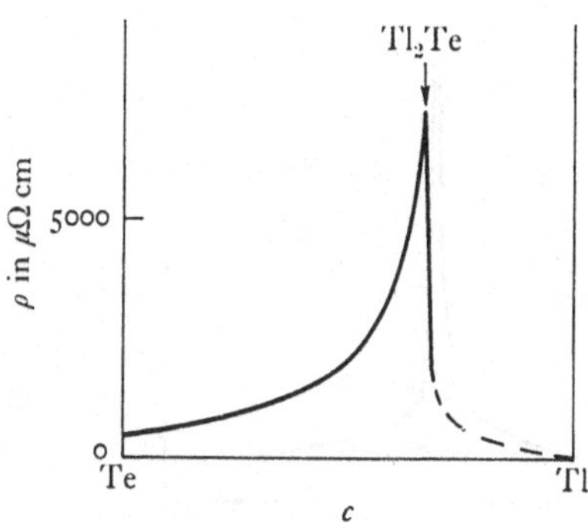

Fig. 6.33. Resistivity isotherm for liquid Te–Tl at 535 °C.

ρ is always negative. It is in liquid Cu_2Te, Ag_2Te, Tl_2Te and perhaps In_2Te_3, however, that the resistivity becomes especially large; for each of these compounds it lies in the range 2000 $\mu\Omega$ cm to 10 000 $\mu\Omega$ cm or more, depending on temperature. These materials must be classified as semiconductors rather than metals.

The variation of resistivity with concentration for the Tl–Te system is shown in fig. 6.33, which makes it clear that the Tl_2Te composition is a critical one, as suggested already by the Hall data in fig. 6.26. The isotherms for Cu–Te and Ag–Te are not dissimilar in appearance. If the Hall coefficient is taken to provide an accurate measure of the carrier density n it becomes meaningful to calculate the *Hall mobility* μ_H $(= cR_H/\rho = e\tau/m?)$. In the Tl–Te system, μ_H appears to be almost independent of T and to vary in a linear fashion with concentration, for $c_{Te} > \frac{1}{3}$. It should perhaps be pointed out that measurements for $c_{Te} < \frac{1}{3}$, whether of R_H, σ or the thermo-electric power P, are impeded by the occurrence of a 'miscibility gap' in the Tl–Te phase diagram; the liquid tends to separate into two layers, one of them quite close to Tl_2Te in composition and the other consisting mainly of Tl. A similar separation is observed for liquid Cu–Te and Ag–Te. To surmount the miscibility gap requires inconveniently high temperatures, which accounts for the limited range covered by the data in figs. 6.26, 6.33 and 6.38 below.

The resistivity of the Mg–Bi system was first studied in the liquid state by Ilschener & Wagner (1958). They found that it rises to a very sharp maximum of at least 1000 $\mu\Omega$ cm^{-1} at the composition

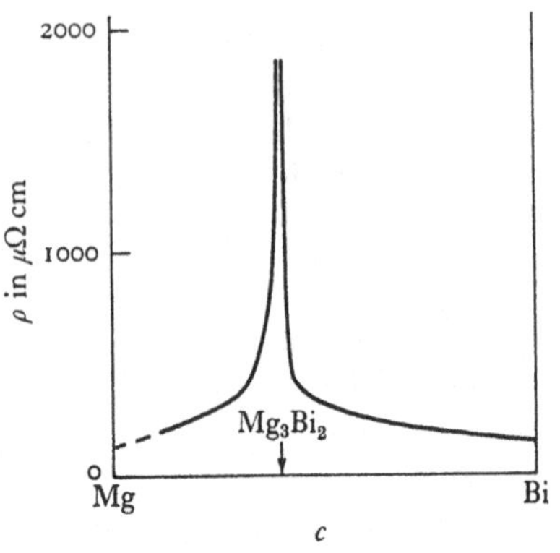

Fig. 6.34. Resistivity isotherm for liquid Mg–Bi at 900 °C.

corresponding to Mg_3Bi_2. Their results have been confirmed by Enderby & Collings (1970), whose curve is sketched in fig. 6.34; according to these authors the peak at 900 °C is in fact about 2×10^4 $\mu\Omega$ cm. Ferrier & Herrell (1969) have reported similar behaviour for amorphous Mg–Bi in the solid state at liquid nitrogen temperatures, and in this case the maximum resistivity is at least 10^5 $\mu\Omega$ cm. Ferrier & Herrell found $\rho^{-1}(\partial\rho/\partial T)_p$ to be strongly negative ($\sim 10^{-2}$ °C^{-1}) at the compound composition. It is of interest to note that this temperature dependence in the frozen amorphous state cannot possibly be attributed to variation in the partial interference functions; thermal excitation of electrons must in some way be responsible for it. The properties of these semiconducting liquid compounds are further discussed in §6.28.

6.27. Thermo-electric power

The variation of the thermo-electric power with concentration should be particularly straightforward for liquid alloy systems in which the variation of k_F may be ignored; the quantity r defined by (6.52) should deviate from linearity, according to (6.54), by a small amount determined solely by $\bar{a}(2k_F)$. To test this conclusion Enderby *et al.* (1968) have investigated five homovalent systems for which Ω_0 and Ω_1 are not very different, and their results are reproduced in fig. 6.35. Although the quantity plotted here is P rather than r the full curves correspond to equation (6.54). The overall agreement is not unsatisfactory. The excellent fit for

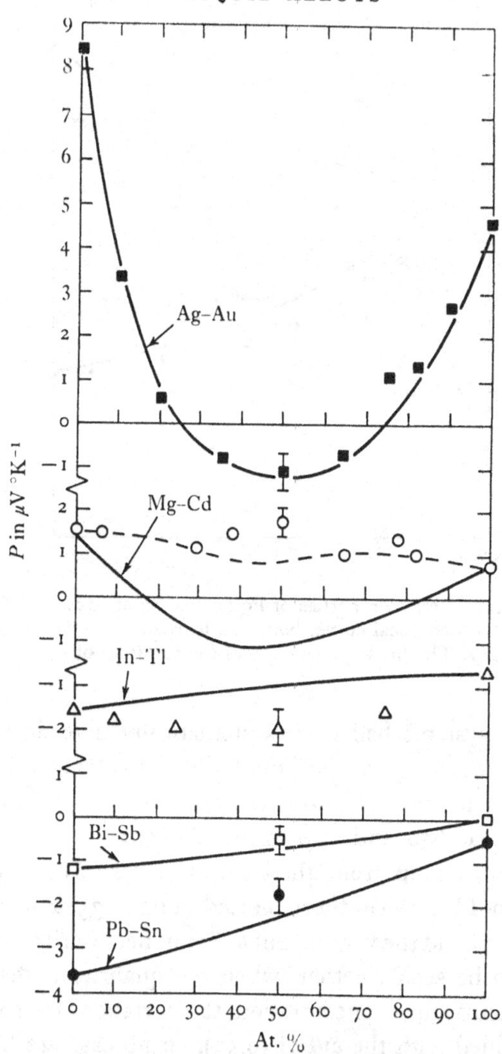

Fig. 6.35. The thermo-electric power for homovalent alloys. (Redrawn by permission from Enderby, Van Zytveld, Howe & Mian, 1968.)

Ag–Au has been secured by choosing $\bar{a}(2k_F)$ to be 0.7, but this does seem to be consistent with the X-ray diffraction data. Otherwise no adjustable parameters are involved. The deviations observed for Mg–Cd and In–Tl have been interpreted by Enderby *et al.* as a sign that the pseudo-potentials are not independent of k as the elementary theory assumes.

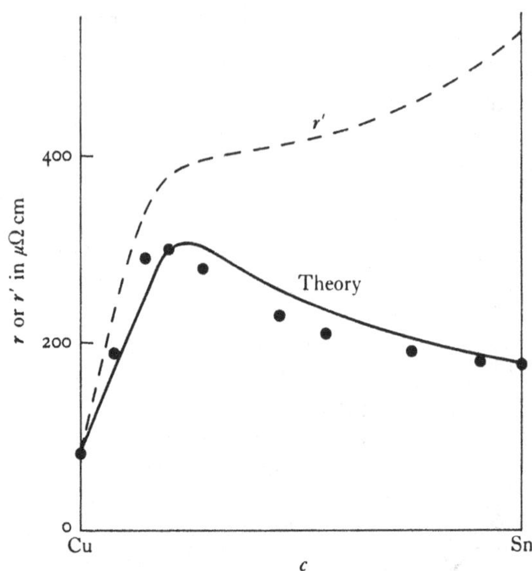

Fig. 6.36. Thermo-electric properties of liquid Cu–Sn at 1100 °C. The experimental points and theoretical curve, both due to Enderby & Howe (1968), are for the parameter r. The broken curve shows the variation of r'.

Few other systems had been systematically investigated until the work of Tougas (1970) on liquid alloys involving Cd, In, Tl, Sn, Pb or Bi; he has now provided data for P as a function of concentration at 630 and 830 °K in all fifteen binary systems that can be made up from these constituents. The results for In–Tl and Sn–Pb confirm those plotted in fig. 6.35. For the other systems $z_1 \neq z_0$ and the variation of k_F is not necessarily negligible. It remains to be seen whether, when the quantity r' defined by (6.57) is plotted against concentration, the deviations from linearity can be explained with the aid of (6.53). In no case are they particularly marked.

Enderby & Howe (1968) have studied the Cu–Sn system, which is of interest because of the large difference between z_0 and z_1 and because this system is known to depart from the rigid-sphere model (see p. 466). Fig. 6.36 shows the experimental points for r and a broken curve derived from these to show the variation of r'; in computing the latter, equation (6.4) has been used to evaluate k_F. The shape of the curve corresponds qualitatively to what is

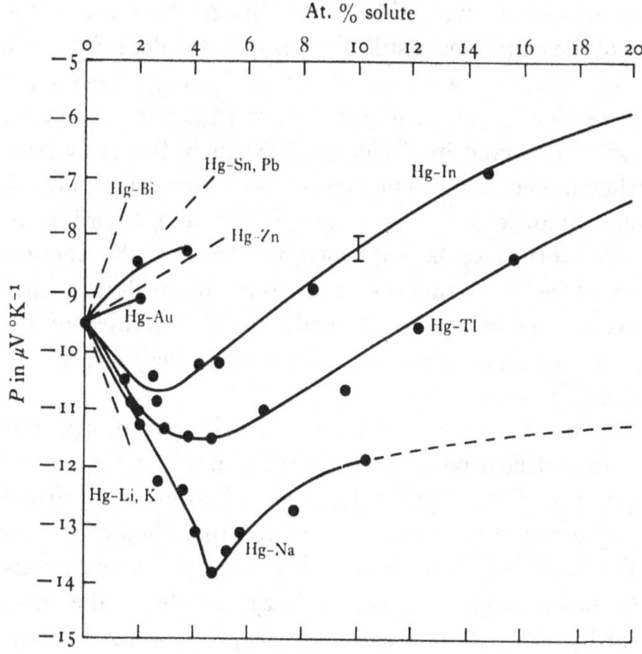

Fig. 6.37. The thermo-electric power of dilute amalgams at 500 °K. (Redrawn by permission from Fielder, 1967, with broken lines superimposed to represent the results of Takeuchi & Noguchi, 1966.)

expected according to (6.53); in particular, its steep initial rise can be attributed largely to the variation of $\bar{a}(2k_F)$, which may be expected to climb from 0.7, say, up to a peak value of perhaps 2 as the mean valence increases from 1.0 to only 1.5. In an attempt to fit their results in a quantitive fashion, Enderby & Howe assumed $u_0(2k_F)$ and $u_1(2k_F)$ to be constants. They achieved remarkable success, as shown in the figure, when they evaluated $\bar{a}(2k_F)$ from the partial interference functions determined experimentally for Cu_6Sn_5 by Enderby *et al.* (1966); when they used interference functions derived from the Percus–Yevick rigid-sphere theory, the fit was not nearly so good.

Thermo-electric power results are available for three series of dilute liquid alloys: for **Na**–Ag, Cd, In, Sn (Davies, 1968), for **Hg**–Li, Na, K, Au, Zn, Cd, In, Tl, Sn, Pb, Bi (Takeuchi & Noguchi, 1966; Fielder, 1967), and for **Bi**–Mg, Hg, As, Se, Te, S (Ichikawa & Shimoji, 1970*a*,*b*). Davies has shown that his data are reasonably consistent with the Faber–Ziman theory, but the behaviour of the amalgams is still unexplained. Fielder's data for a temperature of 500 °K are reproduced in fig. 6.37, with some broken lines superimposed to indicate roughly the rate of variation of P at 400 °K observed by Takeuchi & Noguchi for six systems which Fielder did not study. The thermo-electric power of pure Hg is of course anomalously large in magnitude, and according to Evans (1970) this is to be explained in terms of the energy-dependence of the Hg pseudo-potential. But why small quantities of In, say, should increase $|P|$ while small quantities of Sn decrease it, though their effect on the resistivity is very similar (see fig. 6.30), remains mysterious.

Tamaki has studied not only the resistivity (see p. 518) but also the thermo-electric power of dilute solutions of Fe, Co, Ni and Mn in liquid Sn. If the impurity ions are such as to create virtual bound states near the Fermi level, of course, then their scattering cross-section is liable to vary rapidly with energy; in these circumstances the derivative which occurs in (5.23), and hence the effect of the impurity on the thermo-electric power, may be anomalously large. Quite spectacular effects were indeed reported by Tamaki (1968*b*) in the case of Fe, Co and Ni, but later work has rendered the published data suspect.

There seem to be no obvious anomalies in the thermo-electric behaviour of liquid In–Bi or In–Sb which might be associated with compound formation in these systems (Tougas, 1970; Blakeway, 1969). Of the more tenacious and interesting compounds discussed in previous sections, only ZnSb, CdSb, Mg_3Bi_2 and a number of tellurides have yet been studied in the liquid phase. Liquid Zn–Sb is unusual in that P, which is positive for all concentrations, varies with c in a sigmoid fashion between about 2 and about 6 $\mu V\ °K^{-1}$ (Asanovich, Kozlov & Sryvalin, 1969); at the compound concentration, it is about 3 $\mu V\ °K^{-1}$ but it seems to decrease on heating (Enderby & Walsh, 1966). In liquid Te P is about $+20\ \mu V\ °K^{-1}$, which is outside the normal metallic range, and the rate of fall on heating is quite marked. In liquid alloys which have Te as one constituent, and especially in those whose resistivity is large, P can rise to $+100\ \mu V\ °C^{-1}$ or more, though in all cases heating tends to reduce it. But in such systems the sigmoid tendency shown by

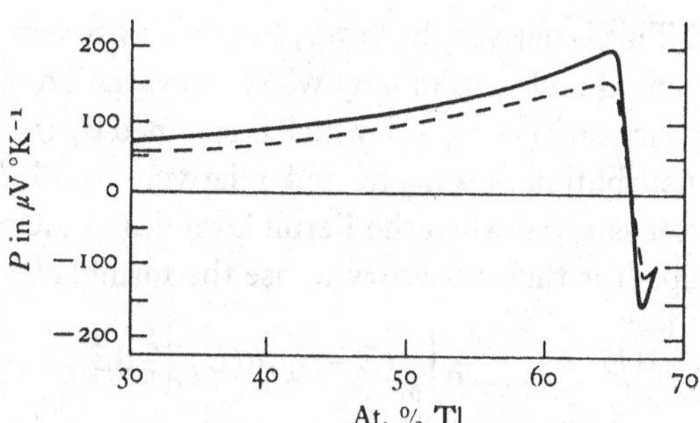

Fig. 6.38. The thermo-electric power of liquid Te–Tl at 425 °C (full curve) and 525 °C (broken curve). (Redrawn by permission from Cutler & Mallon (1966), *Phys. Rev.* **144**, 642.)

Zn–Sb is exaggerated, for as the stoichiometric composition is approached, P usually falls abruptly and may reverse in sign. The best documented example of such behaviour is provided by Tl_2Te and Cutler & Mallon's (1966) curves for the Tl–Te system are reproduced in fig. 6.38; for detailed information about other tellurides the reader should consult the papers already listed on p. 528. Curves similar to those in fig. 6.38 have been reported for liquid Mg–Bi by Enderby & Collings (1970) and for Mg–Bi in the solid amorphous state by Ferrier & Herrell (1969); in this case it is on the Bi-rich side that P is positive.

6.28. Liquid semiconductors

The study of liquid semiconductors such as Se is of respectable antiquity (Ioffe & Regel, 1960; Glazov *et al.*, 1969; Allgaier, 1969) but it is only recently that intermetallic compounds such as Tl_2Te and Mg_3Bi_2 have received much attention. It is still uncertain, therefore, which of the models that have been put forward to explain their intriguing transport properties is nearest to the truth.

Before we examine these models, a brief reminder is required about the significance of the thermo-electric power in semiconductors. For metals we have the formula (see (5.23))

$$P = \Pi/T = \frac{\pi^2 k_B T}{3e}\left(\frac{\partial \log(\sigma(E))}{\partial E}\right)_F, \tag{6.87}$$

where $\sigma(E)$ is the conductivity that the speciman would have if the distribution function of the electrons dropped abruptly from 1 to 0 at the energy E. This is derived, however, from a Taylor expansion of the heat current $\mathbf{Q}$ and is valid only when the variation of $\sigma(E)$ over the energy range of order $4k_B T$ which is spanned by the actual Fermi–Dirac distribution is smooth and relatively small. These conditions are not satisfied when the Fermi level lies in the middle of an energy gap. It is then necessary to use the formula

$$P = \Pi/T = -\frac{1}{e\sigma T}\int_0^\infty (E-E_F)\sigma(E)\frac{df}{dE}dE, \qquad (6.88)$$

which requires only that the electrons contributing to conduction be governed by a single distribution function $f(E)$.† In a typical n-type semiconductor the current is all carried by electrons whose energy is substantially greater than E_F. Then

$$P = \frac{E_c + ak_B T}{eT},$$

where E_c is the gap separating the bottom of the conduction band from the Fermi level and a is a constant, of order unity, whose magnitude depends upon the variation of $\sigma(E)$ with E above the gap. In a p-type semiconductor, where the current is all carried by electrons in the valence band whose energy is at least E_v *below* the Fermi level, we expect

$$P = -\frac{E_v + a'k_B T}{eT}.$$

Because of the negative sign of e, the thermo-electric power is negative in the first case and positive in the second. Clearly its magnitude is liable to be large by metallic standards ($\simeq \pm 10^6/T$ μV per degree if E_c or E_v is 1 eV). Moreover, it tends to decrease like $1/T$ on heating, whereas in metals it tends to increase like T.

In liquid Tl–Te the thermo-electric power is observed to vary more or less like $1/T$, at any rate on the Te-rich side of the compound Tl_2Te, and this encouraged Cutler & Mallon (1966) to treat the system as a semiconducting one of a quite conventional sort. The compound itself they envisaged as an *intrinsic* semiconductor with a density of states as shown schematically in fig.

† Care may be needed if part of the current is due to hopping of electrons between localised states. The distribution functions for localised and non-localised electrons in the same specimen are not necessarily identical (see p. 348).

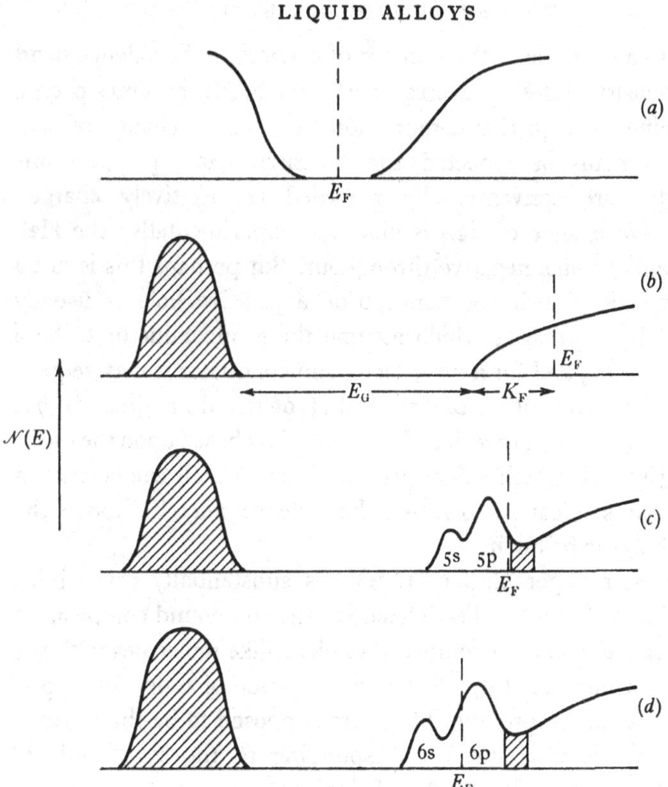

Fig. 6.39. Some models for the density of states in liquid Te–Tl. Hatching indicates localisation.

6.39(*a*), a value of about 0.6 eV being tentatively assigned to the energy gap. They supposed excess Tl atoms to act as *donor* centres and excess Te atoms as *acceptors*, thus explaining the increase of σ which is observed on moving away from the compound composition and the change of sign of P which is observed on passing through it. If the donors are fully ionised at the temperature of measurement, whereas the ionisation energy of the acceptors is about 0.25 eV, the asymmetrical curves for Tl–Te in figs. 6.33 and 6.38 can be fitted by the conventional formulae of semiconductor physics with fair success.

Confidence in this interpretation was momentarily shaken when the Hall coefficient for liquid Tl–Te was measured, with the results displayed in fig. 6.26. The rise in $|R_H|$ as the compound composition is approached from the Tl-rich side can readily be explained as due to a decrease in the number of carriers in the valence band as the density of donors decreases. If the material becomes p-type on passing through this composition, however, a change of sign would normally be expected, for the carriers in a p-type semiconductor are conventionally regarded as positively charged 'holes'. No change of sign is observed experimentally; the Hall coefficient remains negative throughout. But perhaps this is not a fatal objection since the concept of a positive hole is usually justified by arguments which assume the semiconductor to be a regular crystal, and for a disordered semiconductor it may require modification. Matsubara & Kaneyoshi (1968) and Friedman (1971) have now developed theories of the Hall effect based upon the tight-binding model, which safely predict negative Hall coefficients in disordered semiconductors, whether the carriers lie above the Fermi level or below it.

In a later paper Cutler (1971) has substantially revised his original model. On the Te-rich side of the compound composition he pictures the alloy as composed of chain-like molecules with the general formula Tl–Te_n–Tl, where the value of n depends upon the amount of Te present. There are supposed to be three bands of valence electron states corresponding to the σ, π and σ^* molecular orbitals. When the chains are complete the σ and π bands are completely full and the σ^* band completely empty, but holes may arise in the π band if some of the chains rupture under the influence of thermal agitation. The picture is based, of course, upon the idea that chains are favoured in pure liquid Te, and this has been questioned already on p. 82.

Enderby & Simmons (1969) have put forward a quite different model which is illustrated schematically by fig. 6.39(*b*). The liquid is supposed to consist of Tl_2Te molecules in equilibrium with separate Tl and Te atoms. The valence electrons in the molecules are trapped in low-lying states that are probably localised and they play no part in the conduction process. The valence electrons contributed by the separate atoms, however, occupy a conduction band and can be described by NFE theory. Since a small fraction of the molecules is bound to be dissociated, even when the composition corresponds nominally to pure Tl_2Te, there are always some electrons in the conduction band, but their density may be expected to increase with the addition of either Tl or Te. The Hall data are explained in this way. For the purposes of quantitative calculations it is assumed that a reasonably accurate value for n may be obtained from the measured R_H by using the NFE formula. The mobility μ_H which determines the magnitude of the conductivity may then be discussed in terms of the modification of the Faber–Ziman theory which Schaich & Ashcroft have analysed (see p. 461), i.e. in terms of the scattering of conduction electrons by Tl or Te ions on the one hand, or by Tl_2Te molecules. There is reason to hope that the concentration- and temperature-dependence of μ_H can be accounted for by this model.

The problem now is to explain the thermo-electric power. Enderby & Simmons point out that unless m^*, the effective mass for the conduction band, is substantially bigger than m the energy K_F which is indicated in the diagram should be large compared with $k_B T$, except perhaps in the immediate vicinity of the compound composition. The conduction electrons must therefore be regarded as *degenerate* and it is *a priori* reasonable to apply equation (6.87). When P is as large as 200 μV °K^{-1}, we know from (6.88) that the bulk of the current is being carried by electrons with energies

$$\left(\frac{2\times 10^{-4}\times 1.6\times 10^{-12}}{1.4\times 10^{-16}}\right) k_B T \simeq 2k_B T$$

below the Fermi level, but this is not so far below, compared with the spread of the Fermi–Dirac distribution function, as to make (6.87) clearly inappropriate. To explain such a large value of P with its aid, however, we must clearly suppose the energy-dependence of $\sigma(E)$ to be extremely rapid. Enderby & Simmons postulate energy-dependent scattering by a virtual bound state. Enderby & Collings (1970) suggest that the virtual state in question may be a property of the Tl_2Te molecules. The energy is a minimum at the compound composition but the energy denoted by E_G in the diagram is liable to vary monotonically, and it may do so at such a rate that $(E_G + K_F)$ is also a monotonic function. In that case the Fermi level may pass through resonance with the molecular virtual state as the concentration is varied, $(\partial\sigma/\partial E)_F$ being positive on one side and negative on the other. The change of sign of P is tentatively explained in this way.

The explanation is not entirely convincing. For one thing, it does not include the temperature-dependence of P. For another, it makes it seem largely accidental that P changes sign at almost exactly the concentration which corresponds to the compound; why should this happen not only for the Tl–Te system but apparently for the Mg–Bi system as well?

A modification of the Enderby–Simmons scheme which enables these criticisms to be avoided is illustrated by figs. 6.39(*c*) and (*d*). The molecular levels are the same as in (*b*) but the structure of the putative conduction band is significantly different. Diagram (*c*) is supposed to apply to an alloy which is slightly on the Te-rich side of the Tl_2Te. It is worth noting that, according to the law of mass action which is used by chemists to describe the equilibrium of chemical reactions, the quantity

$$N_{\mathrm{Tl}}^2 N_{\mathrm{Te}} / N_{\mathrm{Tl_2Te}}$$

should be independent of concentration at a fixed temperature. This means that the addition of extra Te, which is bound to increase the number N_{Te} of separate Te atoms in the specimen, will tend to suppress N_{Tl} and vice versa. Diagram (*c*) is appropriate in circumstances where N_{Te} is still small compared with $N_{\mathrm{Tl_2Te}}$ but where N_{Tl} has been suppressed to almost zero.

What the diagram is intended to suggest is the existence of two overlapping bands of non-localised states, based upon the 5s and 5p valence states of the free Te atom, at an energy somewhat below that of the conduction band proper; they resemble the sort of impurity bands that can occur in solid semiconductors which are heavily doped. Between them the bands contain eight states for each Te atom so that they should be about $\frac{3}{4}$ filled, and this determines the placing of the Fermi level. As the Te concentration increases, the 'impurity' bands and the conduction band presumably merge together, though since pure liquid Te does not obey the NFE model there may always be some structure left. At the concentration to which the diagram applies the pseudo-gap between the 5p and conduction bands is supposed to be still a pronounced one. Shading indicates that the states in the pseudo-gap itself could be localised (see p. 63).

According to this model it is the rapid fall of $\mathcal{N}(E)$ in the neighbourhood of the Fermi level, and perhaps the localisation of states in the pseudo-gap, which is primarily responsible for the positive thermopower of liquid Tl_2Te+Te. The bulk of the current is carried by electrons with energies near the peak of the 5p band, which must therefore lie about $2k_B T_M$ or 0.1 eV below the Fermi level. So long as this energy separation does not vary with temperature, and there is no obvious reason why the dissociation of a few more Tl_2Te molecules should make it do so, a variation of P like $1/T$ is to be expected.

Fig. 6.39(*d*) may describe the situation in liquid Tl_2Te containing a small excess of Tl, where the thermopower suggests that the bulk of the current is carried by electrons with energies about 0.1 eV *above* the Fermi level. Rather similar impurity bands are drawn, based upon the 6s and 6p valence states of the free Tl atom, but the Fermi level lies to the left of the 6p peak because the valency of Tl is 3 rather than 6.

Variants of this model seem to be capable of explaining the thermo-electric properties of liquid Mg–Bi, and of several liquid Te alloys besides Tl–Te. In the systems studied by Dancy (1965), for example, P does not change sign until c_{Te} is distinctly less than the value which corresponds both to the compound formula and to the minimum in σ (i.e. $< \frac{1}{3}$ for Cu–Te and Ag–Te, or $< \frac{1}{2}$ for Sn–Te). Perhaps this is because molecules of Cu_2Te, Ag_2Te and SnTe dissociate more readily than molecules of Tl_2Te and Mg_3Bi_2 so that a considerable excess of Cu, Ag or Sn is needed to suppress the Te impurity bands. It is altogether too easy, however, to indulge in speculations of this sort.

Naturally there should be a link between the transport properties of these liquid tellurides and the Knight shift data of Warren (1970*a*,*b*) and Brown *et al.* (1971) presented on p. 507. The link has been explored by these authors themselves and by Mott (1971). Their arguments depend, however, upon the theory of conductivity, based upon the tight-binding model, which Mott has pioneered, and the reader who wishes to understand this is recommended to consult Mott & Davis (1971) instead. The present author has already strayed further from the NFE model than he feels is altogether safe.

INDEX OF PRINCIPAL SYMBOLS

Symbols which occur only in a limited context and which are clearly defined where they are used have been omitted from this list. The figures in italics indicate pages of the text where the symbols are further defined or discussed.

Roman

Chemical elements are denoted throughout by the conventional symbols in roman type.

Italic

a, $a(q)$, $a_0(q)$ The interference function [*110*]
$\bar{a}$, $\bar{a}_{XR}$ mean interference functions for alloys [*456*, *462*]
$a_{\alpha\beta}$, a_{00}, etc. partial interference functions [*122*]
a_c, a_s, a_t coefficients for ψ_c, ψ_s, ψ_t [*34*, *245*]
a_H Bohr radius
A_l well depth for model potential [*37*]
A_s, A_t, $A_{\mathbf{k}}$ second-order energy shift associated with a particular $\mathbf{k}$ [*248*]
A_R, A_L amplitude of wave function in one-dimensional chain [*232*]
b spacing in one-dimensional chain [*234*]
b bulk viscosity [*207*]
c wave velocity, normally of light
c_α, c_β, c_0, etc. atomic concentration in alloy [*121*]
c_s, c_t coefficients for ψ_s, ψ_t [*245*]
C normalisation constant for pseudo-waves [*36*]
C_p, C_Ω specific heat at constant pressure or volume
$C(q)$ Fourier transform of direct correlation function [*141*]
C_n coefficient for ψ_n [*354*]
d separation between molecules
D self-diffusion coefficient
D_1, D_{01} diffusion constants in alloys
e electronic charge
E, E_n, E_m electronic energy, especially an eigenvalue
E_c energy of core state
$E_{\mathbf{k}}$, E_s, E_t first-order energy associated with a particular $\mathbf{k}$
E_F Fermi energy
f attenuation factor in one-dimensional chain [*242*]
f, $f(q)$ form factor in scattering theory
$f(E)$ Fermi–Dirac distribution function [*7*]
$f(\mathbf{k})$ electronic distribution function in $\mathbf{k}$-space (*285*]
$f(y)$ function describing equation of state of rigid spheres [*96*]
$f(R)$ direct correlation function [*140*]
$F(\mathbf{q})$ structure factor [*45*]
$F(\mathbf{q}, \omega)$, $F'(\mathbf{q}, \omega)$ frequency spectrum of structure factor [*184*]
$\mathbf{F}$ drag force exerted by current on ions [*487*]

g	acceleration due to gravity
g	shear modulus of elasticity
g, $g(R)$	pair distribution function [*106*]
g_3, g_4	higher-order distribution functions [*107*]
G_s, G_d	time-dependent distribution functions [*188*]
$\mathbf{G}$, $\mathbf{G}_1$	reciprocal lattice vector
h, $\hbar$	Planck and Dirac constants
H	Hamiltonian operator
I	nuclear spin quantum number
I	ionisation energy [*347*]
I, I_∞	intensity
$\mathbf{j}$, j	current density
$\mathbf{k}$, k	wave vector of electron
k_B	Boltzmann constant
K, K_F	kinetic energy of electron [*7*]
l	electronic coherence length [*255*, *322*]
l	diffusion step [*160*]
l	angular momentum quantum number [*23*]
L	electronic mean free path [*322*]
L_0	length of one-dimensional chain
L_M, L_V	latent heat of melting or vaporisation
m	electronic mass
m^*	effective mass [*48ff.*]
m^*_{XC}	effective mass due to exchange and correlation alone [*13*]
m^*_1	effective mass omitting second-order effects [*50*]
m_{opt}	optical effective mass [*380*]
m_A	atomic mass
m	number of molecules [*107*]
$m_\mathbf{k}$	coefficient for $\psi_\mathbf{k}$ [*356*]
M	gram atomic weight
n	density of conduction electrons [*7*]
n^*	effective density of conduction electrons [*364*, *370*]
n_c	density of core electrons [*376*]
n	quantum number [*24*]
$n_\mathbf{k}$	coefficient for $\psi_\mathbf{k}$ [*356*]
N	number of molecules or ions in specimen
N_e	number of electrons
p	pressure
p	strength of δ-function [*232*]
P	thermo-electric power
P_{12}	transition probability [*209*]
P_F	peaking factor relevant to Knight shift [*303*]
P_l	spherical harmonic
$\mathbf{q}$, q	scattering vector [*31*]
q_s, q'_s	screening parameter [*15*, *21*]
q_{max}	position of main peak in $a(q)$ [*136*]
Q	electric charge
$\mathbf{Q}$, Q	phonon wave vector
$\mathbf{Q}$	energy flux [*334*]

Q_D apparent activation energy for diffusion [*159*]
r_s separation between electrons in Bohr units [*10*]
r, r' parameters defining thermo-electric power of alloys [*457, 460*]
$\mathbf{r}$ position vector
$\mathbf{r}_i, \mathbf{r}_j$ position vectors defining sites of molecules or ions
$\mathbf{R}, R$ position vector measured from centre of molecule or ion
R gas constant
R_A atomic radius [*44*]
R_H Hall coefficient [*409*]
R_M model radius [*37*]
R_0 radius of specimen
s size of crystallites [*83*]
s time interval [*170*]
S entropy
S_E excess entropy [*80*]
$S(\mathbf{q}, \omega), S_{inc}, S_{coh}$ frequency spectrum of scattered radiation [*187*]
$\tilde{S}(\mathbf{q}, \omega), \tilde{S}_{inc}, \tilde{S}_{coh}$ symmetrised version of S [*184*]
t time
T temperature
T_D temperature corresponding to Q_D
T_M, T_C melting point; critical point
T_1, T_2 relaxation times in n.m.r. [*214–15*]
$u(\mathbf{r}), u(\mathbf{q})$ ionic pseudo-potential [*45*]
u_0, u_1 pseudo-potentials in alloys [*426ff.*]
u' effective pseudo-potential [*48*]
$U(\mathbf{r}), U(\mathbf{q})$ total pseudo-potential [*45*]
U' effective pseudo-potential [*32, 47*]
U^{XC} contribution of exchange and correlation to U [*14*]
U_{st} matrix element of pseudo-potential [*245*]
v velocity of molecule
v_D drift velocity of electron or molecule
v_F Fermi velocity [*317*]
$v_{\parallel}$ sound velocity for longitudinal wave
$V(\mathbf{r}), V(\mathbf{q})$ screened potential
V_b unscreened potential
$w(R), w(q)$ pair potential between molecules or ions [*63ff.*]
w', w'' indirect and direct contributions to w
$W(R)$ effective pair potential [*138*]
X root mean square displacement of molecule or ion [*103*]
X_{nm}, X'_{mn} matrix element [*353, 354*]
y packing fraction for rigid spheres [*71, 93*]
y $\equiv q/2k_F$
z valency
z^* effective valency [*39*]
z dimensionless parameter describing electronic energy [*251*]
$Z(\omega)$ frequency spectrum of diffusive motion [*177*]

Script

$\mathscr{D}$	thermal diffusivity
ℓ	surface tension
$\mathscr{F}$	free energy
$\mathscr{G}$	Gibbs function
$\mathscr{K}$	Knight shift
$\mathscr{N}(E)$, $\mathscr{N}(E_{\mathbf{k}})$, $\mathscr{N}(\omega)$	Densities of states
$\mathscr{S}$	shear stress
$\mathscr{U}$	internal energy

Sans Serif

A	vector potential
B	magnetic induction
D	electric induction
E	electric field
H	magnetic field
I	magnetisation
M	magnetic moment
P	electric polarisation
q	electric field gradient
Q	quadrupole moment

Greek

α	thermal expansion coefficient
α	factor in the modified Korringa relation [214]
α	misfit ratio for mixtures of rigid spheres [435]
$\alpha_{\mathbf{k}}$	angle between **k** and electric field
β, β_T	isothermal compressibility
β_S	adiabatic compressibility
β	critical exponent [525]
β	parameter describing energy dependence of Γ [264]
γ	electronic specific heat coefficient
γ	specific heat ratio C_p/C_Ω
γ	attenuation factor in one-dimensional chain [236]
$\gamma(\mathbf{r})$, $\gamma(\mathbf{R})$	modulation factor [31, 276]
γ_e, γ_n	gyromagnetic ratio of electron or nucleus
Γ_s, Γ_t, $\Gamma_{\mathbf{k}}$	imaginary part of second-order energy shift associated with a particular **k** [248]
$\delta_{\alpha\beta}$, $\delta(\mathbf{r})$	Kronecker delta, Dirac function
δ	skin depth [366]
δ	departure from equilibrium distribution function [209]
δ	frequency difference [355]
δ	dilatation in alloys [438]
$\Delta\Omega$, ΔS	change of volume or entropy on melting
$\underline{\Delta\Omega}$, $\underline{\Delta S}$, etc.	change of volume, entropy, etc. on mixing for alloys
$\overline{\Delta S}$, etc.	partial entropy, etc. of mixing [474]

ΔV	perturbation of potential etc. in jellium [14ff.]
$\epsilon(q)$, $\epsilon'(q)$	dielectric constant describing screening [15, 17]
$\epsilon(\omega)$	dielectric constant measured in optical experiment [361]
$\tilde{\epsilon}_{\perp}(q, \omega)$, $\tilde{\epsilon}_{\parallel}(q, \omega)$	generalised dielectric constant for transverse or longitudinal disturbance [361, 368]
ϵ_c	core contribution to dielectric constant [381]
ζ	friction coefficient [163]
ζ_R, ζ_L	phase of wave function in one-dimensional chain [232]
η	viscosity
η_l	phase shift [23]
θ	scattering angle [33]
θ	angle of incidence
θ	shear strain
Θ_D	Debye temperature
κ	thermal conductivity
κ_e	electronic contribution to κ
λ	wavelength
μ_B	Bohr magneton
μ_H	Hall mobility
ν_S, ν_L	frequency, e.g. of vibration in solid and liquid
ν	exponent describing temperature dependence of packing fraction [71]
ξ	thermo-electric parameter [335]
ξ	fudge factor for Knight shift [303]
$\boldsymbol{\xi}$, ξ	displacement of molecule or ion
$\boldsymbol{\xi}_Q$, ξ_Q	amplitude of vibration mode
π	Poisson ratio
Π	Peltier coefficient
ρ	charge density
ρ, $\rho(\mathbf{r})$	density
ρ	resistivity
σ	rigid-sphere diameter
σ^*	closest distance of approach of molecules [128]
σ	d.c. conductivity
$\sigma(\omega)$	high frequency conductivity measured in optical experiment [361]
$\tilde{\sigma}(q, \omega)_{\perp}$, $\tilde{\sigma}(q, \omega)_{\parallel}$	generalised conductivity for transverse or longitudinal disturbance [361, 365–6]
σ_v, σ_c	contributions to $\sigma(\omega)$ due to valence and core electrons [379, 381]
Σ_s, Σ_t, $\Sigma_{\mathbf{k}}$	complex second-order energy shift associated with particular $\mathbf{k}$ [246]
τ	electronic relaxation time [210]
τ_l	electronic lifetime [208]
τ_D	Drude relaxation time [364, 370]
τ_L	Langevin relaxation time [171]
τ_M	Maxwell relaxation time [205]
τ_q	quadrupole relaxation time [218]
ϕ	angle, especially Hall angle

ϕ	work function
ϕ_i	inner work function [13]
ϕ	pseudo wave function [34]
χ	total susceptibility
χ_i	ionic contribution to χ
χ_p, χ_d	electronic contributions to χ [294, 300]
ψ	partial wave [23]
ψ, ψ_k	true as opposed to pseudo wave function for electron
ψ_c	eigenfunction for core electron
$\psi(s)$	function describing decay of velocity auto-correlation [170]
Ψ	wave group, especially eigenfunction [245]
ω	angular frequency
ω	element of volume of specimen [107]
ω_0, ω_1	volume occupied by atom in alloys [445]
ω_D	angular frequency associated with Debye temperature [173]
ω_c	angular frequency at which core excitation begins [378]
ω_C	cyclotron frequency [407]
ω_p	free electron plasma frequency
Ω	volume of specimen
Ω_{CP}	close-packed volume of rigid spheres [93]
Ω	phonon angular frequency

BIBLIOGRAPHY AND AUTHOR INDEX

The figures in italics indicate pages of the text on which the reference is quoted.

Aarnodt, R., Case, K. M., Rosenbaum, M. & Zweifel, P. F. (1962), *Phys. Rev.* **126**, 1165. [*187*]

Abarenkov, I. *see* Heine, V.

Abowitz, G. & Gordon, R. B. (1963*a*), *Trans. Met. Soc. AIME*, **227**, 51. [*472*]

Abowitz, G. & Gordon, R. B. (1963*b*), *Acta Met.* **10**, 671. [*472*]

Abragam, A. (1961), *The Principles of Nuclear Magnetism*, Oxford: Clarendon Press. [*218*]

Abrahams, E. *see* Miller, A.

Adams, J. E., Berry, L. & Hewitt, R. R. (1966), *Phys. Rev.* **143**, 164. [*307*]

Adams, P. D. (1968), *Phys. Rev. Lett.* **20**, 537. [*522*]

Adams, P. D. (1970), *Phys. Rev. Lett.* **25**, 1012. [*525*]

Agacy, R. L. & Borland, R. E. (1964), *Proc. Phys. Soc.* **84**, 1017. [*225*]

Aldea, A. (1967), *Phys. Stat. Sol.* *22*, 377. [*411*]

Aldea, A. *see* Banyai, L.

Alder, B. J. (1964), *Phys. Rev. Lett.* **12**, 317. [*139*]

Alder, B. J. & Wainwright, T. E. (1957), *J. Chem. Phys.* **27**, 1208. [*74, 92*]

Alder, B. J. & Wainwright, T. E. (1967), *Phys. Rev. Lett.* **18**, 988. [*174, 176*]

Alder, B. J. *see* Ross, M.

Aldridge, R. V. (1968), *Phil. Mag.* **18**, 1. [*410, 511*]

Alekse'ev, N. V. & Evse'ev, A. M. (1959), *Sov. Phys. Cryst.* **4**, 323. [*462*]

Alekseyeva, T. O., Nikitin, B. S., Khabibullin, B. M. & Kharakhashyan, E. G. (1968), *Phys. Met. Metallography*, **26** (4), 66. [*493*]

Alexandropoulos, N. G. *see* Brumberger, H.

Allen, P. B. & Cohen, M. L. (1969), *Phys. Rev.* **187**, 525. [*53*]

Allen, P. B., Cohen, M. L., Falicov, L. M. & Kasowski, R. V. (1968), *Phys. Rev. Lett.* **21**, 1794. [*51, 308*]

Allen, P. S. & Seymour, E. F. W. (1965), *Proc. Phys. Soc.* **85**, 509. [*506*]

Allgaier, R. S. (1969), *Phys. Rev.* **185**, 227. [*535*]

Allgaier, R. S. *see* Mott, N. F.

Anderson, P. D. *see* Hultgren, R. R.

Anderson, P. W. (1958), *Phys. Rev.* **109**, 1492. [*230*]

Anderson, P. W. (1961), *Phys. Rev.* **124**, 41. [494]

Anderson, P. W. (1970), *Comments on Sol. State Phys.* **2**, 193. [*230*]

Anderson, P. W. & McMillan, W. L. (1967), 'Enrico Fermi', *Proc. Int. School of Physics*, course 37 (ed. Marshall), New York & London: Academic Press. [*56, 58*]

Andrade, E. N. da C. (1934), *Phil. Mag.* **17**, 497 & 698. [*167*]

Andre'ev, A. A. & Regel, A. R. (1966), *Sov. Phys. Sol. State*, **7**, 2076. [*511*]

Andre'ev, A. A. & Regel, A. R. (1967), *Sov. Phys. Sol. State*, **8**, 2950. [*511*]
Angus, J. C. & Hücke, E. E. (1961), *J. Phys. Chem.* **65**, 1549. [*488*]
Animalu, A. O. E. (1965), *Phil. Mag.* **11**, 379. [*40, 325*]
Animalu, A. O. E. (1967), *Adv. Phys.* **16**, 605. [*340*]
Animalu, A. O. E. & Heine, V. (1965), *Phil. Mag.* **12**, 1249. [*12*]
Armstrong, J. E. (1935), *Phys. Rev.* **47**, 391. [*414*]
Asanovich, V. Y., Kozlov, V. A. & Sryvalin, I. I. (1969), *Phys. Met. Metallography*, **28** (2), 195. [*534*]
Ascarelli, P. (1966), *Phys. Rev.* **143**, 36. [*129, 132, 146, 147, 219*]
Ascarelli, P. (1968), *Phys. Rev.* **173**, 371. [*149*]
Ascarelli, P. & Caglioti, G. (1966), *Nuovo Cim.* **43** B, 375. [*186*]
Ascarelli, P., Harrison, R. J. & Paskin, A. (1967), *Adv. Phys.* **16**, 717. [*144*]
Ascarelli, P. & Paskin, A. (1968), *Phys. Rev.* **165**, 222. [*175*]
Ashcroft, N. W. (1966*a*), *Phys. Lett.* **23**, 48. [*38*]
Ashcroft, N. W. (1966*b*), *Phys. Lett.* **23**, 529. [*331*]
Ashcroft, N. W. (1968), *Proc. Phys. Soc.* **1**, 232. [*38*]
Ashcroft, N. W. & Langreth, D. C. (1967*a*), *Phys. Rev.* **155**, 682. [*38, 48, 149, 152*]
Ashcroft, N. W. & Langreth, D. C. (1967*b*), *Phys. Rev.* **159**, 500. [*38, 67, 435, 442, 516, 520, 522*]
Ashcroft, N. W. & Lawrence, W. E. (1968), *Phys. Rev.* **175**, 938. [*51*]
Ashcroft, N. W. & Lekner, J. (1966), *Phys. Rev.* **145**, 83. [*132, 136, 326*]
Ashcroft, N. W. & March, N. H. (1967), *Proc. Roy. Soc.* A **297**, 336. [*132*]
Ashcroft, N. W. & Schaich, W. (1970), *Phys. Rev.* B **1**, 1370 (*for errata see* (1971), *Phys. Rev.* B **3**, 1511). [*281, 501*]
Ashcroft, N. W. & Wilkins, J. W. (1965), *Phys. Lett.* **14**, 185. [*51*]
Ashcroft, N. W. see Koyama, R. Y.
Ashcroft, N. W. *see* Schaich, W.
Ashley, E. J. *see* Bennett, H. E.
Averbach, B. L. *see* Fessler, R. R.
Averbach, B. L. *see* Kaplow, R.
Ballay, M. (1928), *Rev. de Mét.* **25**, 427 & 509. [*492*]
Ballentine, L. E. (1965), *Ph.D. Thesis*, Cambridge University. [*231*]
Ballentine, L. E. (1966*a*), *Proc. Phys. Soc.* **89**, 689. [*42*]
Ballentine, L. E. (1966*b*), *Canad. J. Phys.* **44**, 2533. [*268–70, 292, 300*]
Ballentine, L. E. (1968), *Canad. J. Phys.* **46**, 2567. [*40*]
Ballentine, L. E. & Heine, V. (1964), *Phil. Mag.* **9**, 617. [*231*]
Ballentine, L. E. *see* Chan, T.
Bambakidis, G. *see* Taylor, P. L.
Ban, N. T., Randall, C. M. & Montgomery, D. J. (1962), *Phys. Rev.* **128**, 6. [*169*]
Banyai, L. & Aldea, A. (1966), *Phys. Rev.* **143**, 652. [*411*]
Bardeen, J. (1938), *J. Chem. Phys.* **6**, 367 & 372. [*51*]
Bardeen, J. & Pines, D. (1955), *Phys. Rev.* **99**, 1140. [*149*]
Barker, J. A. (1963), *Lattice Theories of the Liquid State*, Oxford: Pergamon Press. [*77, 91*]

Barnes, R. G. *see* Jones, W. H.
Bates, L. F. & Somekh, E. M. (1944), *Proc. Phys. Soc.* **56**, 182. [*496*]
Baun, W. L. *see* Fischer, D. W.
Beaglehole, D. & Hunderi, O. (1970), *Phys. Rev.* B **2**, 309. [*407*]
Becker, G. *see* Odle, R. L.
Becker, M. *see* Gebhardt, E. M.
Beeby, J. L. (1964), *Proc. Roy. Soc.* A **279**, 82. [*231*]
Beeby, J. L. & Edwards, S. F. (1963), *Proc. Roy. Soc.* A **274**, 395. [*230*]
Benkirane, M. & Robert, J. (1967), *C.R. Acad. Sci.* B **264**, 470 & 1584. [511]
Bennett, H. E., Bennett, J. M., Ashley, E. J. & Motyka, R. J. (1968), *Phys. Rev.* **165**, 755. [*385*]
Bennett, J. M. *see* Bennett, H. E.
Bennett, L. H. *see* Watson, R. E.
Ben-Yosef, N. & Rubin, A. G. (1969), *Phys. Rev. Lett.* **23**, 289. [*350*]
Berger, A. G. *see* Knight, W. D.
Berglund, C. N. & Spicer, W. E. (1964), *Phys. Rev.* **136** A 1030. [*401*]
Bernal, J. D. (1965), *Liquids: Structure, Properties, Solid Interactions* (ed. Hughel), p. 25, Amsterdam: Elsevier. [*78*]
Bernal, J. D. & Finney, J. L. (1967), *Disc. Farad. Soc.* **43**, 62. [*78*]
Berne, A., Boato, G. & de Paz, M. (1962), *Nuovo Cim.* **24**, 1179. [*161*]
Berne, B. J., Boon, J. P. & Rice, S. A. (1966), *J. Chem. Phys.* **45**, 1086. [*178*]
Berry, L. *see* Adams, J. E.
Berry, M. V. *see* Lloyd, P.
Bhatia, A. B. & Krishnan, K. S. (1948), *Proc. Roy. Soc.* A **194**, 185. [*313*]
Bhatia, A. B. & Thornton, D. E. (1970). *Phys. Rev.* B **2**, 3004. [455]
Bhatia, A. B. & Thornton, D. E. (1971), *Phys. Rev.* B **4**, 2325. [*455*]
Bird, N. F. *see* Borland, R. E.
Black, P. J. & Cundall, J. A. (1965), *Acta Cryst.* **19**, 807. [*132*]
Blair, D. G. (1967), *Proc. Phys. Soc.* **91**, 8 & 736. [*226*]
Blakeway, R. (1969), *Phil. Mag.* **20**, 965. [*512, 528, 534*]
Blandin, A., Daniel, E. & Friedel, J. (1959), *Phil. Mag.* **4**, 180. [*503*]
Blatt, F. J. (1957), *Phys. Rev.* **108**, 285. [*440*]
Bloch, A. & Rice, S. A. (1969), *Phys. Rev.* **185**, 933. [*391, 393*]
Blodgett, J. A. & Flynn, C. P. (1969), *Phil. Mag.* **20**, 917. [*498, 499*]
Blodgett, J. A. *see* Rigney, D. A.
Bloembergen, N. *see* Hanabusa, M.
Bloembergen, N. *see* Rimai, L.
Blough, J. L. *see* Olson, D. L.
Boato, G. *see* Berne, A.
Bohm, B. & Klemm, W. (1939), *Z. anorg. Chem.* **243**, 69. [*497*]
Bohm, D. & Staver, T. (1951), *Phys. Rev.* **84**, 836. [*149*]
Boiani, J. & Rice, S. A. (1969), *Phys. Rev.* **185**, 931. [*391*]
Bonch-Bruevich, V. L. (1965), *Phys. Lett.* **18**, 260. [*230*]
Bonch-Bruevich, V. L. (1968), *The Theory of Condensed Matter*, p. 989, Vienna: International Atomic Energy Comm. [*230*]
Bonilla, C. F. *see* Hockman, J. M.
Boon, J. P. *see* Berne, B. J.

Borland, R. E. (1961*a*), *Proc. Phys. Soc.* **77**, 705. [*229*]
Borland, R. E. (1961*b*), *Proc. Phys. Soc.* **78**, 926. [*224, 226*]
Borland, R. E. (1963), *Proc. Roy. Soc.* A **274**, 529. [*228, 243, 244*]
Borland, R. E. & Bird, N. F. (1964), *Proc. Phys. Soc.* **83**, 23. [*227–9, 243*]
Borland, R. E. *see* Agacy, R. L.
Borland, R. E. *see* West, R. N.
Born, M. (1943), *Rep. Progr. Phys.* **9**, 294. [*118*]
Bornemann, K. & von Rauschenplatt, G. (1912), *Metallurgie* **9**, 473 & 505. [*516*]
Bornemann, K. & Wagenmann, G. (1913), *Ferrum*, **11**, 276, 289 & 330. [*516*]
Borsa, F. & Rigamonti, A. (1967), *Nuovo Cim.* **488**, 144. [*220*]
Bortolani, V. & Magnaterra, A. (1968), *Phys. Lett.* **28**A, 316. [*67*]
Bortolani, V. & Calandra, C. (1970), *Phys. Rev.* B **1**, 2405. [*40*]
Bosio, L. & Defrain, A. (1964), *C. R. Acad. Sci.* **258**, 4929. [*80*]
Bradley, C. C. (1963), *Phil. Mag.* **8**, 1533. [*337, 342*]
Bradley, C. C. (1966), *Phil. Mag.* **14**, 953. [*521*]
Bradley, C. C., Faber, T. E., Wilson, E. G. & Ziman, J. M. (1962), *Phil. Mag.* **7**, 865. [*314*]
Brandt, G. B. & Rayne, J. A. (1966), *Phys. Rev.* **148**, 644. [*328*]
Brandt, W. & Waung, H. F. (1968), *Phys. Lett.* **27**A, 700. [*292*]
Bratby, P., Gaskell, T. & March, N. H. (1970), *Phys. Chem. Liq.* **2**, 53. [*86*]
Braunbek, W., Gahn, U., Sommer, F., Werber, K. & Willée Ch. (1970), *Phys. Lett.* **31**A, 237. [*67*]
Breuil, M. *see* Tourand, G.
Bridgman, P. W. (1958), *The Physics of High Pressure*, London: Bell. [*339, 343*]
Brockhouse, B. N. & Pope, N. K. (1959), *Phys. Rev. Lett.* **3**, 259. [*180, 189, 198*]
Brouers, F. (1970), *J. Non-Cryst. Sol.* **4**, 428. [*230*]
Brown, D., Moore, D. S. & Seymour, E. F. W. (1971), *Phil. Mag.* **23**, 1249. [*496, 497, 507, 541*]
Brown, R. C. & March, N. H. (1968), *Phys. Chem. Liq.* **1**, 129. [*169*]
Brumberger, H., Alexandropoulos, N. G. & Claffey, W. (1967), *Phys. Rev. Lett.* **19**, 555. [*524*]
Buff, F. P. *see* Kirkwood, J. G.
Bundy, F. P. (1959), *Phys. Rev.* **148**, 644. [*343*]
Busch, G. & Guggenheim, J. (1968), *Helv. Phys. Acta*, **41**, 1301. [*514*]
Busch, G. & Güntherodt, H-J. (1967), *Phys. kond. Mat.* **6**, 325. [*410, 510, 516*]
Bush, G. & Güntherodt, H-J. (1968), *Phys. Lett.* **27**A, 110. [*498*]
Busch, G., Güntherodt, H-J. & Künzi, H. U. (1970), *Phys. Lett.* **32**A, 376. [*312, 414*]
Busch, G., Güntherodt, H-J. & Künzi, H. U. (1971), *Phys. Lett.* **34**A, 309. [*414*]
Busch, G., Güntherodt, H-J, Künzi, H. U. & Schapbach, L. (1970), *Phys. Lett.* **31**A, 191. [*312, 414*]

Busch, G. & Tièche, Y. (1963), *Phys. kond. Mat.* **1**, 78. [*512*]
Busch, G. & Yuan, S. (1963), *Phys. kond. Mat.* **1**, 37. [*299, 301*]
Buschert, R. C. *see* Henninger, E. H.
Butler, J. A. V. (1932), *Proc. Roy. Soc.* A **135**, 348. [*479*]
Cabane, B. & Friedel, J. (1971), *J. de Physique*, **32**, 73. [*82, 129*]
Cabane, B. & Froidevaux, C. (1969), *Phys. Lett.* **29A**, 512. [*303*]
Caglioti, G., Corchia, M. & Rizzi, G. (1967), *Nuovo Cim.* **49B**, 222. [*132*]
Caglioti, G., Corchia, M. & Rizzi, G. (1969), *Nuovo Cim.* **60B**, 48. [*146*]
Caglioti, G. *see* Ascarelli, P.
Calandra, C. *see* Bortolani, V.
Carbotte, J. & Kahana, S. (1965), *Phys. Rev.* A **139**, 213. [*290*]
Carlson, F. F. *see* McMillan, R. G.
Carnahan, N. F. & Starling, K. E. (1969), *J. Chem. Phys.* **51**, 635. [*96*]
Case, K. M. *see* Aarnodt, R.
Catterall, J. A. & Trotter, J. (1963), *Phil. Mag.* **8**, 897. [*397, 398*]
Catterall, J. A. *see* Kubaschewski, O.
Cech, R. E. *see* Turnbull, D.
Chan, T. & Ballentine, L. E. (1971*a*), *Phys. Lett.* **35A**, 385. [*268, 270, 328*]
Chan, T. & Ballentine, L. E. (1971*b*), *Phys. Chem. Liq.* **2**, 165. [*268*]
Chapman, S. & Cowling, T. G. (1939), *The Mathematical Theory of Non-Uniform Gases*, Cambridge: University Press. [*172*]
Chen, S. H., Eder, O. J., Egelstaff, P. A., Haywood, B. C. G. & Webb, F. J. (1965), *Phys. Lett.* **19**, 269. [*203*]
Chizhevskaya, S. N. *see* Glazov, V. M.
Chollet, L. F. & Templeton, I. M. (1968), *Phys. Rev.* **170**, 656. [*432*]
Christensen, N. E. (1971), *Phys. Lett.* **35A**, 206. [*403*]
Christman, J. R. (1967*a*), *Phys. Rev.* **153**, 217 & 225. [*443, 470*]
Christman, J. R. (1967*b*), *Phys. Rev.* **159**, 108. [*443*]
Christman, J. R. & Huntington, H. B. (1965), *Phys. Rev.* **139**, A 83. [*443, 470*]
Claffey, W. *see* Brumberger, H.
Clark, W. G. *see* Warren, W. W.
Clay, J. (1940), *Physica*, **7**, 838. [*521*]
Clayton, G. T. *see* Kim, Y. S.
Cocking, S. J. (1967), *Adv. Phys.* **16**, 189. [*203, 206*]
Cocking, S. J. (1969), *J. Phys.* C **2**, 2047. [*180, 197*]
Cocking, S. J. & Egelstaff, P. A. (1965), *Phys. Lett.* **16**, 130. [*202*]
Cocking, S. J. & Egelstaff, P. A. (1968), *J. Phys.* C **1**, 507. [*203*]
Cocking, S. J. & Heard, C. R. T. (1965), *AERE Report, R 5016*, London: H.M. Stationery Office. [*125*]
Cohen, M. H. (1962), *J. Phys. Rad.* **23**, 643. [*38*]
Cohen, M. H. & Heine, V. (1970), *Solid State Physics* (ed. Ehrenreich, Seitz & Turnbull) **24**, 37, New York: Academic Press. [*34, 37*]
Cohen, M. H. & Turnbull, D. (1959), *J. Chem. Phys.* **31**, 1164. [*162*]
Cohen, M. L. *see* Allen, P. B.
Colburn, R. P. *see* Kanda, F. A.
Coles, B. R. (1958), *Adv. Phys.* **7**, 40. [*330*]
Collings, E. W. (1965), *Phys. kond. Mat.* **3**, 335. [*299*]

Collings, E. W. (1967), *Adv. Phys.* **16**, 459. [*495*]
Collings, E. W. *see* Enderby, J. E.
Collings, E. W. *see* Van Zytveld, J. B.
Collins, R. (1965), *Proc. Phys. Soc.* **86**, 199. [*84*]
Colver, C. P. *see* Vadovic, C. J.
Comins, N. R. (1972), *Phil. Mag.* **25**, 817. [*386, 392, 514*]
Cooper, J. R. A. *see* West, R. N.
Corchia, M. *see* Caglioti, G.
Cordes, M. R. *see* Shyu, W.-M.
Cornell, D. A. (1967), *Phys. Rev.* **153**, 208. [*216, 218, 305*]
Cornell, E. K. & Slichter, C. P. (1969), *Phys. Rev.* **180**, 358. [*493*]
Cotts, R. M. *see* Murday, J. S.
Covington, E. J. & Montgomery, D. J. (1957), *J. Chem. Phys.* **27**, 1030. [*168*]
Cowling, T. G. *see* Chapman, S.
Crisp, V. H. C. & Cusack, N. E. (1969), *Phys. Lett.* **28A**, 712. [*342, 351*]
Crisp, V. H. C., Cusack, N. E. & Kendall, P. W. (1970), *J. Phys.* C **3** (Metal Phys. Suppl.) S102. [*342*]
Croxton, C. A. (1969), *Ph.D. Thesis*, Cambridge University. [*90*]
Cubiotti, G., Giuliano, E. & Ruggeri, R. (1971*a*), *Nuovo Cim.* **3B**, 193. [*327, 337*]
Cubiotti, G., Giuliano, E. & Ruggeri, R. (1971*b*), *Nuovo Cim.* **6B**, 37. [*327, 330, 337*]
Cumming, P. A. *see* Mott, B. W.
Cundall, J. A. *see* Black, P. J.
Curien, H., Rimsky, A. & Defrain, A. (1962), *Bull. Soc. Franç. Minér. Crist.* **84**, 260. [*80*]
Cusack, N. E. (1963), *Rep. Progr. Phys.* **26**, 361. [*312*]
Cusack, N. E. & Kendall, P. W. (1963), *Phil. Mag.* **8**, 157. [*511*]
Cusack, N. E., Kendall, P. W. & Marwaha, A. S. (1962), *Phil. Mag.* **7**, 1745. [*337*]
Cusack, N. E. *see* Crisp, V. H. C.
Cusack, N. E. *see* Postill, D. R.
Cusack, N. E. *see* West, R. N.
Cutler, M. (1971), *Phil. Mag.* **24**, 381. [*538*]
Cutler, M. & Field, M. B. (1968), *Phys. Rev.* **169**, 642. [*528*]
Cutler, M. & Mallon, C. E. (1966), *Phys. Rev.* **144**, 642. [*528, 535, 536*]
Cutler, M. & Peterson, R. L. (1970), *Phil. Mag.* **21**, 1033. [*528*]
Cutler, M. *see* Donally, J. M.
Damle, P. S., Sjölander, A. & Singwi, K. S. (1968), *Phys. Rev.* **165**, 277. [*178*]
Dancy, E. A. (1965), *Trans. Met. Soc. AIME*, **233**, 270. [*528, 541*]
Daniel, E. (1960), *J. Phys. Chem. Sol.* **13**, 353. [*503*]
Daniel, E. *see* Blandin, A.
Dasannacharya, B. A. & Rao, K. R. (1965), *Phys. Rev.* **137**, A 417. [*180, 190*]
Dasannacharya, B. A. *see* Venkataraman, G.
Dashevskii, M. Ya., Kukuladze, G. V., Lasarev, V. B. & Mirgalovskii, M. S. (1967), *Inorganic Materials*, **3**, 1360. [*480*]

Davies, H. A. (1968), *Phys. Chem. Liq.* **1**, 179. [*533*]
Davies, H. A. & Leach, J. S. Ll. (1969), *Phil. Mag.* **19**, 1271. [*468, 520*]
Davies, H. A., Leach, J. S. Ll. & Draper, P. H. (1971), *Phil. Mag.* **23**, 1163. [*511*]
Davies, W. J. & Evans, E. J. (1930), *Phil. Mag.* **10**, 569. [*521*]
Davis, C. M. *see* Jarsynski, J.
Davis, E. A. *see* Mott, N. F.
Davis, K. G. & Fryzuk, P. (1965), *Trans. Met. Soc. AIME*, **233**, 1662. [*482*]
Davis, L. A. & Gordon, R. B. (1967), *J. Chem. Phys.* **46**, 2650. [*148*]
Dean, P. *see* Eisenschitz, R.
De Dycker, E. & Phariseau, P. (1965), *Physica*, **31**, 1337. [*226*]
De Dycker, E. & Phariseau, P. (1967), *Adv. Phys.* **16**, 401. [*231*]
Defrain, A. *see* Bosio, L.
Defrain, A. *see* Curien, H.
Degenkolbe, J. & Sauerwald, F. (1952), *Z. anorg. Chem.* **270**, 317. [*484, 485*]
De Paz, M. *see* Berne, A.
Desai, R. C. & Yip, S. (1968), *Phys. Rev.* **166**, 129. [*178*]
Desai, R. C. & Yip, S. (1969), *Phys. Rev.* **180**, 299. [*178*]
Devine, R. A. B. & Dupree, R. (1970*a*), *Phil. Mag.* **21**, 787. [*298*]
Devine, R. A. B. & Dupree, R. (1970*b*), *Phil. Mag.* **22**, 1069. [*298*]
Devine, R. A. B. *see* Helman, J. S.
Dickey, J. M., Meyer, A. & Young, W. H. (1966), *Phys. Rev. Lett.* **16**, 727. [*427, 517*]
Dickey, J. M., Meyer, A. & Young, W. H. (1967), *Proc. Phys. Soc.* **92**, 460. [*327, 329, 345*]
Dickey, J. M. *see* Epstein, S. G.
Dickson, E. M. *see* Rossini, F. A.
Dilger, H. *see* Steeb, H.
Dingle, R. B. (1953), *Physica*, **19**, 729. [*384*]
Döge, G. (1965), *Z. Naturforsch.* **20A**, 634. [*482*]
Donally, J. M. & Cutler, M. (1968), *Phys. Rev.* **176**, 1003. [*512*]
Doniach, S. (1970), *Adv. Phys.* **18**, 819. [*62*]
Dorner, B., Plesser, Th. & Stiller, H. (1965), *Physica*, **31**, 1537. [*202*]
Dorner, B., Plesser, Th. & Stiller, H. (1967), *Disc. Farad. Soc.* **43**, 160. [*203*]
Downey, M. E. *see* Mott, B. W.
Draisma, G. G. *see* Van der Molen, S. B.
Drakin, S. I. & Maltsev, A. K. (1957), *Zhur. Fiz. Khim.* **31**, 2036. [*488*]
Drakin, S. I. *see* Mikhaylov, V. A.
Draper, P. H. *see* Davies, H. A.
Drickamer, H. G. *see* Winter, F. R.
Duffill, C. *see* Egelstaff, P. A.
Dugdale, J. S., Gugan, D. & Okumara, K. (1961), *Proc. Roy. Soc.* A **263**, 407. [*169*]
Duggin, M. J. (1969), *Phys. Lett.* **29A**, 470. [*338*]
Dupree, R. & Seymour, E. F. W. (1970), *Phys. kond. Mat.* **12**, 97. [*299*]
Dupree, R. *see* Devine, R. A. B.

Dutchak, Ya. I., Stets'kiv, O. P. & Klyus, I. P. (1966), *Sov. Phys. Sol. State*, **8**, 455. [*511*]

Dworin, L. (1965), *Phys. Rev.* **138**, 1121. [*226*]

Eastman, D. E. (1971), *Phys. Rev. Lett.* **26**, 1108. [*403*]

Ebisawa, H. *see* Fukuyama, H.

Eder, O. J. *see* Chen, S. H.

Edwards, S. F. (1958), *Phil. Mag.* **3**, 1020. [*357*]

Edwards, S. F. (1961), *Phil. Mag.* **6**, 617. [*229, 253*]

Edwards, S. F. (1962), *Proc. Roy. Soc.* A **267**, 518. [*230, 253, 264, 320*]

Edwards, S. F. (1965), *Proc. Phys. Soc.* **85**, 1. [*253*]

Edwards, S. F. *see* Beeby, J. L.

Egelstaff, P. A. (1963), *Inelastic Scattering of Neutrons in Solids and Liquids*, **1**, 65, Vienna: International Atomic Energy Authority. [*195*]

Egelstaff, P. A. (1965), (editor) *Thermal Neutron Scattering*, London: Academic Press. [*156, 198, 206*]

Egelstaff, P. A. (1966), *Rep. Progr. Phys.* **29**, 333. [*180, 191, 197*]

Egelstaff, P. A. (1967), *An Introduction to the Liquid State*, London: Academic Press. [*120*]

Egelstaff, P. A., Duffill, C., Rainey, V., Enderby, J. E. & North, D. M. (1966), *Phys. Lett.* **21**, 286. [*146*]

Egelstaff, P. A., Page, D. I. & Heard, C. R. T. (1969), *Phys. Lett.* **30A**, 376. [*109*]

Egelstaff, P. A. *see* Chen, S. H.

Egelstaff, P. A. *see* Cocking, S. J.

Egelstaff, P. A. *see* Enderby, J. E.

Egelstaff, P. A. *see* North, D. M.

Egelstaff, P. A. *see* Page, D. I.

Egelstaff, P. A. *see* Wignall, G. D.

Ehrenreich, H. *see* Schwartz, L.

Eisenschitz, R. & Dean, P. (1957), *Proc. Phys. Soc.* A **70**, 713. [*230*]

Eisenschitz, R. *see* Sah, P.

Eldridge, H. B. *see* Ritchie, R. H.

Eldridge, J. M., Miller, E. & Komarek, K. L. (1967), *Trans. Met. Soc. AIME*, **239**, 570 and 775. [*474*]

El-Hanany, U. & Zamir, D. (1969), *Phys. Rev.* **183**, 809. [*303*]

Enderby, J. E. & Collings, E. W. (1970), *J. Non-Cryst. Sol.* **4**, 161. [*530, 535, 539*]

Enderby, J. E., Hasan, S. B. & Simmons, C. J. (1967), *Adv. Phys.* **16**, 667. [*510, 512, 514, 528*]

Enderby, J. E. & Howe, R. A. (1968), *Phil. Mag.* **18**, 923. [*532*]

Enderby, J. E. & March, N. H. (1965), *Adv. Phys.* **14**, 453. [*137, 138, 143, 144*]

Enderby, J. E. & March, N. H. (1966), *Proc. Phys. Soc.* **88**, 717. [*106*]

Enderby, J. E. & North, D. M. (1968), *Phys. Chem. Liq.* **1**, 1. [*435, 436, 463*]

Enderby, J. E., North, D. M. & Egelstaff, P. A. (1966), *Phil. Mag.* **14**, 961. [*463, 466, 467, 533*]

Enderby, J. E. & Simmons, C. J. (1969), *Phil. Mag.* **20**, 125. [*512, 513, 528, 538*]

Enderby, J. E., Titman, J. M. & Wignall, G. D. (1964), *Phil. Mag.* **10**, 633. [*298*]
Enderby, J. E. & Walsh, L. (1966), *Phil. Mag.* **14**, 991. [*410, 512, 527, 528, 534*]
Enderby, J. E., Van Zytveld, J. B., Howe, R. A. & Mian, A. J. (1968), *Phys. Lett.* **28A**, 144. [*530, 531*]
Enderby, J. E., *see* Egelstaff, P. A.
Enderby, J. E. *see* Howe, R. A.
Enderby, J. E. *see* North, D. M.
Enderby, J. E. *see* Page, D. I.
Enderby, J. E. *see* Van Zytveld, J. B.
Enderby, J. E. *see* Wingfield, B. F.
Endo, H. *see* Honda, K.
Endo, H. (1963), *Phil. Mag.* **8**, 1403. [*134, 148, 339*]
Endo, H. *see* Takeuchi, S.
Endo, H. *see* Tsuji, K.
Endo, H. *see* Watabe, M.
Entress, H. *see* Steeb, S.
Epstein, S. G. (1967), *Adv. Phys.* **16**, 325. [*486*]
Epstein, S. G. (1968), *Phys. Chem. Liq.* **1**, 109. [*486*]
Epstein, S. G. & Dickey, J. M. (1970), *Phys. Rev.* B **1**, 2442. [*486*]
Epstein, S. G. & Paskin, A. (1967), *Phys. Lett.* **24A**, 309. [*489*]
Eshelby, J. D. (1954), *J. Appl. Phys.* **25**, 255. [*445*]
Evans, E. J. *see* Davies, W. J.
Evans, R. (1970), *J. Phys.* C **3**, S137. [*327, 337, 340, 343, 523, 534*]
Evans, R. (1971). *Phys. Chem. Liq.* **2**, 249. [*327, 330, 337*]
Evans, R., Greenwood, D. A. & Lloyd, P. (1971), *Phys. Lett.* **35A**, 57 [*330*]
Evans, R., Greenwood, D. A., Lloyd, P. & Ziman, J. M. (1969), *Phys. Lett.* **30A**, 313. [*40*]
Evans, W. A. B. (1966), *Proc. Phys. Soc.* **88**, 723. [*411*]
Even, U. & Jortner, J. (1972*a*), *Phys. Rev. Lett.* **28**, 31. [*413*]
Even, U. & Jortner, J. (1972*b*), *Phil. Mag.* **25**, 715. [*413*]
Evse'ev, A. M. *see* Alekse'ev, N. V.
Faber, T. E. (1963), *Sol. State Comm.* **1**, 41. [*216*]
Faber, T. E. (1966), *Adv. Phys.* **15**, 547. [*229, 253, 316, 320, 321, 357, 359*]
Faber, T. E. (1967), *Adv. Phys.* **16**, 637. [*459, 501, 517*]
Faber, T. E. & Smith, N. V. (1968), *J. Opt. Soc. Am.* **58**, 102. [*383, 392*]
Faber, T. E. & Ziman, J. M. (1965), *Phil. Mag.* **11**, 153. [*438, 441, 455, 515*]
Faber, T. E. *see* Bradley, C. C.
Fadikov, I. *see* Kikoin, I.
Falicov, L. M. *see* Allen, P. B.
Falicov, L. M. *see* Stark, R. W.
Fan, H. Y. (1951), *Phys. Rev.* **82**, 900. [*235*]
Faulkner, J. S. (1964), *Phys. Rev.* **135**, A 124. [*229*]
Faulkner, J. S. & Korringa, J. (1961), *Phys. Rev.* **122**, 390. [*225*]
Fenninger, H. (1914), *Die Electrische Leitfähigkeit und innere Reibung Verdünntner Amalgame*, University of Freiburg. [*521*]
Ferrell, R. P. (1958), *Phys. Rev.* **111**, 1214. [*405*]

Ferrier, R. P. & Herrell, D. J. (1969), *Phil. Mag.* **19**, 853. [*530*, *535*]

Fessler, R. R., Kaplow, R. & Averbach, B. L. (1966), *Phys. Rev.* **150**, 34. [*110*, *133*]

Field, M. B. *see* Cutler, M.

Fielder, M. L. (1967), *Adv. Phys.* **16**, 681. [*533*]

Fiks, V. B. (1964), *Sov. Phys. Sol. State*, **5**, 2549. [*492*]

Filippov, L. P. (1968), *Internat. J. Heat Mass Transfer*, **11**, 331. [*338*]

Finney, J. L. (1970), *Proc. Roy. Soc.* A **319**, 479 & 495. [*78*]

Finney, J. L. *see* Bernal, J. D.

Fiorani, M. *see* Oleari, L.

Fischer, D. W. & Baun, W. L. (1965), *Phys. Rev.* **138**, A 1047. [*398*, *399*]

Fisher, I. Z. (1964), *Statistical Theory of Liquids* (transl. Switz, suppl. by Rice, S. A. & Gray, P.), Chicago: University Press. [*107*, *109*, *138*]

Fletcher, N. H. (1967*a*), *Adv. Phys.* **16**, 703. [*231*]

Fletcher, N. H. (1967*b*), *Proc. Phys. Soc.* **91**, 724. [*231*]

Flynn, C. P. & Odle, R. L. (1963), *Proc. Phys. Soc.* **81**, 412. [*29*]

Flynn, C. P., Rigney, D. A. & Gardner, J. A. (1967), *Phil. Mag.* **15**, 1255. [*497*, *499*]

Flynn, C. P. & Seymour, E. F. W. (1959), *Proc. Phys. Soc.* **73**, 945. [*303*]

Flynn, C. P. & Seymour, E. F. W. (1960), *Proc. Phys. Soc.* **76**, 301. [*303*]

Flynn, C. P. *see* Blodgett, J. A.

Flynn, C. P. *see* Gardner, J. A.

Flynn, C. P. *see* Odle, R. L.

Flynn, C. P. *see* Rigert, J. A.

Flynn, C. P. *see* Rigney, D. A.

Flynn, C. P. *see* Stupian, G. W.

Foley, W. T. & Liu, M. T. H. (1964), *Canad. J. Chem.* **42**, 2607. [*483*]

Foley, W. T. & Reid, L. E. (1963), *Canad. J. Chem.* **41**, 1782. [*483*]

Fornazero, J. & Mesnard, G. (1967), *J. Phys.* **28**, 221. [*226*]

Forstmann, F. (1967), *Z. Phys.* **203**, 495. [*385*]

Franck, E. U. & Hensel, F. (1966), *Phys. Rev.* **147**, 109. [*351*]

Franck, E. U. *see* Hensel, F.

Franck, E. U. *see* Renkert, H.

Freedman, J. F. & Robertson, W. D. (1961), *J. Chem. Phys.* **34**, 769. [*517*]

Freeman, A. J. *see* Watson, R. E.

Friedel, J. (1954), *Adv. Phys.* **3**, 446. [*22*, *427*, *429*, *444*]

Friedel, J. (1969), *Comments on Sol. State Phys.* **2**, 21. [*394*]

Friedel, J. *see* Blandin, A.

Friedel, J. *see* Cabane, B.

Friedman, L. (1971), *J. Non-Cryst. Sol.* **6**, 329. [*414*, *538*]

Frisch, H. & Lloyd, S. P. (1960), *Phys. Rev.* **120**, 1175. [*229*]

Froidevaux, C. *see* Cabane, B.

Frolova, G. M. *see* Mikhaylov, V. A.

Frost, B. R. T. (1954), *Progress in Metal Physics*, **5**, 96, Oxford: Pergamon Press. [*132*]

Frost, B. R. T. *see* Smallman, R. E.

Fryzuk, P. *see* Davis, K. G.

Fuchs, R. *see* Kliewer, K. L.

Fukui, Y. & Morita, T. (1970), *J. Phys.* C **3**, 1839. [*178*]
Fukushima, J. *see* Tsuji, K.
Fukuyama, H., Ebisawa, H. & Wada, Y. (1939), *Progr. Theor. Phys.* **42**, 494. [*411*]
Furukawa, K. (1960), *Sci. Rep. Tohoku Univ.* **12**, 368. [*132, 162*]
Furukawa, K. (1962), *Rep. Progr. Phys.* **25**, 395. [*132*]
Gahn, U. *see* Braunbek, W.
Gardner, J. A. & Flynn, C. P. (1967), *Phil. Mag.* **15**, 1233. [*497, 499*]
Gardner, J. A. *see* Flynn, C. P.
Gaskell, T. & March, N. H. (1963), *Phys. Lett.* **7**, 169. [*29, 286*]
Gaskell, T. *see* Bratby, P.
Gaspari, G. D. *see* Shyu, W-M.
Gebhardt, E. M., Becker, M. & Sebastian, H. (1955*a*), *Z. Metallk.* **46**, 669. [*484*]
Gebhardt, E. M., Becker, M. & Tragner, E. (1955*b*), *Z. Metallk.* **46**, 90. [*484*]
Gebhardt, E. M., Becker, M. & Kostlin, K. (1956), *Z. Metallk.* **47**, 684. [*485*]
Geissler, E. *see* Rossini, F. A.
Gel'man, E. V. *see* Kikoin, I. K.
Germer, D. & Mayer, H. (1968), *Z. Phys.* **210**, 391. [*87*]
Gerstenkorn, H. (1952), *Ann. Phys.* **10**, 49. [*188, 313*]
Gingrich, N. S. & Heaton, L. (1961), *J. Chem. Phys.* **34**, 873. [*129*]
Ginsbarg, A. *see* James, H.
Giraud, H. J. *see* Orr, R. L.
Giuliano, E. *see* Cubiotti, G.
Glagoleva, N. N. *see* Glazov, V. M.
Glazov, V. M., Chizhevskaya, S. N. & Glagoleva, N. N. (1969), *Liquid Semiconductors*, New York: Plenum Press. [*82, 535*]
Glazov, V. M., Glagoleva, N. N. & Romantseva, L. A. (1966), *Inorganic Materials*, **2**, 1498. [*484*]
Glazov, V. M., Krestovnikov, A. N. & Glagoleva, N. N. (1965), *Dokl. Akad. Nauk. SSSR*, **161**, 629. [*497*]
Godel, D. *see* Steeb, S.
Golovinskiy, L. P. *see* Rudenko, A. G.
Gordon, R. B. (1961), *Physical Chemistry of Process Metallurgy*, part I, p. 461, New York: Interscience Publishers. [472]
Gordon, R. B. *see* Abowitz, G.
Gordon, R. B. *see* Davis, L. A.
Gordon, W. L. *see* Shepherd, J. P. G.
Graham, T. P. *see* Jones, W. H.
Gray, P. (1964), *Molec. Phys.* **7**, 235. [*173*]
Gray, P. *see* Rice, S. A.
Green, M. S. (1960), *Handbuch der Physik*, **10**, 1, Berlin: Springer-Verlag. [*139*]
Greenfield, A. J. (1964), *Phys. Rev.* **135**, A 1589. [*411*]
Greenfield, A. J. (1966), *Phys. Rev. Lett.* **16**, 6. [*134, 136, 331*]
Greenfield, A. J., Wellendorf, J. & Wiser, N. (1971), *Phys. Rev.* A **4**, 1607 [*127, 146*]

Greenwood, D. A. (1958), *Proc. Phys. Soc.* **71**, 585. [*353*]
Greenwood, D. A. (1966), *Proc. Phys. Soc.* **87**, 775. [*280*]
Greenwood, D. A. *see* Evans, R.
Greenwood, D. A. *see* Ross, R. G.
Grosse, A. V. (1964), *J. Inorg. Nucl. Chem.* **26**, 1349. [*87*]
Grosse, A. V. (1966), *J. Inorg. Nucl. Chem.* **28**, 795. [*338, 350*]
Gubanov, A. I. (1965), *Quantum Electronic Theory of Amorphous Conductors*, New York: Consultants Bureau. [*3, 229*]
Gugan, D. *see* Dugdale, J. S.
Guggenheim, E. A. (1960), *Elements of the Kinetic Theory of Gases*, Oxford: Pergamon Press. [*92, 163, 166*]
Guggenheim, J. (1970), *Phil. Mag.* **22**, 833. [*515*]
Guggenheim, J. *see* Busch, G.
Guinier, A. (1963), *X-Ray Diffraction*, San Francisco: W. H. Freeman. [*118*]
Güntherodt, H-J. & Künzi, H. U. (1969), *Phys. kond. Mat.* **10**, 285. [*510*]
Güntherodt, H-J., Menth, A. & Tièche, Y. (1966), *Phys. kond. Mat.* **5**, 392. [*495, 510, 511*]
Güntherodt, H-J & Thièche, Y. (1968), *Helv. Phys. Acta*, **41**, 855. [*516*]
Güntherodt, H-J. *see* Busch, G.
Gupta, Y. P. (1966), *Acta Met.* **14**, 297. [*482*]
Gupta, Y. P. (1967), *Adv. Phys.* **16**, 333. [*482*]
Gurney, R. W. *see* Mott, N. F.
Gustafson, D. R., Mackintosh, A. R. & Zaffarano, D. J. (1963), *Phys. Rev.* **130**, 1455. [*292*]
Gyorffy, B. L. (1970), *Phys. Rev.* B **1**, 3290.
Haeberlin, H. *see* Menth, A.
Haeffner, E. (1953), *Nature*, **172**, 775. [*490*]
Halder, N. C. (1969), *Phys. Rev.* **177**, 471. [*502*]
Halder, N. C. (1970), *J. Phys. Chem. Sol.* **31**, 2281. [*503*]
Halder, N. C., Metzger, R. J. & Wagner, C. N. J. (1966), *J. Chem. Phys.* **45**, 1259. [*462, 519*]
Halder, N. C. & Wagner, C. N. J. (1966), *J. Chem. Phys.* **45**, 482. [325]
Halder, N. C. & Wagner, C. N. J. (1967*a*), *Z. Naturforsch.* **22A**, 1489. [*462*]
Halder, N. C. & Wagner, C. N. J. (1967*b*), *J. Chem. Phys.* **47**, 4385. [*468, 519*]
Halder, N. C. & Wagner, C. N. J. (1968), *Z. Naturforsch.* **23A**, 992. [*520*]
Halder, N. C. *see* Jena, P.
Halder, N. C. *see* Wagner, C. N. J.
Halperin, B. I. (1965), *Phys. Rev.* **139**, A 104. [*229*]
Halperin, B. I. (1967), *Adv. Chem. Phys.* **13**, 123. [*226, 272*]
Halperin, B. I. & Lax, M. (1966), *Phys. Rev.* **148**, 722. [*229*]
Halperin, B. I. & Lax, M. (1967), *Phys. Rev.* **153**, 802. [*229*]
Ham, F. S. (1962), *Phys. Rev.* **128**, 82 and 2524. [*51*]
Hanabusa, M. & Bloembergen, N. (1966), *J. Phys. Chem. Sol.* **27**, 363. [*216, 499*]

Harrison, M. J. *see* Melnyk, A. R.
Harrison, R. J. *see* Ascarelli, P.
Harrison, W. A. (1958), *Phys. Rev.* **110**, 514. [*440*]
Harrison, W. A. (1960), *Acta Met.* **8**, 168. [*440*]
Harrison, W. A. (1966), *Pseudopotentials in the Theory of Metals*, New York: Benjamin. [*29, 63, 67*]
Harrison, W. A. (1969), *Phys. Rev.* **181**, 1036. [*40*]
Harrison, W. A. *see* Shaw, R. W.
Hartmann, W. M. (1971), *Phys. Rev. Lett.* **26**, 1640. [*106*]
Hasan, S. B. (1969), *Ph.D. Thesis*, Sheffield University. [*410*]
Hasan S. B. *see* Enderby, J. E.
Hasegawa, M. & Watabe, M. (1972), *J. Phys. Soc. Jap.* **32**, 14. [*67, 153*]
Havill, R. L. (1967), *Proc. Phys. Soc.* **92**, 945. [*499, 505*]
Haywood, B. C. G. *see* Chen, S. H.
Heard, C. R. T. *see* Cocking, S. J.
Heard, C. R. T. *see* Egelstaff, P. A.
Heaton, L. *see* Gingrich, N. S.
Heaton, L. *see* Henninger, E. H.
Hedin, L. & Lundqvist, H. (1969), *Solid State Physics* (ed. Seitz & Turnbull), **23**, 1, New York: Academic Press. [*19, 216, 295, 297*]
Heer, C. V. *see* Martin, B. D.
Heighway, J. & Seymour, E. F. W. (1971*a*), *Phys. kond. Mat.* **13**, 1. [*303*]
Heighway, J. & Seymour, E. F. W. (1971*b*), *J. Phys.* F **1**, 138. [*499, 506*]
Heine, V. (1969), *The Physics of Metals* (ed. Ziman), p. 41, Cambridge: University Press. [*48*]
Heine, V. (1970), *Solid State Physics* (ed. Ehrenreich, Seitz & Turnbull), **24**, 1, New York: Academic Press. [*34, 35*]
Heine, V. & Abarenkov, I. (1964), *Phil. Mag.* **9**, 451. [*336*]
Heine, V, Nozières, P. & Wilkins, J. W. (1966), *Phil. Mag.* **13**, 741. [*18*]
Heine, V. & Weaire, D. (1966), *Phys. Rev.* **152**, 603. [*18, 63, 69*]
Heine, V. & Weaire, D. (1970), *Solid State Physics* (ed. Ehrenreich, Seitz & Turnbull), **24**, 249, New York: Academic Press. [*18, 19, 21, 40, 64, 69*]
Heine, V. *see* Animalu, A. O. E.
Heine, V. *see* Ballentine, L. E.
Heine, V. *see* Cohen, M. H.
Heine, V. *see* Knight, W. D.
Helland, J. *see* Landauer, R.
Helman, J. S. & Devine, R. A. B. (1971), *Phys. Rev.* B **4**, 1153 & 1156. [*493*]
Henninger, E. H., Buschert, R. C. & Heaton, L. (1966), *J. Chem. Phys.* **44**, 1758. [*466*]
Hensel, F. (1970), *Phys. Lett.* **31A**, 88. [*352*]
Hensel, F. & Franck, E. U. (1968), *Rev. Mod. Phys.* **40**, 697. [*351*]
Hensel, F. *see* Franck, E. U.
Hensel, F. *see* Renkert, H.
Hensel, F. *see* Schmutzler, R.

Henshaw, D. G. (1956), *Phys. Rev.* **105**, 976. [*129*]
Herrell, D. J. *see* Ferrier, R. P.
Herring, C. (1950), *J. Appl. Phys.* **21**, 437. [*154*]
Herring, C. (1960), *J. Appl. Phys.* **31**, 1939. [*509*]
Hewitt, R. R. & Taylor, T. T. (1962), *Phys. Rev.* **125**, 524. [*219*]
Hewitt, R. R. & Williams, B. F. (1963), *Phys. Rev.* **129**, 1188. [*219*]
Hewitt, R. R. & Williams, B. F. (1964), *Phys. Rev. Lett.* **12**, 216. [*219, 308*]
Hewitt, R. R. *see* Adams, J. E.
Hewitt, R. R. *see* Knight, W. D.
Hezel, R. *see* Steeb, S.
Hietel, B. *see* Mayer, H.
Hill, J. E. & Ruoff, A. L. (1965), *J. Chem. Phys.* **43**, 2150. [*472*]
Hiroike, K. (1965), *Phys. Rev.* **138**, A 422. [*226*]
Hoar, T. P. & Melford, D. A. (1957), *Trans. Farad. Soc.* **53**, 315. [*479, 480*]
Hockman, J. M. & Bonilla, C. F. (1964), *Trans. Nucl. Soc.* **7**, 101. [*345*]
Hodgson, J. N. (1959), *Phil. Mag.* **4**, 183. [*392*]
Hodgson, J. N. (1960), *Phil. Mag.* **5**, 272. [*386*]
Hodgson, J. N. (1961), *Phil. Mag.* **6**, 509. [*386*]
Hodgson, J. N. (1962), *Phil. Mag.* **7**, 229. [*386*]
Hodgson, J. N. (1966), *Optical Properties and Electronic Structure of Metals and Alloys* (ed. Abelès), p. 60, Amsterdam: North-Holland. [*297, 380*]
Hodgson, J. N. (1967), *Adv. Phys.* **16**, 675. [*514*]
Höhler, J. *see* Steeb, S.
Holcomb, D. F. & Norberg, R. E. (1955), *Phys. Rev.* **98**, 1074. [*215*]
Holstein, T. (1952), *Phys. Rev.* **88**, 1427. [*384*]
Honda, K. & Endo, H. (1927), *J. Inst. Met.* **37**, 29. [*496*]
Hoover, W. G. & Ree, F. H. (1968), *J. Chem. Phys.* **49**, 3609. [*94*]
Hopfield, J. J. (1965), *Phys. Rev.* **139**, A 419. [*368*]
Hopfield, J. J. (1965), *Comments on Sol. State Phys.* **2**, 40. [*394*]
Hori, J. (1964), *Progr. Theor. Phys.* **32**, 371 & 471. [*226*]
Hori, J. (1966), *Suppl. Progr. Theor. Phys.* **36**, 3. [*226*]
Hori, J. (1968*a*), *Spectral Properties of Disordered Chains and Lattices*, Oxford: Pergamon. [*226*]
Hori, J. (1968*b*), *J. Phys.* A **1**, 314. [*226*]
Hori, J. & Matsuda, H. (1964), *Progr. Theor. Phys.* **32**, 183. [*226*]
Howe, R. A. & Enderby, J. E. (1967), *Phil. Mag.* **16**, 467. [*336*]
Howe, R. A. *see* Enderby, J. E.
Hücke, E. E. *see* Angus, J. C.
Hultgren, R. R. & Orr, R. L. (1967), *Rév. Int. Hautes Temp. et Réfr.* **4**, 123. [*452, 476*]
Hultgren, R. R., Orr, R. L., Anderson, P. D. & Kelley, K. K. (1963), *Selected Values of Thermodynamic Properties of Metals and Alloys*, New York: Wiley. [*79, 469*]
Hultgren, R. R. *see* Orr, R. L.
Hunderi, O. *see* Beaglehole, D.
Huntington, H. B. *see* Christman, J. R.

Hutchinson, P. *see* Johnson, M. D.

Hutner, R. A. *see* Saxon, S. D.

Ibramigov, K. I. *see* Pokrovsky, N. L.

Ichikawa, K. & Shimoji, M. (1969), *Phil. Mag.* **19**, 33. [*518*]

Ichikawa, K. & Shimoji, M. (1970*a*), *Phil. Mag.* **22**, 873. [*518, 534*]

Ichikawa, K. & Shimoji, M. (1970*b*), *Phys. Chem. Liq.* **2**, 115. [*518, 534*]

Igumnova, A. A. *see* Korol'kov, A. M.

Ilschener, B. B. & Wagner, C. N. J. (1958), *Acta Met.* **6**, 712. [*529*]

Ioffe, A. F. & Regel, A. R. (1960), *Progress in Semiconductors*, **4**, 239, New York: Wiley. [*535*]

Isherwood, S. P. & Orton, B. R. (1968), *Phil. Mag.* **17**, 561. [*132, 325*]

Isherwood, S. P. & Orton, B. R. (1969), *J. Appl. Cryst.* **2**, 219. [*462*]

Isherwood, S. P. & Orton, B. R. (1970), *Phys. Lett.* **31A**, 164. [*463*]

Itami, T. & Shimoji, M. (1972), *Phil. Mag.* **25**, 229. [*432*]

Jalickee, J. E., Morika, T. & Tanaka, T. (1965), *Phil. Mag.* **12**, 209. [*231*]

James, H. & Ginsbarg, A. (1953), *J. Phys. Chem.* **57**, 840. [*227*]

James, R. W. (1945), *The Optical Principles of the Diffraction of X-Rays*, London: G. Bell. [*118*]

Jarayaman, A., Newton, R. C. & McDonough, J. M. (1967), *Phys. Rev.* **159**, 527. [*343, 344*]

Jarzynski, J. & Litowitz, T. A. (1964), *J. Chem. Phys.* **41**, 1290. [*472*]

Jarzynski, J., Smirnow, J. R. & Davis, C. M. (1969), *Phys. Rev.* **178**, 288. [*149*]

Jean, P. (1965), *C.R. Acad. Sci.* **260**, 2465. [*404*]

Jena, P. & Halder, N. C. (1971), *Phys. Rev. Lett.* **26**, 1024. [*268, 270, 300, 308*]

Johnson, M. D., Hutchinson, P. & March, N. H. (1964), *Proc. Roy. Soc.* A **282**, 283. [*144, 145*]

Jones, B. K. *see* Watabe, M.

Jones, H. *see* Mott, N. F.

Jones, P. L. *see* Roberts, A. P.

Jones, W. H., Graham, T. P. & Barnes, R. G. (1960), *Acta Met.* **8**, 663. [*303*]

Jortner, J. *see* Even, U.

Joshi, M. L. & Wagner, C. N. J. (1965), *Z. Naturforsch.* **20A**, 564. [*463*]

Joshi, M. L. *see* Wagner, C. N. J.

Kaeck, J. A. (1968), *Phys. Rev.* **175**, 897. [*499*]

Kahana, S. *see* Carbotte, J.

Kanda, F. A. & Colburn, R. P. (1968), *Phys. Chem. Liq.* **1**, 159. [*484*]

Kaneyoshi, T. *see* Matsubara, T.

Kaplan, M. *see* Kleppa, O. J.

Kaplow, R., Strong, S. L. & Averback, B. L. (1965*a*), *Phys. Rev.* **138**, A 1336. [*132*]

Kaplow, R., Strong, S. L. & Averbach, B. L. (1965*b*), *M.I.T. Report DSR-7954*. [*118, 464*]

Kaplow, R. *see* Fessler, R. R.

Kasowski, R. V. (1969), *Phys. Rev.* **187**, 891. [*308*]

Kasowski, R. V. *see* Allen, P. B.

Kaufman, S. M. & Whalen, T. J. (1965), *Acta Met.* **13**, 797. [*480*]

Kawakami, S. (1951), *Nippon Kinzoku Gakkai-Si* (B) **15**, 59, 303, 528 & 620. [*492*]
Kawakami, S. (1955), *Nippon Kinzoku Gakkai-Si* (B) **19**, 322. [*492*]
Keating, D. J. (1963), *J. Appl. Phys.* **34**, 923. [*465*]
Keeton, S. C. & Loucks, T. L. (1966), *Phys. Rev.* **152**, 548. [*328*]
Keller, J. (1971), *J. Phys.* C **4**, 3143. [*231*]
Kelley, K. K. *see* Hultgren, R.
Kellington, S. H. & Titman, J. M. (1967), *Phil. Mag.* **15**, 1045. [*499*]
Kellington, S. H. *see* Titman, J. M.
Kendall, P. W. (1968), *Phys. Chem. Liq.* **1**, 33. [*336*]
Kendall, P. W. *see* Crisp, V. H. C.
Kendall, P. W. *see* Cusack, N. E.
Kent, C. V. (1919), *Phys. Rev.* **14**, 459. [*514*]
Khabibullin, B. M. *see* Alekseyeva, T. O.
Kharakhashyan, E. G. *see* Alekseyeva, T. O.
Khar'kov, Ye. I. (1966), *Ukr. Fiz. Zh.* **11**, 677. [*489*]
Khar'kov, Ye. I. *see* Rudenko, A. G.
Khodov, Z. L. (1960), *Phys. Met. Metallography*, **10** (5), 129. [*472*]
Khov, K. E. *see* Roberts, A. P.
Kikoin, I. & Fadikov, I. (1935), *Phys. Z. Sowjet.* **7**, 507. [*414*]
Kikoin, I. K., Senchenkov, A. P., Gel'man, E. V., Korsunskii, M. M. & Naurzakov, S. P. (1966), *Sov. Phys. JETP*, **22**, 89. [*351*]
Kikoin, I. K. & Senchenkov, A. P. (1967), *Phys. Met. Metallography*, **24** (5), 74. [*351*]
Kilby, G. E. *see* Young, W. H.
Kim, M. G. & Letcher, S. V. (1971), *J. Chem. Phys.* **55**, 1164. [*472, 524*]
Kim, Y. S., Standley, C. L., Kruh, R. F. & Clayton, G. T. (1961), *J. Chem. Phys.* **34**, 1464. [*462*]
Kindig, N. B. & Spicer, W. E. (1965), *Phys. Rev.* **128**, 1622. [*401*]
King, G. J. *see* McMillan, R. G.
Kirkwood, J. G. & Buff, F. P. (1951), *J. Chem. Phys.* **19**, 774. [*438*]
Kittel, C. (1958), *Elementary Statistical Physics*, New York: Wiley. [*187*]
Kittel, C. (1963), *Quantum Theory of Solids*, New York: Wiley. [*26*]
Kjeldaas, T. & Kohn, W. (1957), *Phys. Rev.* **105**, 806. [*300*]
Klauder, J. R. (1961), *Annals of Phys.* **14**, 43. [*229*]
Kleinman, L. *see* Phillips, J. C.
Klemm, A. (1958), *Z. Naturforsch.* **13**A, 1039. [*469, 489*]
Klemm, W. (1950), *Landolt-Bornstein, Zahlenwerte & Funktionen* 6. Auflage, Bd. **1**/**1**, 395. [*299*]
Klemm, W. *see* Bomm, B.
Kleppa, O. J. (1958), *Liquid Metals & Solidification*, p. 56, Cleveland, Ohio: American Society for Metals. [*469*]
Kleppa, O. J., Kaplan, M. & Thalmeyer, C. E. (1961), *J. Phys. Chem.* **65**, 843. [*472*]
Kliewer, K. L. & Fuchs, R. (1968), *Phys. Rev.* **172**, 607. [*385*]
Kliewer, K. L. & Fuchs, R. (1969), *Phys. Rev.* **181**, 552. [*286, 368*]
Klyus, I. P. *see* Dutchak, Ya. I.
Knight, W. D. (1956), *Solid State Physics* (ed. Seitz & Turnbull), **2**, 93, New York: Academic Press. [*303*]

Knight, W. D., Berger, A. G. & Heine, V. (1959), *Annals of Phys.* **8**, 173. [*303*]

Knight, W. D., Hewitt, R. R. & Pomeranz, M. (1956), *Phys. Rev.* **104**, 271. [*219*]

Knight, W. D. *see* Rossini, F. A.

Knol, J. S. *see* Van der Lugt, W.

Kohn, W. *see* Kjeldaas, T.

Kohn, W. *see* Lang, N. D.

Komarek, K. L. *see* Eldridge, J. M.

Komarek, K. L. *see* Miller, E.

Kopp, W. V. *see* Wachtel, E.

Korol'kov, A. M. & Igumnova, A. A. (1961), *Izv. Akad. Nauk. SSSR, Met. i Topl.* (6), 95. [*480, 481*]

Korringa, J. *see* Faulkner, J. S.

Korsunskii, M. M. *see* Kikoin, I. K.

Kostlin, K. *see* Gebhardt, E. M.

Koyama, R. Y. & Spicer, W. E. (1971), *Phys. Rev.* B **4**, 4318. [*403, 404*]

Koyama, R. Y., Spicer, W. E., Ashcroft, N. W. & Lawrence, W. E. (1967), *Phys. Rev. Lett.* **19**, 1284. [*403*]

Kozlov, V. A. *see* Asanovich, V. Y.

Krainova, I. F. *see* Shpil'rain, E. E.

Krestovnikov, A. N. *see* Glazov, V. M.

Krishnan, K. S. *see* Bhatia, A. B.

Kruh, R. F. (1962), *Chem. Rev.* **62**, 319. [*132*]

Kruh, R. F. *see* Kim, Y. S.

Kubaschewski, O. (1943), *J. Physik. Chem.* **192**, 292. [*476*]

Kubaschewski, O. & Catterall, J. A. (1956), *Thermochemical Data of Alloys*, London, New York: Pergamon. [*469*]

Kubaschewski, O. & Slough, W. (1969), *Progress in Materials Science* (ed. Chalmers & Hume-Rothery), **14**, 3, Oxford: Pergamon. [*469*]

Kukuladze, G. V. *see* Dashevskii, M. Ya.

Kumar, R. (1965), *Trans. Ind. Inst. Met.* **18**, 131. [*492*]

Kumar, R., Singh, M. & Sivaramakrishnan, C. S. (1967), *Trans. Met. Soc. AIME*, **239**, 1219. [*492*]

Kumar, R. *see* Singh, M.

Künzi, H. U. *see* Güntherodt, H.-J.

Künzi, H. U. *see* Busch, G.

Kurkijärvi, J. (1967), *Physica*, **35**, 143. [*195*]

Kusmiss, J. H. & Stewart, A. T. (1967), *Adv. Phys.* **16**, 471. [*292*]

Lackmann-Cyrot, F. (1964), *Phys. kond. Mat.* **3**, 75. [*307*]

Lambert, M. *see* Taupin, C.

Landau, L. D. & Lifshitz, E. M. (1958), *Statistical Physics*, Oxford: Pergamon. [*107*]

Landauer, R. & Helland, J. (1954), *J. Chem. Phys.* **22**, 1655. [*227*]

Lang, N. D. & Kohn, W. (1970), *Phys. Rev.* B **1**, 4555. [*86*]

Lang, N. D. & Kohn, W. (1971), *Phys. Rev.* B **3**, 1215. [*13, 404*]

Langreth, D. C. *see* Ashcroft, N. W.

Lasarev, V. B. (1964), *Russ. J. Phys. Chem.* **38**, 172. [*480*]

Lasarev, V. B. & Malov, Yu. I. (1965*a*), *Dokl. Akad. Nauk. SSSR*, **161**, 875. [*481*]

Lasarev, V. B. & Malov, Yu. I. (1965 *b*), *Dokl. Akad. Nauk. SSSR*, **164**, 846. [*481*]

Lasarev, V. B. *see* Dashevskii, M. Ya.

Lasarev, V. B. *see* Malov, Yu. I.

Lawrence, W. E. *see* Ashcroft, N. W.

Lawrence, W. E. *see* Koyama, R. Y.

Lax, M. & Phillips, J. C. (1958), *Phys. Rev.* **110**, 41. [*229*]

Lax, M. *see* Halperin, B. I.

Leach, J. S. Ll. *see* Davies, H. A.

Lebowitz, J. L. (1964), *Phys. Rev.* **133**, A 895. [*435*]

Lekner, J. *see* Ashcroft, N. W.

Lelyuk, L. G., Shklyarevskii, I. N. & Yarovaya, R. G. (1964), *Opt. & Spectr.* **16**, 263. [*386, 392*]

Letcher, S. V. *see* Kim, M. G.

Levesque, D. & Verlet, L. (1970), *Phys. Rev.* A **2**, 2514. [*92, 174*]

Lien, W. H. & Phillips, N. E. (1964), *Phys. Rev.* **133**, A 1370. [*52*]

Lien, S. Y. & Sivertsen, J. M. (1969), *Phil. Mag.* **20**, 759. [*331, 339*]

Lifshitz, E. M. *see* Landau, L. D.

Lifshitz, I. M. (1964), *Adv. Phys.* **13**, 483. [*229*]

Lindemann, F. A. (1910), *Phys. Zeits.* **11**, 609. [*103*]

Litowitz, T. A. *see* Jarsynski, J.

Liu, M. T. H. *see* Foley, W. T.

Lloyd, P. (1967), *Proc. Phys. Soc.* **90**, 207 and 217. [*231*]

Lloyd, P. (1969), *J. Phys.* C **2**, 1717. [*230*]

Lloyd, P. & Berry, M. V. (1967), *Proc. Phys. Soc.* **91**, 678. [*231*]

Lloyd, P. *see* Evans, R.

Lloyd, S. P. *see* Frisch, H.

Lodding, A. R. E. (1963), *J. Chim. Phys.* **60**, 254. [*490*]

Lodding, A. R. E. (1966), *Z. Naturforsch.* **21A**, 1348. [*492*]

Lodding, A. R. E. (1967), *J. Phys. Chem. Sol.* **28**, 557. [*490*]

Lodding, A. R. E., Mundy, J. N. & Ott, A. (1970), *Phys. Stat. Solidi*, **38**, 559. [*169*]

Lodding, A. R. E. & Ott, A. (1966), *Z. Naturforsch.* **21A**, 1344. [*492*]

Lodding, A. R. E. *see* Löwenberg, L.

Lodding, A. R. E. *see* Ott, A.

Longuet-Higgins, H. C. & Pople, J. A. (1956), *J. Chem. Phys.* **25**, 884. [*163, 165, 171, 207*]

Longuet-Higgins, H. C. & Widom, B. (1964), *Molec. Phys.* **8**, 549. [*94, 95*]

Loucks, T. L. *see* Keeton, S. C.

Love, H. M. *see* McCracken, G. M.

Löwenberg, L. & Lodding, A. R. E. (1967), *Z. Naturforsch.* **22A**, 2077. [*169*]

Löwenberg, L. Nordén-Ott, A. & Lodding, A. R. E. (1968), *Z. Naturforsch.* **23A**, 1771. [*492*]

Lukes, T. (1965), *Phil. Mag.* **12**, 719. [*231*]

Lundén, A. *see* Ott, A.

Lundén, A. *see* Randsalu, A.
Lundqvist, H. *see* Hedin, L.
Luttinger, J. M. (1951), *Philips Res. Rep.* **6**, 303. [*225*]
McCracken, G. M. & Love, H. M. (1960), *Phys. Rev. Lett.* **5**, 201. [*169*]
MacDonald, D. K. C. (1959), *Phil. Mag.* **4**, 1283. [*312*]
McDonald, I. R. & Singer, K. (1967), *Disc. Farad. Soc.* **43**, 40. [*92*]
McDonough, J. M. *see* Jarayaman, A.
McGervey, J. D. (1967), *Proc. Positron Annihilation Conf., Detroit*, New York: Academic Press. [*292*]
Mackintosh, A. R. *see* Gustafson, D. R.
McMillan, R. G., King, G. J., Miller, B. S. & Carlson, F. F. (1962), *J. Phys. Chem. Sol.* **23**, 1379. [*298*]
McMillan, W. L. *see* Anderson, P. W.
Magnaterra, A. *see* Bortolani, V.
Maier, U. *see* Steeb, S.
Majumdar, C. K. (1965), *Phys. Rev.* A **140**, 237. [*290*]
Makinson, R. E. B. & Roberts, A. P. (1960), *Austr. J. Phys.* **13**, 437. [*227*]
Makinson, R. E. B. *see* Roberts, A. P.
Mallon, C. E. *see* Cutler, M.
Malov, Yu. I. & Lasarev, V. B. (1968), *Phys. Met. Metallography*, **26** (5), 171. [*481*]
Malov, Yu. I. *see* Lasarev, V. B.
Maltsev, A. K. *see* Drakin, S. I.
March, N. M. (1966), *Phys. Lett.* **20**, 231. [*73*]
March, N. M. (1968), *Liquid Metals*, Oxford: Pergamon. [*63, 145*]
March, N. M. *see* Ashcroft, N. W.
March, N. M. *see* Bratby, P.
March, N. M. *see* Brown, R. C.
March, N. M. *see* Enderby, J. E.
March, N. M. *see* Gaskell, T.
March, N. M. *see* Johnson, M. D.
March, N. M. *see* Rousseau, J.
Martin, B. D., Zych, D. A. & Heer, C. V. (1964), *Phys. Rev.* **135**, A 671. [*52*]
Martin, D. L. (1959), *Physica*, **25**, 1193. [*169*]
Martin, D. L. (1961), *Proc. Roy. Soc.* A **263**, 378. [*52*]
Marwaha, A. S. (1967), *Adv. Phys.* **16**, 617. [*336*]
Marwaha, A. S. *see* Cusack, N. E.
Matsubara, T. & Kaneyoshi, T. (1968), *Progr. Theor. Phys.* **40**, 1257. [*414, 538*]
Matsuda, H. (1962), *Progr. Theor. Phys.* **27**, 811. [*226*]
Matsuda, H. (1966), *Suppl. Progr. Theor. Phys.* **36**, 97. [*226*]
Matsuda, H. & Okada, K. (1965), *Progr. Theor. Phys.* **34**, 539. [*226*]
Matsuda, H. *see* Hori, H.
Mattuck, R. D. (1962), *Phys. Rev.* **127**, 738. [*229*]
Matuyama, Y. (1927), *Sci. Rep. Tohoku Univ.* **16**, 447. [*516, 526*]
Mayer, H. & Hietel, B. (1966), *Optical Properties and Electronic Structure of Metals and Alloys* (ed. Abelès), p. 47, Amsterdam: North-Holland. [*386, 387*]

Mayer, H. *see* Germer, D.
Maziers, C. *see* Taupin, C.
Melford, D. A. *see* Hoar, T. P.
Melnyk, A. R. & Harrison, M. J. (1970), *Phys. Rev.* B **2**, 835 & 851. [*385*]
Menth, A. & Haeberlin, H. (1967), *Helv. Phys. Acta*, **40**, 366. [*497*]
Menth, A. & Wullschleger, J. (1967), *Helv. Phys. Acta*, **40**, 820. [*495*]
Menth, A. *see* Güntherodt, H.-J.
Mesnard, G. *see* Fornazero, J.
Metzger, R. J. *see* Halder, N. C.
Meyer, A., Nestor, C. W. & Young, W. H. (1967), *Proc. Phys. Soc.* **92**, 446. [*24, 28, 34, 327, 380*]
Meyer, A. *see* Dickey, J. M.
Meyer, A. *see* Young, W. H.
Mian, A. J. *see* Enderby, J. E.
Micah, E. T., Stocks, G. M. & Young, W. H. (1969), *J. Phys.* C **2**, 1653 & 1661. [*37*]
Mikhaylov, V. A., Polovinkina, R. A., Drakin, S. I. & Frolova, G. M. (1966), *Phys. Met. Metallography*, **22** (6), 63. [*486*]
Mikolaj, P. G. & Pings, C. J. (1967), *J. Chem. Phys.* **46**, 1401 & 1412. [*92*]
Miller, A. & Abrahams, E. (1960), *Phys. Rev.* **120**, 745. [*273*]
Miller, B. S. *see* McMillan, R. G.
Miller, E., Paces, J. & Komarek, K. L. (1964), *Trans. Met. Soc. AIME*, **230**, 1557. [*524, 528*]
Miller, E. *see* Eldridge, J. M.
Miller, J. C. (1969), *Phil. Mag.* **20**, 1115. [*383, 386, 396, 400, 401*]
Mirgalovskii, M. S. *see* Dashevskii, M. Ya.
Montgomery, D. J. *see* Ban, N. T.
Montgomery, D. J. *see* Covington, E. J.
Montgomery, D. J. *see* Robertson, V. M.
Moore, D. S. *see* Brown, D.
Morgan, G. J. (1969), *J. Phys.* C **2**, 1454. [*57*]
Morgan, G. J. & Ziman, J. M. (1967), *Proc. Phys. Soc.* **91**, 689. [*231*]
Moriarty, J. A. (1970), *Phys. Rev.* B **1**, 1363. [*40, 327, 337*]
Morika, T. *see* Jalickee, J. E.
Morita, T. *see* Fukui, Y.
Mott, B. W. (1968), *AERE – Bib 151 A*, London: H.M. Stationery Office. [*421, 494, 515*]
Mott, B. W., Downey, M. E. & Cumming, P. A. (1966), *AERE – Bib 151*, London: H.M. Stationery Office. [*421, 494, 515*]
Mott, N. F. (1934), *Proc. Roy. Soc.* A **146**, 465. [*82, 312*]
Mott, N. F. (1962), *Rep. Progr. Phys.* **25**, 218. [*445*]
Mott, N. F. (1964), *Adv. Phys.* **13**, 325. [*330*]
Mott, N. F. (1966), *Phil. Mag.* **13**, 989. [*328, 392, 522*]
Mott, N. F. (1967), *Adv. Phys.* **16**, 49. [*229, 230, 272, 328*]
Mott, N. F. (1968), *Phil. Mag.* **17**, 1259. [*230, 352*]
Mott, N. F. (1969), *Phil. Mag.* **19**, 835. [*60, 230, 352*]
Mott, N. F. (1970), *Phil. Mag.* **22**, 7. [*230*]

Mott, N. F. (1971), *Phil. Mag.* **24**, 1. [*352, 541*]
Mott, N. F. & Allgaier, R. S. (1967), *Phys. Stat. Solidi*, **21**, 343. [*229*]
Mott, N. F. & Davies, E. A. (1971), *Electronic Processes in Non-crystalline Materials*, Oxford: Clarendon Press. [*230, 310, 424, 541*]
Mott, N. F. & Gurney, R. W. (1939), *Trans. Farad. Soc.* **35**, 364. [*83*]
Mott, N. F. & Jones, H. (1936), *The Theory of the Properties of Metals and Alloys*, Oxford: Clarendon Press. [*82, 321, 380*]
Mott, N. F. & Twose, W. D. (1961), *Adv. Phys.* **10**, 107. [*228, 244, 274*]
Motyka, R. J. *see* Bennett, H. E.
Moulson, D. J. & Seymour, E. F. W. (1967), *Adv. Phys.* **16**, 449. [*499, 501, 502, 504, 505*]
Moulson, D. J. & Styles, G. A. (1967), *Phys. Lett.* **24A**, 438. [*499*]
Mueller, W. E. (1969), *J. Opt. Soc. Am.* **59**, 1246. [*391*]
Mueller, W. E. & Thompson, J. C. (1969), *Phys. Rev. Lett.* **23**, 1037. [*398*]
Müller, P. (1910), *Metallurgie*, **7**, 730. [*527*]
Mundy, J. N. *see* Lodding, A.
Mungurwadi, B. D. *see* Radhakrishna Setty, D. L.
Murday, J. S. & Cotts, R. M. (1968), *J. Chem. Phys.* **48**, 4938. [*159, 213*]
Muto, T. & Oyama, S. (1950), *Progr. Theor. Phys.* **5**, 833. [*235*]
Muto, T. & Oyama, S. (1951), *Progr. Theor. Phys.* **6**, 61. [*235*]
Nabarro, F. R. N. (1948), *Report on a Conference on the Strength of Solids*, p. 75, London: Physical Society. [*154*]
Nachtrieb, N. H. (1962), *J. Phys. Chem.* **66**, 1163. [*299*]
Nachtrieb, N. H. & Petit, J. (1956), *J. Chem. Phys.* **24**, 746. [*162, 168*]
Naghizadeh, J. & Rice, S. A. (1962), *J. Chem. Phys.* **36**, 2710. [*159*]
Nakagawa, Y. (1956), *J. Phys. Soc. Jap.* **11**, 855. [*301*]
Nakagawa, Y. (1969), *Phys. Lett.* **28A**, 494. [*498*]
Nakahara, Y. & Takahashi, H. (1966), *Proc. Phys. Soc.* **89**, 747. [*178*]
Nakamura, Y. & Shimoji, M. (1969), *Trans. Farad. Soc.* **65**, 1509. [*528*]
Naumov, A. A. & Ryskin, G. Ya. (1965), *Sov. Phys. Sol. State*, **7**, 558. [*169*]
Naurzakov, S. P. *see* Kikoin, I. K.
Nestor, C. W. *see* Meyer, A.
Nettel, S. (1966), *Phys. Rev.* **150**, 421. [*375*]
Newton, R. C. *see* Jarayaman, A.
Nikitin, B. S. *see* Alekseyeva, T. O.
Noguchi, S. *see* Takeuchi, T.
Norberg, R. E. *see* Holcomb, D. F.
North, D. M., Enderby, J. E. & Egelstaff, P. A. (1968), *J. Phys.* C **1**, 784 & 1075. [*126, 132, 134, 146, 186, 325, 330*]
North, D. M. & Wagner, C. N. J. (1969), *Phys. Lett.* **30A**, 440. [*325*]
North, D. M. & Wagner, C. N. J. (1970), *Phys. Chem. Liq.* **2**, 87. [*468*]
North, D. M. *see* Egelstaff, P. A.
North, D. M. *see* Enderby, J. E.
North, D. M. *see* Wagner, C. N. J.

Nozières, P. *see* Heine, V.
Nozières, P. *see* Pines, D.
Ocken, H. & Wagner, C. N. J. (1966), *Phys. Rev.* **149**, 122. [*325*]
Ocken, H. *see* Wagner, C. N. J.
Odle, R. L., Becker, G. & Sotier, S. (1969), *J. Phys. Chem. Sol.* **30**, 2479. [*497*, *499*]
Odle, R. L. & Flynn, C. P. (1965), *J. Phys. Chem. Sol.* **26**, 1685. [*303*]
Odle, R. L. & Flynn, C. P. (1956), *Phil. Mag.* **13**, 699. [*499*]
Odle, R. L. *see* Flynn, C. P.
Okada, K. *see* Matsuda, H.
Okumara, K. *see* Dugdale, J. S.
Oleari, L. & Fiorani, M. (1959), *Ric. Sci.* **29**, 2589. [*526*]
Olson, D. L., Blough, J. L. & Rigney, D. A. (1972), *Acta Met.* **20**, 305. [*489*]
Ookawa, A. (1960), *J. Phys. Soc. Jap.* **15**, 2191. [*84*]
Oriani, R. A. & Webb, M. B. (1959), *Acta Met.* **7**, 63. [*499*]
Orr, R. L., Giraud, H. J. & Hultgren, R. R. (1962), *Trans. Quarterly ASM*, **55**, 853. [*476*]
Orr, R. L. *see* Hultgren, R. R.
Orton, B. R. & Street, R. (1970), *J. Phys.* C **3**, L 143. [*462*]
Orton, B. R. & Williams, G. I. (1960), *Fulmer Institute Report R. 129/5*. [*463*]
Orton, B. R., Williams, G. I. & Shaw, B. A. (1960), *Acta Met.* **8**, 177. [*462*]
Orton, B. R. *see* Isherwood, S. P.
Orton, B. R. *see* Tsuji, K.
Ott, A. & Lodding, A. R. E. (1965), *Z. Naturforsch.* **20**A, 1578. [*159*]
Ott, A. & Lundén, A. (1964), *Z. Naturforsch.* **19**A, 822. [*492*]
Ott, A. *see* Lodding, A. R. E.
Ott, A. *see* Löwenberg, L.
Otter, M. (1961), *Z. Phys.* **5**, 539. [*396*]
Oyama, S. *see* Muto, T.
Paalman, H. H. & Pings, C. J. (1963), *Rev. Mod. Phys.* **35**, 389. [*127*]
Paasch, G. & Trepte, P. (1971), *Phys. Stat. Solidi.* (b) **44**, K 37. [*327*]
Paces, J. *see* Miller, E.
Page, D. I., Egelstaff, P. A., Enderby, J. E. & Wingfield, B. R. (1969), *Phys. Lett.* **29**A, 296. [*133*]
Page, D. I. *see* Engelstaff, P. A.
Palmer, R. E. & Schnatterly, S. E. (1971), *Phys. Rev.* B **4**, 2329. [*297*]
Paskin, A. (1967), *Adv. Phys.* **16**, 223. [*179*]
Paskin, A. & Rahman, A. (1966), *Phys. Rev. Lett.* **16**, 300. [*92*, *145*, *174*]
Paskin, A. *see* Ascarelli, P.
Paskin, A. *see* Epstein, S. G.
Pells, G. P. *see* Shiga, M.
Pendry, J. B. (1971), *J. Phys.* C **4**, 427. [*35*]
Peterson, R. L. *see* Cutler, M.
Petit, J. *see* Nachtrieb, N. H.
Phariseau, P. & Ziman, J. M. (1963), *Phil. Mag.* **8**, 1487. [*231*]
Phariseau, P. *see* De Dycker, E.

Phillips, J. C. & Kleinman, L. (1959), *Phys. Rev.* **116**, 287. [*34*]
Phillips, J. C. *see* Lax, M.
Phillips, N. E. *see* Lien, W. H.
Pines, D. & Nozières, P. (1966), *The Theory of Quantum Liquids*, New York: Benjamin. [*19, 300, 317, 371*]
Pines, D. *see* Bardeen, J.
Pings, C. J. *see* Mikolaj, P. G.
Pings, C. J. *see* Paalman, H. H.
Pings, C. J. *see* Rodriguez, S. E.
Piroue, P. A. *see* Wannier, G. H.
Plesser, Th. *see* Dorner, B.
Pokrovsky, N. L., Pugachevich, P. P. & Ibramigov, K. I. (1967), *Dokl. Akad. Nauk. SSSR*, **172**, 829. [*480*]
Polovinkina, R. A. *see* Mikhaylov, V. A.
Pomeranz, M. *see* Knight, W. D.
Pope, N. K. *see* Brockhouse, B. N.
Pople, J. A. *see* Longuet-Higgins, H. C.
Postill, D. R., Ross, R. G. & Cusack, N. E. (1967), *Adv. Phys.* **16**, 493. [*351*]
Postill, D. R., Ross, R. G. & Cusack, N. E. (1968), *Phil. Mag.* **18**, 519. [*148, 351*]
Powell, C. J. (1968), *Phys. Rev.* **175**, 972. [*401, 406*]
Powell, R. W. (1953), *Phil. Mag.* **44**, 772. [*330*]
Pratt, J. N. (1967), *Rév. Int. des Hautes Temp. et Réfr.* **4**, 97. [*469*]
Pratt, J. N. *see* Sellors, R. G. R.
Price, D. L. (1971), *Phys. Rev.* A **4**, 358. [*153*]
Pugachevich, P. P. *see* Pokrovsky, N. L.
Radhakrishna Setty, D. L. & Mungurwadi, B. D. (1971), *Phys. Lett.* **35**A, 11. [*499*]
Rahman, A. (1964*a*), *Phys. Rev.* **136**, A 405. [*92, 174, 177, 190, 195*]
Rahman, A. (1964*b*), *Phys. Rev. Lett.* **12**, 575. [*139*]
Rahman, A. (1966), *J. Chem. Phys.* **45**, 2585. [*92, 174, 176*]
Rahman, A. *see* Paskin, A.
Rainey, V. *see* Egelstaff, P. A.
Ramakrishnan, T. V. (1971), *Ph.D. Thesis*, Columbia University. [*411*]
Randall, C. M. *see* Ban, N. T.
Randolph, P. D. (1964), *Phys. Rev.* **134**, 1483. [*180, 197*]
Randolph, P. D. & Singwi, K. S. (1966), *Phys. Rev.* **152**, 99. [*198*, 202]
Randsalu, A. & Lundén, A. (1965), *Z. Naturforsch.* **20**A, 1081. [*336*]
Rao, K. R. *see* Dasannacharya, B. A.
Rao, K. R. *see* Venkataraman, G.
Rapoport, E. (1967), *Phys. Rev. Lett.* **19**, 345. [*345*]
Rayne, J. A. *see* Brandt, G. B.
Ree, F. H. *see* Hoover, W. G.
Regel, A. R. *see* Andre'ev, A. A.
Regel, A. R. *see* Ioffe, A. F.
Reid, L. E. *see* Foley, W. T.
Renkert, H., Hensel, F. & Franck, E. U. (1969), *Phys. Lett.* **30**A, 494. [*345*]

Reuter, G. E. H. & Sondheimer, E. H. (1948), *Proc. Roy. Soc.* A **195**, 336. [*366*]

Rice, M. J. (1970), *Phys. Rev.* B **2**, 4800. [*338*]

Rice, S. A. & Gray, P. (1965), *The Statistical Mechanics of Simple Liquids*, New York: Interscience. [*138, 141, 142, 164, 173, 174*]

Rice, S. A. *see* Berne, B. J.

Rice, S. A. *see* Bloch, A.

Rice, S. A. *see* Boiani, J.

Rice, S. A. *see* Naghizadeh, J.

Rice, S. A. *see* Wilson, E. G.

Ricker, T. & Schaumann, G. (1966), *Phys. kond. Mat.* **5**, 31. [*336*]

Rigamonti, A. *see* Borsa, F.

Rigert, J. A. & Flynn, C. P. (1971), *Phys. Rev. Lett.* **26**, 1177. [*429, 493*]

Rigney, D. A. & Blodgett, J. A. (1969), *J. Phys. Chem. Solids*, **30**, 2247. [*303*]

Rigney, D. A., Blodgett, J. A. & Flynn (1969), *Phil. Mag.* **20**, 907. [*498, 499*]

Rigney, D. A. & Flynn, C. P. (1967), *Phil. Mag.* **15**, 1213. [*499*]

Rigney, D. A. *see* Flynn, C. P.

Rigney, D. A. *see* Olson, D. L.

Rimai, L. & Bloembergen, N. (1960), *J. Phys. Chem. Sol.* **13**, 257. [*499, 505*]

Rimsky, A. *see* Curien, M.

Ritchie, R. H. & Eldridge, H. B. (1962), *Phys. Rev.* **126**, 1935. [*407*]

Rivlin, V. G., Waghorne, R. M. & Williams, G. I. (1966), *Phil. Mag.* **13**, 1169. [*132*]

Rivlin, V. G. *see* Waghorne, R. M.

Rizzi, G. *see* Caglioti, G.

Robert, J. *see* Benkirane, M.

Roberts, A. P. (1963), *Proc. Phys. Soc.* **81**, 990. [*227*]

Roberts, A. P. & Makinson, R. E. B. (1962), *Proc. Phys. Soc.* **79**, 630. [*226, 228*]

Roberts, A. P., Jones, P. L., Khov, K. E. & Smith, P. V. (1969), *J. Phys.* C **2**, 1502. [*229*]

Roberts, A. P., Jones, P. L. & Smith, P. V. (1968), *J. Phys.* C **1**, 549. [*229*]

Roberts, A. P. *see* Makinson, R. E. B.

Robertson, V. M. & Montgomery, D. J. (1960), *Phys. Rev.* **117**, 440. [*168*]

Robertson, W. D. & Uhlig, H. H. (1949), *Trans. Met. Soc. AIME*, **180**, 345. [*452*]

Robertson, W. D. *see* Friedman, J. F.

Rodriguez, S. E. & Pings, C. J. (1965), *J. Chem. Phys.* **42**, 2435. [*132*]

Romantseva, L. A. *see* Glazov, V. M.

Rosenbaum, M. *see* Aarnodt, R.

Ross, M. & Alder, B. J. (1966), *Phys. Rev. Lett.* **16**, 1077. [*94*]

Ross, R. G. & Greenwood, D. A. (1969), *Progress in Materials Science* (ed. Chalmers & Hume-Rothery), **14**, 3, Oxford: Pergamon. [*76, 345*]

Ross, R. G. (1971), *Phys. Lett.* **34**A, 183. [*345*]
Ross, R. G. *see* Postill, D. R.
Rossini, F. A., Geissler, E., Dickson, E. M. & Knight, W. D. (1967), *Adv. Phys.* **16**, 287. [*216*]
Rossini, F. A. & Knight, W. D. (1969), *Phys. Rev.* **178**, 641. [*216, 220*]
Rousseau, J., Stoddart, J. C. & March, N. H. (1970), *Proc. Roy. Soc.* A **317**, 211. [*231*]
Rowlinson, J. S. (1965), *Rep. Progr. Phys.* **28**, 169. [*141, 142*]
Rubin, A. G. *see* Ben-Yosef, N.
Rudenko, A. G., Golovinskiy, L. M. & Khar'kov, Ye. I. (1968), *Phys. Met. Metallography*, **25** (3), 193. [*486, 489*]
Ruggeri, R. *see* Cubiotti, G.
Ruoff, A. L. *see* Hill, J. E.
Ruppersberg, H. & Winterberg, K. H. (1971), *Phys. Lett.* **34**A, 11. [*330*]
Ryskin, G. Ya. *see* Naumov, A. A.
Sah, P. & Eisenschitz, R. (1960), *Proc. Phys. Soc.* **75**, 700. [*229*]
Samarin, A. M. *see* Vertman, A. A.
Sauerwald, F. *see* Degenkolbe, J.
Saxon, S. D. & Hutner, R. A. (1949), *Philips Res. Rep.* **4**, 81. [*225*]
Schaich, W. & Ashcroft, N. W. (1970), *Phys. Lett.* **31**A, 174. [*461, 539*]
Schaich, W. *see* Ashcroft, N. W.
Schaumann, G. *see* Ricker, T.
Schiff, D. (1969), *Phys. Rev.* **186**, 151. [*92, 133, 174, 180*]
Schiff, L. I. (1955), *Quantum Mechanics*, 2nd edition, New York: McGraw-Hill. [*56*]
Schmidt, H. (1957), *Phys. Rev.* **105**, 425. [*224*]
Schmutzler, R. & Hensel, F. (1968), *Phys. Lett.* **27**A, 587. [*342, 351*]
Schmutzler, R. & Hensel, F. (1971), *Phys. Lett.* **35**A, 55. [*342, 351*]
Schnatterly, S. E. *see* Palmer, R. E.
Schneider, T. & Stoll, E. (1966), *Phys. kond. Mat.* **5**, 331 & 364. [*38, 327*]
Schneider, T. & Stoll, E. (1967*a*), *Sol. State Comm.* **5**, 455. [*200*]
Schneider, T. & Stoll, E. (1967*b*), *Sol. State Comm.* **5**, 837. [*67*]
Schneider, T. & Stoll, E. (1967*c*), *Adv. Phys.* **16**, 731. [*268, 270*]
Schneider, T. *see* Stoll, E.
Schulz, L. G. (1957), *Adv. Phys.* **6**, 102. [*382, 386, 390, 514*]
Schuhmann, H. (1962), *Z. anorg. Chem.* **317**, 204. [*465*]
Schürmann, H. K. & Parks, R. D. (1971*a*), *Phys. Rev. Lett.* **26**, 367. [*525*]
Schürmann, H. K. & Parks, R. D. (1971*b*), *Phys. Rev. Lett.* **26**, 835. [*525*]
Schürmann, H. K. & Parks, R. D. (1971*c*), *Phys. Rev. Lett.* **27**, 1790. [*525*]
Schürmann, H. K. & Parks, R. D. (1972). To be published. [*525*]
Schwartz, L. & Ehrenreich, H. (1971), *Annals of Phys.* **64**, 100. [*57, 58*]
Sears, V. F. (1965), *Proc. Phys. Soc.* **86**, 953. [*178*]
Sebastian, H. *see* Gebhardt, E. M.
Sellors, R. G. R. & Pratt, J. N. (1970), *Phys. Chem. Liq.* **2**, 19. [*486*]

Semenchenko, V. K. (1961), *Surface Phenomena in Metals and Alloys*, Oxford: Pergamon. [*86, 90, 477, 480*]
Senchenkov, A. P. *see* Kikoin, I. K.
Seymour, E. F. W. & Styles, G. A. (1964), *Phys. Lett.* **10**, 269. [*303*]
Seymour, E. F. W. & Styles, G. A. (1966), *Proc. Phys. Soc.* **87**, 473. [*499*]
Seymour, E. F. W. see Allen, P. S.
Seymour, E. F. W. *see* Brown, D.
Seymour, E. F. W. *see* Dupree, R.
Seymour, E. F. W. *see* Flynn, C. P.
Seymour, E. F. W. *see* Heighway, J.
Seymour, E. F. W. see Moulson, D. J.
Shackle, P. W. (1970), *Phil. Mag.* **21**, 987. [*410*]
Sham, L. J. & Ziman, J. M. (1963), *Solid State Physics* (ed. Seitz & Turnbull), **2**, 221, New York: Academic Press. [*236*]
Sharma, K. C. (1968), *Phys. Rev.* **174**, 309. [*207*]
Sharma, P. K. *see* Srivastava, S. K.
Shaw, B. A. *see* Orton, B. R.
Shaw, R. W. (1968), *Phys. Rev.* **174**, 769. [*38*]
Shaw, R. W. (1969), *J. Phys.* C **2**, 2335 & 2350. [*51, 67, 68*]
Shaw, R. W. & Harrison, W. A. (1967), *Phys. Rev.* **163**, 604. [*40*]
Shaw, R. W. & Smith, N. V. (1969), *Phys. Rev.* **178**, 985. [*51, 266, 268–71, 298, 300, 308*]
Shepherd, J. P. G. & Gordon, W. L. (1968), *Phys. Rev.* **169**, 541. [*432*]
Shiga, M. & Pells, G. P. (1969), *J. Phys.* C **2**, 1835. [*401*]
Shiota, I. *see* Tamaki, S.
Shimoji, M. *see* Ichikawa, K.
Shimoji, M. *see* Itami, T.
Shimoji, M. *see* Nakumura, Y.
Shklyarevskii, I. N. *see* Lelyuk, L. G.
Sholl, C. A. (1967), *Proc. Phys. Soc.* **91**, 130. [*220*]
Shpil'rain, E. E. & Krainova, I. F. (1967), *High Temp.* (U.S.A.) **5**, 50. [*338*]
Shyu, W.-M. & Gaspari, G. D. (1967), *Phys. Rev.* **163**, 667. [*67*]
Shyu, W.-M. & Gaspari, G. D. (1968), *Phys. Rev.* **170**, 687. [*67*]
Shyu, W.-M. & Gaspari, G. D. (1969), *Phys. Lett.* **30A**, 53. [*67*]
Shyu, W.-M., Singwi, K. S. & Tosi, M. P. (1971*a*), *Phys. Rev.* B **3**, 237. [*67*]
Shyu, W.-M., Wehling, J. H., Cordes, M. R. & Gaspari, G. D. (1971*b*), *Phys. Rev.* B **4**, 1802. [*67*]
Simmons, C. J. *see* Enderby, J. E.
Singer, K. *see* McDonald, I. R.
Singh, M. & Kumar, R. (1966), *Trans. Ind. Inst. Met.* **19**, 117. [*492*]
Singh, M. *see* Kumar, R.
Singwi, K. S. (1964), *Phys. Rev.* **136**, A 969. [*195*]
Singwi, K. S. *see* Damle, P. S.
Singwi, K. S. *see* Randolph, P. D.
Singwi, K. S. *see* Shyu, W.-M.
Sinha, O. P. (1972) *Phys. Lett.* **38A**, 193. [*489*]

Sivaramakrishnan, C. S. *see* Kumar, R.
Sivertsen, J. M. *see* Lien, S. Y.
Sjölander, A. *see* Damle, P. S.
Skaupy, F. (1914), *Verhandl. deut. Phys. Ges.* **16**, 156. [*488*]
Skaupy, F. (1920), *Phys. Zeits.* **21**, 597. [*521*]
Skinner, H. W. B. (1946), *Phil. Trans.* A **239**, 95. [*395, 398*]
Skold, K. (1967), *Phys. Rev. Lett.* **19**, 1023. [*195*]
Slichter, C. P. (1963), *Principles of Magnetic Resonance*, New York: Harper & Row. [*213*]
Slichter, C. P. *see* Cornell, E. K.
Slough, W. *see* Kubaschewski, O.
Smallman, R. E. & Frost, B. R. T. (1956), *Acta Met.* **4**, 611. [*462*]
Smirnow, J. R. *see* Jarzynski, J.
Smit, W. *see* Van der Molen, S. B.
Smith, L. E. & Stromberg, R. T. (1966), *J. Opt. Soc. Am.* **56**, 1539. [*392*]
Smith, N. V. (1967*a*), *Phys. Rev.* **163**, 553. [374]
Smith, N. V. (1967*b*), *Adv. Phys.* **16**, 629. [*386*]
Smith, N. V. (1968), *Phys. Lett.* **26**A, 126. [*392*]
Smith, N. V. (1969), *Phys. Rev.* **183**, 634. [*375, 387, 388*]
Smith, N. V. & Spicer, W. E. (1969), *Phys. Rev.* **188**, 593. [*407*]
Smith, N. V. & Traum, M. M. (1970), *Phys. Rev. Lett.* **25**, 1017. [*403*]
Smith, N. V. *see* Faber, T. E.
Smith, N. V. *see* Shaw, R. W.
Smith, P. V. *see* Roberts, A. P.
Smith, R. A. (1964), *Semiconductors*, Cambridge: University Press. [*346*]
Smith, T. (1967), *J. Opt. Soc. Am.* **57**, 1207. [*383, 392*]
Somekh, E. M. *see* Bates, L. F.
Sommer, F. *see* Braunbek, W.
Sondheimer, E. H. *see* Reuter, G. E. H.
Sotier, S. *see* Odle, R. L.
Spicer, W. E. *see* Berglund, C. N.
Spicer, W. E. *see* Kindig, N. B.
Spicer, W. E. *see* Koyama, R. Y.
Spicer, W. E. *see* Smith, N. V.
Springer, B. (1964), *Phys. Rev.* A **136**, 115. [*281, 328*]
Springer, B. (1967), *Phys. Rev.* A **154**, 614 & 622. [*411*]
Srivastava, S. K. & Sharma, P. K. (1969*a*), *Indian J. Pure Appl. Phys.* **7**, 644. [*268*]
Srivastava, S. K. & Sharma, P. K. (1969*b*), *Nuovo Cim. Lett.* **1**, 698. [*330*]
Sryvalin, I. I. *see* Asanovich, V. F.
Standley, C. L. *see* Kim, Y. S.
Stark, R. W. & Falicov, L. M. (1967), *Phys. Rev. Lett.* **19**, 795. [*38*]
Starling, K. E. *see* Starling, N. F.
Staver, T. *see* Bohm, D.
Steeb, S., Dilger, H. & Höhler, J. (1969), *Phys. Chem. Liq.* **1**, 235. [*465*]
Steeb, S. & Entress, H. (1966), *Z. Metallk.* **57**, 803. [*464, 465, 528*]
Steeb, S. & Hezel, R. (1966), *Z. Metallk.* **57**, 374. [*465*]

Steeb, S., Maier, U. & Godel, D. (1969), *Phys. Chem. Liq.* **1**, 221. [*515*]
Steeb, S. & Woerner, S. (1965), *Z. Metallk.* **56**, 771. [*462*]
Stephens, R. W. B. (1963), *Dispersion and Absorption of Sound by Molecular Processes*, p. 693, New York: Academic Press. [*207*]
Stephens, R. W. B. *see* Webber, G. M. B.
Stern, E. A. (1967), *Phys. Rev.* **162**, 556. [*384*]
Stets'kiv, O. P. *see* Dutchak, Ya. I.
Stewart, A. T. (1964), *Phys. Rev.* **133A**, 1651. [*432*]
Stewart, A. T. *see* Kusmiss, J. H.
Stiller, H. *see* Dorner, B.
Stocks, G. M. *see* Micah, E. T.
Stoddart, J. C. *see* Rousseau, J.
Stoll, E., Szabo, N. & Schneider, T. (1971), *Phys. kond. Mat.* **12**, 279. [*268*]
Stoll, E. *see* Schneider, T.
Stoneburner, D. F. (1965), *Trans. Met. Soc. AIME*, **233**, 153. [*528*]
Street, R. *see* Orton, B. R.
Stromberg, R. T. *see* Smith, L. E.
Strong, S. L. *see* Kaplow, R.
Stupian, G. W. & Flynn, C. P. (1968), *Phil. Mag.* **17**, 295. [*498, 499*]
Styles, G. A. (1967), *Adv. Phys.* **16**, 275. [*503, 506*]
Styles, G. A. *see* Moulson, D. J.
Styles, G. A. *see* Seymour, E. F. W.
Sundström, L. H. (1965), *Phil. Mag.* **11**, 657. [*336*]
Suzuki, K. & Uemura, O. (1971), *J. Phys. Chem. Sol.* **32**, 1801. [*305*]
Suzuki, K. *see* Waseda, Y.
Szabo, N. (1971), *Phys. kond. Mat.* **13**, 118. [*320*]
Szabo, N. *see* Stoll, E.
Takahashi, H. *see* Nakahara, Y.
Takeuchi, S. & Endo, H. (1961), *Trans. Jap. Inst. Met.* **2**, 246. [*299*]
Takeuchi, S. & Endo, H. (1962), *Trans. Jap. Inst. Met.* **3**, 35. [*528*]
Takeuchi, T. (1971), *J. Phys. Soc. Jap.* **30**, 995. [*524*]
Takeuchi, T. & Noguchi, S. (1966), *J. Phys. Soc. Jap.* **21**, 2222. [*533*]
Tamaki, S. (1968*a*), *J. Phys. Soc. Jap.* **25**, 379. [*497, 534*]
Tamaki, S. (1968*b*), *J. Phys. Soc. Jap.* **25**, 1596. [*518, 534*]
Tamaki, S. & Shiota, I. (1966), *Phys. Lett.* **23**, 543. [*444, 470*]
Tanaka, M. *see* Watabe, M.
Tanaka, T. *see* Jalickee, J. E.
Taupin, C., Lambert, M. & Mazieres, C. (1971). *J. Phys. Chem. Sol.* **32**, 2045. [*298*]
Taylor, P. L. (1966*a*), *Proc. Phys. Soc.* **88**, 753. [*226*]
Taylor, P. L. (1966*b*), *Proc. Phys. Soc.* **90**, 233. [*226*]
Taylor, P. L. & Bambakidis, G. (1967), *Adv. Phys.* **16**, 409. [*230*]
Taylor, T. T. *see* Hewitt, R. R.
Templeton, I. M. *see* Chollet, L. F.
Thalmeyer, C. E. *see* Kleppa, O. J.
Thompson, J. C. *see* Mueller, W. E.
Thornton, D. E., Young, W. H., Van der Molen, S. B. & Van der Lugt, W. (1968), *Phys. Lett.* **27A**, 396. [*499, 503*]

Thornton, D. E. *see* Bhatia, A. B.
Tièche, Y. *see* Busch, G.
Tièche, Y. *see* Güntherodt, H.-J.
Timbie, J. P. & White, R. M. (1970), *Phys. Rev.* B **1**, 2409. [*300*]
Titman, J. M. & Kellington, S. H. (1967), *Proc. Phys. Soc.* **90**, 499. [*499*]
Titman, J. M. *see* Enderby, J. E.
Titman, J. M. *see* Kellington, S. H.
Tong, B. Y. & Tong, S. Y. (1969), *Phys. Rev.* **180**, 739. [*225*]
Tong, S. Y. *see* Tong, B. Y.
Toombs, G. A. (1965), *Phys. Lett.* **15**, 222. [*320*]
Tosi, M. P. *see* Shyu, W.-M.
Tougas, R. (1970), *Phys. Chem. Liq.* **2**, 13. [*532, 534*]
Tourand, G. & Breuil, M. (1971), *J. Physique*, **32**, 813. [*129*]
Tragner, E. *see* Gebhardt, E. M.
Traum, M. M. *see* Smith, N. V.
Trepte, P. *see* Paasch, G.
Trotter, J. *see* Catterall, J. A.
Tsuji, K., Fukushima, J., Oshima, R. & Endo, H. (1969), *Phys. Lett.* **30A**, 173. [*432*]
Turnbull, D. (1950), *J. Chem. Phys.* **18**, 769. [*91*]
Turnbull, D. & Cech, R. E. (1950), *J. Appl. Phys.* **21**, 804. [*78*]
Turnbull, D. *see* Cohen, M. H.
Turner, G. (1970), *Phil. Mag.* **21**, 257. [*481*]
Twose, W. D. *see* Mott, N. F.
Ubbelohde, A. R. (1966), *Proc. Roy. Soc.* A **293**, 291. [*336*]
Uemura, O. *see* Suzuki, K.
Uhlig, H. H. *see* Robertson, W. D.
Upthegrove, W. R. *see* Walls, H. A.
Vadovic, C. J. & Colver, C. P. (1970*a*), *Phil. Mag.* **21**, 971. [*175*]
Vadovic, C. J. & Colver, C. P. (1970*b*), *Phys. Rev.* B **1**, 4850. [*175*]
Vadovic, C. J. & Colver, C. P. (1971), *Phil. Mag.* **24**, 509. [*177*]
Van der Lugt, W. & Knol, J. S. (1967), *Phys. Stat. Sol.* **23**, K83. [*303*]
Van der Lugt, W. & Van der Molen, S. B. (1967), *Phys. Stat. Sol.* **19**, 327 [*499*]
Van der Lugt, W. *see* Thornton, D. E.
Van der Lugt, W. *see* Van der Molen, S. B.
Van der Molen, S. B., Van der Lugt, W., Draisma, G. G. & Smit, W. (1968), *Physica*, **38**, 275. [*499*]
Van der Molen, S. B. *see* Thornton, D. E.
Van der Molen, S. B. *see* Van der Lugt, W.
Van Zytveld, J. B., Enderby, J. E. & Collings, E. W. (1972), *J. Phys.* F **2**, 73. [*329*]
Van Zytveld, J. B. *see* Enderby, J. E.
Varley, J. H. O. (1954), *Phil. Mag.* **45**, 887. [*444*]
Venkataraman, G., Dasannacharya, B. A. & Rao, K. R. (1967), *Phys. Rev.* **161**, 133. [*195*]
Verhoeven, J. (1963), *Metallurgical Reviews*, **8**, 311. [*486, 489, 490*]
Verlet, L. (1967), *Phys. Rev.* **159**, 98. [*92*]
Verlet, L. (1968), *Phys. Rev.* **165**, 201. [*92, 142*]

Verlet, L. *see* Levesque, D.

Vertman, A. A., Samarin, A. M. & Yakobson, A. M. (1960), *Izv. Akad. Nauk. SSSR, Met. i Topl.* (3), 17. [*492*]

Von Rauschenplat, G. *see* Bornemann, K.

Wachtel, E. & Kopp, W. V. (1969), *Phys. Lett.* **29A**, 164. [*498*]

Wada, Y. *see* Fukuyama, H.

Wagenmann, G. *see* Bornemann, K.

Waghorne, R. M., Rivlin, V. G. & Williams, G. I. (1967), *Adv. Phys.* **16**, 215. [*464, 465*)

Waghorne, R. M., *see* Rivlin, V. G.

Wagner, C. N. J. & Halder, N. C. (1967), *Adv. Phys.* **16**, 241. [*519*]

Wagner, C. N. J., Halder, N. C. & North, D. M. (1967), *Phys. Lett.* **25A**, 663. [*468, 520*]

Wagner, C. N. J., Ocken, H. & Joshi, M. L. (1965), *Z. Naturforsch.* **20A**, 325. [*132*]

Wagner, C. N. J. *see* Halder, N. C.

Wagner, C. N. J. *see* Ilschener, B. B.

Wagner, C. N. J. *see* Joshi, M. L.

Wagner, C. N. J. *see* North, D. M.

Wagner, C. N. J. *see* Ocken, H.

Wainwright, T. E. *see* Alder, B. J.

Wallace, P. R. (1960), *Solid State Physics* (ed. Seitz & Turnbull), New York: Academic Press. [*288*]

Wallace, W. E. (1955), *J. Chem. Phys.* **23**, 2281. [*522*]

Walls, H. A. & Upthegrove, W. R. (1964), *Acta Met.* **12**, 461. [*159*]

Walsh, L. *see* Enderby, J. E.

Wannier, G. H. & Piroue, P. A. (1956), *Helv. Phys. Acta*, **29**, 221. [*85*]

Warren, W. W. (1970*a*), *J. Non-Cryst. Sol.* **4**, 168. [*506, 508, 541*]

Warren, W. W. (1970*b*), *Sol. State Comm.* **8**, 1269. [*506, 508, 541*]

Warren, W. W. & Clark, W. G. (1969), *Phys. Rev.* **177**, 600. [*217, 303, 506*]

Warren, W. W. & Wernick, J. H. (1971), *Phys. Rev.* B **4**, 1401. [*506*]

Waseda, Y. & Suzuki, K. (1970*a*), *Phys. Lett.* **31A**, 573. [*132*]

Waseda, Y. & Suzuki, K. (1970*b*), *Phys. Stat. Solidi*, **40**, 183. [*136*]

Waseda, Y. & Suzuki, K. (1971*a*), *Phys. Lett.* **34A**, 69. [*146*]

Waseda, Y. & Suzuki, K. (1971*b*), *Phys. Lett.* **35A**, 315. [*330*]

Watabe, M. & Tanaka, M. (1964), *Progr. Theor. Phys.* **31**, 525. [*268, 270*]

Watabe, M., Tanaka, M., Endo, H. & Jones, B. K. (1965), *Phil. Mag.* **12**, 347. [*305*]

Watabe, M. *see* Hasegawa, M.

Watson, R. E., Bennett, L. H. & Freeman, A. J. (1968), *Phys. Rev. Lett.* **20**, 653. [*503*]

Waung, H. F. *see* Brandt, W.

Weaire, D. (1967), *Proc. Phys. Soc.* **92**, 956. [*49, 51, 298*]

Weaire, D. (1968), *J. Phys.* C **1**, 210. [*69, 133*]

Weaire, D. *see* Heine, V.

Webb, F. J. *see* Chen, S. H.

Webb, M. B. *see* Oriani, R. A.

Webber, G. M. B. & Stephens, R. W. B. (1968), *Physical Acoustics* (ed. Mason), **4B**, 53, New York: Academic Press. [*149, 207*]
Wehling, J. H. *see* Shyu, W.-H.
Weiss, R. J. (1966), *X-Ray Determination of Electron Distributions*, Amsterdam: North Holland. [*294*]
Wellendorf, J. *see* Greenfield, A. J.
Werber, K. *see* Braunbek, W.
Wernick, J. H. *see* Warren, W. W.
West, R. N., Borland, R. E., Cooper, J. R. A. & Cusack, N. E. (1967), *Proc. Phys. Soc.* **92**, 195. [*291*]
Whalen, T. J. *see* Kaufman, S. M.
White, D. W. G. (1966), *Trans. Met. Soc. AIME*, **236**, 796. [*87*]
White, D. W. G. (1968), *Metallurgical Reviews*, **13**, 124. [*87*]
White, R. M. *see* Timbie, J. P.
Widom, B. *see* Longuet-Higgins, H. C.
Wignall, G. D. & Egelstaff, P. A. (1968), *J. Phys.* C **1**, 519 & 1088. [*202, 524*]
Wignall, G. D. *see* Enderby, J. E.
Wilkins, J. W. *see* Ashcroft, N. W.
Wilkins, J. W. *see* Heine, V.
Willée, Ch. *see* Braunbek, W.
Williams, B. F. *see* Hewitt, R. R.
Williams, E. J. (1925), *Phil. Mag.* **50**, 27. [*414*]
Williams, G. I. *see* Orton, B. R.
Williams, G. I. *see* Rivlin, V. G.
Williams, G. I. *see* Waghorne, R. M.
Wilson, E. G. & Rice, S. A. (1966), *Phys. Rev.* **145**, 55. [*391, 398, 406*]
Wilson, E. G. *see* Bradley, C. C.
Wilson, J. R. (1965), *Metallurgical Reviews*, **10**, 381. [*76, 87, 159, 312, 421, 469, 477, 482, 484, 515*]
Wilson, J. R. (1966), *Phys. Lett.* **20**, 561. [*528*]
Wingfield, B. R. & Enderby, J. E. (1968), *Phys. Lett.* **27A**, 704. [*330*]
Wingfield, B. R. *see* Page, D. I.
Winter, F. R. & Drickamer, H. G. (1955), *J. Phys. Chem.* **59**, 1229. [*492*]
Winterberg, K. H. *see* Ruppersberg, H.
Wiser, N. *see* Greenfield, A. J.
Woerner, S. *see* Steeb, S.
Wood, W. W. (1968), *Physics of Simple Liquids* (ed. Temperley, Rowlinson & Rushbrooke), p. 115, Amsterdam: North-Holland. [*91*]
Wullschleger, J. *see* Menth, A.
Yakobson, A. M. *see* Vertman, A. A.
Yarovaya, R. G. *see* Lelyuk, L. G.
Yip, S. *see* Desai, R. C.
Yonezawa, F. (1964), *Progr. Theor. Phys.* **31**, 357. [*229*]
Young, W. H., Meyer, A. & Kilby, G. E. (1967), *Phys. Rev.* **160**, 482. [*327, 337*]
Young, W. H. *see* Dickey, J. M.
Young, W. H. *see* Meyer, A.
Young, W. H. *see* Micah, E. T.

Young, W. H. *see* Thornton, D. E.
Yuan, S. *see* Busch, G.
Yul'met'ev, R. M. (1968), *Izv. Vuz. Fiz. USSR*, **8**, 28. [*220*]
Zaffarano, D. J. *see* Gustafson, D. R.
Zamir, D. *see* El-Hanany, U.
Ziman, J. M. (1960), *Electrons and Phonons*, Oxford: Clarendon Press. [*315*]
Ziman, J. M. (1961), *Phil. Mag.* **6**, 1013. [*314, 325, 329*]
Ziman, J. M. (1964), *Adv. Phys.* **13**, 89. [*2, 32, 45*]
Ziman, J. M. (1965), *Proc. Phys. Soc.* **86**, 337. [*56, 231*]
Ziman, J. M. (1966), *Proc. Phys. Soc.* **88**, 387. [*231*]
Ziman, J. M. (1967), *Adv. Phys.* **16**, 421. [*304, 320, 325, 328, 342, 411*]
Ziman, J. M. (1968), *J. Phys.* C **1**, 1532. [*271*]
Ziman, J. M. (1969*a*), *J. Phys.* C **2**, 1230. [*230*]
Ziman, J. M. (1969*b*), *J. Phys.* C **2**, 1740. [*230, 271*]
Ziman, J. M. (1970), *J. Non-Cryst. Sol.* **4**, 426. [*230*]
Ziman, J. M. *see* Bradley, C. C.
Ziman, J. M. *see* Evans, R.
Ziman, J. M. *see* Faber, T. E.
Ziman, J. M. *see* Morgan, G. J.
Ziman, J. M. *see* Phariseau, P.
Ziman, J. M. *see* Sham, L. J.
Zweifel, P. F. *see* Aarnodt, R.
Zych, D. A. *see* Martin, B. D.

SUBJECT INDEX

Page references in italics are to tables or figures in which experimental data or the results of theoretical calculations are presented. Chemical elements are listed, under their conventional symbols, only if they are the subject of special comment in the text. The same is true of binary alloy systems, which are listed alphabetically, e.g. as Te–Tl rather than Tl–Te.

Abé approximation, 138, 280
absorption, *see* interband absorption
absorption sum rule, 369, 375–81
activation energy, 159–62, 176
Ag, 152, 328, 381, 388–90, *389*, 396
Ag–Sn, 463, 468, *497*, 510, *519*
Al, *41*, 67, *68*, 72, 110, *145*, *397*, *399*
alkali metals, *see* Cs, K, Li, Na, Rb
alkaline earth metals, 329
alloys, *see* homovalent alloys, semiconducting alloys
amalgams, *see* Hg amalgams, Hg–K, Hg–Na
amorphous films, 274
anomalous skin effect, 365–7
Ar, 68–9, 97–8, 120, 128, *129*, 136, *137*, 139, *143*, *145*, 161, 180
attenuation, *see* ultrasonic attenuation
Auger effect, 371, 395
Au–Sn, 464, *468*, *475*, 480, 510–11, 520
autocorrelation function for electrons, 230, 255–6, 260, 271, 350
autocorrelation function for molecular velocity, 155–6, 169–74; frequency spectrum of, 156, 177–80, 196, 200

back-scatter correction, 174–7
band structure of conduction electrons, 1, 3; *see also* Brillouin zone, effective mass, energy gaps, Fermi surface, rigid band model
bare potential, pseudo-potential, *see* potential, pseudo-potential
Be, 231, 303
Bi, 52, 53, 81, 132, 301, 308
Bi–In, 465, 503, *504*, 527–8, 534
Bi–Mg, *475*, 529–30, 535
Bloch wave, 1, 221, 235–6, 262, 275, 318, 320
Bohr units, 10
Boltzmann equation, 210–11, 353
bonding in alloys, 423, 453, 522
Born approximation, 31–2, 42, 47, 181, 248, 250, 262, 328, 332; improvements on, 48, 276–85, 358
Born–Green theory, 137–9, 144–5
bound states, in alloys, 427, 428, 453, 493, 508, 511, 538, 540; *see also* virtual bound states
Bragg reflection, 182, 192
Brillouin zone, 58–9, 117–18, 135, 198, 389
bulk modulus, *see* compressibility
bulk viscosity, 207

Cd, 53, 88, 270, 304, 307–8
Cd–Sb, *475*, *515*, 526–7, *528*
centrifugal barrier, 27–8
centrifuge, 492
charging, in alloys, 444
Clausius–Clapeyron equation, 343
coherence length, *see* conduction electrons
coherent scattering, *see* scattering
compound formation, in alloys, 448–55, 474–6
compressibility, 63, 97–103, 112, 119, 134, 147–53, *150*, 438, *470*, 472
Compton scattering, 127, 293–4
concentration fluctuations, *see* fluctuations
conduction electrons: coherence length (or range of coherence), 255, 260, 322; density, 7; density of states, 7, 13, 48–63, 251–3, 258, 268–71, 298, 402; Drude relaxation time, 364, 370; effective density, in Drude equations, 364, 370, 379–81, *387*, 388, *514*; lifetime, 208, 259, 400; mean free path, *5*, 322, 350, 358; relaxation time, *5*, 210, 315; spin relaxation time, 298, 493; *see also* band structure, effective mass, eigenfunctions, free-electron model, magnetic susceptibility, many-body effects, momentum distribution,

conduction electrons (*cont.*)
nearly-free-electron model, scattering, specific heat
conductivity, *see* electrical conductivity, thermal conductivity
configurational entropy, *see* entropy
contact potential, 481
convolution approximation, 195
coordination number, 77–8, 128–9, 135, 451
core electrons, 25, 291–2, 376–9, 386, 394–6; diamagnetism of, 299; electric polarisability of, 381
core states, 34–6, 380, 395–7
correlation, 12–14, 17, 149; *see also* exchange + correlation hole
covalent compounds, 451–2
creep, 154
critical exponent, 525
critical opalescence, 524
critical point, behaviour near, 351, 353, 524
critical temperature, 85, 345
crystallites, 83
Cs, 19, *26*, 40, *41*, 343–50, *344*, *346*
Cs–Na, 472, 524
Cu, 40, *41*, 88, 152, 328, 381, 388–9, 396
Cu–Sn, 460, *463*, 466–8, *467*, 510–11, *514*, *516*, 532
Curie temperature, 301, 401, 498
cyclotron frequency, 407, 409

d band *see* d electrons
Debye model, 102, 103, 115–20, 135–6, 172, 179–80, 193
Debye temperature, 53, *104*, 169
Debye–Waller factor, 116, 226, 234
degeneracy, of conduction electrons, 8, 539
de Haas–van Alphen effect, 3, 51, 432
d electrons, 40, 58, 152, 328–9, 396–8, 401
density, *see* fluctuations, volume
density matrix, 253, 260
density of states, *see* band structure, conduction electrons, effective mass
detailed balance, 186, 211
diamagnetism, *see* Landau diamagnetism, magnetic susceptibility
dielectric constant: longitudinal, 14–19, 368, 404–5; transverse, 361, 363, 365, 368
diffraction, *see* electron diffraction, neutron diffraction, X-ray diffraction
diffuse reflection, 384
diffusion, 154–5, 158–65, 168–9, 179, 220, 482–3
diffusion coefficient, *159*, *164*, 171, 175–6, *483*
dilatation, of solvent by solute ion, 438–9, 445, 517
dipole layer, *see* surface dipole layer
direct correlation function, 140–4
dispersion curves, *see* phonons
dispersion theory, 54–6
displacement, *see* mean square displacement
distribution functions, *see* higher-order distribution functions, pair distribution function, three-body distribution function
divalent metals, 324, 332, 457, 521
Doppler effect, 181–8
Drude equations, 361, 364, 377, 386, 390, 513
dynamic screening, *see* screening

effective mass of conduction electrons, 1, 48–58, 266, 295–300, 317–22, 379–81, *387*; for alkali metals, *297*; for jellium, 13, *296*; for solid metals, 51–3, *52*; optical, *297*, 380
effective potential, pseudo-potential, *see* potential, pseudo-potential
effective rigid-sphere diameter, 70–2, 104
effective valency, 39, 40; in electromigration, 488, *490*
eigenfunctions for conduction electrons, 222, 233, 236, 239, 241, 245–54; *see also* Bloch waves, localisation
eigenstates, *see* eigenfunctions
elastic continuum model, 444–6, 450, 471
elasticity, *see* compressibility, shear modulus
elastic modes, *see* longitudinal compression waves, shear waves
elastic scattering, *see* scattering
electrical conductivity: DC, 353–60; *see also* electrical resistivity; longitudinal, 366–8; transverse, 272–3, 360–7
electrical resistivity, 313–6, 322–30, *326*; dependence on temperature, 312, *314*, 321, 330–4, *331*, *516*, 527,

electrical resistivity (*cont.*)
530; dependence on volume, 339–53, *340*; in alloys, 455–61, 515–30
electrodiffusion, *see* electromigration
electrolysis, 521; *see also* electromigration
electromigration, 485–92
electron compounds, 450–1
electron diffraction, 90
electro-negativity, 423
electronic, *see* specific heat, susceptibility, etc.
electrons, *see* conduction electrons, core electrons, d electrons
electron spin resonance, 296–8, 493
electron states, *see* eigenfunctions
electro-osmosis, 491
electrotransport, *see* electromigration
ellipsometry, 382, 391
empty-core model, 38–9, 66–7, 151–2, 434–5
energy: of jellium, 11–13; of real metal, 63; *see also* internal energy
energy bands, 1, 58–63; *see also* d electrons
energy gaps, 38, 58–63, 225–6, 228, 233–4, 236–40, 262; *see also* pseudogap
energy shift due to perturbation: to first order, 48; to second order, 50, 249, 266–7; to third order, 283–4
enthalpy, 443; *see also* internal energy
entropy, 79–85, 100–2; excess, 74, *80*, 85, 101, 109–10, 132, 139; of melting, 74, *80*, 95, 100, *101*, 452; of mixing, in alloys, 446–8, *470*, *473*, *475*, *476*; *see also* fluctuations, surface properties
equation of state, 96, 97–103, 131
eutectic, 450
excess entropy, *see* entropy
exchange, 8–14, 17, 149; enhancement of spin susceptibility due to, 214, 295, *296*
exchange + correlation hole, 9
excitons, 396
expansion coefficient, *see* thermal expansion coefficient

Faber–Ziman theory, 455–61, 515–20
Fermi–Dirac distribution, 7, 211, 294, 333, 348–9, 536, 539
Fermi energy, 7, 18
Fermi level, 270, 348–9, 425, 536–41
Fermi radius, 8, 323, 429–32, 508
Fermi surface, 1, 3, 8, 38, 51, 59, 409
Fermi velocity, 317, 320, 409
ferromagnetism, 301, 401, 498
fluctuations: of concentration, 437, 524–6; of density, 107, 120, 397; of entropy, 120, 193; of pressure, 120, 193
form factor, 46, 122–4, 127; *see also* pseudo-potential
Fourier component, how defined, 31
free-electron model, 6–8
free energy, 70, 442, 448
frequency spectrum, *see* autocorrelation function for molecular velocity, structure factor
friction coefficient, 162–3, 166, 171
Friedel oscillations, 28–30, 69, 144, 146, 425
Friedel sum rule, 22–5, 32, 55–6, 429–30
F-sum rule, 377

Ga, 52, 53, 80, 81, *129*, 132, *143*, 147, 218
Gaussian approximation, 190, 196
Ge, 3, 81
geometrical approximation, 279
geometrical neighbours, 78, 93

Haeffner effect, 490
halides, 518
Hall angle, 408–9, 412
Hall coefficient, 3, 409, *410*; dependence on effective mass, 409, 411, 413; of alloys, 508–13; positive, 414, 513
Hall effect, 407–14
Hartree, Hartree–Fock potential, *see* potential
heat of mixing, *see* internal energy
Hermitian, 248, 277, 278, 415
Hg, 40, *41*, 60, 81, 90, 110, 132–3, 148, 217, 328–9, 331, 351–3, *351*, 360, 390–4, *391*, 398, 413, 496, 520
Hg amalgams, 472, 474, 481, *483*, 505, 514–15, 520–4, *522*, *523*, *533*, 534
Hg–In, *495*, 504, *505*, *510*, 511, 514, *522*, *523*
Hg–K, 465, *475*, 477, 481, 483, *485*, 511, 526
Hg–Na, 465, *475*, 481, 483, *485*, 488, 511, *523*, 526, *527*
higher-order distribution functions, 75, 107, 433

hole model, for liquid structure, 161–2, 173, 176
hole conduction, 414, 536, 538
homovalent alloys: Knight shift, 499, *501*; resistivity, 457, *516*; thermoelectric power, 457–8, 531
hopping, 273–4, 536
Hume–Rothery rules, 432, 451
hypernetted chain theory, 141, 144

ideal law of mixing, 442, 448, 478
incoherent scattering, *see* scattering
inelastic scattering, *see* scattering
inner work function, 13, 425
In–Sb, *475*, 497, 506, 527–8, 534
interband absorption, 3, 352, 371–5, 378, 386–7, *389*, 392, 394, 396
interference function, 50, 60, 75, 100, 110–14, *113*, 124–37, *129*, *137*, 142; for solid, 114–21; frequency spectrum of, 184–204; mean, for alloys, 456, 462; sum rule, 113, 115, 118, 121, 261; *see also* partial interference functions
internal energy, 93, 97–8, 107; of mixing, in alloys, 443, 451–2, *470*, *475*; of solution, in dilute alloys, 450, *471*
ionic compounds, 451–2, 497
ionisation energy, 347–8, 353, 537
ions, *see* core electrons, core states, pair potential, pseudo-potential, valency
isotope effects, 52, 168–9
isotopic replacement, 191, 466
itinerant oscillator model, 178–9

jellium, 1, 6–22, 63–4, 148–9, 379

K, 13, *297*, 388
K–Na, 462, 469, *470*, 472, 484, 488, 493, 520
Knight shift, 214, 302–8, *304*, 499–507, *500*, 541
Koopman's theorem, 10, 395
Kopp's law, 476
Korringa–Kohn–Rostoker (KKR) method, 231
Korringa relation, 214–8, 505–7
Kramers–Kronig relations, 363, 376–7, 382, 393
Kronig–Penney model, 224

Landau diamagnetism, 299, 300, 493
Landau theory, *see* quasi-particles
Langevin theory, 162, 171, 179
latent heat of melting, 79, 91, 107, 162; *see also* entropy
latent heat of vaporisation, *87*, 161
Lennard–Jones potential, 92
Li, 28, 168–9, 212, 270–1, *297*, 298, 322, 341, 360, 373, 380–1
lifetime, *see* conduction electrons
lifetime broadening, effects of, 4, 29–30, 264–6, 285–7, 290, 334, 339, 358–60, 398–400
Lindemann correlation, 75, 103–6, *104*
Linde's rule, 517
liquid–solid interface, *see* surface properties
liquidus, 449
localisation, of electrons, 60–3, 229–30, 241–4, 256, 271–4, 348–9, 352, 394; *see also* bound states, in alloys
localised moments, 494, 497–8, 499, 519
longitudinal compression waves, 63, 115–21, 151, 201
Lorenz number, *338*

magnetic susceptibility, 214, 294–301, *299*, 492–8, *498*
magneto-resistance, 414
many-body effects, *see* correlation, exchange, phonon enhancement, quasi-particles
mass action, 540
matrix elements, between eigenstates, 353–4, 356–7, 363, 365, 415–20
matrix elements of scattering potential, 245–6, 248, 415–16; between Bloch states, 235–6, 318–20
Matthiessen's rule, 461
Maxwell relaxation time, 205–7
mean free path, *see* conduction electrons, lifetime broadening
mean square displacement, of molecule in liquid, 103–5, 116, 160, 170–3, 180, 196, 219
melting, *for change on melting see under property concerned*
melting pressure, 95, 99
melting temperature, 94, 103–6, 343
Mg–Pb, 465, *475*, *481*, 484, 528
Mg–Sn, *464*, 465, *475*, *481*, 484, *485*, 528
misfit ratio, 435, 442
miscibility gap, 451, 524–6, 529
mobility, 529

models, of liquid structure, 74, 76–9; *see also* elastic continuum model, quasi-crystalline model, random close-packed model, rigid-sphere model, Percus–Yevick rigid-sphere model, substitutional model
modulation factor, 31, 47, 51, 277, 279–81, 287, 306, 427
molecular dynamics, 91–6, 131, 133–4, 139, 142, 145, 174–6, 179–80
molecules, formation of, in liquid alloys, 453–4, 461, 539–41
moments, of structure factor, 187, 191–7
momentum distribution, of conduction electrons, 285, 288–94
Monte Carlo methods, 91–2, 139, 227
Mott transition, 62, 274, 349
muffin tin, *see* potential
multi-phonon processes, 116, 201
multiple scattering, 47, 125, 127

Na, 13, *85*, 134, *136*, 145, 153, 180, *297*, 298, 322, 375, 388
nearly-free-electron (NFE) model, 1, 4, 221–2, 309, 428–32
neutron diffraction, 124–7, 181–6, 190–1, 198–200, 461–9
noble metals, *see* Ag, Au, Cu
node counting, 227
non-direct transitions, 402–3
non-local operators, 33, 36
normalisation corrections, 36, 251, 316, 321, 357–8
nuclear magnetic resonance, 213–20, 302–8, 499–508

optical effective mass, *see* effective mass
optical properties, 382–401, 513–15

packing fraction, 71, 74, 93, 96, 327, 339, 435, 442, 520, 522; dependence on temperature, 71–2, 103, 136–7, 148, 176
pair correlation function, 75, 140, 266
pair distribution function, 75–6, 106–10, 127–30, 137–41; for mixtures, 122, 433; *see also* time-dependent pair distribution functions
pair potential, between ions, 63–72, 133–4, 144–7, 151–3, 179–80, 284, 433–5; between rare gas atoms, 69, 92, 145
partial interference functions, 76, 122–4, 435–41, 455, 465–9, *467*, *468*, 525
partial thermodynamic parameters, for alloys, 474–5
partial waves, 23, 35
Pb, 135–6, *137*, *143*, *145*, 180, *202*
Peltier coefficient, 334
Percus–Yevick rigid-sphere model, 131–2, 136, 142, 326–7, 331, 339; for alloys, 435–6, 463, 482, 520
Percus–Yevick theory, 131, 140–7
phase diagrams, for alloys, 448–51
phase, of wave function, 232–3, 238–40
phase shift, 23–8, 54–7, 428, 430–1, 518
phonons, 38, 53, 135, 157, 183, 186, 193, 198–204, *202*, 206
phonon enhancement, of electronic specific heat, *52*, 53
photo-emission, 401–4
plasma frequency, 367, 405
plasma oscillations, 404–7
plasma term in resistivity, 325
plasmons, 405–6
polycrystalline solids, 118, 154, 275–6
positron annihilation, 288–93, 432
potential: bare, 14, 13; effective, 32, 41; Hartree or Hartree–Fock, 6, 8–9, 25, 27; muffin-tin zero, 61, 230, 350; *see also* pair potential, pseudo-potential
pressure, 93–103, 107
profile, of eigenfunction in **k**-space, 252–6, 259–60, 261, 263, 373
pseudo-gap, 60, 274, 284, 328, 352, 414, 522, 540–1
pseudo-potential: effective, 47–8, 50, 276–82, 287, 440; of assembly of ions, 42–8; of single ion, 33–42, *41*
pseudo wave function, 34

quadrupole effects, 158, 213–20, 506
quadrupole moment, *215*
quasi-crystalline model, 74, 77–8, 79, 82–4, 107, 133, 154
quasi-elastic scattering, 194
quasi-particles, Landau theory of, 21, 317–18, 321–2, 370
quasi-potential, 42

radial distribution function, 128, 130
random close-packed (RCP) model, 78–9, 84, 93, 107

range of coherence, *see* conduction electrons
rare earth metals, 330, 414, 498
rare gases, 86; *see also* Ar
Rb, *129*, 136, *145*, 343
reciprocal lattice vector, 58, 70, 114–18, 198–202, 226, 235–6, 338–9
reflectivity, 382, 391, 394
relaxation time, *see* conduction electrons, Langevin theory, Maxwell relaxation time
relaxation times, of nuclear spin system, 213–22, *215*, 505–8
resistivity, *see* electrical resistivity
rigid band model, 432
rigid-sphere model, 70–2, 92–105, 129–131, 149, 162–8, 171, 174–7, 433–41, 482, 484, 520; *see also* Percus–Yevick rigid-sphere model
ripples, 88–91
roughness, of liquid surface, 91, 385

Saxon–Hutner theorem, 225–6, 275
Sb, 3, 81, 132, 218
scattering: coherent and incoherent, 123, 125, 157, 185, 189–91, 198, 462; inelastic, 30, 124, 156, 181–6, 210–12, 325, 490; of conduction electrons, 30–3, 42–8, 207–13, 245–8 319–20; of neutrons, *see* neutron diffraction; of X-rays, *see* X-ray diffraction; *see also* Born approximation, Compton scattering, critical opalescence, form factor, matrix elements, multiple scattering, structure factor
scattering vector, 31, 183, 185
screening, 14–22, 25, 213, 285–8; dynamic, 367–70
screening length, 20
screening parameter, 15
secular equation, 250
self transport, 490
semiconducting alloys; Hall coefficient, 512–13, 537–8; Knight shift, 506–8, *507*; magnetic susceptibility, *494*, *498*; resistivity, 452, 528–30; thermodynamic properties, *475*; thermo-electric power, 534–41
semi-metals, *see* Bi, Ga, Sb
shear modulus, 119, 204
shear waves, 84, 117, 204–7
Si, 81
skin depth, 366–7, 383
Sn, *41*, 132, *386*
solidification, *see* melting
solubility, 450, 453
solution, heat of, *see* internal energy
Soret effect, 492
sound, *see* ultrasonic attenuation, velocity of sound
specific heat, 3, *52*, 53, 76, 84–6, *85*, 136, 476
specular reflection, 384
spherical approximation, for effective mass, 51, 56–7, 266
spin-orbit interaction, 398, 411, 519
spin relaxation or resonance, *see* conduction electrons, nuclear magnetic resonance, relaxation times
spin susceptibility, *see* magnetic susceptibility
Stokes law, 166
structural neighours, 78
structure of solids, 69
structure factor, 2, 45, 110, 114–16, 226, 235, 278; frequency spectrum of, 184–204
substitutional model, 122–3, 183, 430, 450
sum rule, *see* absorption sum rule, F-sum rule, Friedel sum rule, interference function
supercooling, 78, 91
superposition approximation, 138–9, 220
surface-active impurities, 383, 477
surface scattering, of conduction electrons, 384
surface dipole layer, 10, 13, 73, 404
surface-inactive impurities, 477
surface properties, 73–91, *87*, *88*, 477–81
surface states, 384
surface structure, 90, 384, 393–4
susceptibility, *see* magnetic susceptibility

Te, 81, 128, 285, 512, 534, 538
tellurides, *475*, *494*, 506–8, *512*, 528–9, 534–5; *see also* Te-Tl
Te–Tl, 454, 497, 507–8, *512*, *513*, *529*, *535*, 536–41
thermal conductivity, 168, 337–9
thermal expansion coefficient, 72, 85, 95, *96*, 103
thermo-electric parameter, 335–7, *335*, *340*, 342

thermo-electric power, 335, 457–60, 530–41
Thomas–Fermi approximation, 15, 18, 20–1, 67
three-body distribution function, 75, 107, 109, 138–9, 220, 278–80, 284, 288
tight-binding model, 1, 61, 230, 274, 310, 414, 541
time-dependent pair distribution functions, 156, 188–90, 194–8
T-matrix, 42, 282
total correlation function, 140
transition metals, 57–8, 301, 330, *400*, 401, 414, 497–8, 518, 534
transverse modes, *see* shear waves

ultrasonic attenuation, 193, 203–4, 206–7, 472, 524
umklapp processes, 158, 199, 200, 206, 235
uncertainty principle, *see* lifetime broadening

V, 125, 185
vacancies, 72–3, 115, 154, 161, 173
valency, 7; *see also* effective valency
van der Waals's equation, 97
van der Waals force, 152
vaporisation, *see* latent heat of vaporisation
velocity of sound, 76, 117, 119, 135
virial, 107, 132
virtual bound states, 26–8, 57, 494, 518, 534, 539
visco-elasticity, 204–7
viscosity, 154, 165–9, *168*, 177, 484–5
volume: change of, on melting, 75, 79, 95, 100, *101*; of mixing, in alloys, 441–6, 448, 470–4, *470*, *471*, *475*
Voronoi polyhedra, 78, 105

wave groups, *see* eigenfunctions
Wiener–Khintchine theorem, 187–9, 255
work function, 10, 403–4, 481

X-ray absorption and emission, 394–400
X-ray diffraction, 124–7, 461–9

Ziman theory, 309, 314–15, 322–5, 349–52
Zn, 88, 132
Zn–Sb, *475*, 479, *512*, 527, 534